Handbook of
Psychopharmacology

Volume 17

Biochemical Studies of CNS Receptors

Handbook of Psychopharmacology

SECTION I: BASIC NEUROPHARMACOLOGY

SECTION II: BEHAVIORAL PHARMACOLOGY IN ANIMALS

SECTION III: HUMAN PSYCHOPHARMACOLOGY

SECTION IV: BASIC NEUROPHARMACOLOGY: AN UPDATE

Volume 17

Biochemical Studies of CNS Receptors

Edited by

Leslie L. Iversen
Department of Pharmacology
University of Cambridge

Susan D. Iversen
Department of Psychology
University of Cambridge

and

Solomon H. Snyder
Departments of Neuroscience, Pharmacology, and Psychiatry
The Johns Hopkins University
School of Medicine

PLENUM PRESS • NEW YORK AND LONDON

Library of Congress Cataloging in Publication Data

Main entry under title:

Biochemical studies of CNS receptors.

(Handbook of psychopharmacology; v. 17)
Includes bibliographical references and index.
1. Neurotransmitter receptors. 2. Drug receptors. 3. Neurochemistry. I. Iversen, Leslie Lars. II. Iversen, Susan D., 1940– . III. Snyder, Solomon H., 1938– IV. Title: Biochemical studies of C.N.S. receptors. V. Series. [DNLM: 1. Receptors, Sensory. 2. Central nervous system—Physiology. 3. Central nervous system—Drug effects. 4. Receptors, Drug. QV 77 H236 sect. 4 v. 17]
QP364.7.B56 1983 612'.82 82-22376
ISBN 978-1-4684-4363-9 ISBN 978-1-4684-4361-5 (eBook)
DOI 10.1007/978-1-4684-4361-5

Softcover reprint of the hardcover 1st edition 1983
A Division of Plenum Publishing Corporation
233 Spring Street, New York, N.Y. 10013

CONTRIBUTORS

CLAUS BRAESTRUP, *Research Laboratories, A/S Ferrosan, DK-2860 Soeborg, Denmark; and Skt. Hans Mental Hospital, DK-4000 Roskilde, Denmark*

IAN CREESE, *Department of Neurosciences, University of California, San Diego, School of Medicine, La Jolla, California 92093*

FREDERICK J. EHLERT, *Department of Pharmacology, University of Arizona Health Sciences Center, Tucson, Arizona 85724*

G. FILLION, *Pharmacology Department, Pasteur Institute, 75015 Paris, France*

JACK PETER GREEN, *Department of Pharmacology, Mount Sinai School of Medicine, The City University of New York, New York, New York 10029*

MARK W. HAMBLIN, *Department of Neurosciences, University of California, San Diego, School of Medicine, La Jolla, California 92093*

HANS W. KOSTERLITZ, *Unit for Research on Addictive Drugs, University of Aberdeen, Marischal College, Aberdeen AB9 1AS, Scotland*

STUART E. LEFF, *Department of Neurosciences, University of California, San Diego, School of Medicine, La Jolla, California 92093*

MOGENS NIELSEN, *Psychopharmacological Research Laboratories, Department E, Skt. Hans Mental Hospital, DK-4000 Roskilde, Denmark*

STEWART J. PATERSON, *Unit for Research on Addictive Drugs, University of Aberdeen, Marischal College, Aberdeen AB9 1AS, Scotland*

LINDA E. ROBSON, *Unit for Research on Addictive Drugs, University of Aberdeen, Marischal College, Aberdeen AB9 1AS, Scotland*

WILLIAM R. ROESKE, *Department of Internal Medicine, University of Arizona Health Sciences Center, Tucson, Arizona 85724*

DAVID R. SIBLEY, *Department of Neurosciences, University of California, San Diego, School of Medicine, La Jolla, California 92093*

S. Robert Snodgrass, *Department of Neurology, University of Southern California, Children's Hospital of Los Angeles, Los Angeles, California 90027*

Solomon H. Snyder, *Departments of Neuroscience, Pharmacology, and Experimental Therapeutics, and Department of Psychiatry and Behavioral Sciences, The Johns Hopkins University of Medicine, Baltimore, Maryland 21205*

Henry I. Yamamura, *Departments of Pharmacology, Psychiatry, and Biochemistry, University of Arizona Health Sciences Center, Tucson, Arizona 85724*

PREFACE

It is now eight years since the first *Handbook* volumes on Basic Neuropharmacology were published, and there have been many important advances. As in many other areas in science, progress in this field has depended to a considerable extent on the availability of new experimental methods, and Volume 15 reviews some major recent developments, including new autoradiographic techniques that allow direct visualization of drug and transmitter receptors in the nervous system, and the pinpointing of the precise locations of the changes in brain metabolism elicited by various drug treatments. Volume 16 and 17 cover two of the most active areas for basic research in psychopharmacology at the moment: the characterization of drug and transmitter receptors in brain by radioligand binding techniques, and studies of the role of small peptides in brain function. The latter area, in particular, illustrates how rapidly progress continues to be made in basic research on the mechanisms of chemical communication within the nervous system. Eight years ago when the *Handbook* first appeared none of the opioid peptides (enkephalins and endorphins) had yet been identified. Since then a whole new area of basic biological research has focused on these substances, and in addition we know of more than thirty other neuropeptides with putative CNS transmitter functions.

We hope that these new volumes will help to keep the *Handbook of Psychopharmacology* abreast of the most recent advances in the field, and continue to make it a valuable reference work for all who are involved in research in this increasingly active field of science. The response to earlier volumes has been remarkably positive, and we remain indebted to the publishers for conceiving the original idea, and to the many contributors who have labored long and hard to bring it to fruition.

L.L.I.
S.D.I.
S.H.S.

CONTENTS

CHAPTER 3

CNS Dopamine Receptors

IAN CREESE, MARK W. HAMBLIN, STUART E. LEFF, AND DAVID R. SIBLEY

CHAPTER 6

The Nature of Muscarinic Receptor Binding

FREDERICK J. EHLERT, WILLIAM R. ROESKE, AND HENRY I. YAMAMURA

Chapter 7

Benzodiazepine Receptors

Claus Braestrup and Mogens Nielsen

CHAPTER 8

Histamine Receptors in Brain

JACK PETER GREEN

1

MOLECULAR ASPECTS OF NEUROTRANSMITTER RECEPTORS: AN OVERVIEW

Solomon H. Snyder

1. INTRODUCTION

It is axiomatic that following synaptic release neurotransmitters interact with specific sites on adjacent cellular membranes. Molecular approaches to neurotransmitter receptors are surprisingly recent, with most research on brain neurotransmitters having taken place within the past 5–8 years. During this relatively brief period, biochemical studies of receptors have burgeoned so rapidly that they form the most common theme of the chapters in this volume.

Much of the research described in these chapters deals with neurotransmitter receptor binding studies. Frequently, reports dealing with receptor binding are titled "Such and Such Neurotransmitter Receptor." There ensues discussion as to just what is meant by a neurotransmitter receptor. Some argue that binding studies do not deal with receptors at all but merely with binding sites which may not be associated with a physiological response and hence may not be functional. Accordingly, such entities do not deserve the name receptor.

Because of these controversies, it might be best to commence with some definitions. The concept of hormone or neurotransmitter receptors

---

Solomon H. Snyder • Department of Neuroscience, Pharmacology, and Experimental Therapeutics, and Department of Psychiatry and Behavioral Sciences, The Johns Hopkins University School of Medicine, Baltimore, Maryland 21205.

is largely a pharmacological one. Prior to the advent of molecular probes, pharmacologists measured physiological responses to drugs, hormones, and neurotransmitters, often monitoring contractions of smooth muscles. The "receptor" was that entity which was responsible for the "response." It was generally assumed that the total receptor apparatus must include a recognition site, some transducing mechanism, and an effector device. According to this view, one does not study receptors unless one investigates in each experiment all three elements. Thus, the only legitimate receptor research is that which measures smooth muscle contraction, secretion of a bodily substance, or some other physiological response.

Of course, this caricature applies only to an extreme point of view. However, it does address an important issue. In some instances, receptor recognition sites may not be linked to an effector apparatus and thus fit the designation "silent receptors." On the other hand, when we cannot detect a response associated with a receptor site, we may not be justified in concluding that there is no response but merely that we have been monitoring the wrong measure. This issue comes up in different experimental situations. For instance, the regional distribution of some neurotransmitter receptors does not parallel precisely the distribution of neuronal systems containing the transmitter. Thus, beta-adrenergic receptor binding sites are widely distributed throughout the brain and are highly concentrated in areas such as the corpus striatum which contains little or no norepinephrine (Bylund and Snyder, 1976). One may conclude either the beta-adrenergic receptor recognition sites are meaningless entities or that we are missing a subtlety of nature. For instance, perhaps there exists a hitherto unrecognized neurotransmitter other than norepinephrine which interacts with beta-receptors. Recently it has been possible to monitor by autoradiography the distribution of cholecystokinin (CCK) receptor sites using [^{125}I]-CCK-33, the 33 amino acid form of the peptide (Zarbin *et al.*, 1981). A number of very discrete localizations of the autoradiographically visualized grains, including discrete clusters in the optic tract and superior colliculus, differ markedly from the localization of CCK-containing neuronal systems mapped with antibodies to CCK. The rapidly accumulating evidence for multiple forms of CCK stored in distinct neuronal systems (Rehfeld, 1978) suggests that the visualized receptors are associated with neurons containing a form of CCK other than that which has been visualized in immunohistochemical studies.

It has been generally thought that the transducing element connecting neurotransmitter recognition to alterations in cellular function involves either changes in ion permeability or alterations in cyclic AMP formation. Recent evidence has greatly increased the number of possible second messenger or transducing mechanisms. Studies, especially from Axelrod's laboratory, suggest that in some instances adenylate cyclase may be a third messenger whose enhancement is provoked by an initial stimulation of phospholipid methylation evoked by the hormone or neurotransmitter (Hirata and Axelrod, 1980). Beginning with the pioneering work of the Hokins, there has been a progressive accumulation of data linking formation of phosphatidylinositol with cholinergic, histamine, and substance P-associated synaptic activity. Alterations in calcium permeability may be important and linked to various types of messages. For instance, cyclic GMP formation is related to changes in calcium disposition. Calcium is also associated with specific binding proteins, most notably calmodulin, which in turn modulates numerous cellular responses. Calcium is also crucial for the activity of numerous enzymes, such as a protease which may be linked to glutamate receptors (Baudry and Lynch, 1980; Vargas *et al.*, 1980). Also, calcium is required for the activity of phospholipase A_2 which generates arachidonic acid, the precursor of prostaglandins which in turn mediate many receptor-associated responses.

How all these "second messengers" interact is unclear. One model holds that the rapid opening and closing of ion channels is responsible for classical excitation and inhibition in synaptic transmission. Biochemical alterations, such as changes in cyclic nucleotide, phospholipid, or prostaglandin formation may mediate slower changes. Some of these latter activities may be long-lasting "trophic" alterations involved in synaptic plasticity and other events associated with memory processes. Clearly one of the tasks of neurotransmitter research is to work out the molecular events which link receptor recognition to subsequent cellular events.

2. RECEPTOR RECOGNITION: ION AND NUCLEOTIDE REGULATION ASSOCIATED WITH "SECOND MESSENGERS"

The first biochemical approaches to receptor labeling for neurotransmitters dealt with the nicotinic cholinergic receptor in the electric organs of invertebrate fishes. These studies took advantage of the extremely high receptor content of these organs, amounting to 20% of membrane protein in some instances. Additionally, in these studies potent snake toxins were utilized which bound with high affinity and virtually irreversibly to the receptors. Such approaches could not be readily applied to the brain where receptor density in most cases is only about one-millionth by weight of brain tissue and in which unique irreversible, yet neurotransmitter-specific, toxins do not exist. Early studies using [^{3}H]atropine and intestinal smooth muscle indicated the feasibility of labeling receptors with reversible ligands, but technical difficulties, such as low signal–noise ratio, precluded any widespread development of these limited initial studies.

Simple, sensitive, and specific techniques for labeling opiate receptors with good signal–noise ratios permitted an extension of these general approaches to receptors for almost all the known neurotransmitters in the brain (Snyder and Goodman, 1980).

In initial studies it was assumed that the binding sites being examined involved only the neutrotransmitter recognition moiety, a membrane protein which was quite distinct from the macromolecular entities involved in further translation of the neurotransmitter recognition information. Very soon, however, numerous findings suggested that receptor binding sites were linked allosterically to ion conductance channels and other modulating entities. Some of the first evidence derived from studies of the opiate receptor and the observation that sodium ion could discriminate between receptor interactions of agonists and antagonists (Pert *et al.*, 1973). The key difference between an agonist and antagonist is that receptor recognition of an agonist triggers a change in cellular function while no such change follows antagonist binding to the receptor. These findings, therefore, suggested that sodium might have some second messenger role, perhaps as an ion whose conductance is altered upon opiate synaptic transmission. Sodium ion is required for some neurophysiologic actions of opiates (Zieglgansberger and Bayerl, 1976), for pharmacologic influences of opiates on smooth muscle (Enero, 1971), and for the regulation of cyclic AMP formation in neuroblastoma clones (Lichtshtein *et al.*, 1979).

Sodium also regulates receptor binding differentially for agonists and antagonists at alpha-adrenergic receptors (Greenberg *et al.*, 1978), bradykinin (Innis *et al.*, 1981), histamine H_1 (Chang and Snyder, 1980), and adenosine (Bruns *et al.*, 1980; Goodman *et al.*, 1981) receptors. Just how sodium ion regulates synaptic transmission at all of these receptors is quite unclear.

Cellular hyperpolarization by glycine or GABA involves increases in chloride ion permeability. Interestingly, receptor interactions of the glycine antagonist, strychnine and of glycine itself are regulated by chloride ion which decreases the binding of strychnine and decreases the affinity of glycine for the receptor (Young and Snyder, 1974; Muller and Snyder, 1978). Evidence that this effect reflects a link of the receptor recognition protein to the chloride channel derives from findings that the relative potencies of a series of anions in influencing glycine receptor binding fit extremely closely with their influences on synaptic hyperpolarization.

Benzodiazepines appear to act pharmacologically by facilitating GABA neurotransmission, with the molecular counterpart that GABA enhances benzodiazepine receptor binding. Accordingly, the enhancement of benzodiazepine binding by chloride ion suggests a link to the chloride ion channel (Costa *et al.*, 1979).

Guanine nucleotides influence hormone and neurotransmitter receptors by selectively decreasing the affinity of agonists. Such effects were first described by Rodbell and associates at the glucagon receptor (Rodbell *et al.*, 1971) but have been subsequently extended to beta-adrenergic and many other receptors (Snyder and Goodman, 1980). Purification of the GTP binding protein and studies of its interaction with receptor and cyclase, represent one of the leading points of neurotransmitter receptor research and have been most extensively exploited at beta-adrenergic receptors in red cell preparations.

One of the major questions addressed is how a decrease in agonist binding elicited by GTP can be responsible for the GTP enhancement of hormone- and neurotransmitter-stimulated adenylate cyclase. One possibility is that decreases in agonist binding reflect an acceleration of the kinetics of receptor interaction with somewhat greater speeding up of the dissociation than of the association rate. Indeed, such effects have been directly demonstrated at opiate receptors (Childers and Snyder, 1980). In many instances of neurotransmitter receptor binding, agonists have nanomolar potency with dissociation rates much too slow to support synaptic transmission. Perhaps the membrane preparations generally used involved desensitized receptors with the desensitization being caused by the unnaturally high affinity of transmitter for receptor. According to this model, GTP "resensitizes" the receptor by accelerating both the association and dissociation rates.

Divalent cations regulate numerous receptors including opiate (Pasternak *et al.*, 1975), alpha-adrenergic (U'Prichard and Snyder, 1980), histamine H_1 (Chang and Snyder, 1980), adenosine (Bruns *et al.*, 1980), bradykinin (Innis *et al.*, 1981), dopamine (Usdin *et al.*, 1980), and CCK (Innis and Snyder, 1980) receptors. Unlike the effects of sodium, which decreases agonist but not antagonist binding, the divalent cations selectively increase agonist binding. The relative potencies of various divalent cations vary with different receptors. At opiate, alpha-adrenergic, dopamine, CCK, and histamine H_1 receptors manganese is either the most potent or is similar in effect to magnesium while calcium is substantially less effective. This pattern resembles the influence of these ions upon adenylate cyclase and its coupling to the GTP-binding protein which links receptors to the cyclase. Accordingly, in these instances the divalent cation effect might be related to guanine nucleotide effects which presumably are involved in linking receptor, GTP-binding protein, and cyclase.

In terms of differential actions on agonist and antagonist binding, divalent cations seem to exert effects opposite to those of guanine nucleotides. At receptors such as the alpha-adrenergic receptor, interactive receptor influences have been described which support the notion that the divalent cation effects are linked somehow to the guanine nucleotide actions (U'Prichard and Snyder, 1980).

One very different pattern of divalent cation influence has been observed recently in our laboratory at bradykinin receptors (Innis *et al.*, 1981). Here physiological concentrations of calcium are more potent than magnesium or manganese. Moreover, the divalent cations decrease agonist binding, opposite to their effects on other receptors. Bradykinin elicits many of its actions through an enhancement of prostaglandin synthesis via the stimulation of phospholipase A_2. Conceivably, the calcium effects on binding reflect an allosteric link in tissue membranes between the bradykinin recognition site and the phospholipase enzyme with its calcium-recognizing site. Alternatively, the effect of calcium might indicate an association with calcium conductance channels.

3. SOLUBILIZED AND PURIFIED RECEPTORS CLARIFY SYNAPTIC MECHANISMS

The most straightforward goal of receptor solubilization and purification is to isolate the receptor macromolecule, insert it back into membranes, and attempt to reconstitute physiological function. At nicotinic cholinergic receptors in invertebrate electric organs, substantial progress toward this goal has been made. With the brain such a goal seems still far away, at least using conventional techniques. However, the limited studies of brain receptor solubilization and purification have already provided some initial insights into receptor functioning.

One of the first tasks is to develop procedures which will detect receptor binding in the solubilized state. By using a variety of detergents it is now possible to monitor reliably, receptor binding for various brain receptors. Ligand-bound receptor can be quantified by precipitating the ligand–receptor complex with agents such as polyethylene glycol or ammonium sulfate or binding the unreacted ligand to charcoal. In several instances one can obtain ratios of specific to nonspecific binding at least as good in the solubilized as in the membrane-bound state. While it has proved extremely difficult to purify these detergent-solubilized receptors, considerable information about receptor properties has been obtained even with partially purified or unpurified receptors.

The question of the relationship of the receptor-recognition protein to other proteins involved in the synaptic response has been studied. How closely are recognition and GTP-binding proteins related? Recently we were able to demonstrate a retention of guanine nucleotide regulation of adenosine receptors in the soluble state (Fig. 1) (Gavish *et al.*, 1982). The fact that the complex between receptor and GTP protein remains intact throughout the solubilization procedure may permit a molecular dissection of forces that determine the interactions between these proteins. For

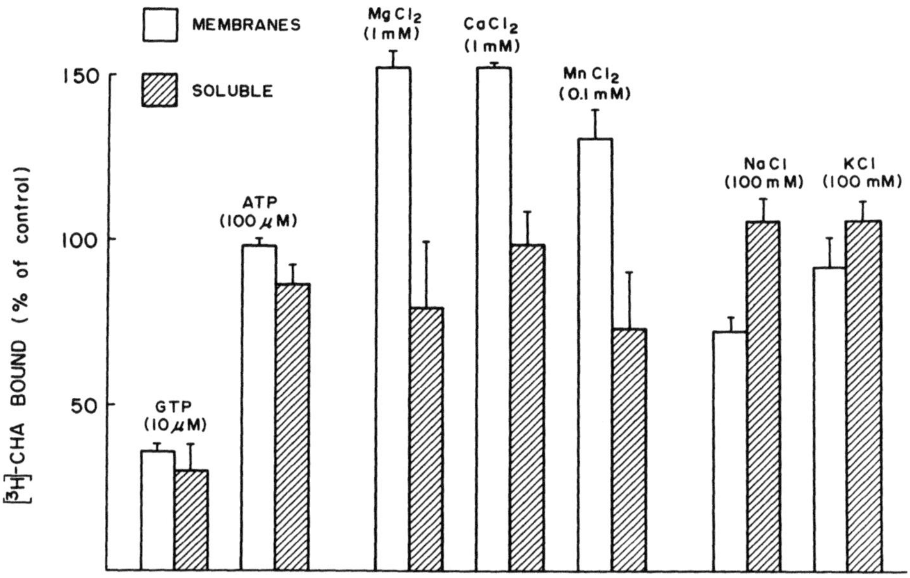

FIG. 1. Differential effects of guanine nucleotides and divalent cations on [^{3}H]cyclohexyladenosine ([^{3}H]-CHA) binding to membrane-bound and soluble adenosine receptors. Note that GTP reduction of binding is retained in the soluble state while the divalent-cation enhancement and sodium-elicited reductions are not detected with soluble receptors. (Data adapted from Gavish *et al.*, 1982.)

instance, it may prove feasible to dissociate the two proteins and then reunite them.

A rather different pattern has been described for solubilized opiate receptors. Guanine nucleotide reduction of ligand binding to the receptors is retained in the soluble state (Koski *et al.*, 1981). However, solubilization of opiate receptors abolishes the differentiation by guanine nucleotides of agonist and antagonist receptor interactions.

Interestingly, divalent cations and sodium, which regulate adenosine receptors in the membrane-bound state, have no effect on soluble adenosine receptors (Gavish *et al.*, 1982). This suggests that their actions do not involve a GTP-binding protein or that solubilization inactivates the cation-recognizing portion of the protein. Unlike the results with adenosine receptors, the regulation of histamine H_1 receptors by sodium is retained when receptors are solubilized, whereas GTP regulation is lost (Toll and Snyder, 1982).

Benzodiazepine and GABA receptors appear to be related somehow, and both are associated with chloride ion conductance. Receptor solubilization has facilitated studies of these interactions. When receptors are solubilized with Triton, GABA stimulation of benzodiazepine binding is retained (Gavish and Snyder, 1980), while solubilization with Lubrol abolishes the influences of GABA (Yousufi *et al.*, 1979). A close link of GABA and benzodiazepine receptor proteins is indicated by the finding that muscimol binding to GABA receptors and benzodiazepine receptor binding copurify via several procedures (Gavish and Snyder, 1981). Moreover, after extensive purification by several techniques, GABA stimulation of benzodiazepine binding is retained. Interestingly, the chloride stimulation of benzodiazepine binding also remains after extensive purification. Thus, it may be conceivable to obtain extensively purified and perhaps homogeneous preparations of a GABA–benzodiazepine–chloride ion channel complex.

4. NEUROTRANSMITTER UPTAKE RECEPTORS

The ligand binding procedures that have facilitated studies of synaptic neurotransmitter receptors may be useful for labeling other synaptic sites. Biogenic amine uptake has been traditionally studied by the accumulation of radiolabeled neurotransmitter into synaptosomes. Recently several laboratories have labeled serotonin uptake sites in brain membranes and platelets using [^{3}H]imipramine, the tricyclic antidepressant which is

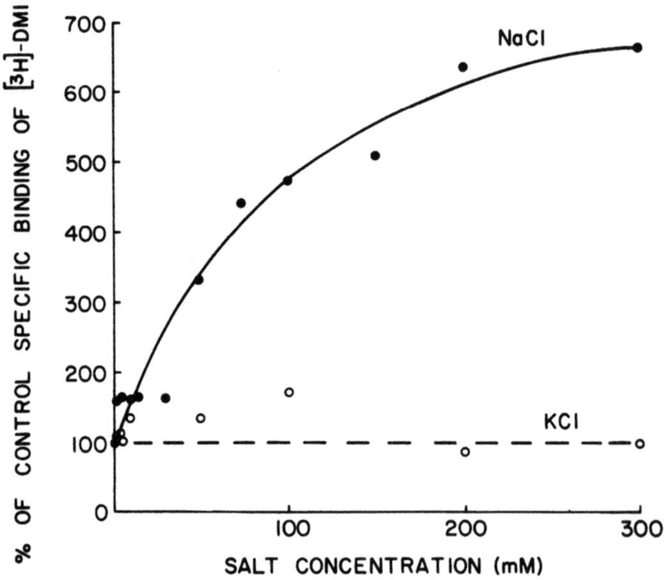

FIG. 2. Sodium dependence of [^{3}H]-DMI binding to rat cerebral cortical membranes. Specific binding of [^{3}H]-DMI was measured at 2 nM [^{3}H]-DMI in the presence of various concentrations of sodium (●) or potassium (○) chloride. Specific binding of [^{3}H]-DMI in the absence of sodium was 19.2 ± 2.7 fmol/mg protein. Results are expressed as percent of binding relative to that of the no-sodium control and are the means of three experiments. (Data adapted from Lee and Snyder, 1981.)

TABLE 1
Drug Inhibitory Potencies on [^{3}H]-NE Uptake and [^{3}H]-DMI Binding in the Rat Cerebral Cortex[a]

Drugs	IC$_{50}$[b] (μM) [^{3}H]-NE Uptake	IC$_{50}$[b] (μM) [^{3}H]-DMI Binding
Tricyclic antidepressants		
Desipramine	0.02	0.03
Imipramine	0.79	0.36
Doxepin	0.31	1.0
Chlorimipramine	1.8	1.4
Atypical antidepressants		
Mianserin	0.88	2.25
Iprindole	12.0	17.0
Others		
Chlorpromazine	1.8	0.13
Benztropine	2.7[c]	4.0
(+)-Amphetamine	0.07	12.0
(−)-Amphetamine	0.12	10.0
Cyproheptidine	2.7	3.0
Atropine	—	195.0
Neurotransmitters		
NE	0.4[d]	>1000
DA	—	>1000
5-HT	—	>1000

[a] GABA had no effect on [^{3}H]-DMI binding even when tested at 1 mM. The following drugs had less than 50% inhibition of [^{3}H]-DMI binding when tested at 1 mM: yohimbine, phenylephrine, clonidine, and naloxone. [^{3}H]-NE uptake was assayed in rat cerebral cortical homogenates by minor modifications of the techniques of Coyle and Snyder (1969). (Data adapted from Lee and Snyder, 1981.)
[b] IC$_{50}$ is the concentration for 50% inhibition of [^{3}H]-NE uptake or [^{3}H]-DMI binding. Values are means of three independent determinations which varied less than 15%.
[c] From Horn *et al.* (1971).
[d] From Coyle and Snyder (1969).

a potent inhibitor of serotonin uptake (Langer and Briley, 1981). In our own (Lee and Snyder, 1981) and other (Rehavi *et al.*, 1981) laboratories, it has proved feasible to label norepinephrine uptake sites with [^{3}H]-desipramine, which is a substantially more selective inhibitor of norepinephrine than serotonin uptake (Table 1). [^{3}H]Desipramine binds to more than one membrane entity. However, the highest affinity component which is selectively influenced by norepinephrine is specifically associated with norepinephrine uptake sites.

Biogenic amine uptake is absolutely dependent upon sodium. It has been assumed that this reflects a link between the uptake process and the sodium-potassium ATPase. Sodium exerts a striking influence upon [^{3}H]-desipramine binding to norepinephrine uptake sites potently stimulating binding. The effect is selective as potassium is inactive (Fig. 2).

It appears unlikely that the influence of sodium on [^{3}H]desipramine binding is due to a link to the sodium–potassium ATPase, since oubain, a potent inhibitor of this enzyme, does not affect binding or its regulation by sodium. Thus, there may be a unique sodium-recognizing site associated with the norepinephrine uptake process. Studies of this regulation by sodium in the soluble state may help clarify its mechanism. The sodium effect appears to be a general one, since [^{3}H]imipramine binding to serotonin uptake sites is also regulated in a similar fashion by sodium.

There are numerous other ways in which neurotransmitter receptor binding studies have enhanced our understanding of synaptic transmission and had other ramifications. The ability to localize receptors microscopically by autoradiography (Young and Kuhar, 1979) has been a giant step forward in explaining how drugs exert their pharmacological effects. Thus, the numerous pharmacological actions of opiates are readily explained simply by knowing sites in the brain where opiate receptors are highly concentrated. Similarly, the localization of histamine H_1 and benzodiazepine receptors can explain most of their pharmacological actions.

Neurotransmitter receptor binding has proved a valuable tool in drug development. Screening studies in intact animals do not permit systematic structure–activity analysis since drugs may differ in their pharmacological potencies because of differences in bioavailability or metabolism as well as differences in receptor affinity. Besides focusing selectively upon the receptor, binding studies can amplify the productivity of drug research, since the chemist need synthesize only small quantities of test compounds and only a few test tubes need be used for the receptor assay rather than large numbers of experimental animals.

As with most new areas of research, one cannot predict all the future applications of molecular research into receptors. The chapters of this volume point to many possibilities.

Acknowledgments

This work was supported by USPHS grants MH-18501, NS-16375, DA-00266, RSA Award DA-00074, and a grant from the McKnight Foundation. I thank Dawn C. Hanks for the preparation of the manuscript.

5. REFERENCES

Baudry, M., and Lynch, G., 1980, Regulation of hippocampal glutamate receptors: evidence for the involvement of calcium-activated protease, *Proc. Natl. Acad. Sci. USA* **77**:2298–2302.

Bruns, R. F., Daly, J. W., and Snyder, S. H., 1980, Adenosine receptors in brain membranes:

binding of N^6-cyclohexyl[^{3}H]adenosine and 1,3-diethyl-8-[^{3}H]phenylxanthine, *Proc. Natl. Acad. Sci. USA* **77:**5547–5551.

BYLUND, D. B., and SNYDER, S. H., 1976, Beta adrenergic receptor binding in membrane preparations from mammalian brain, *Mol. Pharmacol.* **12:**568–680.

CHANG, R. S. L., and SNYDER, S. H., 1980, Histamine H_1 receptor binding sites in guinea pig brain membranes: regulation of agonist interactions by guanine nucleotides and cations, *J. Neurochem.* **34:**916–922.

CHILDERS, S. R., and SNYDER, S. H., 1980, Differential regulation by guanine nucleotides of opiate agonist and antagonist receptor interactions, *J. Neurochem.* **34:**583–593.

COSTA, T., RODBARD, and PERT, C. B., 1979, Is the benzodiazepine receptor coupled to a chloride anion channel? *Nature (London)* **277:**315–317.

COYLE, J. T., and SNYDER, S. H., 1969, Catecholamine uptake by synaptosomes in homogenates of rat brain: stereospecificity in different areas, *J. Pharmacol. Exp. Ther.* **170:**221–231.

ENERO, M. A., 1971, Properties of the peripheral opiate receptors in the cat nictitating membrane, *Eur. J. Pharmacol.* **45:**349–356.

GAVISH, M., and SNYDER, S. H., 1980, Soluble benzodiazepine receptors: GABAergic regulation, *Life Sci.* **26:**579–582.

GAVISH, M., and SNYDER, S. H., 1981, Gamma-aminobutyric acid and benzodiazepine receptors: copurification and characterization, *Proc. Natl. Acad. Sci. USA* **78:**1939–1942.

GAVISH, M., GOODMAN, R. R., and SNYDER, S. H., 1982, Solubilized adenosine receptors in the brain regulated by guanine nucleotides, *Science* **215:**1633–1635.

GOODMAN, R. R., COOPER, M. J., GAVISH, M., and SNYDER, S. H., 1982, Guanine nucleotide and cation regulation of the binding of [^{3}H]cyclohexyladenosine and [^{3}H]diethylphenylxanthine to adenosine A_1 receptors in brain membranes, *Mol. Pharmacol.* **21:**329–335.

GREENBERG, D. A., U'PRICHARD, D. C., SHEEHAN, P., and SNYDER, S. H., 1978, Alpha-Noradrenergic receptors in the brain: differential effects of sodium on binding of [^{3}H]-agonists and [^{3}H]antagonists, *Brain Res.* **140:**378–384.

HIRATA, F., and AXELROD, J., 1980, Phospholipid methylation and biological signal transmission, *Science* **209:**1082–1090.

HORN, A. S., COYLE, J. T., and SNYDER, S. H. 1971, Catecholamine uptake by synaptosomes from rat brain. Structure–activity relationships of drugs with differential effects on dopamine and norepinephrine neurons, *Mol. Pharmacol.* **7:**66–80.

INNIS, R. B., and SNYDER, S. H., 1980, Distinct cholecystokinin receptors in brain and pancreas, *Proc. Natl. Acad. Sci. USA* **77:**6917–6921.

INNIS, R. B., MANNING, D. C., STEWART, J. M., and SNYDER, S. H., 1981, [^{3}H]Bradykinin receptor binding in mammalian tissue membranes, *Proc. Natl. Acad. Sci. USA* **78:**2630–2634.

KOSKI, G., SIMONDS, W. F., and KLEE, W. A., 1981, Guanine nucleotides inhibit binding of agonists and antagonists to soluble opiate receptors, *J. Biol. Chem.* **256:**1536–1538.

LANGER, S. Z., and BRILEY, M. S., 1981, High-affinity [^{3}H]imipramine binding: a new biological tool for studies in depression, *Trends Neuroscie.* **4:**28–31.

LEE, C.-M., and SNYDER, S. H., 1981, Norepinephrine neuronal uptake binding sites in rat brain membranes labeled with [^{3}H]desipramine, *Proc. Natl. Acad. Sci. USA* **78:**5250–5254.

LICHTSHTEIN, D., BOONE, G., and BLUME, A. J., 1979, A physiological requirement of Na^+ for the regulation of cAMP levels in intact NG108–15 cells. *Life Sci.* **25:**985–992.

MULLER, W. E., and *Snyder*, S. H., 1978, Strychnine binding associated with synaptic glycine receptors in rat spinal cord membranes: ionic influences, *Brain Res.* **147:**107–116.

PASTERNAK, G. W., SNOWMAN, A. M., and SNYDER, S. H., 1975, Selective enhancement of [^{3}H]-opiate agonist binding by divalent cations, *Mol. Pharmacol.* **11:**735–744.

PERT, C. B., PASTERNAK, G., and SNYDER, S. H., 1973, Opiate agonists and antagonists discriminated by receptor binding in brain, *Science* **182:**1359–1361.

REHAVI, M., SKOLNICK, P., HULIHAN, B., and PAUL, S. M., 1981, High-affinity binding of [^{3}H]desipramine to rat cerebral cortex: relationship to tricyclic antidepressant-induced inhibition of norepinephrine uptake, *Eur. J. Pharmacol.* **70:**597–599.

REHFELD, J. F., 1978, Immunochemical studies on cholecystokinin. II. Distribution and molecular heterogeneity in the central nervous system and small intestines of man and hog, *J. Biol. Chem.* **253**:4022–4030.

RODBELL, M., KRANS, H. J., POHL, S. L., and BIRNBAUMER. L., 1971, The glucagon-sensitive adenyl cyclase system im plasma membranes of rat liver. IV. effects of guanyl nucleotides on binding of [^{125}I]glucagon, *J. Biol. Chem.* **246**:1872–1876.

SNYDER, S. H., and GOODMAN, R. R., 1980, Multiple neurotransmitter receptors, *J. Neurochem.* **35**:5–15.

TOLL, L., and SNYDER, S. H., 1982, Solubilization and characterization of histamine H_1 receptors in brain, *J. Biol. Chem.*, in press.

U'PRICHARD, D. C., and SNYDER, S. H., 1980, Interactions of divalent cations and guanine nucleotides at α-noradrenergic receptor binding sites in bovine brain, *J. Neurochem.* **34**:385–394.

USDIN, T. B., CREESE, I., and SNYDER, S. H., 1980, Regulation by cations of [^{3}H]spiroperidol binding associated with dopamine receptors of rat brain, *J. Neurochem.* **34:916–922.**

VARGAS, F., GREENBAUM, L., and COSTA, E., 1980, Participation of cysteine proteinase in the high-affinity calcium-dependent binding of glutamate to hippocampal synaptic membranes, *Neuropharmacology* **19**:781–794.

YOUNG, A. B., and SNYDER, S. H., 1974, The glycine synaptic receptor: Evidence that strychnine binding is associated with the ionic conductance mechanism, *Proc. Natl. Acad. Sci. USA* **71**:4002–4005.

YOUNG, W. S., III, and KUHAR, M. J., 1979, A new method for receptor autoradiography: [^{3}H]opioid receptors in rat brain, *Brain Res.* **179**:255–270.

YOUSUFI, M. A. K., THOMAS, J. W., and TALLMAN, J. F., 1979, Solubilization of benzodiazepine binding sites from rat cortex, *Life Sci.* **25**:463–470.

ZARBIN, M. A., INNIS, R. B., WAMSLEY, J. K., and SNYDER, S. H., 1981, Autoradiographic localization of CCK receptors in guinea pig brain, *Eur. J. Pharmacol.* **71**:349–350.

ZIEGLGANSBERGER, W., and BAYERL, H., 1976, The mechanism of inhibition of neuronal activity by opiates in the spinal cord of the cat, *Brain Res.* **115**:111–123.

2

OPIATE RECEPTORS

Linda E. Robson, Stewart J. Paterson, and Hans W. Kosterlitz

1. INTRODUCTION

A fundamental concept of modern pharmacology is that the biological activity of a drug results from its interaction with a macromolecular tissue component. Such a receptor must be defined in terms of its pharmacology: it encompasses both a recognition site to which a drug binds, and a factor which translates binding into the events which lead ultimately to a biological response. A drug which binds specifically to the recognition site has an affinity for that receptor, usually expressed as the equilibrium dissociation constant, K_D, which is the ratio of the association and dissociation rate constants. Agonists have, in addition to affinity for the recognition site, the ability to trigger the translational phase to produce a biological effect, while antagonists fail to elicit a biological response. This chapter will deal primarily with the biochemical characterization of the opiate receptors. Since binding studies yield information only about the recognition site of the receptor, they cannot distinguish between the agonist or antagonist nature of the interactions. We shall therefore correlate the biochemical data with results from pharmacological assays.

The theory that opiates produce their effects by interaction with a specific receptor developed from observations of the rigid stereochemical and structural requirements for their analgesic action (Beckett and Casy, 1954; Beckett *et al.*, 1956; Braenden *et al.*, 1955; Portoghese, 1965). The

Linda E. Robson, Stewart J. Paterson, and Hans W. Kosterlitz • Unit for Research on Addictive Drugs, University of Aberdeen, Marischal College, Aberdeen AB9 1AS, Scotland.

investigation of the mode of action of opiates in the central nervous system was facilitated by the use of the guinea pig ileum. In this relatively simple model, the potency of opiates to inhibit neurotransmission correlated well with their potency as analgesic agents; furthermore, these effects were competitively reversed by the narcotic antagonists naloxone and naltrexone (Paton, 1957; Kosterlitz and Watt, 1968; Kosterlitz and Waterfield, 1975).

Early attempts at biochemical demonstration of the existence of specific binding sites for opiates were unsuccessful due to the difficulty of distinguishing between specific and nonspecific binding. In 1971, Goldstein and co-workers showed that, while [^{3}H]levorphanol was bound stereospecifically to mouse brain homogenates, the stereospecific binding amounted to only 2% of the total binding. However, the use of low concentrations of labeled ligand of high specific activity allowed unequivocal demonstration of sites to which opiates such as dihydromorphine, etorphine, and naloxone are bound in a saturable and specific manner (Pert and Snyder, 1973; Simon *et al.*, 1973; Terenius, 1973*a*).

In view of the existence of specific opiate receptors in nervous tissue and the observation that electrically induced analgesia is antagonized by naloxone (Akil *et al.*, 1972), it seemed likely that an endogenous ligand was present. In 1975, Hughes *et al.* (1975*b*) identified two pentapeptides with opiate activity, [Met]enkephalin and [Leu]enkephalin, from extracts of pig brain. When the amino acid sequence of [Met]enkephalin was found to be identical to residues 61–65 of the pituitary hormone β-lipotropin, it was quickly established that the sequence 61–91 of β-lipotropin, C-fragment or β-endorphin, also had potent opiate activity (Bradbury *et al.*, 1976, 1977; Cox *et al.*, 1976; Waterfield *et al.*, 1977). One of the more important developments in our understanding of the opiate receptors has been the characterization of the binding of these and other opioid peptides to the recognition sites of the receptor. These studies have led to the hypothesis that the opioid peptides interact with a δ-binding site in addition to the μ-binding site characteristic for morphine.

Several reviews and reports of symposia on this subject have been published (Snyder *et al.*, 1975; Goldstein and Cox, 1978; Simon and Hiller, 1978; Kosterlitz and Paterson, 1980; Kosterlitz and McKnight, 1980, 1981; Snyder and Goodman, 1980).

2. SCOPE AND LIMITATIONS OF METHODS

2.1. Pharmacological Methods *in Vitro* and *in Vivo*

The earliest approach to the investigation of opiate receptor interactions was the study of their effect, usually their antinociceptive effect, in the whole animal; systematic modification of the molecular structure of a prototype drug and comparison of the potencies of congeners to produce a certain effect allows inferences to be drawn as to the receptor structure (Beckett and Casy, 1954). The antinociceptive action of opiates may be determined by several methods, e.g., the hot plate test, the tail flick test, and several writhing tests. The results obtained with the different methods may be inconsistent, particularly with analgesics that have both agonist and antagonist actions. When a heterogeneous population of receptors is involved, interpretation is difficult unless specific antagonists for each subtype of receptor are available. The concentration of drug at the receptor site, which is affected by absorption, distribution, metabolism, and excretion, is difficult to assess. This problem can be minimized by application of the drug close to its site of action. For instance, the effects on the central nervous system of drugs that do not pass the blood–brain barrier can be investigated after administration into the cerebral ventricles. Furthermore, since the observed antinociceptive response is a consequence of a complex sequence of events initiated by the drug–receptor interaction, it may be modified by many factors.

The influences of distribution and metabolism are reduced by the use of isolated tissues in which neurotransmission is sensitive to inhibition by opiates. In the last decade three preparations, the guinea pig ileum (Kosterlitz and Watt, 1968), the mouse vas deferens (Hughes *et al.*, 1975*a*, and the rat vas deferens (Lemaire *et al.*, 1978) have been extensively used to study the interactions of opiates and opioid peptides with their receptors. However, before one can accurately determine the potencies of opioid peptides, it is necessary to prevent degradation by peptidases in the biophase. In this context, it has been shown that the use of thiorphan, an enkephalinase inhibitor, together with bestatin, an aminopeptidase inhibitor, and captopril, an inhibitor of the angiotensin-converting enzyme, increases the potency of [Met]enkephalin by about 10-fold in the guinea pig ileum, 25-fold in the rat vas deferens, and 8-fold in the mouse vas deferens (Corbett *et al.*, 1982).

The *in vitro* pharmacology of opioids is usually studied in tissues from the peripheral nervous system, while the biochemical characterization of the binding sites is most commonly investigated in the central nervous system. It would be useful, wherever possible, to establish the pharmacological properties and the biochemical characteristics of the binding sites in the same tissue. This would require the development of pharmacological assay systems in the CNS.

2.2. Biochemical Methods; Binding Assays *in Vivo* and *in Vitro*

The analysis of the interaction of radiolabeled opioids with their recognition site can be performed *in vivo* or *in vitro*. In studies where a labeled drug is administered to the whole animal and its interaction with binding sites in various isolated tissues is determined after sacrifice, the results are modified by the various factors already discussed in Section 2.1. The specific binding of a labeled ligand to a receptor site is saturable and is the difference between that observed in the presence and absence of an excess of unlabeled opioid. When pairs of stereoisomers are tested, the (−)-isomers have greater affinity for the binding sites than the (+)-isomers. It is important that the chemical structure of the unlabeled ligand used to differentiate between specific and nonspecific binding be not too closely related to that of the labeled ligand; it has been found in our laboratory that compounds of the benzomorphan group such as bremazocine, ethylketazocine, and Mr 2266 bind with low affinity to a site from which they are not displaced by etorphine, diprenorphine, and naloxone. Thus, this site cannot be defined as an opiate binding site. Similar observations were obtained in NCB-20 cells (McLawhon *et al.*, 1981) and in guinea pig brain (Su, 1981).

The kinetic parameters, equilibrium dissociation constant (K_D), and maximal number of binding sites can be estimated directly from saturation experiments using the methods of Scatchard (1949) and Hill (1910). The affinity of unlabeled ligands can be estimated indirectly by the inhibition constant (K_I) using the equation

$$K_I = \frac{IC_{50}}{1 + [L]/K_D}$$

where [L] is the concentration of labeled ligand, K_D is its equilibrium dissociation constant, and IC_{50} is the concentration of unlabeled ligand required to displace 50% of the binding of labeled ligand (Cheng and Prusoff, 1973). Unfortunately, most of the ligands available are not selective and bind to more than one site with varying affinities. To resolve this problem, the binding of a nonselective labeled ligand can be restricted to, for example, the κ-binding site if its binding to additional sites is displaced with unlabeled ligands which have negligible affinity for the κ-site (Magnan *et al.*, 1982). An alternative method is the use of computer programs for nonlinear curve fitting which allow simultaneous analysis of saturation curves for several labeled ligands or of displacement curves for several unlabeled ligands to obtain estimates of their common parameters (Munson and Rodbard, 1980). An ideal situation would be the availability of selective ligands for each subtype of binding site.

Binding assays provide information only about the interaction of opioids with the recognition site of the receptor since by definition agonism and antagonism are pharmacological parameters. It is therefore necessary to attempt to correlate binding data with those from pharmacological assays.

2.3. Histochemical Methods

Autoradiographical visualization of opiate binding sites with labeled opioids of high specific activity and high affinity offers a detailed differential analysis of the distribution of binding sites. As already discussed for pharmacological and biochemical studies, the disposition of a drug after administration *in vivo* is liable to modification. However, the binding sites can be labeled in lightly fixed tissue sections, a procedure that does not appear to alter the characteristics of the binding sites. The use of *in vitro* labeling of tissue sections is more economical than the use of *in vivo* techniques; it also allows studies in human postmortem tissue and makes variation in the conditions of incubation possible. Simultaneous quantitative measurements of different subtypes of binding sites can be obtained in adjacent sections, and the distribution of the binding sites can be correlated with that of endogenous opioid peptides (Young and Kuhar, 1979).

3. HETEROGENEITY OF OPIATE RECEPTORS

3.1. Evidence Based on Pharmacological Assays

The concept that there are subtypes of the opiate receptor was originally suggested by Martin (1967) who observed that in man nalorphine had a dual action, antagonizing the analgesic effect of morphine and also acting as an analgesic in its own right. He concluded that the analgesic effect of nalorphine was mediated by a receptor which was different from the morphine receptor and which he later referred to as the κ-receptor. In a detailed analysis, the pharmacological profiles of several types of opiates in neurophysiological and behavioral tests in the chronic spinal dog were sufficiently different to suggest the existence of three subclasses of receptor designated μ, κ, and σ for which the prototype agonists were morphine, ketazocine, and *N*-allylnormetazocine (SKF 10047), respectively (Gilbert and Martin, 1976; Martin *et al.*, 1976). Thus, morphine induced analgesia, meiosis, bradycardia, hypothermia, and indifference to environmental stimuli. Ketazocine produced meiosis, general sedation, and depression of flexor reflexes but did not alter the skin twitch reflex or pulse rate. *N*-allylnormetazocine caused mydriasis, tachypnoea, tachycardia, and mania. Under this classification, morphine had high affinity for the μ-receptor, medium affinity for the κ-receptor, and low affinity for the σ-receptor. Ketazocine had high affinity for the κ-receptor and low affinity for the μ- and σ-receptors, and nalorphine had high affinity for the μ- and κ-receptors. Cyclazocine and *N*-allylnormetazocine, which are distinguished by their ability to produce dysphoric and psychotomimetic effects, had high affinity for the μ- and σ-receptors. Cyclazocine had high affinity also for the κ-receptor. Furthermore, the withdrawal syndrome in the morphine-dependent dog was suppressed by morphine, but not by ketazocine, ethylketazocine, and pentazocine, which therefore appeared to have little affinity for the μ-receptor. Morphine and ketazocine suppressed the withdrawal syndrome in the cyclazocine-dependent dog.

The use of a variety of *in vivo* tests has allowed similar classifications to be made for the rat, squirrel monkey, rhesus monkey, and pigeon (cf. Adler, 1981; Cowan, 1981; Herling and Woods, 1981). Cyclazocine and *N*-allylnormetazocine shared discriminative effects with morphine and ketazocine as well as with the nonopioids phencyclidine and ketamine. Thus, there is a great deal of evidence for the view that some of the actions of cyclazocine and *N*-allylnormetazocine may be mediated through a nonopioid mechanism, the putative σ-receptor, as will be discussed in Section 7.

It was possible to obtain evidence for the existence of separate μ- and κ-receptors in *in vitro* experiments on the isolated preparations of the guinea pig ileum and mouse vas deferens. Certain benzomorphans, including ketazocine, which are good antinociceptive agents in the rat but do not substitute for morphine in the morphine-dependent monkey (Villarreal and Seevers, 1972; Swain and Seevers, 1974, 1976), could be distinguished by the fact that in the mouse vas deferens their agonist potency relative to normorphine was only 25% of that found in the guinea pig ileum. Furthermore, to antagonize the inhibitory effects of such benzomorphans three to seven times more naloxone was required than for antagonism of normorphine (Hutchinson *et al.*, 1975). A similarly high concentration is required to antagonize the effects of *N*-allylnormetazocine in the guinea pig ileum (Su *et al.*, 1981).

When the enkephalins and β-endorphin were examined in the guinea pig ileum and mouse vas deferens assays, it was found that their action could not be explained solely by interaction with the μ-, κ-, and σ-receptors. Whereas β-endorphin (LPH 61–91) was equipotent in the two assay systems, [Met]enkephalin (LPH 61–65) and [Leu]enkephalin were relatively more potent in the mouse vas deferens than in the guinea pig ileum. In contrast, morphine had greater potency in the guinea pig ileum than in the mouse vas deferens. These observations were mirrored by the binding assays which showed that, while β-endorphin was equipotent in inhibiting the binding of [^{3}H]naloxone and [^{3}H]-[Leu]enkephalin in homogenates of guinea pig brain, morphine was more potent in inhibiting the binding of [^{3}H]naloxone than [^{3}H]-[Leu]enkephalin, and the enkephalins were more potent in inhibiting the binding of [^{3}H]-[Leu]enkephalin than [^{3}H]naloxone (Fig. 1). These observations led to the suggestion that there are two types of receptors for the opioid peptides; the μ-

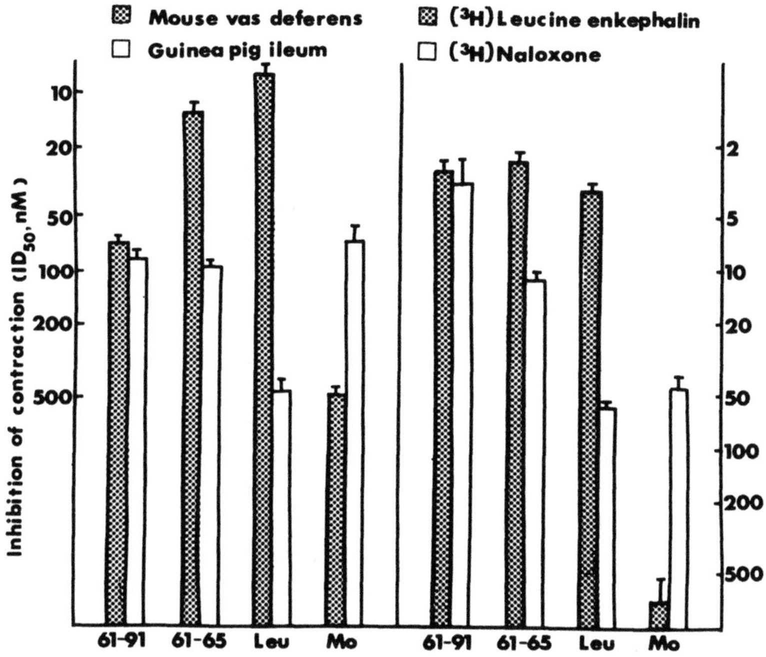

FIG. 1. The agonist activities of various compounds in the mouse was deferens and guinea pig ileum and their potencies to inhibit [^{3}H]-[Leu]enkephalin and [^{3}H]naloxone binding in homogenates of guinea pig brain (pH 7.4 at 0°C, no Na^{+}, 150 min). The numbers on the abscissa indicate the amino acid sequence of β-LPH. 61–91: β-endorphin; 61–65: [Met]-enkephalin; Leu: [Leu]enkephalin; Mo: morphine. (Reprinted with modification from Lord *et al.*, 1977; Copyright © 1977, Macmillan Journals Limited).

receptor with which morphine preferentially interacts appeared to be predominant in the guinea pig ileum, while an enkephalin-preferring δ-receptor, which is different from the μ- or κ-receptor, appeared to predominate in the mouse vas deferens. In agreement with this interpretation, a concentration of naloxone about 10 times higher than that needed to antagonize morphine was required for the opioid peptides in the mouse vas deferens. In the guinea pig ileum assay, the enkephalins appeared to interact with μ-receptors because they did not require more naloxone for antagonism than did morphine (Lord *et al.*, 1976, 1977). Enkephalin analogs such as [D-Ala^{2},L-Met^{5}]enkephalin, [D-Ala^{2},D-Met^{5}]enkephalin, and [D-Ala^{2},D-Leu^{5}]enkephalin, which retain an activity profile more similar to the endogenous enkephalins than to morphine in the pharmacological and binding assays (Section 3.2), also required more naloxone for antagonism than did normorphine in the mouse vas deferens. On the other hand, two analogs, [D-Ala^2,MePhe^4,Met(O)-ol^5]enkephalin and [D-Ala^2,MePhe^4,Gly-ol^5]enkephalin, which have an activity profile more similar to morphine than to [Met]enkephalin, did not. In the guinea pig ileum the K_e values for naloxone antagonism were all between 1.9 and 3.8 nM (Table 1) (Kosterlitz *et al.*, 1980; Gillan *et al.*, 1981).

Hughes *et al.* (1975*a*) reported that, while morphine inhibited the electrically evoked contractions of the mouse vas deferens, the vasa deferentia of the rat, rabbit, guinea pig, cat, hamster, and gerbil were insensitive. More recently it has been shown that, compared with the mouse vas deferens and the guinea pig ileum, the rat vas deferens has a different pattern of sensitivity to the inhibitory actions of opiates and opioid peptides (Lemaire *et al.*, 1978; Schulz *et al.*, 1979*a*; Wüster *et al.*, 1979; Huidobro *et al.*, 1980; Jacquet, 1980; Gillan *et al.*, 1981). The potency of β-endorphin in the rat vas deferens (24–130 nM) is close to that observed in the guinea pig ileum and the mouse vas deferens, but the potencies of [Met]enkephalin, [Leu]enkephalin, [D-Ala^2,L-Leu^5]enkephalin, [D-Ala^2,D-Leu^5]enkephalin, and etorphine are two to three orders of magnitude lower in the rat vas deferens than in the other two preparations. A possible inhibitory effect of morphine cannot be shown because it appears to be masked by an excitatory action which is not mediated by an opiate receptor; however, the highly selective μ-agonist [D-Ala^2,MePhe^4,Gly-ol^5]enkephalin inhibits the twitch with an IC_{50} value of 410 nM (Gillan *et al.*, 1981) compared to 10–30 nM and 20–40 nM in the guinea pig ileum and mouse vas deferens, respectively (Handa *et al.*, 1981; Gillan *et al.*, 1981, and unpublished data). The relatively high potency of β-endorphin in the rat vas deferens prompted the suggestion (Wüster *et al.*, 1979) that its effect in this preparation is mediated by a further subtype of opiate receptor, the ε-receptor. So far, there has been no evidence from binding assays which would support the existence of the ε-subtype; this failure may be due to the fact that the receptor density in the rat vas deferens is very low (J. Magnan, unpublished observations). It is interesting that compounds with known agonist–antagonist activity are pure antagonists in the rat vas deferens (Wüster *et al.*, 1980; Magnan *et al.*, 1982), but, unexpectedly, ketazocine-like compounds such as bremazocine and ethylketazocine, which are potent pure agonists in the mouse vas deferens and guinea pig ileum, are pure antagonists in the rat vas deferens (Gillan *et al.*, 1981). There is so far no explanation for this observation, particularly since there is no direct information about the subtype of the opiate receptor to which the benzomorphans bind in the rat vas deferens. As a working hypothesis, it has been suggested that, since the benzomorphans antagonize the selective μ-agonists normorphine and [D-Ala^2,MePhe^4,Gly-ol^5]enkephalin more readily than the selective δ-agonist [D-Ala^2,D-Leu^5]enkephalin, their antagonist action may be due to an interaction with the μ-site (Gillan *et al.*, 1981).

TABLE 1
Effectiveness of Naloxone (K_e, nM) in Antagonizing the Agonist Actions of Enkephalin Analogs and Normorphine[a]

Agonists	Guinea pig ileum	Mouse vas deferens
[Gly^2,L-Met^5]enkephalin	2.5	23
[D-Ala^2,L-Met^5]enkephalin	2.6	22
[D-Ala^2,D-Met^5]enkephalin	2.4	26
[D-Ala^2,D-Leu^5]enkephalin	1.9	32
[D-Ala^2,MePhe^4,Met(O)-ol^5]enkephalin	3.8	5.7
[D-Ala^2,MePhe^4,Gly-ol^5]enkephalin	3.7	1.6
Normorphine	1.9	1.8

[a] Modified from Lord *et al.* (1977), Kosterlitz *et al.* (1980), and Gillan *et al.* (1981).

The electrically evoked contractions of the vas deferens of the rabbit, but not those of the cat and guinea pig, are inhibited by β-endorphin (Lemaire *et al.*, 1978). In the same preparation, the effects of bremazocine, ethylketazocine, ketazocine, and cyclazocine are inhibitory, but β-endorphin, [D-Ala^2,MePhe^4,Met(O)-ol^5]enkephalin, [Leu]enkephalin, [Met]enkephalin, morphine, and *N*-allylnormetazocine are inactive. Since the inhibitory effects of the benzomorphans were antagonized by naloxone, the lack of effect of the opioid peptides may be indicative of the exclusive presence of κ-receptors (Oka *et al.*, 1981).

While the rabbit ileum is insensitive to morphine (Cowie *et al.*, 1978), a recent report shows that the enkephalins inhibit the responses of the ilea of rabbit, rat, and mouse to electrical stimulation (Oka, 1980). In the mouse ileum [Met]enkephalin was 240 times more potent than the μ-selective agonist [D-Ala^2,MePhe^4,Gly-ol^5]enkephalin (Handa *et al.*, 1981). In this connection, it is of interest that the flux of Na^+ and K^+ across the mucous membrane of guinea pig small intestine is decreased by [D-Ala^2,D-Leu^5]enkephalin but not by morphine (Kachur *et al.*, 1980).

Further evidence for multiple opiate receptors has been derived from the observation that the relative potencies of normorphine and [Met]-enkephalin are different in the vasa deferentia from different strains of mice (Berti *et al.*, 1978; Waterfield *et al.*, 1978; Szerb and Vohra, 1979). In seven strains of mice the IC_{50} values of normorphine varied between 46 and 620 nM while those of [Met]enkephalin varied only between 2.6 and 11.4 nM.

In addition, it has been shown that, compared with those from naive mice, the vasa deferentia from mice infused subcutaneously with [D-Ala^2,D-Leu^5]enkephalin (5 μg/hr, 6 days) were less sensitive to the inhibitory effects of the δ-selective ligands; [D-Ala^2,D-Leu^5]enkephalin and [Leu]-enkephalin had dose ratios of 800 and 250, respectively, while the sensitivities of the mouse vas deferens to the μ-selective agonists normorphine and [D-Ala^2,MePhe^4,Met(O)-ol^5]enkephalin and to ketazocine-like compounds were not affected. In contrast, infusion with sufentanyl (10 μg/hr, 6 days) resulted in selective tolerance to μ-agonists while the sensitivities to δ-selective agonists and to ketazocine remained unchanged (Schulz *et al.*, 1980*a,b*; Wüster *et al.*, 1981). Chronic exposure of rats to etorphine resulted in losses of sensitivity of the vas deferens to etorphine and sufentanyl, giving dose ratios of >250 and >1250, respectively, but in a smaller decrease in sensitivity to β-endorphin with a dose ratio of only 7 (Schulz *et al.*, 1980*b*). A similar approach used in the guinea pig (Su *et al.*, 1981) showed that ileum preparations from animals chronically treated with morphine exhibit a greater loss of sensitivity to morphine than to ketazocine or *N*-allylnormetazocine.

3.2. Binding Assays

3.2.1. Evidence for μ- and δ-Binding Sites

Evidence for the existence of multiple opiate binding sites was first obtained from the comparison of the inhibitory effects of unlabeled opiates and opioid peptides on the binding of tritiated opiates and opioid peptides. In homogenates of guinea pig brain, [Leu]enkephalin was 25 times more potent at displacing the binding of [^{3}H]-[Leu]enkephalin than of [^{3}H]naloxone (Fig. 1). In contrast, morphine preferentially displaced the binding of [^{3}H]naloxone rather than of [^{3}H]-[Leu]enkephalin while β-endorphin was equipotent at displacing both tritiated ligands (Lord *et al.*, 1976, 1977). Similarly, in homogenates of rat brain, [Met]enkephalin and [Leu]enkephalin competed more readily for the binding sites of labeled enkephalins than for the binding sites of labeled opiates, while the converse was true for the competition by nonpeptide opiates (Terenius, 1977; Law and Loh, 1978; Simantov *et al.*, 1978; Chang and Cuatrecasas, 1979; Simon *et al.*, 1980). These findings, together with the results obtained in the guinea pig ileum and mouse vas deferens (Section 3.1), have led to the conclusion that in the brain there are at least two types of binding sites for opioid peptides; as a first approximation, the main part of [^{3}H]naloxone binding may be to the μ-site, while the tritiated enkephalins bind mainly to a site which is different from that of the κ-receptor and which has been assigned to the δ-receptor.

In contrast to β-endorphin, the enkephalins are rapidly degraded by aminopeptidases, carboxypeptidases, and enkephalinase. Thus, stable analogs are essential for use *in vivo*. Although such analogs may be useful for the characterization of receptors in *in vitro* assay systems, it is important to establish that their binding profiles are similar to those of [Met]-

enkephalin and [Leu]enkephalin, both of which, and particularly the latter, exhibit much higher affinity for the δ-binding site than for the μ-binding site. As expected, changes in the structure of the enkephalins induced changes in the pattern of activity (Kosterlitz *et al.*, 1980). Analogs obtained by the substitution of D-Ala for Gly^2 and D-Met or D-Leu for L-Met^5 or L-Leu^5 exhibited the same pattern of inhibition of the binding of [^{3}H]naltrexone and [^{3}H]-[Leu]enkephalin as [Met]enkephalin and [Leu]-enkephalin (Table 2). However, in the bioassays, substitution of Gly^2 by D-Ala in [Leu]enkephalin increased the potency in the mouse vas deferens 5-fold and in the guinea pig ileum 15-fold. The same substitution in [Met]-enkephalin increased the activity in both assays 5- to 6-fold. This increase in potency in the pharmacological assays, in the absence of corresponding potency changes in the binding assays, is probably due to a decreased degradation in the biophase of the tissue. When, in addition to the substitution of D-Ala for Gly^2, the C-terminal amino acid was replaced by its D-isomer, [D-Ala^2,D-Met^5]enkephalin showed no change in the mouse vas deferens but a 70% loss of activity in the guinea pig ileum. The corresponding [Leu]enkephalin analog, [D-Ala^2,D-Leu^5]enkephalin, showed the same loss of activity in the guinea pig ileum but was more potent in the mouse vas deferens.

In contrast, the analog Tyr-D-Ala-Gly-MePhe-Met(O)-ol, FK 33-824 (Römer *et al.*, 1977), exhibited a very different pattern; while its affinity to the μ-binding site was unchanged, that to the δ-binding site was reduced to 6.4% of that of its parent compound, [Met]enkephalin. In the bioassays, [D-Ala^2, MePhe^4,Met(O)-ol^5]enkephalin, which had the same activity as

TABLE 2
Relative Agonist Potencies of Analogs of [Met]enkephalin and [Leu]enkephalin[a]

	Mean inhibition of contraction (36°C; 1/IC$_{50}$ nM)			Mean inhibition of binding in brain homogenates (0°C; 1/K_I nM)		
Substitution in enkephalin	Guinea pig ileum	Mouse vas deferens	G.p.i./ M.v.d.	[^{3}H]-Naltrexone	[^{3}H]-[Leu]-enkephalin	Nal/Leu
[Gly^2,L-Met^5]	0.006	0.067	0.09	0.16	0.91	0.18
[Gly^2,L-Met-NH$_2$^{5}]	0.009	0.048	0.19	0.10	0.30	0.33
[D-Ala^2,L-Met^5]	0.036	0.33	0.11	0.08	1.11	0.07
[D-Ala^2,D-Met^5]	0.011	0.40	0.03	0.030	0.50	0.06
[D-Ala^2,MePhe^4,L-Met(O)-ol^5]	0.13	0.062	2.1	0.20	0.06	3.4
[Gly^2,L-Leu^5]	0.002	0.11	0.02	0.040	0.77	0.05
[D-Ala^2,L-Leu^5]	0.034	0.62	0.06	0.042	0.50	0.08
[D-Ala^2,D-Leu^5]	0.021	2.0	0.01	0.033	0.26	0.13

[a] Modified from Lord *et al.* (1977) and Kosterlitz *et al.* (1980).

[Met]enkephalin in the mouse vas deferens, was 20 times more active than [Met]enkephalin in the guinea pig ileum. Since FK 33-824 is a much more potent antinociceptive agent than [D-Ala^2,D-Leu^5]enkephalin, the μ-receptor may be more important than the δ-receptor for this action. Other analogs with high antinociceptive activities also showed preferential interaction with the μ-binding site, for instance Tyr-D-Met-Gly-Phe-Pro-NH_2 (Székely *et al.*, 1977).

Amidation of the C-terminal carboxyl group of [Met]enkephalin led to a marked decrease in the inhibition of the binding of [^{3}H]-[Leu]-enkephalin but only to a smaller extent of the binding of [^{3}H]naloxone (Table 2). Thus, the affinity to the δ-binding site had decreased while that to the μ-binding site was less affected (Lord *et al.*, 1977). These and other experiments (Kosterlitz *et al.*, 1980) pointed to the importance of a free carboxyl group for the activity at the δ-binding site.

Highly selective ligands are required to obtain further information on the biochemistry and pharmacology of the different binding sites. Dihydromorphine, although it is 29 times more active at the μ-binding site than at the δ-site, is not the most selective ligand available but has nevertheless frequently been used to characterize the μ-site (Gillan *et al.*, 1980; Leslie *et al.*, 1980; Pfeiffer and Herz, 1981). Similar considerations hold for the use of [D-Ala^2,MePhe^4,Met(O)-ol^5]enkephalin (Chang and Cuatrecasas, 1979; Chang *et al.*, 1979; Kream and Zukin, 1979). Normorphine is more selective, being 72 times more active at the μ-binding site than at the δ-binding site, and the corresponding value for Tyr-D-Ala-Gly-MePhe-Gly-ol is 220 (Handa *et al.*, 1981; Kosterlitz and Paterson, 1981). A further μ-selective but less potent peptide is Tyr-Pro-Phe-Pro-NH_2 or morphiceptin (Chang *et al.*, 1981*b*). [D-Ala^2,D-Leu^5]enkephalin, the most commonly used ligand for the characterization of the δ-binding site (Chang *et al.*, 1979; Gillan *et al.*, 1979; Fields *et al.*, 1980; Gillan *et al.*, 1980; Leslie

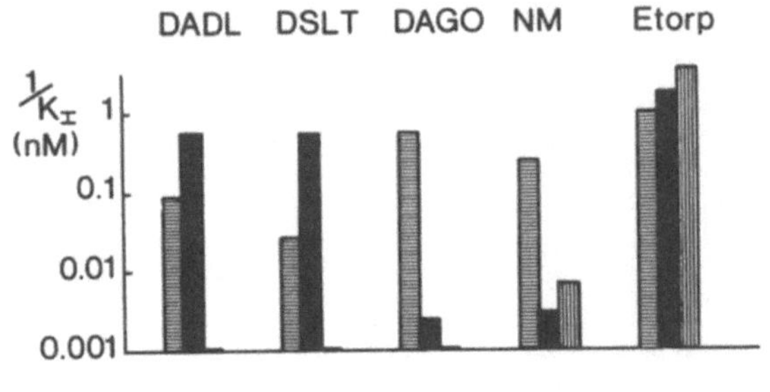

FIG. 2. Interaction of several opiates and opioid peptides at the μ-, δ-, and κ-binding sites. Inhibition of the binding of [^{3}H]-[D-Ala^2,MePhe^4,Gly-ol^5]enkephalin (1 nM; first column, μ), of [^{3}H]-[D-Ala^2,D-Leu^5]enkephalin (1–1.8 nM; second column, δ), and of [^{3}H]-(±)-ethylketazocine after suppression of μ- and δ-binding (0.65 nM; third column, κ). Ordinate: log of the reciprocal K_I (nM). DADL: [D-Ala^2,D-Leu^5]enkephalin; DSLT: [D-Ser^2,L-Leu^5]enkephalyl-Thr; DAGO: [D-Ala^2,MePhe^4,Gly-ol^5]enkephalin; NM: normorphine; Etorp: etorphine.

TABLE 3
Maximal Number of Binding Sites for Tritiated Opioid Peptides and Opiates in Guinea Pig Brain at 0 and 25°C[a]

Tritiated ligand	Maximal number of binding sites (pmol/g brain)	
	0°C	25°C
[Leu]enkephalin	4.9	—
[Met]enkephalin	6.1	—
[D-Ala^2,D-Leu^5]enkephalin	6.4	6.1
[D-Ala^2,MePhe^4,Gly-ol^5]enkephalin	—	3.0
Dihydromorphine	2.8	3.8
[D-Ala^2,L-Leu-NH_2^{5}]enkephalin	10.6	13.0
[D-Ala^2,L-Met-NH_2^{5}]enkephalin	12.7	13.9
Etorphine	—	14.6

[a] Modified from Gillan *et al.* (1980) and Kosterlitz and Paterson (1981).

et al., 1980), is less selective, being only 12 times more active at the δ-binding site than at the μ-binding site. [D-Ser^2,L-Leu^5]enkephalyl-Thr (Gacel *et al.*, 1980) is twice as selective as [D-Ala^2,D-Leu^5]enkephalin for the δ-binding site (Kosterlitz *et al.*, 1982*b*).

The relative affinities of various compounds for the different binding sites is most effectively demonstrated in a histogram (Fig. 2). The logarithms of the reciprocal inhibition constants (K_I) for the binding of [^{3}H]-[D-Ala^2,MePhe^4,Gly-ol^5]enkephalin (μ), [^{3}H]-[D-Ala^2, D-Leu^5]enkephalin (δ), and [^{3}H]ethylketazocine, after suppression of μ- and δ-binding (κ), are given on the ordinate. This histogram illustrates that the peptide analogs [D-Ala^2,D-Leu^5]enkephalin, DADL, and [D-Ser^2,L-Leu^5]enkephalyl-Thr, DSLT, are fairly selective for δ-binding sites; normorphine, NM, and particularly [D-Ala^2,MePhe^4,Gly-ol^5]enkephalin, DAGO, are selective for the μ-binding sites. Etorphine, Etorp, may be considered to be a universal ligand.

Binding assays in which saturation is achieved yield information on the maximum number of the different binding sites provided the ligands used are selective (Table 3). In homogenates of guinea pig brain, [^{3}H]-[Met]enkephalin and [^{3}H]-[Leu]enkephalin have a maximum binding of 5–6 pmol/g tissue at 0°C. The more stable analog [^{3}H]-[D-Ala^2,D-Leu^5]-enkephalin has a similar number of binding sites (6 pmol/g tissue), while the μ-ligands, [^{3}H]dihydromorphine and [^{3}H]-[D-Ala^2,MePhe^4,Gly-ol^5]enkephalin, have 4 and 3 pmol/g tissue, respectively (Gillan *et al.*, 1980; Kosterlitz and Paterson, 1981). Amidation of the carboxyl group of both

[^{3}H]-[D-Ala^2,L-Met^5]enkephalin and [^{3}H]-[D-Ala^2,L-Leu^5]enkephalin increased the maximum number of binding sites to 10–14 pmol/g tissue, a value somewhat lower than that found with the universal ligand etorphine. This indicates that these compounds interact at least with both the μ- and δ-binding sites. In the rat, the relative proportions of μ- and δ-binding sites may be different from the guinea pig. By saturation analysis it was found that δ-binding sites were predominant in the corpus striatum (Fields *et al.*, 1980). However, when the maximum binding was calculated indirectly either from observations with single concentrations of labeled ligands (Chang *et al.*, 1979) or from computer analysis of a series of competitive displacement experiments (Pfeiffer and Herz, 1981), equal numbers of μ- and δ-binding sites were found in the striatum and frontal cortex whereas μ-binding sites predominated in the thalamus and hypothalamus. Equal numbers of μ- and δ-binding sites are found in the bovine corpus striatum (Leslie *et al.*, 1980).

Further evidence for the existence of separate μ- and δ-binding sites is the finding that certain neuroblastoma × glioma hybrid cells possess binding sites with characteristics similar to those of the δ-binding sites in the brain, but do not appear to have μ-binding sites (Chang and Cuatrecasas, 1979) (Section 4.3).

In a recent report, it has been found that straight-chain alcohols up to pentan-1-ol cause a selective inhibition of the binding of the δ-ligand [^{3}H]-[D-Ala^2,D-Leu^5]enkephalin in rat brain homogenate without affecting the binding of [^{3}H]dihydromorphine or [^{3}H]-[D-Ala^2,MePhe^4,Met(O)-ol^5] enkephalin (Hiller *et al.*, 1981). This is the first report of an *in vitro* treatment which discriminates between the two binding sites.

3.2.2. Evidence for the Putative κ-Binding Site

The initial evidence for a κ-receptor based on behavioral and neurophysiological investigations has already been discussed (Section 3.1).

In homogenates of guinea pig brain, the maximal number of binding sites for [^{3}H]ethylketazocine was 13.9 pmol/g tissue, a value close to that observed for the nonselective ligand [^{3}H]etorphine (Kosterlitz *et al.*, 1981). Since ethylketazocine and other ketazocine-like compounds have a high degree of cross-reactivity to the μ-binding sites of [^{3}H]dihydromorphine and [^{3}H]-[D-Ala^2,MePhe^4,Gly-ol^5]enkephalin and, to a lesser extent, to the δ-binding site of [^{3}H]-[D-Ala^2,D-Leu^5]enkephalin (Kosterlitz *et al.*, 1981; Magnan *et al.*, 1982; Section 7), the evidence for a separate κ-binding site is based on the observation that μ-selective and δ-selective agonists have low potency to displace the binding of [^{3}H]ethylketazocine (Table 4). The inhibition curve for the highly selective μ-agonist [D-Ala^2,MePhe^4,Gly-ol^5] enkephalin was biphasic and could be separated into two Hill plots with slopes of unity (Fig. 3). Biphasic inhibition curves observed for other μ-

TABLE 4
Inhibitory Effects of Agonist Opiates and Opioid Peptides on the Binding of [^{3}H]-(±)-Ethylketazocine (0.65 nM) at 25°C[a]

Unlabeled compounds	Inhibition of [^{3}H]-(±)-ethylketazocine binding: K_I (nM)	Hill coefficient	Separation of the two components of the Hill plot, K_I (nM)	Inhibition of [^{3}H]-dihydromorphine binding, K_I (nM)
μ-Selective agonists				
Normorphine	83	0.60	6.3 520	5.0
[D-Ala^2,MePhe^4,Met(O)-ol^5]enkephalin	165	0.46	2.2 1340	1.3
[D-Ala^2,MePhe^4,Gly-ol^5]enkephalin	784	0.35	4.6 4960	3.6
δ-Selective agonists				
[D-Ala^2,D-Leu^5]enkephalin	5230	0.42	38 27,900	16.8
Nonselective agonists				
Etorphine (μ + δ + κ)	0.48	1.11	—	1.1

[a] When the Hill coefficient was 0.6 or less, the K_I values for the two components were calculated separately; the Hill coefficients for the separated components were between 0.8 and 1.2. The readily displaced component was 37 (35–40)% for the μ-agonists and 31% for the δ-agonists. Modified from Kosterlitz *et al.* (1981).

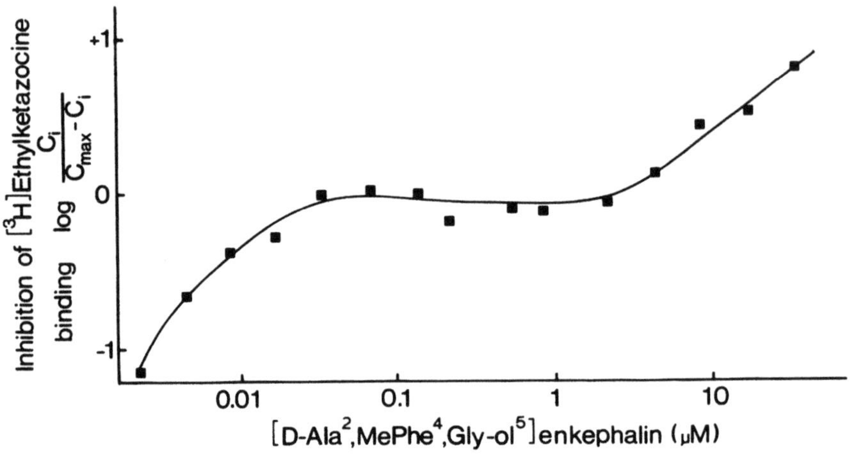

FIG. 3. The inhibitory effect of [D-Ala^2,MePhe^4,Gly-ol^5]enkephalin, a μ-selective ligand, on the binding of [^{3}H]-(±)-ethylketazocine (0.65 nM) in a homogenate of guinea pig brain. C_i: inhibited counts; C_{max}: maximal counts. (From Kosterlitz *et al.*, 1981.)

selective agonists and the δ-selective agonist [D-Ala^2,D-Leu^5]enkephalin could also be separated into two components (Table 4). The K_I values for the readily displaced portion of [^{3}H]ethylketazocine binding were sufficiently close to those for the inhibition of [^{3}H]dihydromorphine binding to suggest that they represented displacement of [^{3}H]ethylketazocine from the μ-binding site. Much higher concentrations were required to displace the remainder of the binding, which was to the putative κ-binding site. Etorphine may be considered to be a universal ligand, the K_I values being similarly low (0.44–1 nM) against the binding of [^{3}H]dihydromorphine, [^{3}H]-[D-Ala^2,D-Leu^5]enkephalin, and [^{3}H]ethylketazocine.

In a further study, the κ-binding site was characterized by measuring the binding of [^{3}H]ethylketazocine in the presence of unlabeled [D-Ala^2,MePhe^4,Gly-ol^5]enkephalin (100 nM) and unlabeled [D-Ala^2,D-Leu^5] enkephalin (100 nM) (Magnan *et al.*, 1982). These concentrations of the unlabeled ligands are 100 times larger than their K_D values for the μ- and δ-binding sites, respectively, and therefore suppress the μ- and δ-binding of [^{3}H]ethylketazocine without affecting the κ-binding. Under these conditions, the maximum number of binding sites for ethylketazocine and etorphine is 5–6 pmol/g tissue compared with 12–13 pmol/g tissue in the absence of the unlabeled μ- and δ-ligands (Fig. 4).

After suppression of μ- and δ-binding, [D-Ala^2,D-Leu^5]enkephalin, [D-Ser^2,L-Leu^5]enkephalyl-Thr, and [D-Ala^2,MePhe^4,Gly-ol^5]enkephalin were found to have negligible affinity for the κ-binding site, while the universal ligand etorphine was more potent at the κ-binding site than at either the μ- or δ-sites (Fig. 2).

[Met]enkephalin and [Leu]enkephalin have very low affinity for the κ-binding site, the K_I values against [^{3}H]bremazocine binding after suppression of μ- and δ-binding being 2980 nM and 5540 nM, respectively (Kosterlitz *et al.*, 1982*a*). This finding indicates that neither [Met]enkephalin nor [Leu]enkephalin is the endogenous ligand for the κ-binding site. The activity of β-endorphin at the κ-binding site is only 2–3% of that at the μ- and δ-binding sites. It is therefore unlikely to be the endogenous ligand for the κ-receptor.

There are species differences in the distribution and density of κ-binding sites. In both guinea pig and rat brain, it has been demonstrated in several laboratories that ketazocine-like compounds have high affinities for the μ-binding sites and somewhat lower affinities for δ-binding sites (Hiller and Simon, 1979, 1980; Chang *et al.*, 1980, 1981*a*; Harris and Sethy, 1980; Römer *et al.*, 1980; Snyder and Goodman, 1980). In contrast, observations of the potencies of μ- and δ-ligands for inhibition of [^{3}H]-ethylketazocine binding were inconsistent. However, it would appear that the density of κ-binding sites in guinea pig brain is greater than in rat brain. In guinea pig brain, of a total of 13 pmol/g tissue approximately 25% have been assigned to the μ-binding sites, 45% to the δ-binding sites,

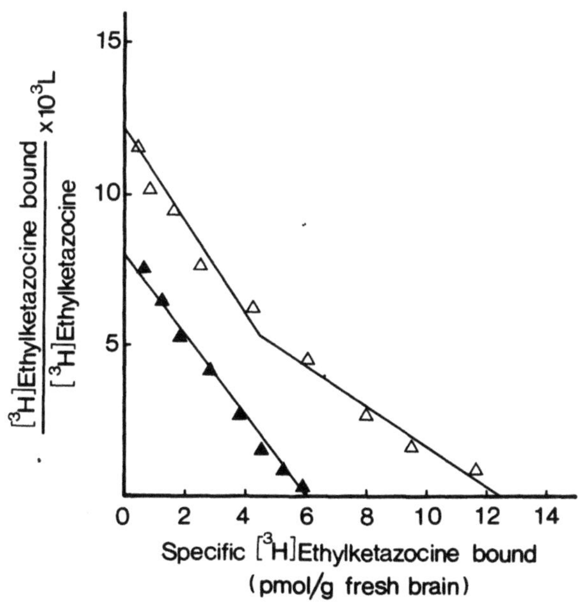

FIG. 4. Scatchard plot of the specific binding of [^{3}H]-(±)-ethylketazocine in the absence (△) and presence (▲) of 100 nM unlabeled [D-Ala^2,D-Leu^5]enkephalin (δ-ligand) and 100 nM unlabeld [D-Ala^2,MePhe^4,Gly-ol^5]enkephalin (μ-ligand). Unsuppressed and suppressed binding were determined in the same homogenate. K_D: unsuppressed 0.84 and 1.6 nM, suppressed 0.74 nM. (From Magnan *et al.*, 1982.)

TABLE 5
Relative Inhibitory Potencies of Prototype Ligands at μ-, δ-, and κ-Binding Sites[a]

Prototype unlabeled ligands	Inhibition of binding of triated ligands: relative potencies at		
	μ	δ	κ
μ: [D-Ala^2,MePhe^4,Gly-ol^5]enkephalin	1	<0.01	<0.01
δ: [D-Ala^2,D-Leu^5]enkephalin	0.1	1	<0.01
κ: (−)-Ethylketazocine	0.5	0.1	1

[a] The potency of each unlabeled ligand in inhibiting binding at its own site was taken to be 1. The tritiated ligands used were: for the μ-site, [D-Ala^2,MePhe^4,Gly-ol^5]enkephalin, for the δ-site, [D-Ala^2,D-Leu^5]-enkephalin, and for the κ-site, (±)-ethylketazocine after suppression of μ- and δ-binding with 100 nM unlabeled [D-Ala^2,MePhe^4,Gly-ol^5]enkephalin (μ) and 100 nM [D-Ala^2,D-Leu^5]enkephalin (δ). Modified from Kosterlitz *et al.* (1981).

and 30% to the κ-binding sites. In rat brain, of 18.5 pmol/g tissue 40% are μ-binding sites, 46% are δ-binding sites, and 14% are κ-binding sites (M. G. C. Gillan, unpublished observations).

To summarize (Table 5), the μ-binding site is selectively activated by the μ-ligand, [D-Ala^2,MePhe^4,Gly-ol^5]enkephalin, which affects the δ- and κ-binding sites only to a very small extent. The κ-binding site responds only to κ-ligands, the δ-binding site to δ-ligands and also to ethylketazocine, and the μ-binding site to μ-ligands, to ethylketazocine, and, to a smaller extent, to [D-Ala^2,D-Leu^5]enkephalin. It is interesting to speculate as to whether the μ-binding site is less demanding than the κ-binding site as far as ligand structure is concerned; alternatively, the available κ-ligands may not be sufficiently selective to avoid cross-reactivity with the μ-binding site. In this context, it must be stressed that analysis at the binding site does not indicate whether a ligand has agonist or antagonist effects. Such considerations may be of importance in solving the dilemma that ketazocine-like compounds cannot substitute for morphine in morphine-dependent rhesus monkeys (Villarreal and Seevers, 1972; Swain and Seevers, 1974, 1976).

At present there is no evidence from binding studies for the existence of additional subtypes of the opiate receptor. Specific binding sites for [^{3}H]phencyclidine have been found in rat brain (Vincent *et al.*, 1979; Zukin and Zukin, 1979). This binding is displaced by certain benzomorphans, for example by the postulated σ-agonist *N*-allylnormetazocine (SKF 10047), but not by selective μ- or δ-ligands (Zukin and Zukin, 1979; Quirion *et al.*, 1981). Phenylcyclidine and some of its analogs will displace that portion of [^{3}H]-*N*-allylnormetazocine (SKF 10047) binding that is not displaced by high concentrations of morphine (Quirion *et al.*, 1981) or etorphine (IC_{50} = 2.5 μM) (Su, 1981), but are less potent at displacing the binding of [^{3}H]dihydromorphine (IC_{50} = 26 μM) (Vincent *et al.*, 1978).

Furthermore, the low-affinity binding of [^{3}H]cyclazocine is only partly displaced by unlabeled naloxone or morphine although it is readily inhibited by cyclazocine or SKF 10047 (Zukin and Zukin, 1981). These findings along with reports that the behavioral effects of σ-agonists and phenylcyclidine are not reversed by naloxone (Holtzman, 1980) indicate that part of the actions of the σ-agonists is not mediated by opiate receptors.

3.3. Selective Protection Assays

While the evidence from pharmacological and binding assays for a heterogeneous population of opiate receptors is compelling, it is essentially indirect. More direct evidence for the existence of μ- and δ-receptors has been obtained from experiments investigating selective protection of the binding sites. In our laboratory, use was made of the observation that phenoxybenzamine causes an irreversible inactivation of opiate binding sites; protection from this alkylating effect could be achieved by simultaneous incubation with levallorphan (Cicero *et al.*, 1974, 1975; Spiehler *et al.*, 1978). In homogenates of guinea pig brain which were exposed to phenoxybenzamine for 15 min at 37°C, the binding of [^{3}H]-dihydromorphine, [^{3}H]-[D-Ala^2,D-Leu^5]enkephalin, or [^{3}H]ethylketazocine, determined after two washes, was found to be inhibited in a dose-related fashion (IC_{50} 0.3–1.3 μM) (Robson and Kosterlitz, 1979; Kosterlitz *et al.*, 1981). The prediction (Robson and Kosterlitz, 1979) that simultaneous incubation with a ligand with a high affinity for the δ-binding site, such as [D-Ala^2,D-Leu^5]enkephalin, should protect the binding sites of the δ-ligand [^{3}H]-[D-Ala^2,D-Leu^5]enkephalin against the alkylating action of phenoxybenzamine more readily than those of the μ-ligand [^{3}H]dihydromorphine was borne out by the experimental results; conversely, unlabeled dihydromorphine protected the binding sites of [^{3}H]dihydromorphine more readily than those of [^{3}H]-[D-Ala^2,D-Leu^5]enkephalin (Fig. 5). Thus, in guinea pig brain, dihydromorphine was six times more potent than [D-Ala^2, D-Leu^5]-enkephalin in protecting the μ-binding of [^{3}H]dihydromorphine but had only 5% of its potency in protecting the binding of [^{3}H]-[D-Ala^2,D-Leu^5]-enkephalin. Similar results were obtained in homogenates of rat brain by Smith and Simon (1980) who used *N*-ethylmaleimide as an irreversible inhibitor of binding (Table 6). Morphine was eight times more potent at protecting the μ-binding of [^{3}H]naltrexone than in protecting the δ-binding of [^{3}H]-[D-Ala^2,D-Leu^5]enkephalin, while [D-Ala^2,D-Leu^5]enkephalin protected the binding sites of [^{3}H]-[D-Ala^2,D-Leu^5]enkephalin more readily than those of [^{3}H]naltrexone. The nonselective ligand [D-Ala^2,L-Leu-NH_2^{5}]enkephalin, and also [D-Ala^2,L-Met-NH_2^{5}]enkephalin, were equipotent in protecting the binding to μ- and δ-sites.

In other protection experiments in guinea pig brain the μ-selective

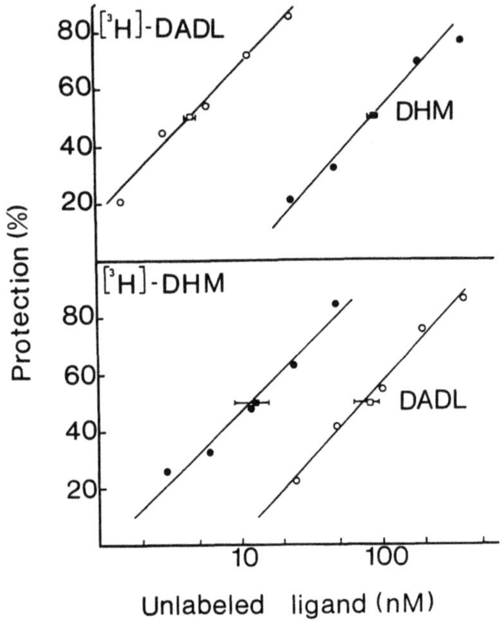

FIG. 5. Protection of the binding in homogenates of guinea pig brain of [^{3}H]-[D-Ala^2,D-Leu^5]enkephalin (1.4 nM; DADL) and [^{3}H]dihydromorphine (0.6 nM; DHM) from the inhibitory effect of phenoxybenzamine (2.4 μM). The protecting effects of the unlabeled ligands were always tested in the same homogenate. Each point is the mean of three to five observations. Abscissa: concentration of unlabeled ligand (nM). Ordinate: percentage of binding protected from inactivation by phenoxybenzamine. (●) Dihydromorphine; (○) [D-Ala^2,D-Leu^5]enkephalin. The mean concentrations of unlabeled ligands required to protect 50% of the binding are indicated by ■ and □. The inhibition of binding by phenoxybenzamine was 77.7 ± 2.1% for [^{3}H]-[D-Ala^2,D-Leu^5]-enkephalin and 78.2 ± 2.7% for [^{3}H]-dihydromorphine. (From Robson and Kosterlitz, 1979.)

ligand dihydromorphine was much less potent than unlabeled ethylketazocine in protecting the binding of [^{3}H]ethylketazocine from inactivation by phenoxybenzamine, while concentrations up to 5000 nM of the δ-ligand [D-Ala^2,D-Leu^5]enkephalin effected less than 40% protection. On the other hand, as was to be expected from the high affinities of ethylketazocine to the μ- and δ-binding sites, unlabeled ethylketazocine readily protected the binding of [^{3}H]ethylketazocine, [^{3}H]dihydromorphine, and [^{3}H]-[D-Ala^2,D-Leu^5]enkephalin (Kosterlitz *et al.*, 1981).

The technique of selective protection has been used to differentiate opiate receptors for [D-Ala^2,D-Leu^5]enkephalin and dynorphin$_{(1-13)}$ in the

TABLE 6

Selective Protection of Opiate Binding Sites from Inactivation by N-Ethylmaleimide (0.5 nM) in Rat Brain[a]

Protecting ligand	Protection of the binding of [^{3}H]Naltrexone (1.1 nM)	[^{3}H]-[D-Ala^2,D-Leu^5]enkephalin (1 nM)
Morphine	100	800
[D-Ala^2,L-Leu^5]enkephalin	400	60
[D-Ala^2,L-Met-NH$_2$^{5}]enkephalin	70	40

[a] The values are the concentration (nM) of protecting agents required to produce 50% of the maximal protection achieved for each labeled ligand. Modified from Smith and Simon (1980).

guinea pig ileum (Chavkin and Goldstein, 1981). Incubation of myenteric plexus–longitudinal muscle preparations for 20 min with the irreversible opiate antagonist chlornaltrexamine (3 nM) reduced the inhibitory potency, after washing, of [Leu]enkephalin, normorphine, and dynorphin$_{(1-13)}$, giving dose ratios of 14, 12, and 17, respectively. When [D-Ala^2,D-Leu^5]-enkephalin (10 μM) was present during exposure to chlornaltrexamine, the dose ratios for [Leu]enkephalin and normorphine were reduced to 3 and 2, respectively, while that for dynorphin$_{(1-13)}$ was unchanged. In contrast, protection with dynorphin$_{(1-13)}$ (20 nM) decreased the dose ratio for dynorphin$_{(1-13)}$ to 5, while that for [Leu]enkephalin was unchanged.

3.4. Effects of Irreversible Agonists and Antagonists

While phenoxybenzamine and *N*-ethylmaleimide cause irreversible inactivation at the opiate binding site, this effect is not specific (Section 3.3) (Cicero *et al.*, 1974, 1975; Simon and Groth, 1975; Simon *et al.*, 1975*b*; Spiehler *et al.*, 1978). Recently, several agents which selectively form covalent bonds with opiate receptors have been described. Chloroxymorphinone, prepared by substitution with a bis(α-chloroethyl) amino group on the C_6 group of oxymorphone, exhibits irreversible pure agonist opiate activity in the guinea pig ileum (Caruso *et al.*, 1979). Its effect, which persisted after 10 washes, was prevented by pretreatment with naloxone. Its antinociceptive activity in the mouse tail flick test was blocked by naloxone and persisted for four times longer than that of morphine. An identical modification of the narcotic antagonist naltrexone results in an irreversible antagonist, chlornaltrexamine (Portoghese *et al.*, 1978, 1979; Caruso *et al.*, 1979). Chlornaltrexamine inhibited the binding of [^{3}H]-naloxone, an effect which in contrast to that of naltrexone and levorphanol was only partly reduced after three washes of the homogenate. Similarly, 2 hr after the i.c.v. administration of chlornaltrexamine to mice a significant dose-related inhibition of the antinociceptive effect of morphine was observed. Pretreatment of mice with naloxone prevented the action of chlornaltrexamine. Spontaneous recovery of the antinociceptive effect of morphine was slow. In mice treated with chlornaltrexamine (1.2 nmol/mouse), the dose of morphine required to produce the same antinociceptive effect as in control animals was 66 times larger after 2 hr, six times larger after 24 hr, and twice as large after 72 hr.

An enkephalin-based reagent, Tyr-D-Ala-Gly-Phe-Leu-chloromethyl ketone, is an irreversible inhibitor of electrically stimulated contractions of the mouse vas deferens; its inhibitory effect is only partly lost by washing. Subsequent addition of naloxone reverses the inhibitory effect of this agonist, but when naloxone is removed the inhibitory effect is reestablished. In binding studies, [D-Ala^2,L-LeuCH_2Cl^5]enkephalin prevented the binding of [^{3}H]-[Leu]enkephalin and [^{3}H]etorphine even after extensive washing; the binding in P_2 homogenates of rat brain pretreated with 20 μM [D-Ala^2,L-LeuCH$_2$Cl^5]enkephalin was reduced by 100% for [^{3}H]-[D-Ala^2,D-Leu^5]enkephalin but only by 50% for [^{3}H]etorphine. The quantitative difference in the effects of the binding of the two ligands may be accounted for by the nonselective nature of etorphine compared to the δ-selective agonist [D-Ala^2,D-Leu^5]enkephalin (Venn and Barnard, 1981).

Naloxazone, naltrexazone, and oxymorphazone are C_6-substituted hydrazone derivatives of naloxone, naltrexone, and oxymorphone; they produce a long-lasting inhibition of the binding of [^{3}H]naltrexone and [^{3}H]-[D-Ala^2,L-Met-NH$_2$^{5}]enkephalin in homogenates of rat brain. The effect of the hydrazones, in contrast to that of the parent compounds, was not reversed by washing (Pasternak and Hahn, 1980). Twenty-four hours after treatment of rats or mice with naloxazone (200–400 mg/kg), but not after treatment with naloxone (200 mg/kg), high-affinity binding sites for tritiated naloxone, naltrexone, dihydromorphine, ethylketazocine, [Met]-enkephalin, [Leu]enkephalin, [D-Ala^2,D-Leu^5]enkephalin, and [D-Ala^2,L-Met-NH$_2$^{5}]enkephalin were inhibited; binding to α- and β-adrenergic, benzodiazepine, or muscarinic binding sites was not affected (Pasternak *et al.*, 1980*a,b*; Pasternak and Hahn, 1980; Hazum *et al.*, 1981; Zhang and Pasternak, 1981). These changes in receptor binding were accompanied by a reduction of the antinociceptive potency of morphine, [D-Ala^2,D-Leu^5]enkephalin, [D-Ala^2,L-Met-NH$_2$^{5}]enkephalin, and β-endorphin, but the lethal effects of morphine were not changed. Binding returned to control levels after three days and was accompanied by recovery of the antinociceptive properties of opioids. It was therefore suggested that the antinociceptive activity, but not the lethal effects, of morphine are mediated by an interaction with the high-affinity sites that are inactivated by naloxazone.

Several lines of evidence indicate that the high-affinity binding sites which are preferentially inhibited by naloxazone treatment are associated with a subgroup of the μ-binding site; they have been termed μ_1 sites. The inhibition by the selective μ-agonist morphiceptin, Tyr-Pro-Phe-Pro-NH$_2$, of the binding of tritiated dihydromorphine (0.3 nM), [D-Ala^2,D-Leu^5]enkephalin (0.4 nM), [D-Ala^2,L-Met-NH$_2$^{5}]enkephalin (0.4 nM), ethylketazocine (0.6 nM), or *N*-allylnormetazocine (0.4 nM) in rat brain was biphasic; between 10 and 60% of the binding was readily displaced (IC_{50} <10 nM), while the remainder of the binding was less sensitive to inhibition by morphiceptin. After pretreatment of the homogenates with naloxazone, the first phase of inhibition of binding by morphiceptin was lost, while the second phase was unaffected (Zhang *et al.*, 1981). The ability of unlabeled [D-Ala^2,D-Leu^5]enkephalin to displace the binding of [^{125}I]-[D-Ala^2,D-Leu^5]enkephalin was unchanged after naloxazone treatment (Hazum *et al.*, 1981). In addition, the binding of [^{3}H]-[D-Ala^2,Met-NH$_2$^{5}]-

enkephalin in homogenates prepared from discrete areas of rat brain was more sensitive to inhibition by naloxazone; the binding in the hypothalamus and spinal cord was most sensitive, that of the corpus striatum and the thalamus less sensitive, while that of the frontal cortex, midbrain, and pons medulla was insensitive (Zhang and Pasternak, 1980).

4. DISTRIBUTION OF OPIATE RECEPTORS

4.1. Central Nervous System

The measurement of specific binding of opiates to synaptosomal membranes obtained from discrete parts of the brain was the first approach used to determine the regional distribution of receptors in the central nervous system (Pert and Snyder, 1973; Terenius, 1973*a*). However, in early studies it was not possible to differentiate between the subtypes of the opiate receptor.

Detailed studies of the regional distribution of opiate receptors in grossly dissected areas have been undertaken for human brain using [^{3}H]-etorphine (Hiller *et al.*, 1973) and for monkey and human brains using [^{3}H]dihydromorphine (Kuhar *et al.*, 1973). In retrospect, the results obtained with [^{3}H]etorphine are of particular interest; the universal nature of this ligand allows it to label all the subtypes. Generally, high specific binding was found in areas related to, or associated with, the limbic system, such as amygdala, thalamus, hypothalamus, caudate nucleus, and periaqueductal gray. In the cortex, with the exception of the frontal lobe, and in the cerebellum, binding was low. In the spinal cord of monkey (Lamotte *et al.*, 1976), binding of [^{3}H]naloxone was detected in all spinal gray areas but was predominantly localized in laminae I and II of the dorsal horn. The reduction in binding following deafferentation suggests its association with primary afferent terminals.

Another approach has been the autoradiographic determination of binding sites after administration of tritiated opiates *in vivo*. Such experiments using etorphine and diprenorphine, which also label μ-, δ-, and κ-binding sites, have yielded useful results in the rat brain (Pert *et al.*, 1976; Atweh and Kuhar, 1977*a,b,c*; Pearson *et al.*, 1980). Results obtained with this method are qualitatively similar to those observed in homogenates. In the spinal cord, high densities of binding sites were observed in lamina I and II of the dorsal horn, in the substantia gelatinosa of spinal trigeminal nucleus, in various components of the vagal system, and in the area postrema. In the brainstem, areas showing high densities of opiate receptors included the parabrachial nuclei, the superior colliculus, the ventral median raphe nucleus, components of the accessory optic system, portions of the habenulo-interpeduncular complex, the pretectal nuclei and the ventral lateral geniculate, the infundibulum, and the medial thalamus. In the telencephalon, higher densities of binding sites were associated with parts of the presubiculum and amygdala, certain areas in the caudate putamen and accumbens, the subfornical organ, the interstitial nucleus of the striae terminalis, and the anterior olfactory nucleus (Atweh and Kuhar, 1977*a,b,c*). Thus opiate binding sites are widely and unevenly distributed in the central nervous system. Of interest is the observation that, in contrast to primate and rodent brain, the cerebellum of lagomorphs has a high density of binding sites for etorphine and diprenorphine (Meunier and Zajac, 1979; Zajac and Meunier, 1981).

Analysis of the regional distribution of the subtypes of the opiate receptor will have value in the understanding of functional differences. In the rat, Chang *et al.*, (1979) using [^{125}I]-[D-Ala^2,MePhe^4,Met(O)-ol^5]-enkephalin to characterize the "morphine" or μ-binding site and [^{125}I]-[D-Ala^2,D-Leu^5]enkephalin for the "enkephalin" or δ-binding site, estimated the maximum binding capacities indirectly from observations with single concentrations of unlabeled ligand. In the cow, Ninkovic *et al.* (1981) compared the binding of 0.5 nM [^{3}H]morphine (μ-binding site) and 0.5 nM [^{3}H]-[D-Ala^2,D-Leu^5]enkephalin (δ-binding site) in homogenates obtained from frozen slices of different brain regions. The ratio of μ-binding to δ-binding in the various regions showed differences in the two species (Table 7). In the rat, the frontal cortex contained an equal number of μ- and δ-binding sites, whereas the μ-binding site was predominant in the brainstem, striatum, hypothalamus, and especially in the thalamus. In contrast, in bovine brain δ-binding sites were predominant in the hippocampus and frontal cortex, while μ-binding sites were progressively more prominent in the caudate, hypothalamus, periaqueductal gray, thalamus, and substantia nigra. From competition studies in the rat with unlabeled naloxone, it was concluded that in the frontal cortex about 80% of the binding sites were δ while in the thalamus 70% of the binding sites were μ. In the rat spinal cord, μ-binding sites ([^{3}H]morphine) and δ-binding sites ([^{3}H]-[D-Ala^2,D-Leu^5]enkephalin) have been detected; in the dorsal root the ratio of μ- to δ-binding sites was 3.2, and in the dorsal horn it was 1.8. Destruction of small-diameter primary afferents after section of the sciatic nerve led to reduction of both μ- and δ-binding sites in the dorsal root (Fields *et al.*, 1980).

Opiate receptors have also been classified into two groups on the basis of their sensitivity to inhibition by GTP (Pert and Taylor, 1980; Pert *et al.*, 1980). GTP-sensitive receptors (Type 1) predominate in areas which are associated with antinociception, e.g., the periaqueductal gray, nucleus gigantocellularis, reticular formation, and thalamus, while GTP-insensitive receptors (Type 2) predominate in areas of the limbic system, that is, in

TABLE 7

Regional Distribution of μ- and δ-Binding Sites in Cow and Rat Brain[a]

Brain regions	Cow: $[^3H]$Morphine / $[^3H]$-[D-Ala^2,D-Leu^5]enkephalin	Rat: $[^{125}I]$-[D-Ala^2,MePhe^4,Met(O)-ol^5]enkephalin / $[^{125}I]$-[D-Ala^2,D-Leu^5]enkephalin
Substantia nigra	3.7	—
Thalamus	2.1	4.5
Periaqueductal gray	1.8	—
Brainstem	—	2.2
Hypothalamus	1.6	2.8
Caudate	1.6	—
Globus pallidus	1.0	—
Striatum	—	1.4
Hippocampus	0.4	2.3
Frontal cortex	0.5	1.1

[a] The data for cow brain were obtained from Ninkovic *et al.* (1981), and for rat brain, from Chang *et al.* (1979). The derivations of the ratios are described in the text.

the amygdala, hypothalamus, nucleus accumbens, frontal cortex, and the pituitary (see Section 5.3).

Different regional densities of μ- and δ-binding sites have been demonstrated autoradiographically after labeling of rat brain slices *in vitro* with [^{125}I]-[D-Ala^2,MePhe^4,Met(O)-ol^5]enkephalin for μ-sites and [^{125}I]-[D-Ala^2,D-Leu^5]enkephalin, in the presence of unlabeled μ-ligand (1 nM), for the δ-sites (Goodman *et al.*, 1980). The regional distribution (Table 8) indicated that, *inter alia*, the thalamus and hypothalamus are rich in μ-binding sites, and the amygdala, nucleus accumbens, and olfactory tubercle are rich in δ-binding sites. In a series of experiments on primate cerebral cortex (Lewis *et al.*, 1981), [^{3}H]naloxone was used to identify the μ-binding sites, and [^{3}H]-[D-Ala^2,D-Leu^5]enkephalin was used to identify the δ-binding sites. In rat striatum, μ-binding sites occurred in distinct patches, while δ-binding sites were diffusely distributed (Herkenham and Pert, 1981).

Several surgical, chemical, or electrolytic lesions (Table 9) cause degeneration of the dopaminergic terminals in the striatum as measured by the reduction in the activity of dopadecarboxylase or the rate of uptake of [^{3}H]dopamine, but do not alter the rate of uptake of [^{3}H]choline into the cholinergic neurons upon which the dopaminergic neurons project. In homogenates of rat striatum, about one-third of the binding of [^{3}H]-[Leu]enkephalin or [^{3}H]naloxone was lost after the three types of lesions

TABLE 8
Comparison of the Distribution of μ- and δ-Binding Sites in Rat Brain by Autoradiography in Slices[a]

Predominantly μ-sites	Predominantly δ-sites	Mixed μ- and δ-sites
Laminae I and IV of cerebral cortex	Laminae II, III, and V of cerebral cortex	Lamina VI of cerebral cortex
Streaks and clusters in corpus striatum	Diffuse grains in corpus striatum	Nucleus corpus solitarius
Dorsomedial and ventral thalamus	Amygdala	Vagal fibers
Hypothalamus	Nucleus accumbens	Nucleus ambiguus
Hippocampus (pyramidal cell layer)	Olfactory tubercle	Substantia gelatinosa of spinal cord trigeminal
Periaqueductal gray	Pontine nuclei	
Interpeduncular nucleus		
Inferior colliculus		
Midbrain median raphe		

[a] Adjacent sections of rat brain were pre-incubated with GTP (50 μM) and NaCl (100 μM) and after washing were labeled with μ- or δ-ligands (0.1–0.2 nM). For the μ-sites, [^{125}I]-[D-Ala^2,MePhe^1,Met(O)-ol^5]enkephalin was used, and for the δ-sites, [^{125}I]-[D-Ala^2,D-Leu^5]enkephalin in the presence of [D-Ala^2,MePhe^4,Met(O)-ol^5]enkephalin (1 nM) was used (Goodman *et al.*, 1980).

TABLE 9

Effect of Various Lesions on the Binding of [^{3}H]-[Leu]enkephalin (12 nM) and on the Rate of Uptake of [^{3}H]-Dopamine (20 nM) and [^{3}H]Choline (200 nM) in Rat Striatal Homogenates[a]

Lesions	[^{3}H]-[Leu]enkephalin binding (fmol/mg protein)	[^{3}H]Dopamine uptake (fmol/mg protein per 5 min)	[^{3}H]Choline uptake (fmol/mg protein per 5 min)
Control	174	2350	197
Intranigral 6-hydroxydopamine	123	141	N.S.
Hypothalamic hemisection	123	399	—
Intrastriatal kainic acid	96	—	67
Hemisection + kainic acid	49	—	95

[a] The binding of [^{3}H]-[Leu]enkephalin was determined in the presence of 20 μM bacitracin. Uptake of [^{3}H]dopamine and [^{3}H]choline were measured after 5 min incubation in the presence of ascorbic acid (0.02 mg/ml), EDTA (0.01 mg/ml), and nialamide (70 μM). N.S., no significant change. Modified from Pollard *et al.* (1978).

(Pollard *et al.*, 1977*b*, 1978; Carenzi *et al.*, 1978; Schwartz *et al.*, 1978; Reisine *et al.*, 1979*a*). Following the destruction of cell bodies by injection of kainic acid into the striatum, the binding of [^{3}H]dihydromorphine and [^{3}H]-[Leu]enkephalin was reduced by 35–45% and the rate of uptake of [^{3}H]choline was reduced by 66% (Carenzi *et al.*, 1978; Pollard *et al.*, 1978; Schwartz *et al.*, 1978). The effects of dopaminergic denervation of the striatum and intrastriatal injection of kainic acid on the binding of opiates were additive (Table 9). Thus, it appears that about one-third of striatal opiate receptors are located presynaptically on dopaminergic neurons, while the remainder are on neurons intrinsic to, or emanating from, the striatum. After electrocoagulations in the ventral tegmental area, the degeneration of dopaminergic nerve endings was accompanied by a loss of the binding of [^{3}H]naloxone (4 nM), which amounted to 52% in the septum and 27% in the nucleus accumbens (Pollard *et al.*, 1977*a*).

In agreement with the results of the biochemical studies, dopaminergic denervation or intrastriatal injection of kainic acid caused a reduction in the autoradiographic grain densities for [^{3}H]diprenorphine in rat striata (Murrin *et al.*, 1980). About 45% of the binding sites appeared as dense clusters, while the remainder were spread diffusely. Kainic acid reduced the grain densities of the clusters by more than 80%, and at sites outside of the clusters, by 62%. The reduction caused by electrolytic lesions of the medial forebrain bundle or injection of 6-hydroxydopamine into the striatum was 28–30% for the clusters and 35–45% for the diffusely distributed binding sites.

Surgical and 6-hydroxydopamine lesions in different positions in the substantia nigra produced two functionally distinct behavioral syndromes in the rat (Gardner *et al.*, 1980). Rats with medially placed lesions exhibited contraversive rotational behavior after systemic injection of apomorphine; this effect was associated with a decreased density of binding sites for [^{3}H]-[D-Ala^2,L-Leu-NH$_2$^{5}]enkephalin. In contrast, the density of binding sites was increased in rats with laterally placed lesions; these animals displayed ipsiversive rotation. Further experiments with selective ligands would prove interesting.

4.2. Peripheral Nervous System

The wide and uneven distribution of opiate receptors in the central nervous system has been extended to the neurohypophysis (Simantov and Snyder, 1977) and large areas of the autonomic nervous system. Investigations of characteristics of the peripheral binding sites may be particularly useful in elucidating functional correlates for the subtypes.

Stereospecific opiate binding to myenteric plexus–longitudinal muscle preparations of the guinea pig (Pert and Snyder, 1973; Creese and Snyder,

1975; Terenius, 1975; Leslie and Kosterlitz, 1979; Leslie *et al.*, 1980) is restricted to the myenteric plexus and has properties similar to those observed in brain homogenates. [^{3}H]Dihydromorphine and [^{3}H]-[D-Ala^2,D-Leu^5]enkephalin have been used respectively to characterize μ- and δ-binding sites of preparations from guinea pig, rabbit, and rat ilea (Leslie *et al.*, 1980). In preparations from all three species, δ-binding sites were found, while μ-binding sites were present only in preparations from guinea pig. The density of binding sites in guinea pig ileum expressed as unit weight of tissue was about 10% of that observed in brain homogenates. The presence of both μ- and δ-binding sites in the mouse vas deferens (Leslie and Kosterlitz, 1979; Leslie *et al.*, 1980) is not surprising in view of the different concentrations of naloxone that are required to antagonize the effects of morphine and enkephalin (Lord *et al.*, 1977) in electrically stimulated preparations. There are about twice as many δ- as μ-binding sites (Leslie *et al.*, 1980).

Specific binding of [^{3}H]naloxone was observed in bovine adrenal medulla but not in cortical tissue (Chavkin *et al.*, 1979), the density being about 25% of that found in brain homogenates. The medulla contains approximately equal proportions of μ- and δ-binding sites (Leslie *et al.*, 1980).

The demonstration of specific binding of [^{3}H]etorphine in retinal membranes from the rat (Medzihradsky, 1976) was confirmed and extended to include binding of [^{3}H]naloxone and [^{3}H]-[Met]enkephalin. Opiate binding sites have also been detected in retinal membranes from the cow, the toad, and the skate (Howells *et al.*, 1980). The density of binding sites determined per gram of tissue is close to those observed in rat brain; their function has not yet been elucidated.

Placental membranes from humans and rats, but not those from mice, hamsters, and rabbits, have been found to have binding sites for [^{3}H]-etorphine (Valette *et al.*, 1979, 1980). Both morphine and opioid peptides were less potent at displacing the binding of [^{3}H]etorphine to placental than to brain membranes. The binding sites found in human tissue have been further characterized by the use of selective ligands (Porthé *et al.*, 1981). Although no specific binding could be detected with the selective μ-ligand, [^{3}H]dihydromorphine, or the selective δ-ligand, [^{3}H]-[D-Ala^2,D-Leu^5]enkephalin, [^{3}H]etorphine and [^{3}H]ethylketazocine had the same maximum number of binding sites (54–66 fmol/mg protein) and K_D values of 0.69 nM and 0.33 nM, respectively. Opiates and opioid peptides were equipotent at displacing the binding of either tritiated ligand. The binding of [^{3}H]ethylketazocine was readily displaced by cyclazocine, etorphine, naltrexone, and nalorphine (K_I = 1 to 7 nM) but not by the μ-ligand morphine (K_I = 270 nM) or the δ-ligand [D-Ala^2,D-Leu^5]enkephalin (K_I = 8230 nM) (Porthé *et al.*, 1981). These values are in close agreement with the inhibition constants found for the same compounds against the binding of [^{3}H]ethylketazocine, after suppression of the μ- and δ-binding sites, to the κ-site in the guinea pig brain (Magnan *et al.*, 1982). Thus, it would appear that the human placenta contains only κ-binding sites. The functional significance of such binding sites is unknown; in this context it is of interest that the placenta lacks any innervation.

4.3. Cultured Cells

Saturable binding sites have been demonstrated in the membranes of several cell lines (Klee and Nirenberg, 1974; Chang *et al.*, 1978; Garcin *et al.*, 1978; Gerber *et al.*, 1978; Chang and Cuatrecasas, 1979; Law *et al.*, 1979*a*; McLawhon *et al.*, 1981). For instance, in the neuroblastoma × glioma hybrid cell line NG108-15, [^{3}H]dihydromorphine binds to about 300,000 sites per cell, while a sister clone NG108-5 has relatively few binding sites; the parental cell lines N18TG-2 and C6BU-1 lacked binding sites for [^{3}H]dihydromorphine (Klee and Nirenberg, 1974). However, in a later study N18TG-2 cells were shown to possess 60,000 sites per cell as estimated with [^{3}H]naloxone and [^{3}H]dihydromorphine (Law *et al.*, 1979*a*). It was found that the interaction of opioids with the binding sites was stereospecific since levorphanol was more potent than its (+)-isomer dextrorphan in displacing the binding of [^{3}H]dihydromorphine from NG108-15 cells (Klee and Nirenberg, 1974), the binding of [^{125}I]-[D-Ala^2,D-Leu^5]enkephalin from N4TG1 cells (Chang *et al.*, 1978), and the binding of [^{3}H]naloxone from N18TG-2 cells (Law *et al.*, 1979*a*).

In N4TG1 cells, [^{125}I]-[D-Ala^2,D-Leu^5]enkephalin binds to 1800 sites per cell (K_D = 2 nM) (Chang *et al.*, 1978). As found in rat brain homogenates, the δ-selective ligands [Leu]enkephalin and [D-Ala^2,D-Leu^5]-enkephalin were more potent inhibitors of [^{125}I]-[D-Ala^2,D-Leu^5]enkephalin (1 nM) binding than μ-selective ligands such as morphine and naloxone. Since, in contrast to observations in rat brain, [D-Ala^2,D-Leu^5]-enkephalin was also more potent in displacing the binding of [^{3}H]naloxone (0.38 nM), it appears that N4TG1 cells possess mainly δ-binding sites and few, if any, μ-binding sites. Similar results have been observed in NIE 115, N18TG-2, and NG108-15 cell lines (Chang *et al.*, 1978; Chang and Cuatrecasas, 1979).

The maximum binding (B_{max}) of [^{3}H]-[D-Ala^2,D-Leu^5]enkephalin was 400 fmol/mg protein in NCB-20 cells and 90 fmol/mg protein in NC108-15 cells (K_D = 3 nM) (McLawhon *et al.*, 1981). The relative potencies of ligands to displace the binding of [^{3}H]-[D-Ala^2,D-Leu^5]enkephalin (4 nM) was similar in the two cell lines. [D-Ala^2,D-Leu^5]enkephalin was 100 times more potent than morphine and naloxone and 10 times more potent than ethylketazocine, observations which suggested that the binding of [^{3}H]-[D-Ala^2,D-Leu^5]enkephalin was to a δ-site. The Scatchard plots obtained from saturation curves for [^{3}H]ethylketazocine were linear in NG108-15 cells (K_D = 5 nM, B_{max} = 70 fmol/mg protein) but could be resolved into two components (K_D = 4 nM, B_{max} = 300 fmol/mg protein, K_D = 20 nM, B_{max} = 1000 fmol/mg protein) in NCB-20 cells. In competitive inhibition assays, ethylketazocine, bremazocine, and ketazocine were potent inhibitors of the binding of [^{3}H]ethylketazocine (7 nM) (IC_{50} = 20 to 40 nM), but concentrations of up to 100 μM of [D-Ala^2,D-Leu^5]enkephalin, morphine, or etorphine displaced less than 30% of the binding; 100 μM naloxone was required to inhibit 50% of the binding (McLawhon *et al.*, 1981). Thus, it would appear that in NCB-20 cells [^{3}H]ethylketazocine binds to the δ-site and, in addition, to a site for which morphine, [D-Ala^2,D-Leu^5]-enkephalin, and the universal ligand etorphine have no affinity. Since the narcotic antagonist naloxone also has negligible affinity for this site, it is doubtful whether it is associated with the opiate receptor. Similar considerations may apply to the observation that [^{3}H]-*N*-allylnormetazocine bound in NCB-20 cells to a site for which [D-Ala^2,D-Leu^5]enkephalin, morphine, etorphine, and naloxone, as well as ethylketazocine, have negligible activity. This site may be equivalent to the phencyclidine binding site in rat olfactory bulb slices to which *N*-allylnormetazocine and cyclazocine bind but ketazocine, etorphine, morphine, [D-Ala^2,D-Leu^5]enkephalin, and naloxone do not (Quirion *et al.*, 1981).

Image-intensified fluorescence microscopy has been used for the visualization and localization of binding sites in neuroblastoma cells (Hazum *et al.*, 1979*b,c,* 1980). When N4TG1 or NG108-15 cells were incubated with a fluorescent derivative of enkephalin, [D-Ala^2,L-Leu^5]-enkephalyl-Lys-rhodamine, opiate binding sites formed clusters on the cell surface which did not become internalized with incubation of longer duration. The clusters were distributed on the cell body and were highly concentrated in regions where neurite processes appeared to interconnect. Their formation was not prevented by sodium azide or dinitrophenol and therefore appeared to be independent of the generation of metabolic energy. Sulfhydryl and disulfide blocking agents prevented formation of, but did not disperse, clusters in concentrations which did not affect the binding of [^{125}I]-[D-Ala^2,D-Leu^5]enkephalin. It was therefore possible to show that preincubation with the agonists morphine and [Leu]enkephalin, as well as the antagonists naloxone and nalorphine, induced cluster formation. However, clusters induced by agonists, but not those induced by antagonists, can be dispersed by dithiothreitol after removal of the inducing drug, an observation suggesting differences between the conformations of clusters induced by agonists and antagonists.

It is likely that these events are due to redistribution of opiate-receptor complexes since the number and affinity of the binding sites are unchanged. However, the formation of clusters appears to be unrelated to the early effects of opiates; when cluster formation is prevented by treatment with sulfhydryl blocking reagents, the ability of enkephalins to inhibit basal or stimulated adenylate cyclase activity is unchanged (Hazum *et al.*, 1980).

Induction of clusters of opiate receptors subsequent to the ligand binding process may be correlated to the observation that conjugation of enkephalin analogs to the capsomers of the tobacco mosaic virus resulted in a potentiation of the activity of the analog (Kosterlitz and Paterson, 1980; Kriwaczek *et al.*, 1981).

4.4. Subcellular Distribution

High-affinity specific binding sites for [^{3}H]dihydromorphine have been found to be associated with the synaptic membrane fraction (Terenius, 1973*a,b*) and microsomal fraction (Pert and Snyder, 1973) of rat brain. In a differential study of their subcellular distribution, the binding of [^{3}H]dihydromorphine occurred primarily in the synaptosomal fraction and was particularly associated with those fractions enriched with synaptic membranes (Pert *et al.*, 1974*b*). Significant specific binding, however, was found in all membranous fractions. In extensively washed fractions of mouse brain, there were comparable levels of binding in synaptosomal and microsomal fractions (Smith and Loh, 1976). The density of binding sites for [^{3}H]etorphine was highest in the microsomal fraction, but nevertheless substantial in synaptosomal and myelin fractions of ovine corpus striatum (Glasel *et al.*, 1980).

After intracisternal administration of [^{3}H]etorphine, the major part of specific binding was associated with the synaptic plasma membranes (Mulé *et al.*, 1974), while, after intravenous administration, binding was predominantly localized in the microsomal fraction (Cerletti *et al.*, 1978).

Cumulatively, these results indicate that opiate binding sites are distributed diffusely on the surface of nerve cells. However, selective localization of the binding sites on neuronal processes has been observed (Medzihradsky, 1976); isolated neuronal perykarya exhibit considerably greater binding of [^{3}H]etorphine than the glial-enriched fractions.

4.5. Ontogenetic and Phylogenetic Variations

There is general agreement that, in rat brain, receptor density increases during fetal development and continues to do so for several weeks after birth. The increase in binding capacity is not associated with a change in affinity (K_D) (Clendeninn *et al.*, 1976; Coyle and Pert, 1976; Garcin and Coyle, 1976; Auguy-Valette *et al.*, 1978). In brains from Sprague-Dawley rats, there was no significant binding of [^{3}H]naloxone before the 14th or 15th day of gestation; at birth the number of binding sites was 40% of that found in 4-week-old rats (Coyle and Pert, 1976), or 15% of that in 20-week-old rats (Clendeninn *et al.*, 1976). Their density increased until the third week after birth and more slowly until the 20th week. In Wistar rats, there was a similar rate of development of binding sites for [^{3}H]diprenorphine (Kirby, 1981).

Regional differences have been observed in the development of binding sites for [^{3}H]naloxone in Sprague-Dawley rats (Garcin and Coyle, 1976; Bardo *et al.*, 1981). The most striking observation was that the rate of increase in receptor density was greater in the cortex than in the striatum or pons and medulla. The binding capacity for [^{3}H]-[Met]enkephalin in the forebrain of Sprague-Dawley rats increased fivefold between the 5th and 20th day after birth (Tsang and Ng, 1980). In the brainstem, the binding capacity increased at a similar rate until the 15th day but by the 20th day had returned to the level found in newborn rats.

In the guinea pig, in contrast to the rat, there was no difference in the binding capacity for [^{3}H]naltrexone in brain homogenates from late-term fetuses and adults (Clendeninn *et al.*, 1976). The binding capacity found at a fetal age of 4 to 5 weeks was about 50% of that in adult animals and increased linearly until birth. This earlier appearance of adult levels of binding is in agreement with the finding that brain development in the guinea pig is complete by birth, whereas considerable maturation of the CNS occurs in rats during the first three weeks after birth.

In an early study of the phylogenetic distribution of [^{3}H]dihydromorphine binding in nervous tissue, a substantial amount of specific binding was found in the vertebrate species examined, whereas none was detected in invertebrates (Pert *et al.*, 1974*a*). Among the vertebrates there was no evidence of an evolutionary trend; the highest binding, expressed as fmol/mg protein, occurred in the brain of the most primitive of the species, the slime hog.

In invertebrates, the presence of opiate receptors was indicated by the demonstration that exogenously applied opioid peptides can increase dopamine levels in the pedal ganglia of the marine mollusc *Mytilus edulis* (Stefano and Catapane, 1979), in the ganglia of the snail *Helix pomatia* (Osborne and Neuhoff, 1979), and in the fresh water bivalve *Amodonta cygnea* (Stefano and Hiripi, 1979); these effects were reversed by naloxone. Several recent reports have demonstrated the presence of binding sites for [^{3}H]naloxone, [^{125}I]-[D-Ala^2,MePhe^4,Met(O)-ol^5]enkephalin (Stefano *et al.*, 1980), and [^{3}H]etorphine (Kream *et al.*, 1980) in homogenates of the pedal ganglia of *Mytilus edulis*. The characteristics of the binding of [^{125}I]-[D-Ala^2,MePhe^4,Met(O)-ol^5]enkephalin in saturation and competition experiments were similar in *Mytilus edulis* ganglia and in rat brain (Kream and Zukin, 1979).

Since tritiated ligands that exhibit little discrimination between the subtypes of the opiate receptor have so far been used in these experiments, there is no information as to whether there are differences in the ontogenetic and phylogenetic development of the μ-, δ-, and κ-binding sites.

5. PROPERTIES OF THE OPIATE BINDING SITES

5.1. Physicochemical Properties

Information concerning the molecular nature of the opiate receptor is largely based on indirect evidence. Early observations that the binding of opiates is reduced by proteolytic enzymes and other protein-modifying agents interacting preferentially with sulfhydryl groups (Simon *et al.*, 1973; Terenius, 1973*b*; Pasternak and Snyder, 1974, 1975*a*; Pasternak *et al.*, 1975*b*; Simon and Groth, 1975; Wilson *et al.*, 1975) have indicated that protein is part of the receptor molecule. Since the simultaneous incubation with opiates protects the binding sites from the deleterious effects of sulfhydryl agents (Pasternak *et al.*, 1975*b*; Simon and Groth, 1975), it appears that a sulfhydryl group may be located at or very near to the binding site. That such agents exert a greater effect on the binding of agonists than on that of antagonists (Pasternak and Snyder, 1975*a*; Pasternak *et al.*, 1975*b*; Wilson *et al.*, 1975) may indicate that a reactive sulfhydryl group is more important for the binding of agonists than of antagonists.

Similarly, the presence of phospholipids in the receptor molecule has been implied by the fact that the binding of opiates is sensitive to inhibition by the lipolytic enzyme phospholipase A (Pasternak and Snyder, 1974, 1975*a*; Abood *et al.*, 1977; Lin and Simon, 1978). Since binding is much less sensitive to inhibition by phospholipase C and insensitive to phospholipase D, both of which selectively affect the polar groups of phospholipids, these moieties may not be required for binding (Simon *et al.*, 1973; Pasternak and Snyder, 1974, 1975*a*; Abood *et al.*, 1977). The destruction of the specific binding of opiates by ascorbate (Dunlap *et al.*, 1979) may be a consequence of peroxidation of lipid which, while being important for the structural integrity of the opiate binding site, is not in close enough proximity to be protected with an opiate ligand. Neuramidase, which releases membrane-bound sialic acid, does not affect opiate binding (Pasternak and Snyder, 1974, 1975*a*).

It has been shown that the relative affinities for the binding of opiates to cerebroside sulfate parallel their analgesic and antinociceptive potencies in man and rodents (Loh *et al.*, 1974, 1975). A solubilized opiate–macromolecule complex obtained from mouse brain was found to be similar to an opiate–cerebroside sulfate complex (Loh *et al.*, 1974). In addition, mice having either a genetic deficiency in cerebrosides or a deficiency produced by treatment with Azure A were less sensitive to the antinociceptive effect of morphine; the levels of binding of [^{3}H]morphine (2.5 nM) or [^{3}H]naloxone (2.5 nM) in brain synaptosomal plasma membranes were reduced (Law *et al.*, 1978). Furthermore, small amounts of antibodies to cerebroside sulfate reduced the ability of morphine and β-endorphin to inhibit the wet-shake response in rats (Craves *et al.*, 1980).

It was observed that the relative potencies of opiate agonists, antagonists, and mixed agonist–antagonists to induce or prevent transfer of [^{3}H]-cerebroside sulfate from an aqueous to a nonaqueous phase correlated well with their analgesic potencies in man or their antagonist potencies in the guinea pig ileum. It has therefore been suggested that the molecular mechanisms of the opiate interaction with cerebroside sulfate may be a useful model for the interactions of opiates with their binding sites (Cho *et al.*, 1979). The validity of such a model awaits unequivocal determination of the binding-site structure.

The binding of [^{14}C]morphine to phosphatidylserine extracted from bovine cerebral cortex was displaced by levorphanol; this effect was stereospecific since dextrorphan had 10% of the potency of levorphanol (Abood and Hoss, 1975). The hypothesis that phosphatidylserine may be an integral component of the receptor is supported by several observations. Addition of phosphatidylserine to neural membranes enhanced specific opiate binding (Abood and Takeda, 1976; Abood *et al.*, 1977) and partly reversed the inhibition produced by phospholipase A, Triton X-100 (Abood *et al.*, 1978), and ascorbate (Dunlap *et al.*, 1979), while several other fatty acids and cerebroside sulfate did not protect the binding site from destruction by ascorbate. Exposure of membranes to phosphatidylserine decarboxylase partly inhibited opiate binding, which returned to control levels on addition of phosphatidylserine (Abood *et al.*, 1978). The latter observations were complicated by the fact that Triton X-100 is required for optimal activity of the enzyme. Dinitrofluorobenzene, which cross-links the amino groups of phospholipids and proteins, prevented enhancement of binding by phosphatidylserine, possibly by preventing access of the lipid to the opiate binding site (Abood *et al.*, 1977).

Analysis of the fatty acids released by phospholipase A indicated preferential release of polyunsaturated fatty acids, with no evident hydrolysis of cerebroside, cerebroside sulfate, sphingomyelin, or ganglioside (Abood *et al.*, 1978). While saturated forms of fatty acids enhanced opiate binding, unsaturated forms inhibited it (Abood *et al.*, 1977). Therefore, the effect of phospholipase A may be due to the release of inhibitory fatty acids. This possibility is supported by the fact that the effect of phospholipase A can be reversed by albumin, which forms complexes with fatty acids (Lin and Simon, 1978). An alternative explanation for the protective effect of albumin, which is more evident at low than high concentrations of phospholipase A, involves redistribution of the remaining phosphatidylserine to replenish depleted lipid at the receptor site (Abood *et al.*, 1980).

Further evidence for the protein and lipid nature of the opiate binding site is obtained from observations that solubilized opiate–macromolecule complexes are destroyed by heat, various proteases, *N*-ethylmaleimide, phospholipase *A*, and phosphatidylserine decarboxylase (Simon *et al.*, 1975*a*; Zukin and Kream, 1979; Bidlack and Abood, 1980; Rüegg *et al.*, 1980, 1981; Simonds *et al.*, 1980).

5.2. Effects of Ions

It has been shown that Na^+ ions (100 mM) decrease the specific binding of agonists, e.g., [^{3}H]etorphine and [^{3}H]dihydromorphine, and increase that of antagonists, e.g., [^{3}H]naloxone and [^{3}H]naltrexone. In addition, the potencies of unlabeled antagonists to displace the binding of [^{3}H]naltrexone or [^{3}H]naloxone is unchanged by Na^+, while those of compounds with dual agonist and antagonist activity are decreased but to a lesser extent than for pure agonists (Pert and Snyder, 1973, 1974; Pert *et al.*, 1973; Simon *et al.*, 1973, 1975*c*). The degree of depression of the binding of opiates varied considerably within groups of agonists, the ketazocine-like agonists being affected much less than the morphine-like compounds (Kosterlitz and Leslie, 1978). This effect was obtained to a lesser extent with Li^+ but was not mimicked by other monovalent cations (Pert and Snyder, 1974; Simon *et al.*, 1975*c*). The divalent cation Mn^{2+} (1 mM) increased the binding of [^{3}H]dihydromorphine but not that of [^{3}H]naloxone, the increase being larger when Na^+ was also present. There was also a small increase of about 10% in the binding of [^{3}H]dihydromorphine by Mg^{2+} (1 mM) and Ni^{2+} (0.02 mM) (Pasternak *et al.*, 1975*a*).

As far as the opioid peptides are concerned, the binding of [^{3}H]-[Met]-enkephalin, [^{3}H]-[Leu]enkephalin (Morin *et al.*, 1976; Meunier and Moisand, 1977; Simantov *et al.*, 1978), and [^{125}I]-[D-Ala^2,D-Leu^5]enkephalin (Miller *et al.*, 1978) was reduced in the presence of Na^+ (100 mM) by about 60–90% and increased in the presence of Mn^{2+} (1–10 mM) by 70–80%. In competitive inhibition experiments with [^{3}H]naloxone, Na^+ reduced the potencies of [Met]enkephalin, [Leu]enkephalin, [D-Ala^2,L-Leu^5]enkephalin, and β-endorphin between 6- and 26-fold compared with 30- to 60-fold for morphine-like agonists (Simantov and Snyder, 1976; Simantov *et al.*, 1978; Childers *et al.*, 1979). Mn^{2+} (1 mM) increased the potency of [Met]enkephalin and [Leu]enkephalin about twofold (Simantov and Snyder, 1976). The binding of [^{3}H-Tyr^{27}]-β_h-endorphin (1.2 and 4 nM) to rat brain membranes was completely inhibited in the presence of

Na^+ and inhibited by 30–40% in the presence of 10 mM of Li^+, Cs^+, and NH_4^+, while K^+ (10 mM) had no effect. In contrast to observations with labeled opiates and enkephalins, the binding of [^{3}H-Tyr^{27}]-β_h-endorphin was also inhibited by the divalent cations Mn^{2+}, Ca^{2+}, and Mg^{2+}, Mn^{2+} being the most potent (Ferrara *et al.*, 1979; Law *et al.*, 1979*b*; Akil *et al.*, 1980; Ferrara and Li, 1980). The binding of [^{3}H]dimethyllysine-β_p-endorphin was inhibited by Li^+, Na^+, Mn^{2+}, Mg^{2+}, and Ni^{2+} and also by K^+ (Hammonds *et al.*, 1981). When the effects of ions on the binding of [^{125}I]-[D-Ala^2]-β_h-endorphin and [^{125}I]-[D-Ala^2,D-Leu^5]enkephalin were compared in the same series of experiments in rat brain, Na^+ was found to decrease the binding of [^{125}I]-[D-Ala^2]-β_h-endorphin by only 17% compared to 48% of that of [^{125}I]-[D-Ala^2,D-Leu^5]enkephalin; Mn^{2+} (1 mM) and Mg^{2+} (2 mM) increased the binding of [^{125}I]-[D-Ala^2,D-Leu^5]-enkephalin by 70–80% but did not alter the binding of [^{125}I]-[D-Ala^2]-β_h-endorphin. In contrast, K^+ (100 mM) had no effect on the binding of [^{125}I]-[D-Ala^{2}D-Leu^5]enkephalin but caused 50% inhibition of [^{125}I]-[D-Ala^2]-β_h-endorphin binding. It has been suggested that these differences are due to the fact that the binding sites for β-endorphin are not identical to those for the enkephalins (Hazum *et al.*, 1979*a*).

There is some controversy as to the mechanism by which Na^+ discriminates between the binding of agonists and antagonists. Whereas some reports have indicated that Na^+ alters the numbers of binding sites (Pert and Snyder, 1974; Pasternak and Snyder, 1975*b*; Morin *et al.*, 1976), other groups have reported that the effect of Na^+ is due primarily to alterations in affinity (Simon *et al.*, 1975*c*; Lee *et al.*, 1977; Ferrara *et al.*, 1979). Since the rate of inactivation of the binding of [^{3}H]naltrexone by *N*-ethylmaleimide, but not the rate of interaction between glutathione and *N*-ethylmaleimide, was decreased by Na^+, it was suggested that Na^+ caused a conformational change in the binding site which rendered sulfhydryl groups less accessible to alkylation (Simon and Groth, 1975). It has been proposed that the opiate receptor exists in two interconvertible states on which Na^+ acts as an allosteric effector. However, the final interpretation of this phenomenon has to await further investigation with ligands highly selective for the μ-, δ-, and κ-sites.

There are conflicting reports on the effects of ions on the binding of [^{3}H]ethylketazocine. In mouse brain, Na^+ and, to a lesser extent, Li^+, but not K^+, decreased the steady-state binding of [^{3}H]ethylketazocine. These effects were due to a reduction in the number of high-affinity binding sites for [^{3}H]ethylketazocine, while the low-affinity binding sites showed relatively little change (Pasternak, 1980). In rat brain, however, Na^+ and, to a lesser extent Li^+, increased the steady-state binding of [^{3}H]ethylketazocine, while K^+, Cs^+, and Mg^{2+} had little or no effect. While the total number of binding sites remained the same, the relative proportion of high-affinity sites was increased in the presence of sodium (Hiller and

Simon, 1980). In view of the high degree of cross-reactivity of ethylketazocine for μ-, δ-, and κ-type binding sites, these data are difficult to interpret.

5.3. Effects of Nucleotides

It was shown that the binding of [^{3}H]-[Leu]enkephalin (1.2 nM), [^{3}H]-[Met]enkephalin (1.2 nM), and [^{3}H]dihydromorphine (0.41 nM) in rat brain was reduced by 40–50% in the presence of guanosine-5′-triphosphate (GTP) (50 μM), while that of etorphine was reduced by 20%. As far as antagonists were concerned, the binding of [^{3}H]naloxone, but not that of [^{3}H]diprenorphine, was reduced by 20%. When in addition Na^+ (100 mM) was present, the binding of agonists was reduced to 10 to 20% of control levels, while that of the antagonists was not affected (Childers and Snyder, 1978, 1980). Similar observations were made by Blume (1978*a*) who observed that the binding of [^{3}H]dihydromorphine (0.1–5 nM) in rat brain was reduced by about 50% with either Na^+ (25 mM) or with guanylyl-5′-imidophosphate [GMP-P(NH)P] (50 μM) and by about 70% with a combination of the two. The binding of [^{3}H]-[Leu]enkephalin (4 nM) to NG108-15 cells was not affected by 50 μM of GMP-P(NH)P alone but was reduced by 26% by Na^+ (100 mM) and by 65% in the presence of both Na^+ and GMP-(NH)P (Blume, 1978*b*). GTP was more effective in inhibiting the binding of [^{3}H]dihydromorphine than of [^{3}H]-[D-Ala^2,L-Leu-NH_2^{5}]enkephalin in guinea pig brain. It was observed that low concentrations of GTP (0.2 μM) increased rather than decreased the binding of [^{3}H]-[D-Ala^2,L-Leu-NH_2^{5}]enkephalin and [^{3}H]dihydromorphine in homogenates of brain and longitudinal muscle–myenteric plexus of guinea pig (Zukin and Gintzler, 1980). In contrast to observations for membrane-bound receptors, GTP (100 μM) inhibited equally well antagonist ([^{3}H]diprenorphine) and agonist ([^{3}H]etorphine) binding to a solubilized component from NG108-15 cells (Koski *et al.*, 1981). Divalent cations, particularly Mn^{2+}, reversed the inhibition of agonist binding caused by GTP or by the combination of Na^+ and GTP but not that caused by the stable analog GMP-P(NH)P. This effect of Mn^{2+} appeared to be the result of an enhanced hydrolysis of GTP (Childers and Snyder, 1980). GTP, GMP-P(NH)P, and guanosine-5′-diphosphate (GDP) were equally effective in inhibiting the binding; inosinetriphosphate (ITP) and inosyl-5′-imidodiphosphate [IMP-P(NH)P] were not as potent, while adenine nucleotides had no effect. Guanosine-5′-monophosphate (GMP) was considerably less potent than the other guanine nucleotides (Blume, 1978*a,b*; Childers and Snyder, 1980).

The potencies of several opioid agonists, but not those of antagonists, in displacing the binding of [^{3}H]diprenorphine in rat brain was reduced by GTP (50 μM) and even more so by GTP together with Na^+ (100 mM). However, GTP alone had negligible effect on the inhibitory potency of etorphine and phenazocine (Childers and Snyder, 1980).

In one series of experiments, the effect of guanine nucleotides alone or with sodium has been reported to be predominantly due to the reduction of the number of binding sites, particularly the high-affinity sites of [^{3}H]dihydromorphine (Blume 1978*a*; Childers and Snyder, 1978). The number of high-affinity sites for [^{3}H]naloxone was reduced by GTP alone but not by the combination of GTP and Na^+ (Childers and Snyder, 1978). In contrast, in a second series, the inhibition by GTP (50 μM) of the binding of [^{3}H]-[D-Ala^2,L-Leu-NH_2^{5}]enkephalin in rat brain and of [^{3}H]dihydromorphine in homogenates of the guinea pig longitudinal muscle–myenteric plexus was attributed to a reduced affinity for the high-affinity sites; the high-affinity binding sites were not seen when GTP and Na^+ (50 mM) were both present (Zukin and Gintzler, 1980; Zukin *et al.*, 1980).

Analysis of biphasic Scatchard plots in which high- and low-affinity sites may be differentially affected is difficult. A more direct approach has been the investigation of the effects of nucleotides at the kinetic level. GTP (50 μM) reduced the maximum binding of [^{3}H]dihydromorphine by 60% and about doubled the rate of association. In the absence of GTP, the dissociation curve for [^{3}H]dihydromorphine was biphasic but in the presence of GTP it was monophasic, the rate of dissociation being accelerated about fourfold. In the absence of GTP, the calculated K_D values were 0.11 nM for the first phase and 0.82 nM for the second phase; in the presence of GTP the K_D value was 0.38 nM. GTP had no significant effect on the rates of association and dissociation or the maximal binding of [^{3}H]diprenorphine (Childers and Snyder, 1980).

Since GTP (10 μM), but not ATP (50 μM), retarded the rate of inactivation by *N*-ethylmaleimide of [^{3}H]naloxone binding, it may modulate opiate binding by an allosteric mechanism (Zukin *et al.*, 1980). It has been shown that both Na^+ and GTP are required for opiate inhibition of adenylate cyclase in NG108-15 neuroblastoma × glioma cells, and it has been suggested that guanine nucleotides may interact with a membrane unit to couple the binding site to adenylate cyclase (Blume *et al.*, 1979).

Opiate receptors have been classified into two subgroups on the basis of their sensitivity to GTP (Pert and Taylor, 1980; Pert *et al.*, 1980); Type 1 receptors, but not Type 2 receptors, are sensitive to inhibition by GTP. GTP (2 μM) was found to reduce the inhibition of [^{3}H]diprenorphine binding by [Met]enkephalin (100 nM) only in membranes from certain tissues (Table 10). Thus, in the guinea pig ileum, the bovine adrenal medulla, and most areas of rat brain, including the periaqueductal gray, nucleus gigantocellularis, and reticular formation, the inhibition of [^{3}H]-diprenorphine binding by [Met]enkephalin was reduced by GTP. In

TABLE 10
Effect of GTP (2 μM) on the Inhibition of [³H]Diprenorphine (0.98 nM) Binding by [Met]enkephalin (100 nM) in Regions of Rat Brain[a]

Brain regions	Inhibition of [^{3}H]diprenorphine binding by [Met]enkephalin (%)	
	without GTP	with GTP
Periaqueductal gray	52	22
Caudate nucleus	49	36
Thalamus	52	35
Reticular formation	55	35
Nucleus gigantocellularis	43	24
Amygdala	47	47
Hypothalamus	39	39
Frontal cortex	30	33
Nucleus accumbens	40	40

[a] Modified from Pert *et al.* (1980).

contrast, GTP was ineffective in membranes from neuroblastoma × glioma cells, NGH109-15, the pituitary gland, drosophila heads, and four areas of the limbic system. It was suggested that there may be a correlation between Type 1 receptors and the μ-binding sites and Type 2 receptors and the δ-binding sites since the regional distribution of Type 1 receptors was similar to that of the μ-binding sites and the distribution of Type 2 receptors was similar to that of the δ-binding sites (see Section 4.1) (Pert and Taylor, 1980; Pert *et al.*, 1980). The validity of this suggestion cannot be evaluated until the effects of GTP on the binding of ligands selective for the μ-, δ-, and κ-binding sites in the various regions have been investigated.

6. ISOLATION OF OPIATE RECEPTORS

6.1. Inactive Ligand–Macromolecular Complexes

Early attempts to isolate components from mammalian brain tissue that retain the ability to bind labeled opiates and opioid peptides after solubilization were unsuccessful because of the inactivation of the binding sites by detergents. Labeled ligand–macromolecular complexes can be isolated when homogenates are incubated with labeled opiates or opioid peptides prior to solubilization. At the end of the incubation period, bound ligand is separated from unbound by centrifugation. The pellet is resuspended in buffer containing the desired concentration of the solubilizing agent (Table 11); the solubilized proteins are recovered in the supernatant by centrifugation.

In 1975, Simon and co-workers (Simon *et al.*, 1975*a*) solubilized a [^{3}H]-etorphine–macromolecular complex from rat brain using the nonionic detergent Brij 36T. The free [^{3}H]etorphine was separated from the bound [^{3}H]etorphine by passage over XAD-4. The labeled complex was destroyed by trypsin, pronase, and *N*-ethylmaleimide, and by heating to 50°C for 10 min, as are membrane-bound binding sites. The solubilized complex was further characterized by gel filtration on Sepharose 6B. A single peak was found which had an elution volume corresponding to a molecular weight of 370,000 daltons.

The same solubilization procedure was used to isolate [^{3}H]-[D-Ala^2,L-Met-NH$_2$^{5}]enkephalin and [^{125}I]-[D-Ala^2,MePhe^4,Met(O)-ol^5]enkephalin complexes from rat brain (Zukin and Kream, 1979). The solubilized complexes were separated from unbound ligand on a Sephadex G-25 column. Thereafter, the labeled ligands were covalently bound to the solubilized proteins using dimethyl suberimidate. The molecular weight of the cross-linked [^{3}H]-[D-Ala^2,Met-NH$_2$^{5}]enkephalin complex was 380,000 daltons.

The [^{3}H]-[D-Ala^2,L-Met-NH$_2$^{5}]enkephalin, [^{3}H]naloxone, [^{3}H]etorphine, and [^{3}H]-β_h-endorphin labeled complexes which had been solubilized from mouse brain by Brij 36T consisted of several components with molecular weights in the range 100,000–500,000 daltons. Electrophoresis of the [^{3}H]-β_h-endorphin complex on SDS polyacrylamide gels, after exposure to suberimidate, yielded a large number of covalently linked species with molecular weights in the range 2,000–200,000 daltons (Smith and Loh, 1979).

A covalently labeled complex has been isolated from mouse brain by the use of the irreversible antagonist [^{3}H]chlornaltrexamine (Caruso *et al.*, 1980). The complex was solubilized with 1% Triton X-100, dialyzed against Brij 36T, and then precipitated with ammonium sulfate. When the precipitated material was fractionated on an Ultrogel AcA 22 column, the main radioactive peak had an elution volume which corresponded to a molecular weight of 590,000 daltons; an additional larger-molecular-weight peak (>1,200,000 daltons), which may be an aggregate of the main peak, and a lower-molecular-weight peak were also present.

A [^{3}H]etorphine complex has also been isolated from rat brain using Na cholate (Puget *et al.*, 1980). Subsequent gel filtration of the solubilized complex indicated that the major radioactive peak was associated with a fraction corresponding to a molecular weight of ~500,000 daltons. However, if the solubilized complex was fractionated on a sucrose density gradient, the major radioactive peak cosedimented with aldolase which has a molecular weight of 158,000 daltons. The discrepancy in the molecular weight as determined by the two methods may be due to bound

TABLE 11
Solubilization of Opiate Receptors and Drug–Receptor Complexes

Source	Solubilizing agent	Molecular weight	Reference
Rat brain	1% Brij 36T in 10 mM Tris	370,000	Simon *et al.* (1975*a*)
Rat brain	1% Brij 36T in phosphate buffer	380,000	Zukin and Kream (1979)
Rat brain	10% (w/v) Na cholate in 50 mM Tris	500,000	Puget *et al.* (1980)
Rat brain or NG108-15	10 mM 3-[3-(cholamidopropyl)-diamethylammino]-1-propanesulfonate (CHAPS) in 50 mM Tris	380,000–669,000	Simonds *et al.* (1980)
Rat brain	1% Triton X-100 in 50 mM Tris	>800,000	Bidlack *et al.* (1981)
Mouse brain	1% Triton X-100 in 50 mM Tris	590,000	Caruso *et al.* (1980)
Toad brain	1% (w/v) digitonin in 50 mM Tris	350,000–400,000	Rüegg *et al.* (1981)

molecules of detergents altering the hydrodynamic properties of the complex, as has been found to occur with solubilized acetylcholine receptors (Heidmann and Changeux, 1978).

6.2. Solubilized Macromolecules to Which Opioid Ligands Bind

The isolation of a component from the neuroblastoma × glioma hybrid cell line NG108-15 which, after solubilization with 3-[(3-cholamidopropyl)dimethylammino]-1-propanesulfonate (CHAPS), can bind [^{3}H]-[D-Ala^2,L-Met-NH_2^{5}]enkephalin, was first reported by Simonds *et al.* (1980). Gel filtration of the solubilized material gave a broad peak eluting between the marker proteins thyroglobulin (669,000) and phosphorylase *a* (380,000). CHAPS was also shown to solubilize a similar component from rat and bovine brain which can bind [^{3}H]etorphine. Analysis of the binding of [^{3}H]-[D-Ala^2,L-Met-NH_2^{5}]enkephalin to the solubilized material indicated a K_D of 2.3 nM and a binding capacity of 0.96 pmol/mg protein. Several unlabeled opioids of different potencies were examined for their ability to displace the binding of [^{3}H]-[D-Ala^2,L-Met-NH_2^{5}]enkephalin; it was found that the K_I value of each opioid was nearly identical in the membrane bound and in the solubilized preparations (Table 12) (Simonds *et al.*, 1980). In solubilized, but not in membrane-bound preparations, GTP was equipotent in decreasing the binding of the agonist etorphine, and the antagonist diprenorphine (Koski *et al.*, 1981).

The binding of [^{3}H]diprenorphine had similar characteristics in homogenates of brains from the toad, *Bufo marinus*, and the rat. In both

TABLE 12

Comparison of the Inhibitory Effects (IC_{50}) of Opiates and an Opioid Peptide on the Binding of [^{3}H]-[D-Ala^2,L-Met-NH_2^{5}]enkephalin and [^{3}H]Naltrexone in Homogenates and in Partially Purified Solubilized Preparations of NG108-15 Cells and Toad Brain

	NG108-15 cells,[a] IC_{50} (nM)		Toad brain,[b] IC_{50} (nM)	
Unlabeled compound	P_2 homogenate	Solubilized preparation[c]	P_2 homogenate	Solubilized preparation[d]
Etorphine	3.7	3.7	0.2	0.3
Levorphanol	19	37	5	2
Dextrorphan	1900	9300	—	—
Naloxone	56	93	7	5
[D-Ala^2,D-Leu^5]enkephalin	—	—	500	200

[a] Assayed against [^{3}H]-[D-Ala^2,L-Met-NH_2^{5}]enkephalin (2 nM) (Simonds *et al.*, 1980).
[b] Assayed against [^{3}H]naltrexone (2.5 nM) (Rüegg *et al.*, 1981).
[c] Solubilized with 10 mM CHAPS.
[d] Solubilized with 1% (w/v) digitonin.

species the K_D was 1 nM, but the toad brain contained 40% more binding sites than the rat (Rüegg *et al.*, 1980). After solubilization of toad brain membranes with digitonin, the recovery of [^{3}H]diprenorphine binding sites was between 40 and 50% of that found in fresh homogenate (Rüegg *et al.*, 1981). The solubilized material had a K_D of 0.86 nM for [^{3}H]-diprenorphine and a binding capacity of 0.33 pmol/mg protein. In competitive displacement assays, the order of potency in displacing the binding of [^{3}H]naloxone was etorphine > levorphanol > naloxone > [D-Ala^2,D-Leu^5]enkephalin, suggesting that the solubilized material obtained from the toad brain consists mainly of μ-binding sites. The solubilized material had a molecular weight of 350,000–400,000 daltons which is similar to that found for the [^{3}H]etorphine–macromolecular complex isolated from rat brain (Simon *et al.*, 1975*a*).

A recent development has been to combine isolation by means of Triton X-100 (Bidlack and Abood, 1980) with purification by affinity chromatography of the solubilized proteins (Bidlack *et al.*, 1981). An affinity column was prepared by linking 14-β-bromoacetamidomorphine to ω-aminohexyl-Sepharose via the acetamido group. Although linked to the ω-aminohexyl-Sepharose, the opiate was still able to interact with opiate binding sites. When solubilized material was applied to the column, any proteins which interacted with the 14-β-bromoacetamidomorphine were retained on the column and could subsequently be eluted with a high concentration of unlabeled opiate, e.g., etorphine or levorphanol. This fraction, which appeared in the void volume of a Sephadex G-200 column (molecular weight >500,000), showed by gel electrophoresis three main peaks of 43,000, 35,000, and 23,000 daltons. The solubilized material bound [^{3}H]dihydromorphine with a K_D of 3.8 nM and a capacity of 40 pmol/mg protein. This represents a 2000-fold purification compared with membrane-bound preparations.

The use of this procedure with selective ligands should facilitate the isolation of the different subtypes of the opiate receptor. This could be achieved either by preparing an affinity column with a highly selective ligand which will only retain one subtype or by eluting the different binding proteins sequentially from the column with selective ligands.

7. ASSESSMENT OF THE PHARMACOLOGICAL AND BINDING PROPERTIES OF NARCOTIC ANALGESIC DRUGS

During the last three or four decades a great deal of effort has been concentrated on attempts to obtain new potent narcotic analgesic drugs which would be free from the dangers of tolerance to, and dependence on, opiates. Progress has necessarily been slow because our understanding of the mechanisms underlying the action of opiates has been and still is incomplete. However, the development of the concept of defined subtypes of the opiate receptor and of new receptor-selective models for *in vitro* bioassays has reached a point at which a reassessment of the known facts may be useful in many respects. In particular, a better correlation of the pharmacology of opiates of the alkaloid type with the action of the endogenous opioid peptides may be helpful in the design of new drugs.

In a first step toward this aim, a number of opiates have been examined in three binding assays to assess their inhibitory potencies at the μ-, δ-, and κ-binding sites in homogenates of guinea pig brain, and in three *in vitro* bioassays, the guinea pig ileum, the mouse vas deferens, and the rat vas deferens to determine their relative agonist and antagonist activities (Table 13) (Magnan *et al.*, 1982).

Agonists. The members of the first group are pure agonists in the guinea pig ileum, mouse vas deferens, and rat vas deferens, with the exception of morphine, which is inactive in the rat vas deferens. Many of the compounds are almost pure μ-ligands; for instance, morphine has only 2% cross-reactivity to the δ-binding site and 0.6% to the κ-binding site. Some of the synthetic and semisynthetic surrogates have higher cross-reactivity to the other binding sites; to give an extreme case, compared to the μ-binding site, etorphine is 1.8 times more potent at the δ-binding site and 4 times more potent at the κ-binding site.

No information is so far available which would make possible a correlation of the pattern of the binding spectra of these agonist compounds with their pharmacological properties *in vitro* and *in vivo*. In particular, it would be of interest to compare in detail the pharmacological effects of compounds that bind almost exclusively to the μ-site (e.g., morphine and sufentanyl) with those compounds that exhibit considerable affinities to the δ- and κ-sites (e.g., phenazocine and etorphine).

Agonist–antagonists. The second group consists of natural and synthetic opiates in which the *N*-methyl side chain has been replaced by *N*-allyl, *N*-cyclopropylmethyl, or *N*-dimethylallyl. All these compounds have agonist and antagonist activities in the guinea pig ileum and mouse vas deferens and, as a group, are pure antagonists in the rat vas deferens, as has previously been pointed out (Wüster *et al.*, 1980). The binding of the *N*-allyl derivatives and particularly the *N*-cyclopropylmethyl homologs show high affinities to the μ-binding sites and, compared to the *N*-methyl homologs, they show increased binding to the δ- and κ-binding sites. In particular, cyclorphan, SKF 10047, and cyclazocine exhibit a high relative affinity to the κ-binding site.

Pharmacologically, the members of this group of compounds have certain important properties in common: *in vivo*, they are potent antinociceptive agents but also exhibit antagonist activities; furthermore, they

TABLE 13

Binding at μ-, δ-, and κ-Binding Sites and the Pharmacological Properties of Opiates[a]

	Inhibition (K_I, nM) of binding at the			Inhibition of contractions. (IC_{50}, nM)			Antagonism (K_e, nM)		
	μ-site	δ-site	κ-site	G.p.i.	M.v.d.	R.v.d.	G.p.i.[b]	M.v.d.[b]	R.v.d.[c]
Agonists									
Morphine	1.8	90	317	68	492		None	None	None
Sufentanyl	1.6	23	124	0.65	1.8	28.3	None	None	None
Phenazocine	1.5	3.4	7.7	5.5	23	2380	None	None	None
Etorphine	1.0	0.56	0.23	0.08	0.4	5.0	None	None	None
Agonist–antagonists									
Nalorphine	1.8	7.2	7.7	24	480 shallow	None	4.5	29	41
Cyclorphan	0.25	0.65	0.28	0.28	3.1 shallow	None	None	1.6	2.2
SKF 10047	2.0	9.0	3.2	37	[d]	None	7.5	36	10.4
Cyclazocine	0.29	2.0	0.40	2.0	9.7	None	0.75	4.9	9.8
Ketazocine-like									
Ethylketazocine	1.0	5.5	0.52	0.18	4.4	None	None	None	82
Bremazocine	0.62	0.72	0.41	0.13	2.0	None	None	None	2.7
Antagonists									
Naloxone	1.8	27	17	None	None	None	1.2	3.7	7.5
Diprenorphine	0.84	1.4	2.2	Shallow	None	None	0.13	0.96	0.94
Mr 2266	1.4	6.0	0.69	Shallow	None	None	1.9	1.5	8.8

[a] Binding at the μ-site was measured by [^{3}H]-[D-Ala^2,MePhe^4,Gly-ol^5]enkephalin (1 nM), at the δ-site by [^{3}H]-[D-Ala^2,D-Leu^5]enkephalin (1–1.8 nM), and at the κ-site by [^{3}H]-(±)-ethylketazocine (0.65 nM) after suppression of binding to μ- and δ-sites by 100 nM unlabeled μ- and δ-ligands. G.p.i.: guinea pig ileum. M.v.d.: mouse vas deferens. R.v.d.: rat vas deferens. Modified from Magnan *et al.* (1982).

[b] Antagonism against normorphine.

[c] Antagonism against [D-Ala^2,MePhe^4,Gly-ol^5]enkephalin.

[d] Maximum inhibition of 50% at 1000 nM.

are characterized by their ability to cause dysphoric and psychotomimetic symptoms. There is little doubt that binding of a μ-agonist such as morphine to the μ-site will lead to an effect which is due to activation at the postrecognition stages of the μ-receptor as observed *in vivo* and also in isolated tissues of guinea pig ileum, mouse vas deferens, and rat vas deferens. Whereas little is known about the pharmacological consequences of the binding of the compounds of this group of agonist–antagonists to the δ-site, their interaction with the κ-binding site or the κ-receptor is attracting much interest. In view of their binding spectra, it is not surprising that the pharmacological effects of the compounds which have been postulated to interact with the κ-receptor are complex. The binding assays have shown that they have high affinities at the μ- and κ-binding sites and lower affinities at the δ-binding site. Based on pharmacological observations, Martin and his colleagues have concluded that their action at the μ-receptor is antagonist, and at the κ-receptor agonist. Binding assays do not distinguish between agonist and antagonist effects, but it has been shown (Wüster *et al.*, 1980; Gillan *et al.*, 1981) that, in the rat vas deferens, the agonist–antagonists and also the ketazocine-like compounds are pure antagonists.

It should be noted that the evidence for existence of the μ-, δ-, and κ-binding sites depends on the use of ligands which are selective, such as the μ-ligand [^{3}H]-[D-Ala^2,MePhe^4,Gly-ol^5]enkephalin and, less stringently, the δ-ligand [^{3}H]-[D-Ala^2,D-Leu^5]enkephalin. A nonselective ligand, such as [^{3}H]ethylketazocine, can be converted to a selective κ-ligand by the suppression of μ- and δ-binding with suitable unlabeled μ- and δ-ligands. In these circumstances, there is no unequivocal evidence that, in this group of compounds, there are sites other than the μ-, δ-, and κ-recognition sites which are necessary for opioid activity. However, it has been shown recently that the psychotomimetic effects in man and the characteristic behavioral effects in animals caused by some of these drugs with *N*-allyl or *N*-cyclopropylmethyl side chains (e.g., cyclorphan, SKF 10047, and cyclazocine) may not be due to interaction with μ-, δ-, and κ-binding sites. It is more likely that these actions are due to activation of the relatively low affinity sites of [^{3}H]cyclazocine (Zukin and Zukin, 1981), [^{3}H]allylnormetazocine (SKF 10047) (Su, 1981), or [^{3}H]phenylcyclidine (Quirion *et al.*, 1981). These binding sites have in common that naloxone, dihydromorphine, or etorphine displace these tritiated ligands only at very high concentrations. It has been postulated that the psychotomimetic and behavioral actions may be induced by the σ-receptor which does not belong to the "classical opiate" receptor consisting of μ-, δ-, and κ-subtypes.

Ketazocine-like Compounds. These drugs differ from those with dual agonist and antagonist actions in that they are pure agonists in the guinea pig ileum and mouse vas deferens and pure antagonists in the rat vas deferens (Hutchinson *et al.*, 1975; Gillan *et al.*, 1981). The most striking difference of the binding spectrum is the affinity to the κ-binding site which is 1.5 to 2 times higher than that to the μ-binding site.

The ketazocine-like compounds have a unique pharmacological pattern in that they are potent antinociceptive agents in rodents but do not suppress signs of withdrawal in morphine-dependent rhesus monkeys and have little or no antagonist activity (Villarreal and Seevers, 1972; Swain and Seevers, 1974, 1976). In the chronic spinal dog, behavioral and neurophysiological observations have indicated that the ketazocine-like compounds interact with the κ-receptor rather than the μ-receptor, a view that is in agreement with the observation that the ketazocine-like compounds are antagonists in the rat vas deferens when tested against pure μ-agonists. The ketazocine-like compounds have no agonist effect on the σ-receptor (Gilbert and Martin, 1976; Martin *et al.*, 1976).

Antagonists. The fourth group of compounds are antagonists that have only negligible or no agonist activity. None of them has a binding spectrum which is sufficiently selective for their use as receptor-specific antagonists *in vivo* or *in vitro*. Naloxone is most suitable for the antagonism of μ-ligands since about 15 times higher concentrations are required for the antagonism of δ-ligands than for that of μ-ligands. While Mr 2266 is about equiactive at the μ- and κ-binding sites, the most universal antagonist available at present is diprenorphine.

In conclusion it has been found that, in each of the four groups, the compounds, whether pure agonists, compounds with dual agonist and antagonist actions, ketazocine-like drugs, or pure antagonists, have high affinities to the μ-binding sites and affinities to the κ-binding sites which vary from low to high. Binding to the δ-site also varies but not to the same degree. It is important to note that this variation in affinity to the κ-binding site is not characteristic for any of the four groups of compounds. Thus, pure agonism or antagonism, or the peculiar behavioral effects of the compounds with agonist–antagonist activity, cannot be explained solely by variation in binding affinities to the κ-site. It will be important to develop pharmacological models in the central nervous system which can be used for the study of highly specific antagonists when they become available. Success in this area of research would facilitate our understanding of the basic mechanisms of interaction of opioids with the different subtypes of the opiate receptor and also the design of drugs with novel characteristics.

8. EXCITATION EFFECTOR COUPLING

While the binding of drugs to the recognition sites of the opiate receptor has been extensively studied, information regarding the subsequent events which lead to a pharmacological response is as yet limited.

Cyclic 3′,5′-adenosine monophosphate (cAMP), which has been shown to be involved in the mediation of the effects of a number of hormones, may have a role in the actions of opiates. The literature has been reviewed recently (Wollemann, 1981).

Preparations of certain cultured cell lines have large numbers of opiate receptors and therefore provide a relatively simple system for the study of postrecognition processes. It has been demonstrated that in such cell lines opiates and enkephalins inhibit basal as well as prostaglandin E_1-stimulated adenylate cyclase activity in a stereospecific and naloxone-reversible manner (Sharma *et al.*, 1975*b*; Traber *et al.*, 1975; Klee and Nirenberg, 1976) and that the ability of opioid agonists to inhibit adenylate cyclase activity varies with the number of opiate binding sites present in different cell lines. Thus, morphine has a marked inhibitory effect on prostaglandin-stimulated cAMP formation in the hybrid cell line NG108-15 which possesses a large number of binding sites, but inhibits adenylate cyclase activity only to a small extent in the parental cell line N18TG-2 which has relatively few opiate binding sites; it does not affect adenylate cyclase activity in the parental line C6BU-1 which lacks opiate receptors (Sharma *et al.*, 1975*b*). That these effects are the result of an interaction with the opiate receptor is confirmed by observations that the potency of the agonists morphine, etorphine, levorphanol, and 3-allylprodine to inhibit adenylate cyclase activity is closely correlated with their affinity for the binding sites of [^{3}H]naloxone (Fig. 6). The antagonist naloxone prevented the binding of [^{3}H]naloxone and, as expected, did not inhibit adenylate cyclase activity. The effect of levorphanol was stereospecific since the (+)-isomer dextrorphan was without effect on adenylate cyclase activity and had a very low affinity for the binding site (Sharma *et al.*, 1975*b*). The observation that the interaction between the inhibitory effect of morphine and the stimulatory effect of prostaglandin was noncompetitive indicated that prostaglandins and morphine were not acting at the same receptor site (Traber *et al.*, 1975).

Similar observations have been made with respect to the effect of opiates on adenylate cyclase activity in homogenates of brain tissues. The stimulation of cAMP formation by prostaglandins E_1 and E_2 was inhibited by morphine and other opiates, but, in contrast to observations in cultured cell preparations, the basal activity was not affected. The potencies of opiates to inhibit the stimulatory effect of prostaglandins on adenylate cyclase activity were well correlated with their antinociceptive potencies, with their affinities to the opiate binding sites, and with their potencies in the guinea pig ileum; these effects were antagonized by naloxone (Collier and Roy, 1974*a,b*).

In slices of rat striatum, [Leu]enkephalin, [Met]enkephalin, morphine, and levorphanol, but not dextrorphan, enhanced the accumulation of cyclic 3′,5′-guanosine monophosphate (cGMP) with a concomitant decrease

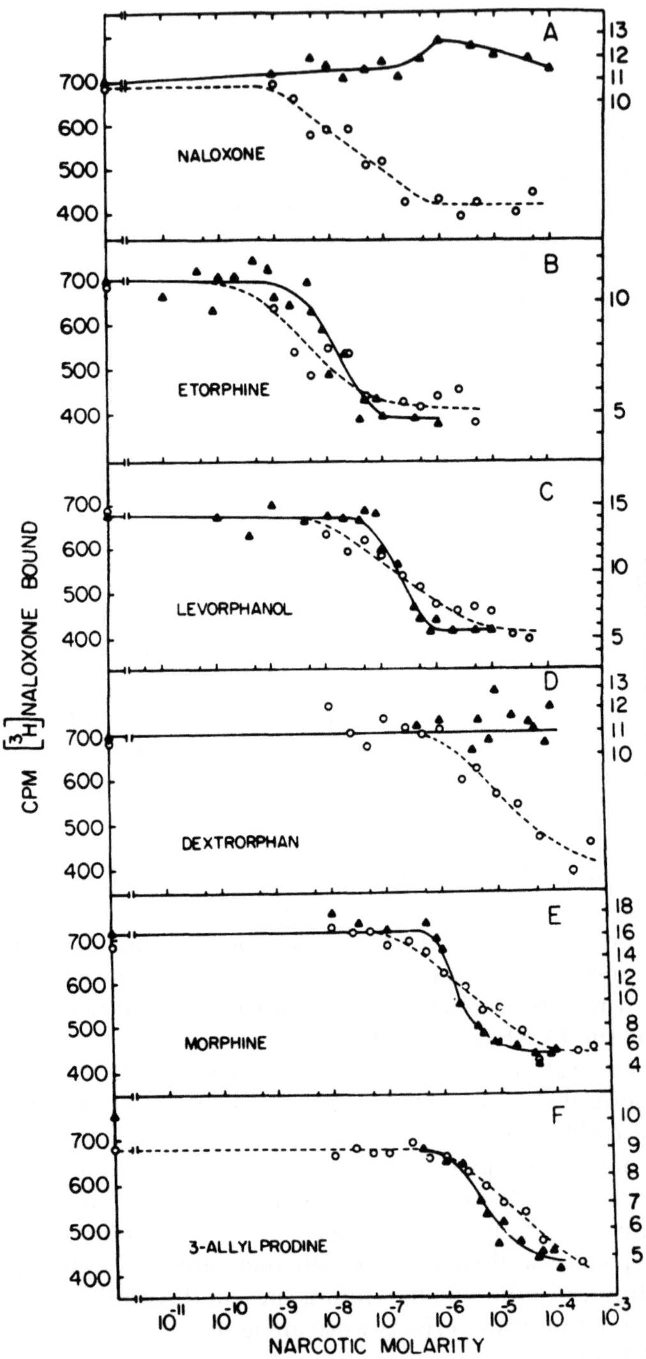

FIG. 6. Comparison of the potencies of opiates to inhibit adenylate cyclase activity with their potencies to displace [^{3}H]naloxone (50 nM) binding in homogenates of NG108-15 cells. Abscissa: concentration of opiate (M). Ordinates: (○) [^{3}H]naloxone bound (counts/min); (▲) rate of formation of [^{32}P]cAMP (pmol/250 μg protein per 5 min). (From Sharma *et al.*, 1975*b*.)

in cAMP levels (Minneman and Iversen, 1976). These effects were antagonized by naloxone. While the effects on cAMP were confirmed in slices of rat striatum, they could not be reproduced in homogenates (Havemann and Kuschinsky, 1978). There may be regional differences in the effects of opioids on cAMP and cGMP contents. After subcutaneous injection of morphine or cyclazocine to the rat the levels of cAMP were increased in the thalamus and cortex. In contrast, in the substantia nigra, levels were decreased by morphine but not changed by cyclazocine. The effects of morphine were not seen after the prior administration of naloxone. As far as cGMP levels were concerned, morphine caused a decrease in the substantia nigra and an increase in the cerebellum. Cyclazocine and naloxone decreased thalamic levels of cGMP, whereas cyclazocine caused an increase in the cerebellum. It remains to be determined by the use of selective ligands whether, as the authors suggested, the different effects of cyclazocine and morphine in the various regions are due to the fact that they interact with different receptor populations (Bonnet, 1975; Bonnet *et al.*, 1978). In another series of experiments, the subcutaneous injection of morphine (30 mg/kg) reduced levels of cGMP by 50% in the periaqueductal gray. Levels of cGMP were decreased to a small extent in the centromedian hypothalamus but were unchanged in the caudate-putamen, in the lateral and ventromedial hypothalamus, and in the pituitary. The effect of morphine could be prevented by prior administration of naloxone and was also seen with levorphanol but not with dextrorphan (O'Callaghan *et al.*, 1979).

Sodium ions and GTP are required for coupling of opiate receptors to adenylate cyclase (Blume *et al.*, 1979; Wilkening *et al.*, 1980). The inhibition by morphine, etorphine, and [D-Ala^2,L-Met-$NH_2{}^5$]enkephalin of basal and prostaglandin-stimulated adenylate cyclase activity was determined in extensively washed nuclei-free homogenates of NG108-15 cells; in the presence of either Na^+ or GTP the inhibition was only 20–27% of that observed when both Na^+ and GTP were present. Furthermore, in intact cells, replacement of Na^+ by sucrose or choline reduced the ability of [D-Ala^2,L-Met-$NH_2{}^5$]enkephalin to decrease cAMP levels (Blume *et al.*, 1979). At high concentrations of Ca^{2+}, the inhibitory effect of morphine on adenylate cyclase activity in washed membrane fractions of NG108-15 cells was reduced (Wilkening *et al.*, 1980).

In NG108-15 cells the hydrolysis of GTP was increased by several opiates, including etorphine, morphine, and [D-Ala^2,L-Met-$NH_2{}^5$]enkephalin in concentrations similar to those required for inhibition of adenylate cyclase activity. This effect exhibited stereospecificity and was seen only in cell lines with opiate binding sites. Sodium ions were required for stimulation of GTPase activity (Koski and Klee, 1981). These observations indicate that inhibition of adenylate cyclase by opiates may be partly due to stimulation of GTP hydrolysis. This hypothesis is supported by the observation that opiates did not inhibit adenylate cyclase activity in the presence of high concentrations (>10 μM) of the stable GTP analog guanylyl-5′-imidodiphosphate [GMP-P(NH)P] (Sharma *et al.*, 1975*b*).

In addition to the readily reversible morphine-induced inhibition of adenylate cyclase activity observed in NG108-15 cells, a second process mediated by opiate receptors has been observed. In cells cultured for up to 4 days in the presence of morphine (10 μM), the level of basal and prostaglandin E_1-stimulated adenylate cyclase activity was at first inhibited but appeared to start to recover after about 12 hr. After 24–48 hr the levels of adenylate cyclase had returned to those observed in untreated cells. Removal of the morphine or application of naloxone resulted in a further increase in adenylate cyclase activity and caused a marked accumulation of cAMP. These events may be related to the phenomena of tolerance and dependence (Sharma *et al.*, 1975*a*).

While the interaction with adenylate cyclase may possibly be one of the initial events following opioid receptor activation, further biochemical changes may be of importance for the pharmacological action of opiates. For instance, in NG108-15 cells, morphine inhibited cAMP-dependent protein kinase and ornithine decarboxylase (Bachrach *et al.*, 1979). In N4TG1 and NG108-15 cells, incubation for 24 hr with morphine, β-endorphin, and [D-Ala^2,D-Leu^5]enkephalin inhibited the biosynthesis of membrane glycosphingolipid and glycoproteins; these effects were stereospecific, dose-dependent, and naloxone-reversible and were not seen in cell lines lacking opiate binding sites (Dawson *et al.*, 1979, 1980).

9. CHANGES IN BINDING SITES INDUCED BY DRUGS OR PATHOLOGICAL CONDITIONS

For an understanding of the phenomena of tolerance to, and dependence on, opiates administered for long periods, it is important to examine whether such treatment causes changes in the affinity and number of opiate binding sites. Several authors have observed that, after chronic treatment with morphine, there are no changes in the binding sites for [^{3}H]dihydromorphine, [^{3}H]etorphine, and [^{3}H]naloxone in homogenates of rat, mouse and guinea pig brain and of guinea pig ileum (Hitzemann *et al.*, 1974; Klee and Streaty, 1974; Davis *et al.*, 1975; Cox and Padhya, 1977). However, in one series of experiments, the binding of [^{3}H]dihydromorphine and [^{3}H]naloxone was enhanced in brains of mice following treatment with morphine under various treatment schedules, but this enhancement was not correlated with the development of tolerance and dependence (Pert and Snyder, 1976). In contrast to these findings, the binding of single concentrations of [^{3}H]morphine, [^{3}H]etorphine, [^{3}H]-[D-Ala^2,L-Met-NH$_2$^{5}]enkephalin, and [^{3}H]naloxone in slices of rat brainstem, measured after rapid homogenization and filtration, was lower in preparations from animals implanted with morphine pellets (150 mg; 72 hr) than in those from naive animals (Davis *et al.*, 1975, 1978, 1979). The affinity for morphine, obtained from Scatchard analysis in brainstem slices from the morphine-treated animals, was lower (K_D = 23 nM) than that in control animals (K_D = 6.2 nM), and the corresponding maximal numbers of binding sites were 0.08 and 0.12 pmol/mg protein. The authors are aware of the possibility that the change in affinity may be due to the difficulty of removing residual morphine from the brain slices. This problem may not arise in homogenates (Davis *et al.*, 1975).

Four hours after cessation of chronic infusion of naloxone (10 mg/kg per day, 2 to 4 weeks) to rats, the antinociceptive action of morphine was increased (Tang and Collins, 1978). Similarly, thoroughly washed ileum preparations from animals treated chronically with naloxone (200 mg pellet; 7 days) were four times more sensitive to the inhibitory effects of normorphine and [Met]enkephalin than those from naive animals (Schulz *et al.*, 1979*b*). These increases in sensitivity may be associated with changes in the receptor population. In the rat brain, the maximal number of binding sites for [^{3}H]naloxone was increased by 40% seven days after chronic naloxone treatment (Lahti and Collins, 1978); the maximal binding for [^{3}H]etorphine was increased by 23% in the longitudinal muscle–myenteric plexus and by 30% in the brainstem of the guinea pig. There were no apparent changes in the binding affinity (Schulz *et al.*, 1979*b*).

Since the discovery of the opioid peptides, pharmacological, electrophysiological, and anatomical observations have suggested that they play an important role in the control of pain. Although significant changes are found in the levels of the endogenous opioid peptides in rats exposed to stress (Madden *et al.*, 1977; Rossier *et al.*, 1978), no difference was found in the maximal binding of [^{3}H]-[Nle^5]enkephalin in control and fear-conditioned rats (Chance *et al.*, 1978). Similarly, a study of the effect of chronic footshock on the postnatal development of [^{3}H]naloxone binding sites found no effect in five brain regions or the spinal cord (Bardo *et al.*, 1981). In rats, experimentally induced polyarthritis caused an increase in [Met]enkephalin-like immunoreactivity in the lumbar spinal cord, but no changes were found in the binding of either [^{3}H]naloxone or [^{3}H]-[Leu]-enkephalin (Cesselin *et al.*, 1980).

In the same species, castration led to a marked increase in the number of [^{3}H]naltrexone binding sites in the brain without altering their affinity (Hahn and Fishman, 1979). Treatment with testosterone restored the number of binding sites. Brattleboro rats, which are homozygous with respect to diabetes insipidus, displayed an impaired development of tolerance to morphine (De Wied and Gispen, 1976). However, a comparison of the binding of [^{3}H]dihydromorphine in seven brain regions obtained from Brattleboro and normal rats did not show any difference

TABLE 14

[^{3}H]Naloxone Binding in Brains from Control Patients and Patients with Parkinson's Disease or Schizophrenia[a]

Patients	Binding of [^{3}H]naloxone (1.6 nM) (fmol/mg protein) in		Binding affinity and capacity for [^{3}H]naloxone in the caudate nucleus	
	Caudate nucleus	Putamen	Equilibrium dissociation constant (K_D, nM)	Maximum number of binding sites (fmol/mg protein)
Control	83.1 ± 9.5(10)	76.7 ± 7.5(7)	1.93 ± 0.1	177 ± 9.7(3)
Parkinson's disease	42.9 ± 7.5(5)	66.1 ± 4.5(7)	1.8 ± 0.1	139 ± 13.6(3)
Control	80.5 ± 11.2(11)	74.2 ± 6.3(11)		
Schizophrenic	45.2 ± 5.8(11)	63.4 ± 6.2(11)	1.56 ± 0.1	73.3 ± 5.6(3)

[a] Modified from Reisine *et al.* (1979*b*, 1980).

in the maximal number of binding sites or in their affinity; the only exception was the amygdala where there was a slight increase in the number of binding sites (Rigter *et al.*, 1979).

To date, only two studies have been performed on the changes in opiate binding associated with disease. In both studies, opiate binding was assessed using [^{3}H]naloxone (1.6 nM) in the caudate nucleus and the putamen (Reisine *et al.*, 1979*b*, 1980). In the caudate nucleus, binding was decreased by 65% in Parkinson's disease and 44% in schizophrenia. No significant changes were found in the putamen (Table 14). Saturation analysis of the binding of [^{3}H]naloxone in the caudate nucleus from three control patients, three patients with schizophrenia, and three patients with Parkinson's disease indicated that the K_D for naloxone was the same in all three groups; however, the maximum number of binding sites was reduced by 21% in Parkinson's disease and by 59% in schizophrenia. Whether these changes in receptor number involve losses of one or other of the subtypes of opiate receptor requires further investigation with selective ligands.

Acknowledgments

This work was supported by grants from the U.S. National Institute on Drug Abuse (DA 00662) and the Medical Research Council.

10. REFERENCES

Abood, L. G., and Hoss, W., 1975, Stereospecific morphine adsorption to phosphatidylserine and other membranous components of brain, *Eur. J. Pharmacol.* **32:**66–75.

Abood, L. G., and Takeda, F., 1976, Enhancement of stereospecific opiate binding to neural membranes by phosphatidylserine, *Eur. J. Pharmacol.* **39:**71–77.

Abood, L. G., Salem, N., MacNeil, M., Bloom, L., and Abood, M. E., 1977, Enhancement of opiate binding by various molecular forms of phosphatidylserine and inhibition by other unsaturated lipids, *Biochim. Biophys. Acta* **468:**51–62.

Abood, L. G., Salem, N., MacNeil, M., and Butler, M., 1978, Phospholipid changes in synaptic membranes by lipolytic enzymes and subsequent restoration of opiate binding with phosphatidylserine, *Biochim. Biophys. Acta* **530:**35–46.

Abood, L. G., Butler, M., and Reynolds, D., 1980, Effect of calcium and physical state of membranes on phosphatidylserine requirement for opiate binding, *Mol. Pharmacol.* **17:**290–294.

Adler, M. W., 1981, Mini symposium I. The *in vivo* differentiation of opiate receptors: introduction, *Life Sci.* **28:**1543–1545.

Akil, H., Mayer, D. J., and Liebeskind, J. C., 1972, Compairison chez le rat entre l'analgésie induite par stimulation de la substance grise péri-aqueducale et l'analgésie morphinique, *C.R. Acad. Sci. Ser. D* **274:**3603–3605.

Akil, H. A., Hewlett, W. A., Barchas, J. D., and Li, C-H., 1980, Binding of [^{3}H]-β-endorphin to rat brain membranes: characterization of opiate properties and interaction with ACTH, *Eur. J. Pharmacol.* **64:**1–8.

Atweh, S. F., and Kuhar, M. J., 1977*a*, Autoradiographical localization of opiate receptors in rat brain. I. Spinal cord and lower medulla, *Brain Res.* **124:**53–67.

Atweh, S. F., and kuhar, M. J., 1977*b*, Autoradiographical localization of opiate receptors in rat brain. II. The brainstem, *Brain Res.* **129:**1–12.

Atweh, S. F., and Kuhar, M. J., 1977*c*, Autoradiographical localization of opiate receptors in rat brain. III. The telencephalon, *Brain Res.* **134:**393–405.

Auguy-Valette, A., Cros, J., Gouarderes, C., Gout, R., and Pontonnier, G., 1978, Morphine analgesia and cerebral opiate receptors: a developmental study, *Br. J. Pharmacol.* **63:**303–308.

Bachrach, U., Benalal, D., and Reches, A., 1979, Morphine inhibits cyclic AMP-dependent protein kinase and ornithine decarboxylase activities in neuroblastoma × glioma hybrid cells, *Life Sci.* **25:**1879–1884.

Bardo, M. T., Bhatnagar, R. K., and Gebhart, G. F., 1981, Opiate receptor ontogeny and morphine-induced effects: influence of chronic footshock stress in preweanling rats, *Dev. Brain Res.* **1:**487–495.

Beckett, A. H., and Casy, A. F., 1954, Synthetic analgesics: stereochemical considerations, *J. Pharm. Pharmacol.* **6:**986–999.

Beckett, A. H., Casy, A. F., and Harper, N. J., 1956, Analgesics and their antagonists: some steric and chemical considerations. Part III. The influence of the basic group on the biological response, *J. Pharm. Sci.* **8:**874–884.

Berti, F., Bruno, F., Omini, C., and Racagni, G., 1978, Genotype-dependent response of morphine and methionine-enkephalin on the electrically induced contractions of the mouse vas deferens, *Naunyn-Schmiedeberg's Arch. Pharmacol.* **305:**5–8.

Bidlack, J. M., and Abood, L. G., 1980, Solubilization of the opiate receptor, *Life Sci.* **27:**331–340.

Bidlack, J. M., Abood, L. G., Osei-Gyimah, P., and Archer, S., 1981, Purification of the opiate receptor from rat brain, *Proc. Natl. Acad. Sci. USA* **78:**636–639.

Blume, A. J., 1978*a*, Interaction of ligands with the opiate receptors of brain membranes: regulation by ions and nucleotides, *Proc. Natl. Acad. Sci. USA* **75:**1713–1717.

Blume, A. J., 1978*b*, Opiate binding to membrane preparations of neuroblastoma × glioma hybrid cells NG 108-15: effects of ions and nucleotides, *Life Sci.* **22:**1843–1852.

Blume, A. J., Lichtshtein, D., and Boone, G., 1979, Coupling of opiate receptors to adenylate cyclase: requirement for Na^+ and GTP, *Proc. Natl. Acad. Sci. USA* **76:**5626–5630.

Bonnet, K. A., 1975, Regional alterations in cyclic nucleotide levels with acute and chronic morphine treatment, *Life Sci.* **16:**1877–1882.

Bonnet, K. A., Gusik, S. A., and Sunshine, A. G., 1978, Multiple opiate receptors reflected in region-specific alterations in brain cyclic nucleotides, in: *Characteristics and Function of Opioids* (J. van Ree and L. Terenius, eds.), pp. 453–464, Elsevier/North Holland Biomedical Press, Amsterdam.

Bradbury, A. F. D., Smyth, D. G., Snell, C. R., Birdsall, N. J. M., and Hulme, E. C., 1976, C-fragment of lipotropin has a high affinity for brain opiate receptors, *Nature (London)* **260:**793–795.

Bradbury, A. F. D., Smyth, D. G., Snell, C. R., Deakin, J. F. W., and Wendlandt, S., 1977, Comparison of the analgesic properties of lipotropin C-fragment and stabilized enkephalins in the rat, *Biochem. Biophys. Res. Commun.* **74:**748–754.

Braenden, O. J., Eddy, N. B., and Halbach, H., 1955, Synthetic substances with morphine-like effects: relationship between chemical structure and analgesic action, *Bull. W.H.O.* **13:**937–998.

Carenzi, A., Frigeni, V., and Della Bella, D., 1978, Synaptic localization of opiate receptors in rat striatum, in: *Advances in Biochemical Psychopharmacology* (E. Costa and M. Trabucchi, eds.), Vol. 18, pp. 265–270, Raven Press, New York.

Caruso, T. P., Takemori, A. E., Larson, D. L., and Portoghese, P. S., 1979, Chloroxymorphamine, an opioid receptor site-directed agent having narcotic agonist activity, *Science (Washington D.C.)* **204:**316–318.

CARUSO, T. P., LARSON, D. L., PORTOGHESE, P. S., and TAKEMORI, A. E., 1980, Isolation of selective [^{3}H]chlornaltrexamine-bound complexes, possible opioid receptor components in brains of mice, *Life Sci.* **27:**2063–2069.

CERLETTI, C., COCCIA, P., MANARA, L., MENNINI, T., and RECCHIA, M., 1978, Subcellular distribution of etorphine in rat brain and evidence for *in vivo* stereospecific binding, *Br. J. Pharmacol.* **62:**31–38.

CESSELIN, F., MONTASTRUC, J. L., GROS, C., BOURGOIN, S., and HAMON, M., 1980, Met-enkephalin levels and opiate receptors in the spinal cord of chronic suffering rats, *Brain Res.* **191:**289–293.

CHANCE, W. T., WHITE, A. C., KRYNOCK, G. M., and ROSECRANS, J. A., 1978, Conditional fear-induced antinociception and decreased binding of [^{3}H]-N-Leu-enkephalin to rat brain, *Brain Res.* **141:**371–374.

CHANG, K.-J., and CUATRECASAS, P., 1979, Multiple opiate receptors. Enkephalins and morphine bind to receptors of different specificity, *J. Biol. Chem.* **254:**2610–2618.

CHANG, K.-J., MILLER, R. J., and CUATRECASAS, P., 1978, Interaction of enkephalin with opiate receptors in intact cultured cells, *Mol. Pharmacol.* **14:**961–970.

CHANG, K.-J., COOPER, B. R., HAZUM, E., and CUATRECASAS, P., 1979, Multiple opiate receptors: different regional distribution in the brain and differential binding of opiates and opioid peptides, *Mol. Pharmacol.* **16:**91–104.

CHANG, K.-J., HAZUM, E., and CUATRECASAS, P., 1980, Possible role of distinct morphine and enkephalin receptors in mediating actions of benzomorphan drugs (putative κ and σ agonists), *Proc. Natl. Acad. Sci. USA* **77:**4469–4473.

CHANG, K.-J., HAZUM, E., and CUATRECASAS, P., 1981*a*, Novel opiate binding sites selective for benzomorphan drugs, *Proc. Natl. Acad. Sci. USA* **78:**4141–4145.

CHANG, K.-J., KILLIAN, A., HAZUM, E., and CUATRECASAS, P., 1981*b*, Morphiceptin (NH_2-Tyr-Pro-Phe-Pro$CONH_2$): a potent and specific agonist for morphine (μ) receptors, *Science (Washington D.C.)* **212:**75–77.

CHAVKIN, C., and GOLDSTEIN, A., 1981, Demonstration of a specific dynorphin receptor in guinea pig ileum myenteric plexus, *Nature (London)* **29:**591–593.

CHAVKIN, C., COX, B. M., and GOLDSTEIN, A., 1979, Stereospecific opiate binding in bovine adrenal medulla, *Mol. Pharmacol.* **15:**751–753.

CHENG, Y.-C., and PRUSOFF, W. H., 1973, Relationship between the inhibition constant (K_I) and the concentration of inhibitor which causes 50 percent inhibition (I_{50}) of an enzymatic reaction, *Biochem. Pharmacol.* **22:**3099–3108.

CHILDERS, S. R., and SNYDER, S. H., 1978, Guanine nucleotides differentiate agonist and antagonist interactions with opiate receptors, *Life Sci.* **23:**759–762.

CHILDERS, S. R., and SNYDER, S. H., 1980, Differential regulation by guanine nucleotides of opiate agonist and antagonist receptor interactions, *J. Neurochem.* **34:**583–593.

CHILDERS, S. R., CREESE, I., SNOWMAN, A. M., and SNYDER, S. H., 1979, Opiate receptor binding affected differentially by opiates and opioid peptides, *Eur. J. Pharmacol.* **55:**11–18.

CHO, T. M., CHO, S. J., and LOH, H. H., 1979, Physicochemical basis of opiate-cerebroside sulphate interaction and its application to receptor theory, *Mol. Pharmacol.* **16:**393–405.

CIČERO, T. J., WILCOX, C. E., and MEYER, E. R., 1974, Effect of α-adrenergic blockers on naloxone binding in brain, *Biochem. Pharmacol.* **23:**2349–2352.

CICERO, T. J., WILCOX, C. E., MEYER, E. R., and MICHAEL, H., 1975, Influence of catecholaminergic agents on narcotic binding in brain, *Arch. Int. Pharmacodyn. Ther.* **218:**221–230.

CLENDENINN, N. J., PETRAITIS, M., and SIMON, E. J., 1976, Ontological development of opiate receptors in rodent brain, *Brain Res.* **118:**157–160.

COLLIER, H. O. J., and ROY, A. C., 1974*a*, Morphine-like drugs inhibit the stimulation by E prostaglandins of cyclic AMP formation by rat brain homogenate, *Nature (London)* **248:**24–27.

COLLIER, H. O. J., and ROY, A. C., 1974*b*, Hypothesis: Inhibition of E-prostaglandin-sensitive adenyl cyclase as the mechanism of morphine analgesia, *Prostaglandins* **7:**361–376.
CORBETT, A. D., KOSTERLITZ, H. W., and MCKNIGHT, A. T., 1982, Inhibition of the breakdown of exogenous and endogenous enkephalins, *Br. J. Pharmacol.* **76:** Proc. Suppl. 26P.
COWAN, A., 1981, Mini symposium III. Simple *in vivo* tests that differentiate prototype agonists at opiate receptors, *Life Sci.* **28:**1559–1570.
COWIE, A. L., KOSTERLITZ, H. W., and WATERFIELD, A. A., 1978, Factors influencing the release of acetylcholine from the myenteric plexus of the ileum of the guinea pig and rabbit, *Br. J. Pharmacol.* **64:**565–580.
COX, B. M., and PADHYA, R., 1977, Opiate binding and effect in ileum preparations from normal and morphine-pretreated guinea pigs, *Br. J. Pharmacol.* **61:**271–278.
COX, B. M., GOLDSTEIN, A., and LI, C. H., 1976, Opioid activity of a peptide, β-lipotropin-(61–91), derived from β-lipotropin, *Proc. Natl. Acad. Sci. USA* **73:**1821–1823.
COYLE, J. T., and PERT, C. B., 1976, Ontogenetic development of [^{3}H]naloxone binding in rat brain, *Neuropharmacology* **15:**555–560.
CRAVES, F. B., ZALC, B., LEYBIN, L., BAUMANN, N., and LOH, H. H., 1980, Antibodies to cerebroside sulfate inhibit the effects of morphine and β-endorphin, *Science (Washington D.C.)* **207:**75–76.
CREESE, I., and SNYDER, S. H., 1975, Receptor binding and pharmacological activity of opiates in the guinea pig intestine, *J. Pharmacol. Exp. Ther.* **194:**205–219.
DAVIS, M. E., AKERA, T., and BRODY, T. M., 1975, Saturable binding of morphine to rat brainstem slices and the effect of chronic morphine treatment, *Res. Commun. Chem. Pathol. Pharmacol.* **12:**409–418.
DAVIS, M. E., AKERA, T., and BRODY, T. M., 1978, [D-Alanine^2]-methionine enkephalinamide binding to rat brain slices: differences from opiate alkaloid binding and reduction by chronic morphine treatment, *Life Sci.* **23:**2675–2680.
DAVIS, M. E., AKERA, T., and BRODY, T. M., 1979, Reduction of opiate binding to brainstem slices associated with the development of tolerance to morphine in rats, *J. Pharmacol. Exp. Ther.* **211:**112–119.
DAWSON, G., MCLAWHON, R., and MILLER, R. J., 1979, Opiates and enkephalins inhibit synthesis of gangliosides and membrane glycoproteins in mouse neuroblastoma cell line N4TG1, *Proc. Natl. Acad. Sci. USA* **76:**605–609.
DAWSON, G., MCLAWHON, R., and MILLER, R. J., 1980, Inhibition of sialoglycosphingolipid (ganglioside) biosynthesis in mouse clonal lines N4TG1 and NG108-15 by β-endorphin, enkephalins, and opiates, *J. Biol. Chem.* **255:**129–137.
DE WIED, D., and GISPEN, W. H., 1976, Impaired development of tolerance to morphine analgesia in rats with hereditary diabetes insipidus, *Psychopharmacologia (Berlin)* **46:**27–29.
DUNLAP, C. E., LESLIE, F. M., RADO, M., and COX, B. M., 1979, Ascorbate destruction of opiate stereospecific binding in guinea pig brain homogenate, *Mol. Pharmacol.* **16:**105–119.
FERRARA, P., and LI, C. H., 1980, β-Endorphin: characteristics of binding sites in rabbit spinal cord, *Proc. Natl. Acad. Sci. USA* **77:**5746–5748.
FERRARA, P., HOUGHTEN, R., and LI, C. H., 1979, β-endorphin: characteristics of binding sites in the rat brain, *Biochem. Biophys. Res. Commun.* **89:**786–792.
FIELDS, H. L., EMSON, P. C., LEIGH, B. K., GILBERT, R. F. T., and IVERSEN, L. L., 1980, Multiple opiate receptor sites on primary afferent fibres, *Nature (London)* **284:**351–353.
GACEL, G., FOURNIE-ZALUSKI, M-C., and ROQUES, B. P., 1980, Tyr-D-Ser-Gly-Phe-Leu-Thr, a highly preferential ligand for δ-opiate receptors, *FEBS Lett.* **118:**245–247.
GARCIN, F., and COYLE, J. T., 1976, Ontogenetic development of [^{3}H]naloxone binding and endogenous morphine-like factor in rat brain, in: *Opiates and Endogenous Opioid Peptides* (H. W. Kosterlitz, ed.), pp. 267–273, Elsevier/North Holland Biomedical Press, Amsterdam.
GARCIN, F., CIESIELSKI-TRESKA, J., ULRICH, G., LOUIS, J-C., and MANDEL, P., 1978, Sites recépteurs des opiacés dans les lignées neuronales et gliales de cellules en culture, *C.R. Acad. Sci. Ser. D* **287:**1023–1026.

GARDNER, E. L., ZUKIN, R. S., and MAKMAN, M. H., 1980, Modulation of opiate receptor binding in striatum and amygdala by selective mesencephalic lesions, *Brain Res.* **194:**232–239.

GERBER, L. D., STEIN, S., RUBINSTEIN, M., WIDEMAN, J., and UDENFRIEND, S., 1978, Binding assay for opioid peptides with neuroblastoma × glioma hybrid cells: specificity of the receptor site, *Brain Res.* **151:**117–126.

GILBERT, P. E., and MARTIN, W. R., 1976, The effects of morhpine- and nalorphine-like drugs in the nondependent, morphine-dependent, and cyclazocine-dependent chronic spinal dog, *J. Pharmacol. Exp. Ther.* **198:**66–82.

GILLAN, M. G. C., KOSTERLITZ, H. W., and PATERSON, S. J., 1979, Comparison of the binding characteristics of tritiated opiates and opioid peptides, *Br. J. Pharmacol.* **66:**86–87P.

GILLAN, M. G. C., KOSTERLITZ, H. W., and PATERSON, S. J., 1980, Comparison of the binding characteristics of tritiated opiates and opioid peptides, *Br. J. Pharmacol.* **70:**481–490.

GILLAN, M. G. C., KOSTERLITZ, H. W., and MAGNAN, J., 1981, Unexpected antagonism in the rat vas deferens by benzomorphans which are agonists in other pharmacological tests, *Br. J. Pharmacol.* **72:**13–15.

GLASEL, J. A., VENN, R. F., and BARNARD, E. A., 1980, Distribution of sterospecific opiate receptor binding activity between subcellular fractions from ovine corpus striatum, *Biochem. Biophys. Res. Commun.* **95:**263–268.

GOLDSTEIN, A., and COX, B. M., 1978, Opiate receptors and their endogenous ligands (endorphins), in: *Progress in Molecular and Subcellular Biology* (F. E. Hahn, H. Kersten, W. Kersten, and W. Szybalski, eds.), Vol. 6, pp. 113–157, Springer-Verlag, Berlin.

GOLDSTEIN, A., LOWNEY, L. I., and PAL, B. K., 1971. Stereospecific and nonspecific interactions of the morphine congener levorphanol in subcellular fractions of mouse brain, *Proc. Natl. Acad. Sci. USA* **68:**1742–1747.

GOODMAN, R. R., SNYDER, S. H., KUHAR, M. J., and YOUNG, W. S., 1980, Differentiation of delta and mu opiate receptor localizations by light microscopic autoradiography, *Proc. Natl. Acad. Sci. USA* **77:**6239–6243.

HAHN, E. F., and FISHMAN, J., 1979, Changes in rat brain opiate receptor content upon castration and testosterone replacement, *Biochem. Biophys. Res. Commun.* **90:**819–823.

HAMMONDS, R. G., LING, N., and PUETT, D., 1981, Interaction of tritiated β-endorphin with rat brain membranes, *Anal. Biochem.* **114:**75–84.

HANDA, B. K., LANE, A. C., LORD, J. A. H., MORGAN, B. A. RANCE, M. J., and SMITH, C. F. C., 1981, Analogues of β-LPH_{61-64} possessing selective agonist activity at μ-opiate receptors, *Eur. J. Pharmacol.* **70:**531–540.

HARRIS, D. W., and SETHY, V. H., 1980, High-affinity binding of [^{3}H]ethylketazocine to rat brain homogenate, *Eur. J. Pharmacol.* **66:**121–123.

HAVEMANN, U., and KUSCHINSKY, K., 1978, Interactions of opiates and prostaglandins E with regard to cyclic AMP in striatal tissue of rats *in vitro, Naunyn Schmiedeberg's Arch. Pharmacol.* **302:**103–106.

HAZUM, E., CHANG, K.-J., and CUATRECASAS, P., 1979*a*, Interaction of iodinated human [D-Ala^2]-β-endorphin with opiate receptors, *J. Biol. Chem.* **254:**1765–1767.

HAZUM, E., CHANG, K.-J., and CUATRECASAS, P., 1979*b*, Opiate (enkephalin) receptors of neuroblastoma cells: occurrence in clusters on the cell surface, *Science (Washington D.C.)* **206:**1077–1079.

HAZUM, E., CHANG, K.-J., and CUATRECASAS, P., 1979*c*, Role of disulphide and sulfhydryl groups in clustering of enkephalin receptors in neuroblastoma cells, *Nature (London)* **282:**626–628.

HAZUM, E., CHANG, K.-J., and CUATRECASAS, P., 1980, Cluster formation of opiate (enkephalin) receptors in neuroblastoma cells: differences between agonists and antagonists and possible relationships to biological functions, *Proc. Natl. Acad. Sci. USA* **77:**3038–3041.

HAZUM, E., CHANG, K.-J., CUATRECASAS, P., and PASTERNAK, G. W., 1981, Naloxazone irreversibly inhibits the high-affinity binding of [^{125}I]-D-ala^2-D-leu^5-enkephalin, *Life Sci.* **28:**2973–2979.

HEIDMANN, T., and CHANGEUX, J.-P., 1978, Structural and functional properties of the acetylcholine receptor protein in its purified and membrane-bound states, *Annu. Rev. Biochem.* **47:**317–357.

HERKENHAM, M., and PERT, C. B., 1981, Mosaic distribution of opiate receptors, parafascicular projections and acetylcholinesterase in rat striatum, *Nature (London)* **291:**415–418.

HERLING, S., and WOODS, J. H., 1981, Mini symposium IV. Discriminative stimulus effects of narcotics: evidence for multiple receptor-mediated actions, *Life Sci.* **28:**1571–1584.

HILL, A. V., 1910, The possible effects of the aggregation of the molecules of haemoglobin on its dissociation curves, *J. Physiol. (London)* **40:**iv–viii.

HILLER, J. M., and SIMON, E. J., 1979, [^{3}H]Ethylketocyclazocine binding: lack of evidence for a separate kappa receptor in rats CNS, *Eur. J. Pharmacol.* **60:**389–390.

HILLER, J. M., and SIMON, E. J., 1980, Specific high-affinity [^{3}H]ethylketocyclazocine binding in rat central nervous system: lack of evidence for kappa receptors, *J. Pharmacol. Exp. Ther.* **214:**516–519.

HILLER, J. M., PEARSON, J., and SIMON, E. J., 1973, Distribution of stereospecific binding of the potent narcotic analgesic etorphine in the human brain: predominance in the limbic system, *Res. Commun. Chem. Pathol. Pharmacol.* **6:**1052–1062.

HILLER, J. M., ANGEL, L. M., and SIMON, E. J., 1981, Multiple opiate receptors: alcohol selectively inhibits binding to delta receptors, *Science (Washington D.C.)* **214:**468–469.

HITZEMANN, R. J., HITZEMANN, B. A., and LOH, H. H., 1974, Binding of [^{3}H]naloxone in the mouse brain: effect of ions and tolerance development, *Life Sci.* **14:**2393–2404.

HOLTZMAN, S. G., 1980, Phencyclidine-like discriminative effects of opioids in the rat, *J. Pharmacol. Exp. Ther.* **214:**614–619.

HOWELLS, R. D., GROTH, J., HILLER, J. M., and SIMON, E. J., 1980, Opiate binding sites in the retina: properties and distribution, *J. Pharmacol. Exp. Ther.* **215:**60–64.

HUGHES, J., KOSTERLITZ, H. W., and LESLIE, F. M., 1975*a*, Effect of morphine on adrenergic transmission in the mouse van deferens. Assessment of agonist and antagonist potencies of narcotic analgesics, *Br. J. Pharmacol.* **53:**371–381.

HUGHES, J., SMITH, T. W., KOSTERLITZ, H. W., FOTHERGILL, L. A., MORGAN, B. A., and MORRIS, H. R., 1975*b*, Identification of two related pentapeptides from the brain with potent agonist activity, *Nature (London)* **258:**577–580.

HUIDOBRO, F., HUIDOBRO-TORO, J. P., and MIRANDA, H., 1980, Interactions between morphine and the opioid-like peptides in the rat vas deferens, *Br. J. Pharmacol.* **70:**519–525.

HUTCHINSON, M., KOSTERLITZ, H. W., LESLIE, F. M., WATERFIELD, A. A., and TERENIUS, L., 1975, Assessment in the guinea pig ileum and mouse vas deferens of benzomorphans which have strong antinociceptive activity but do not substitute for morphine in the dependent monkey, *Br. J. Pharmacol.* **55:**541–546.

JACQUET, Y. F., 1980, Excitatory and inhibitory effects of opiates in the rat vas deferens: a dual mechanism of opiate action, *Science (Washington D.C.)* **210:**95–97.

KACHUR, J. F., MILLER, R. J., and FIELD, M., 1980, Control of guinea pig intestinal electrolyte secretion by a δ-opiate receptor, *Proc. Natl. Acad. Sci. USA* **77:**2753–2756.

KIRBY, M. L., 1981, Development of opiate receptor binding in rat spinal cord, *Brain Res.* **205:**400–404.

KLEE, W. A., and NIRENBERG, M., 1974, A neuroblastoma × glioma hybrid cell line with morphine receptors, *Proc. Natl. Acad. Sci. USA* **71:**3474–3477.

KLEE, W. A., and NIRENBERG, M., 1976, Mode of action of endogenous opiate peptides, *Nature (London)* **263:**609–612.

KLEE, W. A., and STREATY, R. A., 1974, Narcotic receptor sites in morphine-dependent rats, *Nature (London)* **248:**61–63.

KOSKI, G., and KLEE, W. A., 1981, Opiates inhibit adenylate cyclase by stimulating GTP hydrolysis, *Proc. Natl. Acad. Sci. USA* **78:**4185–4189.

KOSKI, G., SIMONDS, W. F., and KLEE, W. A., 1981, Guanine nucleotides inhibit binding of agonists and antagonists to soluble opiate receptors, *J. Biol. Chem.* **256:**1536–1538.

KOSTERLITZ, H. W., and LESLIE, F. M., 1978, Comparison of the receptor binding characteristics of opiate agonists acting with μ or κ-receptors, *Br. J. Pharmacol.* **64**:607–614.

KOSTERLITZ, H. W., and MCKNIGHT, A. T., 1980, Endorphins and enkephalins, in: *Advances in Internal Medicine* (G. H. Stollerman, eds.), Vol. 26, pp. 1–36, Year Book Medical Publishers Inc., Chicago.

KOSTERLITZ, H. W., and MCKNIGHT, A. T., 1981, Opioid peptides and sensory function, in: *Progress in Sensory Physiology* (H. Autrum, D. Ottoson, E. R. Perl, and R. F. Schmidt, eds.), Vol. 1, pp. 31–95, Springer-Verlag, Berlin.

KOSTERLITZ, H. W., and PATERSON, S. J., 1980, Characterization of opioid receptors in nervous tissue, *Proc. R. Soc. London Ser. B* **210**:113–122.

KOSTERLITZ, H. W., and PATERSON, S. J., 1981, Tyr-D-Ala-Gly-MePhe-NH$(CH_2)_2$OH is a selective ligand for the μ-opiate binding site, *Br. J. Pharmacol.* **73**:299P.

KOSTERLITZ, H. W., and WATERFIELD, A. A., 1975, *In vitro* models in the study of structure–activity relationships of narcotic analgesics, *Annu. Rev. Pharmacol.* **15**:29–48.

KOSTERLITZ, H. W., and WATT, A. J., 1968, Kinetic parameters of narcotic agonists and antagonists, with particular reference to *N*-allylnoroxymorphone (naloxone), *Br. J. Pharmacol.* **33**:266–276.

KOSTERLITZ, H. W., LORD, J. A. H., PATERSON, S. J., and WATERFIELD, A. A., 1980, Effects of changes in the structure of enkephalins and of narcotic analgesic drugs on their interactions with μ- and δ-receptors, *Br. J. Pharmacol.* **68**:333–342.

KOSTERLITZ, H. W., PATERSON, S. J., and ROBSON, L. E., 1981, Characterization of the κ-subtype of the opiate receptor in the guinea pig brain, *Br. J. Pharmacol.* **73**:939–949.

KOSTERLITZ, H. W., MAGNAN, J., and PATERSON, S. J., 1982*a*, The interaction of endogenous opioid peptides with the μ-, δ-, and κ-binding sites in the guinea pig, *Br. J. Pharmacol.* **76**:Proc. Suppl. 121P.

KOSTERLITZ, H. W., PATERSON, S. J., and ROBSON, L. E., 1982*b*, Opioid peptides and their receptors, in: *Neuropeptides—Basic and Clinical Aspects. Proceedings of the Eleventh Pfizer Symposium* (G. Fink and L. J. Whalley, eds.), pp. 3–12, Churchill Livingstone, Edinburgh.

KREAM, R. M., and ZUKIN, R. S., 1979, Binding characteristics of a potent enkephalin analog, *Biochem. Biophys. Res. Commun.* **90**:99–109.

KREAM, R. M., ZUKIN, R. S., and STEFANO, G. B., 1980, Demonstration of two classes of opiate binding sites in the nervous tissue of the marine mollusc *Mytilus edulis, J. Biol. Chem.* **255**:9218–9224.

KRIWACZEK, V. M., SCHWYZER, R., GILLAN, M. G. C., PATERSON, S. J., and KOSTERLITZ, H. W., 1981, Tobacco mosaic virus–enkephalin conjugates: potentiation of opioid activity, *Peptides* **2**:89–92.

KUHAR, M. J., PERT, C. B., and SNYDER, S. H., 1973, Regional distribution of opiate receptor binding in monkey and human brain, *Nature (London)* **245**:447–450.

LAHTI, R. A., and COLLINS, R. J., 1978, Chronic naloxone results in prolonged increases in opiate binding sites in brain, *Eur. J. Pharmacol.* **51**:158–186.

LAMOTTE, C., PERT, C. B., and SNYDER, S. H., 1976, Opiate receptor binding in primate spinal cord: distribution and changes after dorsal root section, *Brain Res.* **112**:407–412.

LAW, P. Y., and LOH, H. H., 1978, [^{3}H]-Leu^5-enkephalin specific binding to synaptic membranes—comparison with [^{3}H]-dihydromorphine and [^{3}H]naloxone, *Res. Commun. Chem. Pathol. Pharmacol.* **21**:409–434.

LAW, P. Y., HARRIS, R. A., LOH, H. H., and WAY, E. L., 1978, Evidence for the involvement of cerebroside sulphate in opiate receptor binding: studies with azure A and jimpy mutant mice, *J. Pharmacol. Exp. Ther.* **207**:458.

LAW, P. Y., HERZ, A. A., and LOH, H. H., 1979*a*, Demonstration and characterization of a stereospecific opiate receptor in the neuroblastoma N18TG2 cells, *J. Neurochem.* **33**:1177–1187.

LAW, P.-Y., LOH, H. H., and LI, C. H., 1979*b*, Properties and localization of β-endorphin receptor in rat brain, *Proc. Natl. Acad. Sci. USA* **76**:5455–5459.

Lee, C.-Y., Akera, T., and Brody, T. M., 1977, Effects of Na^+, K^+, Mg^{++}, and Ca^{++} on the saturable binding of [^{3}H]dihydromorphine and [^{3}H]naloxone *in vitro*, *J. Pharmacol. Exp. Ther.* **202**:166–173.

Lemaire, S., Magnan, J., and Regoli, D., 1978, Rat vas deferens: a specific bioassay for endogenous opioid peptides, *Br. J. Pharmacol.* **64**:327–329.

Leslie, F. M., and Kosterlitz, H. W., 1979, Comparison of binding of [^{3}H]methionine-enkephalin, [^{3}H]naltrexone, and [^{3}H]dihydromorphine in the mouse vas deferens and the myenteric plexus and brain of the guinea pig, *Eur. J. Pharmacol.* **56**:379–383.

Leslie, F. M., Chavkin, C., and Cox, B. M., 1980, Opioid binding properties of brain and peripheral tissues: evidence for heterogeneity in opiate ligand binding sites, *J. Pharmacol. Exp. Ther.* **214**:395–402.

Lewis, M. E., Mishkin, M., Bragin, E., Brown, R. M., Pert, C. B., and Pert, A., 1981, Opiate receptor gradients in monkey cerebral cortex: correspondence with sensory processing hierarchies, *Science (Washington D.C.)* **211**:1166–1169.

Lin, H.-K., and Simon, E. J. 1978, Phospholipase A inhibition of opiate receptor binding can be reversed by albumin, *Nature (London)* **271**:383–384.

Loh, H. H., Cho, T. M., Wu, Y.-C., and Way, E. L., 1974, Stereospecific binding of narcotics to brain cerebrosides, *Life Sci.* **14**:2231–2245.

Loh, H. H., Cho, T. M., Wu, Y.-C., Harris, R. A., and Way, E. L., 1975, Opiate binding to cerebroside sulphate: a model system for opiate receptor interaction, *Life Sci.* ***16***:1811–1817.

Lord, J. A. H., Waterfield, A. A., Hughes, J., and Kosterlitz, H. W., 1976, Multiple opiate receptors, in: *Opiates and Endogenous Opioid Peptides* (H. W. Kosterlitz, ed.), pp. 275–280, Elsevier/North Holland Biomedical Press, Amsterdam.

Lord, J. A. H., Waterfield, A. A., Hughes, J., and Kosterlitz, H. W., 1977, Endogenous opioid peptides: multiple agonists and receptors, *Nature (London)* **267**:495–499.

McLawhon, R. W., West, R. E., Miller, R. J., and Dawson, G., 1981, Distinct high-affinity binding sites for benzomorphan drugs and enkephalin in a neuroblastoma–brain hybrid cell line, *Proc. Natl. Acad. Sci. USA* **78**:4309–4313.

Madden, J., Akil, H., Patrick, R. L., and Barchas, J. D., 1977, Stress-induced parallel changes in central opioid levels and pain responsiveness in the rat, *Nature (London)* **265**:358–360.

Magnan, J., Paterson, S. J., Tavani, A., and Kosterlitz, H. W., 1982, The binding spectrum of narcotic analgesic drugs with different agonist and antagnoist properties *Naunyn-Schmiedeberg's Arch. Pharmacol.*, **319**:197–205.

Martin, W. R., 1967, Opioid antagonists, *Pharmacol. Rev.* **19**:463–521.

Martin, W. R., Eades, C. G., Thompson, J. A., Huppler, R. E., and Gilbert, P. E., 1976, The effects of morphine- and nalorphine-like drugs in the nondependent and morphine-dependent chronic spinal dog, *J. Pharmacol. Exp. Ther.* **197**:517–532.

Medzihradsky, F., 1976, Stereospecific binding of etorphine in isolated neural cells and in retina determined by a sensitive microassay, *Brain Res.* **108**:212–219.

Meunier, J.-C., and Moisand, C., 1977, Binding of Leu^5-enkephalin and Met^5-enkephalin to a particulate fraction from rat cerebrum, *FEBS Lett.* **77**:209–213.

Meunier, J.-C., and Zajac, J.-M., 1979, Cerebellar opiate receptors in lagomorphs. Demonstration, characterization, and regional distribution, *Brain Res.* **168**:311–321.

Miller, R. J., Chang, K-J., Leighton, J., and Cuatrecasas, P., 1978, Interaction of iodinated enkephalin analogues with opiate receptors, *Life Sci.* **22**:379–388.

Minneman, K. P., and Iversen, L. L., 1976, Enkephalin and opiate narcotics increase cyclic GMP accumulation in slices of rat neostriatum, *Nature (London)* **262**:313–314.

Morin, O., Caron, M. G., DeLean, A., and Labrie, F., 1976, Binding of the opiate-like pentapeptide methionine-enkephalin to a particulate fraction from rat brain, *Biochem. Biophys. Res. Commun.* **73**:940–946.

Mulé, S. J., Casella, G., and Clouet, D. H., 1974, Localization of narcotic analgesics in synaptic membranes of rat brain, *Res. Commun. Chem. Pathol. Pharmacol.* **9**:55–77.

MUNSON, P. J., and RODBARD, D., 1980, LIGAND: a versatile computerized approach for characterization of ligand-binding systems, *Anal. Biochem.* **107**:220–239.

MURRIN, L. C., COYLE, J. T., and KUHAR, M. J., 1980, Striatal opiate receptors: pre- and postsynaptic localization, *Life Sci.* **27**:1175–1183.

NINKOVIC, M., HUNT, S. P., EMSON, P. C., and IVERSEN, L. L., 1981, The distribution of multiple opiate receptors in bovine brain, *Brain Res.* **214**:163–167.

O'CALLAGHAN, J. P., CHESS, Q., MCKIMMEY, C., and CLOUET, D. H., 1979, The effects of opiates on the levels of cyclic 3':5'-guanosine monophosphate in discrete areas of the rat central nervous system, *J. Pharmacol. Exp. Ther.* **210**:361–367.

OKA, T., 1980, Enkephalin receptor in the rabbit ileum, *Br. J. Pharmacol.* **68**:193–195.

OKA, T., NEGISHI, K., SUDA, M., MATSUMIYA, T., INAZU, T., and UEKI, M., 1981, Rabbit vas deferens: a specific bioassay for opioid κ-receptor agonists, *Eur. J. Pharmacol.* **73**:235–236.

OSBORNE, N. N., and NEUHOFF, V., 1979, Are there opiate receptors in the invertebrates? *J. Pharm. Pharmacol.* **31**:481.

PASTERNAK, G. W., 1980, Multiple opiate receptors: [^{3}H]ethylketocyclazocine receptor binding and ketocyclazocine analgesia, *Proc. Natl. Acad. Sci. USA* **77**:3691–3694.

PASTERNAK, G. W., and HAHN, E. F., 1980, Long-acting opiate agonists and antagonists: 14-hydroxydihydromorphinone hydrazones, *J. Med. Chem.* **23**:674–676.

PASTERNAK, G. W., and SNYDER, S. H., 1974, Opiate receptor binding: effects of enzymatic treatments, *Mol. Pharmacol.* **10**:183–193.

PASTERNAK, G. W., and SNYDER, S. H., 1975*a*, Opiate receptor binding: enzymatic treatments that discriminate between agonist and antagonist interactions, *Mol. Pharmacol.* **11**:478–484.

PASTERNAK, G. W., and SNYDER, S. H., 1975*b*, Identification of novel high-affinity opiate receptor binding in rat brain, *Nature (London)* **253**:563–565.

PASTERNAK, G. W., SNOWMAN, A. M., and SNYDER, S. H., 1975*a*, Selective enhancement of [^{3}H]opiate agonist binding by divalent cations, *Mol. Pharmacol.* **11**:735–744.

PASTERNAK, G. W., WILSON, H. A., and SNYDER, S. H., 1975*b*, Differential effects of protein-modifying reagents on receptor binding of opiate agonists and antagonists, *Mol. Pharmacol.* **11**:340–351.

PASTERNAK, G. W., CHILDERS, S. R., and SNYDER, S. H., 1980*a*, Opiate analgesia: Evidence for mediation by a subpopulation of opiate receptors, *Science (Washington, D.C.)* **208**:514–516.

PASTERNAK, G. W., CHILDERS, S. R., and SNYDER, S. H., 1980*b*, Naloxazone, a long-acting opiate antagonist: effects on analgesia in intact animals and on opiate receptor binding *in vitro*, *J. Pharmacol. Exp. Ther.* **214**:455–462.

PATON, W. D. M., 1957, The action of morphine and related substances on contraction and on acetylcholine output of coaxially stimulated guinea pig ileum, *Br. J. Pharmacol. Chemother.* **12**:119–127.

PEARSON, J., BRANDEIS, L., SIMON, E., and HILLER, J., 1980, Radioautography of binding of tritiated diprenorphine to opiate receptors in the rat, *Life Sci.* **25**:1047–1052.

PERT, C. B., and SNYDER, S. H., 1973, Opiate receptor: demonstration in nervous tissue, *Science (Washington D.C.)* **179**:1011–1014.

PERT, C. B., and SNYDER, S. H., 1974, Opiate receptor binding of agonists and antagonists affected differentially by sodium, *Mol. Pharmacol.* **10**:868–879.

PERT, C. B., and SNYDER, S. H., 1976, Opiate receptor binding-enhancement by opiate administration *in vivo*, *Biochem. Pharmacol.* **25**:847–853.

PERT, C. B., and TAYLOR, D., 1980, Type 1 and Type 2 opiate receptors: a subclassification scheme based on GTP's differential effects on binding, in: *Endogenous and Exogenous Opiate Agonists and Antagonists* (E. L. Way, ed.), pp. 87–90, Pergamon Press, New York.

PERT, C. B., PASTERNAK, G., and SNYDER, S. H., 1973, Opiate agonists and antagonists discriminated by receptor binding in brain, *Science (Washington D.C.)* **182**:1359–1361.

PERT, C. B., APOSHIAN, D., and SNYDER, S. H., 1974*a*, Phylogenetic distribution of opiate receptor binding, *Brain Res.* **75**:356–361.

PERT, C. B., SNOWMAN, A. M., and SNYDER, S. H., 1974*b*, Localization of opiate receptor binding in synaptic membranes of rat brain, *Brain Res.* **70:**184–188.

PERT, C. B., KUHAR, M. J., and SNYDER, S. H., 1976, Opiate receptor: autoradiographic localization in rat brain, *Proc. Natl. Acad. Sci. USA* **73:**3729–3733.

PERT, C. B., DUNCAN, D. P., PERT, A., HERKENHAM, M. A., and KENT, J. L., 1980, Biochemical and autoradiographic evidence for Type 1 and Type 2 opiate receptors, in: *Advances in Biochemical Psychopharmacology* (E. Costa and M. Trabucchi, eds.), Vol. 22, pp. 581–589, Raven Press, New York.

PFEIFFER, A., and HERZ, A., 1981, Demonstration and distribution of an opiate binding site in rat brain with high affinity for ethylketazocine and SKF 10047, *Biochem. Biophys. Res. Commun.* **101:**38–44.

POLLARD, H., LLORENS, C., BONNET, J. J., COSTENTIN, J., and SCHWARTZ, J. C., 1977*a*, Opiate receptors on mesolimbic dopaminergic neurones, *Neurosci. Lett.* **7:**295–299.

POLLARD, H., LLORENS-CORTES, C., and SCHWARTZ, J. C., 1977*b*, Enkephalin receptors on dopaminergic neurones in rat striatum, *Nature (London)* **268:**745–747.

POLLARD, H., LLORENS, C., SCHWARTZ, J. C., GROS, C., and DRAY, F., 1978, Localization of opiate receptors and enkephalins in the rat striatum in relationship with the nigrostriatal dopaminergic system: lesion studies, *Brain Res.* **151:**392–398.

PORTHÉ, G., VALETTE, A., and CROS, J., 1981, Kappa opiate binding sites in human placenta, *Biochem. Biophys. Res. Commun.* **101:**1–6.

PORTOGHESE, P. S., 1965, A new concept on the mode of interaction of narcotic analgesics with receptors, *J. Med. Chem.* **8:**609–616.

PORTOGHESE, P. S., LARSON, D. L., JIANG, J. B., TAKEMORI, A. E., and CARUSO, T. P., 1978, 6β-[*N*,*N*-Bis(2-chloroethyl)amino]-17-(cyclopropylmethyl)-4,5α-epoxy-3,14-dihydroxymorphinan (chlornaltrexamine), a potent opioid receptor alkylating agent with ultralong narcotic antagonist activity, *J. Med. Chem.* **21:**598–599.

PORTOGHESE, P. S., LARSON, D. L., JIANG, J. B., CARUSO, T. P., and TAKEMORI, A. E., 1979, Synthesis and pharmacologic characterization of an alkylating analogue (chlornaltrexamine) of naltrexone with ultralong-lasting narcotic antagonist properties, *J. Med. Chem.* **22:**168–173.

PUGET, A., JAUZAC, P., and MEUNIER, J. C., 1980, Hydrodynamic properties of a [^{3}H]-etorphine macromolecular complex from the rat brain, *FEBS Lett.* **122:**199–202.

QUIRION, R., HAMMER, R. P., HERKENHAM, M., and PERT, C. B., 1981, Phencyclidine (angel dust)/σ "opiate" receptor: visualization by tritium-sensitive film, *Proc. Natl. Acad. Sci. USA* **78:**5881–5885.

REISINE, T. D., NAGY, J. I., BEAUMONT, K., FIBIGER, H. C., and YAMAMURA, H. I., 1979*a*, The localization of receptor binding sites in the substantia nigra and striatum of the rat, *Brain Res.* **177:**241–252.

REISINE, T. D., ROSSOR, M., SPOKES, E., IVERSEN, L. L., and YAMAMURA, H. I., 1979*b*, Alterations in brain opiate receptors in Parkinson's disease, *Brain Res.* **173:**378–382.

REISINE, T. D., ROSSOR, M., SPOKES, E., IVERSEN, L. L., and YAMAMURA, H. I., 1980, Opiate and neuroleptic receptor alterations in human schizophrenic brain tissue, in: *Receptors for Neurotransmitters and Peptide Hormones* (G. Pepeu, M. J. Kuhar, and S. J. Enna, eds.), pp. 443–450, Raven Press, New York.

RIGTER, H., MESSING, R. B., VASQUEZ, B. J., JENSEN, R. A., MARTINEZ, J. L., CRABBE, J. C., and MCGAUGH, J. L., 1979, Regional analysis of brain opiate receptors in rats with hereditary hypothalamic diabetes insipidus, *Life Sci.* **25:**1137–1143.

ROBSON, L. E., and KOSTERLITZ, H. W., 1979, Specific protection of the binding sites of D-Ala^2-D-Leu^5-enkephalin (δ-receptors) and dihydromorphine (μ-receptors), *Proc. R. Soc. London Ser. B* **205:**425–432.

RÖMER, D., BÜSCHER, H. H., HILL, R. C., PLESS, J., BAUER, W., CARDINAUX, F., CLOSSE, A., HAUSER, D., and HUGUENIN, R., 1977, A synthetic enkephalin analogue with prolonged parenteral and oral analgesic activity, *Nature (London)* **268:**547–549.

RÖMER, D., BÜSCHER, H., HILL, R. C., MAURER, R., PETCHER, T. J., WELLE, H. B. A., BAKEL, H. C. C. K., and AKKERMAN, A. M., 1980, Bremazocine: a potent long-acting opiate kappa-agonist, *Life Sci.* **27**:971–978.

ROSSIER, J., GUILLEMIN, R., and BLOOM, F., 1978, Footshock-induced stress decreases Leu^5-enkephalin immunoreactivity in rat hypothalamus, *Eur. J. Pharmacol.* **48**:465–466.

RÜEGG, U. T., HILLER, J. M., and SIMON, E. J., 1980, Solubilization of an active opiate receptor from *Bufo marinus*, *Eur. J. Pharmacol.* **64**:367–368.

RÜEGG, U. T., CUENOD, S., HILLER, J. M., GIOANNINI, T., HOWELLS, R. D., and SIMON, E. J., 1981, Characterization and partial purification of solubilized active opiate receptors from toad brain, *Proc. Natl. Acad. Sci. USA* **78**:4635–4638.

SCATCHARD, G., 1949, The attraction of proteins for small molecules and ions, *Ann. N.Y. Acad. Sci.* **51**:660–674.

SCHULZ, R., FAASE, E., WÜSTER, M., and HERZ, A., 1979*a*, Selective receptors for β-endorphin on the rat vas deferens, *Life Sci.* **24**:843–849.

SCHULZ, R., WÜSTER, M., and HERZ, A., 1979*b*, Supersensitivity to opioids following the chronic blockade of endorphin action by naloxone, *Naunyn Schmiedeberg's Arch. Pharmacol.* **306**:93–96.

SCHULZ, R., WÜSTER, M., KRENSS, H., and HERZ, A., 1980*a*, Selective development of tolerance without dependence in multiple opiate receptors of mouse vas deferens, *Nature (London)* **285**:242–243.

SCHULZ, R., WÜSTER, M., KRENSS, H., and HERZ, A., 1980*b*, Lack of cross-tolerance on multiple opiate receptors in the mouse vas deferens, *Mol. Pharmacol.* **18**:395–401.

SCHWARTZ, J. C., POLLARD, H., LLORENS, C., MALFROY, C., MALFROY, B., GROS, C., PRADELLES, P., and DRAY, F., 1978, Endorphins and endorphin receptors in striatum: relationships with dopaminergic neurones, in: *Advances in Biochemical Psychopharmacology* (E. Costa and M. Trabucchi, eds.), Vol. 18, pp. 245–264, Raven Press, New York.

SHARMA, S. K., KLEE, W. A., and NIRENBERG, M., 1975*a*, Dual regulation of adenylate cyclase accounts for narcotic dependence and tolerance, *Proc. Natl. Acad. Sci. USA* **72**:3092–3096.

SHARMA, S. K., NIRENBERG, M., and KLEE, W. A., 1975*b*, Morphine receptors as regulators of adenylate cyclase activity, *Proc. Natl. Acad. Sci. USA* **72**:590–594.

SIMANTOV, R., and SNYDER, S. H., 1976, Morphine-like peptides, leucine-enkephalin and methionine-enkephalin: interactions with the opiate receptor, *Mol. Pharmacol.* **12**:987–998.

SIMANTOV, R., and SNYDER, S. H., 1977, Opiate receptor binding in the pituitary gland, *Brain Res.* **124**:178–184.

SIMANTOV, R., CHILDERS, S. R., and SNYDER, S. H., 1978, The opiate receptor binding interactions of [^{3}H]methionine enkephalin, an opioid peptide, *Eur. J. Pharmacol.* **47**:319–331.

SIMON, E. J., and GROTH, J., 1975, Kinetics of opiate receptor inactivation by sulfhydryl reagents: evidence for conformational change in presence of sodium ions, *Proc. Natl. Acad. Sci. USA* **72**:2404–2407.

SIMON, E. J., and HILLER, J. M., 1978, The opiate receptors, *Annu. Rev. Pharmacol. Toxicol.* **18**:371–394.

SIMON, E. J., HILLER, J. M., and EDELMAN, I., 1973, Stereospecific binding of the potent narcotic analgesic [^{3}H]etorphine to rat brain homogenate, *Proc. Natl. Acad. Sci. USA* **70**:1947–1949.

SIMON, E. J., HILLER, J. M., and EDELMAN, I., 1975*a*, Solubilization of a stereospecific opiate-macromolecular complex from rat brain, *Science (Washington D.C.)* **190**:389–390.

SIMON, E. J., HILLER, J. M., EDELMAN, I., GROTH, J., and STAHL, K. D., 1975*b*, Opiate receptors and their interactions with agonists and antagonists, *Life Sci.* **16**:1795–1800.

SIMON, E. J., HILLER, J. M., GROTH, J., and EDELMAN, I., 1975*c*, Further properties of stereospecific opiate binding sites in rat brain: on the nature of the sodium effect, *J. Pharmacol. Exp. Ther.* **192**:531–537.

SIMON, E. J., BONNET, K. A., CRAIN, S. M., GROTH, J., HILLER, J. M., and SMITH, J. R., 1980, Recent studies on interaction between opioid peptides and their receptors, in: *Advances in Biochemical Psychopharmacology* (E. Costa and M. Trabucchi, eds.), Vol. 22, pp. 335–346, Raven Press, New York.

SIMONDS, W. F., KOSKI, G., STREATY, R. A., HJELMELAND, L. M., and KLEE, W. A., 1980, Solubilization of active opiate receptors, *Proc. Natl. Acad. Sci. USA* **77:**4623–4627.

SMITH, A. P., and LOH, H. H., 1976, The subcellular localization of stereospecific opiate binding in mouse brain, *Res. Commun. Chem. Pathol. Pharmacol.* **15:**205–219.

SMITH, A. P., and LOH, H. H., 1979, Multiple molecular forms of stereospecific opiate binding, *Mol. Pharmacol.* **16:**757–766.

SMITH, J. R., and SIMON, E. J., 1980, Selective protection of stereospecific enkephalin and opiate binding against inactivation by *N*-ethylmaleimide: evidence for two classes of opiate receptors, *Proc. Natl. Acad. Sci. USA* **77:**281–284.

SNYDER, S. H., and GOODMAN, R. R., 1980, Multiple neurotransmitter receptors, *J. Neurochem.* **35:**5–15.

SNYDER, S. H., PASTERNAK, G. W., and PERT, C. B., 1975, Opiate receptor mechanisms, in: *Handbook of Psychopharmacology* (L. L. Iversen, S. D. Iversen, and S. H. Snyder, eds.) pp. 329–360, Plenum Press, New York and London.

SPIEHLER, V., FAIRHURST, A. S., and RANDALL, L. O., 1978, The interaction of phenoxybenzamine with the mouse brain opiate receptor, *Mol. Pharmacol.* **14:**587–595.

STEFANO, G. B., and CATAPANE, E. J., 1979, Enkephalins increase dopamine levels in the CNS of a marine mollusc, *Life Sci.* **24:**1617–1622.

STEFANO, G. B., and HIRIPI, L., 1979, Methionine-enkephalin and morphine alter monoamine and cyclic nucleotide levels in the cerebral ganglia of the freshwater bivalve *Anodonta cygnea, Life Sci.* **25:**291–297.

STEFANO, G. B., KREAM, R. M., and ZUKIN, R. S., 1980, Demonstration of stereospecific opiate binding in the nervous tissue of the marine mollusc, *Mytilus edulis, Brain Res.* **181:**440–445.

SU, T-P., 1981, Psychotomimetic opioid binding: specific binding of [^{3}H]-SKF-10047 to etorphine-inaccessible sites in guinea pig brain, *Eur. J. Pharmacol.* **75:**81–82.

SU, T.-P., CLEMENTS, T. H., and CERODETZKY, C. W., 1981, Multiple opiate receptors in guinea pig ileum, *Life Sci.* **28:**2519–2528.

SWAIN, H. H., and SEEVERS, M. H., 1974, Evaluation of new compounds for morphine-like physical dependence in the rhesus monkey, *Bull. Probl. Drug Dependence* **36:**Addendum 1168–1195.

SWAIN, H. H., and SEEVERS, M. H., 1976, Evaluation of new compounds for morphine-like physical dependence in the rhesus monkey, *Bull. Probl. Drug Dependence* **38:**Addendum 2, 768–787.

SZÉKELY, J. I., RÓNAI, A. Z., DUNAI-KOVÁCS, Z., MIGLÉGZ, E., BERZÉTEI, I., BAJUSZ, S., and GRÁF, L., 1977, (D-Met^2,Pro^5)-enkephalinamide: a potent morphine-like analgesic, *Eur. J. Pharmacol.* **43:**293–294.

SZERB, J. C., and VOHRA, M. M., 1979, Potencies of normorphine and met-enkephalin in the vas deferens of different strains of mice, *Life Sci.* **24:**1983–1988.

TANG, A. H., and COLLINS, R. J., 1978, Enhanced analgesic effects of morphine after chronic administration of naloxone in the rat, *Eur. J. Pharmacol.* **47:**473–474.

TERENIUS, L., 1973*a*, Stereospecific interaction between narcotic analgesics and a synaptic plasma membrane fraction of rat cerebral cortex, *Acta Pharmacol. Toxicol. (Copenhagen)* **32:**317–320.

TERENIUS, L., 1973*b*, Characteristics of the "receptor" for narcotic analgesics in synaptic plasma membrane fraction from rat brain, *Acta Pharmacol. Toxicol. (Copenhagen)* **33:**377–384.

TERENIUS, L., 1975, Comparison between narcotic "receptors" in the guinea pig ileum and the rat brain, *Acta Pharmacol. Toxicol. (Copenhagen)* **37:**211–221.

TERENIUS, L., 1977, Opioid peptides and opiates differ in receptor selectivity, *Psychoneuroendocrinology* **2:**53–58.

TRABER, J., FISCHER, K., LATZIN, S., and HAMPRECHT, B., 1975, Morphine antagonises action of prostaglandin in neuroblastoma and neuroblastoma × glioma hybrid cells, *Nature (London)* **253:**120–122.

TSANG, D., and NG, S. C., 1980, Development of radioimmunoassayable β-endorphin and methionine-enkephalin binding sites in regions of rat brain, *Can. J. Physiol. Pharmacol.* **58:**947–950.

VALETTE, A., PONTONNIER, G., and CROS, J., 1979, Evidence for a stereospecific [^{3}H]etorphine binding in human placenta, *FEBS Lett.* **103:**362–365.

VALETTE, A., REME, J. M., PONTONNIER, G., and CROS, J., 1980, Specific binding for opiate-like drugs in the placenta, *Biochem. Pharmacol.* **29:**2657–2662.

VENN, R. F., and BARNARD, E. A., 1981, A potent peptide affinity reagent for the opiate receptor, *J. Biol. Chem.* **256:**1529–1532.

VILLARREAL, J. E., and SEEVERS, M. H., 1972, Evaluation of new compounds for morphine-like physical dependence in the rhesus monkey, *Bull. Probl. Drug Dependence* **34:**Addendum 7, 1040–1053.

VINCENT, J. P., CAREY, D., KAMENKA, J. M., GENESTE, P., and LAZDUNSKI, M., 1978, Interaction of phencyclidines with the muscarinic and opiate receptors in the central nervous system, *Brain Res.* **152:**176–182.

VINCENT, J. P., KARTALOVSKI, B., GENESTE, P., KAMENKA, J. M., and LAZDUNSKI, M., 1979, Interaction of phencyclidine ("angel dust") with a specific receptor in rat brain membranes, *Proc. Natl. Acad. Sci. USA* **76:**4678–4682.

WATERFIELD, A. A., SMOKCUM, R. W. J., HUGHES, J., KOSTERLITZ, H. W., and HENDERSON, G., 1977, *In vitro* pharmacology of the opioid peptides, enkephalins and endorphins, *Eur. J. Pharmacol.* **43:**107–116.

WATERFIELD, A. A., LORD, J. A. H., HUGHES, J., and KOSTERLITZ, H. W., 1978, Differences in the inhibitory effects of normorphine and opioid peptides on the responses of the vasa deferentia of two strains of mice. *Eur. J. Pharmacol.* **47:**249–250.

WILKENING, D., SABOL, S. L., and NIRENBERG, M., 1980, Control of opiate receptor–adenylate cyclase interactions by calcium ions and guanosine-5′-triphosphate, *Brain Res.* **189:**459–466.

WILSON, H. A., PASTERNAK, G. W., and SNYDER, S. H., 1975, Differentiation of opiate agonist and antagonist receptor binding by protein-modifying reagents, *Nature (London)* **253:**448–450.

WOLLEMANN, M., 1981, Endogenous opioids and cyclic AMP, *Prog. Neurobiol.* **16:**145–154.

WÜSTER, M., SCHULZ, R., and HERZ, A., 1979, Selectivity of opioids towards the μ-, δ, and ε-opiate receptors, *Neurosci. Lett.* **15:**193–198.

WÜSTER, M., SCHULZ, R., and HERZ, A., 1980, Opioid agonists and antagonists: action on multiple opiate receptors, in: *Endogenous and Exogenous Opiate Agonists and Antagonists* (E. L. Way, ed.), pp. 75–78, Pergamon Press, New York.

WÜSTER, M., SCHULZ, R., and HERZ, A., 1981, Multiple opiate receptors in peripheral tissue preparations, *Biochem. Pharmacol.* **30:**1883–1887.

YOUNG, W. S., and KUHAR, M. J., 1979, A new method for receptor autoradiography: [^{3}H]-opioid receptors in rat brain, *Brain Res.* **179:**255–270.

ZAJAC, J. M., and MEUNIER, J. C., 1981, Opiate receptor sites in the rabbit cerebellum: autoradiographic distribution, *J. Receptor Res.* **1:**403–413.

ZHANG, A.-Z., and PASTERNAK, G. W., 1980, μ- and δ-opiate receptors: correlation with high- and low-affinity binding sites, *Eur. J. Pharmacol.* **67:**323–324.

ZHANG, A.-Z., and PASTERNAK, G. W., 1981, Opiates and enkephalins: a common binding site mediates their analgesic actions in rats, *Life Sci.* **29:**843–851.

ZHANG, A.-Z., CHANG, J.-K., and PASTERNAK, G. W., 1981, The actions of naloxazone on the binding and analgesic properties of morphiceptin (NH_2Tyr-Pro-Phe-Pro-$CONH_2$), a selective mu-receptor ligand, *Life Sci.* **28:**2829–2836.

ZUKIN, R. S., and GINTZLER, A. R., 1980, Guanyl nucleotide interactions with opiate receptors in guinea pig brain and ileum, *Brain Res.* **186:**486–491.

ZUKIN, R. S., and KREAM, R. M., 1979, Chemical cross-linking of a solubilized enkephalin macromolecular complex, *Proc. Natl. Acad. Sci. USA* **76:**1593–1597.

ZUKIN, S. R., and ZUKIN, R. S., 1979, Specific [^{3}H]phencyclidine binding in rat central nervous system, *Proc. Natl. Acad. Sci. USA* **76:**5372–5376.

ZUKIN, R. S., and ZUKIN, S. R., 1981, Demonstration of [^{3}H]cyclazocine binding to multiple opiate receptor sites, *Mol. Pharmacol.* **20:**246–254.

ZUKIN, R. S., WALCZAK, S., and MAKMAN, M. H., 1980, GTP modulation of opiate receptors in regions of rat brain and possible mechanism of GTP action, *Brain Res.* **186:**238–244.

3

CNS DOPAMINE RECEPTORS

Ian Creese, Mark W. Hamblin, Stuart E. Leff, and David R. Sibley

1. INTRODUCTION

The past 20 years has seen our appreciation of the function of dopamine in the brain elevated from that of a precursor for other catecholamines, principally norepinephrine, to a neurotransmitter in its own right. The association of disturbances of dopaminergic neurotransmission with neurological and psychiatric disorders has further emphasized the crucial role of this neurotransmitter in normal brain functioning and stimulated much basic research into dopaminergic neurotransmission. Dopaminergic agonists with the ability to cross the blood–brain barrier now have a firmly established role in the treatment of Parkinson's disease, and may be of value in the therapy of tardive dyskinesia. Dopaminergic antagonists have a longer history in the treatment of schizophrenia, Huntington's disease, and Gilles de la Tourette's syndrome. This pharmacological arsenal, available because of the pharmaceutical industries' search for better therapeutic agents, also provides the major tools for experimental approaches. Both agonists and antagonists are available from diverse structural groups. Some exist as stereo or geometric isomers which differ markedly in their ability to interact with dopaminergic systems. Although agents which alter the synthesis, release, reuptake, or catabolism of dopamine are useful both therapeutically and experimentally, drugs which act directly on dopamine receptors, as agonists or antagonists, have proven

Ian Creese, Mark W. Hamblin, Stuart E. Leff, and David R. Sibley • Department of Neurosciences, University of California, San Diego, School of Medicine, La Jolla, California 92093.

most useful in delineating the biochemical, electrophysiological, and behavioral functioning of the dopamine systems. Most excitingly, recent experiments have clearly divided dopamine receptors into distinct subtypes, much as was done earlier for the alpha and beta adrenergic, and the muscarinic and nicotinic cholinergic, receptors. These findings will have a profound effect on our understanding of dopaminergic pharmacology and neurotransmission, and the role of dopamine in psychiatric and neurological disease.

Since 1975, the elegantly simple radioligand binding technique has allowed direct examination of dopamine agonist and antagonist interactions with dopamine receptors. The simplification thus obtained through elimination of factors both proximal (such as regulation of neurotransmitter synthesis) and distal (such as activation of a second messenger system) to the receptor has been the chief advantage of this approach. From this simplification, however, also comes the chief difficulty of the receptor binding technique—it is a task of utmost importance, and often of considerable difficulty, to demonstrate that the binding sites so identified *in vacuo* can be related back to the real world, that they can be firmly identified with some biological function. Although problems remain, this correspondence between binding sites and their function, on both the behavioral and biochemical level, is steadily being established for the dopamine receptors.

The aim of this review is therefore twofold: to outline the pharmacological characteristics and anatomical localizations of the several distinct dopamine receptor subtypes delineated through radioligand binding, pharmacological, and biochemical studies; and, in addition, to outline what is known concerning the functions of these receptors. By way of introduction, the major dopaminergic agonists and antagonists, along with their chief actions, will first be described.

2. DOPAMINERGIC AGONISTS AND ANTAGONISTS AND THEIR ACTIONS

The major actions of dopaminergic agents will be considered only briefly since they are expanded in many of the other chapters in this Handbook (York, 1975; Iversen, 1977; Creese *et al.*, 1978*a*; Fielding and Lal, 1978). In general, behavioral experiments have concentrated on the effects of dopaminergic agents on motor behavior. Directly acting dopamine agonists act at dopamine receptors in the striatum and nucleus accumbens to promote an increase in locomotor activity and stereotyped behavior in rodents, and these drugs produce turning to the contralateral side when injected unilaterally into the striatum. Peripheral administration of agonists causes a decrease in dopamine turnover in neurons on the nigrostriatal projection by acting at presynaptic dopamine autoreceptors. Biochemically, at least some dopamine receptors in both the CNS and periphery are linked in a stimulatory fashion to the enzyme adenylate cyclase, while others may be linked in an inhibitory fashion. Finally, dopamine agonists act on receptors in the anterior pituitary to decrease prolactin secretion.

The dopamine agonist most frequently used in all types of experiments, whether behavioral, physiological, or biochemical, is apomorphine. This alkaloid of the aporphine class easily crosses the blood–brain barrier although its duration of action is generally found to be short. In rodents it stimulates both locomotor activity and "stereotyped" behavior. Stereotyped behavior consists of repetitious motor behavior such as rearing, sniffing, or gnawing, maintained in one location. In man, dogs, and other animals with a chemoreceptive trigger zone in the area postrema, apomorphine produces nausea and intense vomiting. Interestingly, in low doses apomorphine apparently selectively stimulates autoreceptors and therefore has a sedative action as opposed to its more commonly known stimulant action. Apomorphine acts as a partial agonist of the dopamine-sensitive adenylate cyclase and as a full agonist in decreasing striatal dopamine turnover and pituitary prolactin release.

While dopamine itself is a full agonist in all *in vitro* systems, it is unable to cross the blood–brain barrier. Its levels in the brain can be raised *in vivo*, however, by peripheral administration of its precursor L-DOPA. If a peripheral decarboxylase inhibitor which does not cross the blood–brain barrier is administered, the L-DOPA conversion to dopamine occurs only in the CNS, reducing gastrointestinal side effects. L-DOPA is one of the major therapeutic agents in the treatment of the rigidity, tremor, and bradykinesia of Parkinson's disease, a syndrome in which central dopamine neurons degenerate. It is thought that L-DOPA enhances dopamine production in the remaining intact neurons. The other principal dopaminergic agonist used mainly in biochemical studies is ADTN (2-amino-6,7-dihydroxy-1,2,3,4-tetrahydronaphthalene), an agent which also does not cross the blood–brain barrier. Bromocryptine, an ergot derivative, acts as an agonist in suppressing prolactin secretion, in affecting rotational behavior, and in alleviating Parkinsonian symptoms. However, it acts as an antagonist of the dopamine-stimulated adenylate cyclase.

Dopamine antagonists block the effects of dopamine agonists and may produce effects in the opposite direction if there is a tonic release of dopamine. Among dopamine antagonists two classes of drugs have received the greatest attention. Phenothiazines and butyrophenones are antipsychotic, or neuroleptic, drugs used in the treatment of schizophrenia. These compounds specifically decrease avoidance behavior in animals without affecting escape in response to noxious stimuli. This paradigm is used as a sensitive screen for potential antipsychotic agents. Chlorpromazine (Thorazine), the archetypical phenothiazine, was the first neuroleptic identified, and since its introduction in the early 1950's, many hundreds of analogs have been synthesized. The principal phenothiazines used in clinical practice now include fluphenazine (Prolixin, Permitil) and trifluperazine (Stelazine) which are both about 10 times more potent than chlorpromazine. Within the second major class of dopamine antagonists—the butyrophenones—haloperidol (Haldol) was the first agent identified as being antipsychotic (Janssen and VanBever, 1978). It is about equipotent therapeutically with fluphenazine. The other important butyrophenone for biochemical experiments is spiperone (or spiroperidol). A number of other antipsychotic agents appear to be dopamine antagonists. The thioxanthenes such as flupentixol are closely similar in structure to the phenothiazines. Thioridazine (Melaril), the most widely used thioxanthene, is about equipotent with chlorpromazine. Of major importance is the drug butaclamol (a dibenzocycloheptane derivative) which exists as optical isomers. Only the (+) isomer of butaclamol is active in blocking dopamine-mediated effects *in vivo* and *in vitro*, while the (−) isomer is virtually inactive. The stereospecificity of this compound has been a useful tool in delineating dopamine receptor mechanisms. Another potent dopamine antagonist and antipsychotic agent is pimozide (a diphenylbutylpiperidine), which shares many structural features with the butyrophenones. Domperidone, another butyrophenone-like compound, does not cross the blood–brain barrier and is utilized clinically to increase gastric emptying, probably by acting at gastric dopamine receptors. Domperidone is worthy of note as having, unlike most other dopamine antagonists, almost no ability to block dopamine stimulation of the dopamine-sensitive adenylate cyclase (Laduron and Leysen, 1979). The only other widely used compounds lacking antagonism of the dopamine-sensitive adenylate cyclase while possessing dopamine antagonist activity by many other indices are the substituted benzamides, such as sulpiride, tiapride, and metaclopramide (Jenner and Marsden, 1979).

A number of these compounds in tritiated form have been used as ligands to identify dopamine receptors. This list now includes haloperidol, spiroperidol, domperidone, pimozide, tiapride, sulpiride, flupentixol, lysergic acid diethylamide (LSD), dihydroergocryptine, dopamine, apomorphine, ADTN, and *n*-propylnorapomorphine (NPA). It is now becoming apparent that these structurally diverse ligands can be exploited to preferentially identify distinct dopamine receptors with differences in their binding properties not merely representing quirks in laboratory technique or idiosyncratic ligand/receptor interactions.

3. ANATOMY

Our anatomical knowledge of the dopamine neuronal systems has increased immeasurably since the pioneering studies of Dahlstrom and Fux (1964). The detailed anatomy and histology of the dopamine systems have recently been reviewed in depth (Moore and Bloom, 1978; Lindvall and Bjorklund, 1977) and will be only highlighted here (Table 1). Because the nigrostriatal pathway accounts for about 70% of the total brain content of dopamine, this tract became an obvious focus of research. The existence of the nigrostriatal pathway was strongly indicated by the observations of Hornykiewicz (1966) who demonstrated that patients with Parkinson's disease displayed a concomitant loss of dopamine in the striatum along with the degeneration of the substantia nigra pars compacta. The other major dopamine pathway described at this time originates from a group of cells in the ventrotegmental area surrounding the interpeduncular nucleus and innervates the olfactory tubercle and adjacent limbic and cortical structures. The substantia nigra and ventrotegmental dopamine

TABLE 1
Dopamine Neuron Systems in the Mammalian Brain[a]

System	Nucleus of origin	Site(s) of termination
Nigrostriatal	Substantia nigra, pars compacta; ventral tegmental area	Neostriatum (caudate-putamen), globus pallidus
Mesocortical	Ventral tegmental area; substantia nigra, pars compacta	Isocortex (mesial frontal, anterior cingulate, entorhinal, perirhinal Allocortex (olfactory bulb, anterior olfactory nucleus, olfactory tubercle, piriform cortex, septal area, nucleus accumbens, amygdaloid complex)
Tuberohypophyseal	Arcuate and periventricular hypothalamic nuclei	Neurointermediate lobe of pituitary, median eminence
Retinal	Interplexiform cells, of retina	Inner and outer plexiform layers of retina
Incertohypothalamic	Zona incerta, posterior hypothalamus	Dorsal hypothalamic area, septum
Periventricular	Medulla in area of dorsal motor vagus, nucleus tractus solitarius, periaqueductal and periventricular gray	Periventricular and periaqueductal gray, tegmentum, tectum, thalamus, hypothalamus
Olfactory bulb	Periglomerular cells	Glomeruli (mitral cells)

[a] Modified from Moore and Bloom (1978).

cell groups are frequently referred to as the A-9 and A-10 nuclear groups, respectively, following the original designation of Dahlstrom and Fuxe (1964). Since the introduction of the more sensitive glyoxylic acid fluorescence histochemical method (Lindvall and Bjorklund, 1974), immunohistochemical studies, and orthograde and retrograde nerve circuit tracing (Beckstead *et al.*, 1979), it is becoming clear that the dopamine systems are more complex than originally envisaged.

Ontogenetic and mapping studies have now demonstrated that the A-9 and A-10 nuclear groups are more correctly described as a continuum with the more laterally situated cells predominately innervating the striatum and the more medial cells predominantly innervating the "mesocortical" areas. The striatal projection includes the caudate nucleus, putamen, and globus pallidus, whereas the terminal areas of the mesocortical projection include the medial frontal, anterior cingulate, entorhinal, perirhinal, and piriform cortex. Also apparent is a strong innervation of the olfactory tubercle, septum, nucleus accumbens, and amygdaloid complex. This pathway is frequently referred to as the mesolimbic cortical or limbic dopamine system.

Dopamine cell bodies in the substantia nigra are found only in the pars compacta, although their dendritic trees extend ventrally into the pars reticulata. The cells are of medium size and multipolar. The pronounced varicosities within the dendritic tree are unusual and have been proposed to be indicative of the dendritic release of dopamine. GABAergic and Substance P pathways from the striatum and globus pallidus feed back to the substantia nigra.

In the cortex, it appears that the dopamine innervation is to the deeper layers in contrast to the norepinephrine input which is to the more superficial layers (Berger *et al.*, 1974; Lindvall *et al.*, 1978). A dopaminergic innervation to the spinal cord, probably originating from the substantia nigra, has recently been described (Commissiong and Neff, 1979; Commissiong *et al.*, 1979; Blessing and Chalmers, 1979).

Of the other dopamine pathways, the tuberohypophyseal system has received most attention and will be discussed below. Two other dopamine pathways originate in hypothalamic areas—the incertohypothalamic and periventircular systems (Bjorklund *et al.*, 1975; Lindvall and Bjorklund, 1977). The periglomerular cells of the olfactory bulb (Hökfelt *et al.*, 1975) and the interplexiform cells of the retina (Ehinger, 1976) both also appear to utilize dopamine as a transmitter. In the perhiphery some small, intensely fluorescent cells in sympathetic ganglia are thought to be dopaminergic (Libet, 1976). Dopaminergic nerves have recently been described in the kidney (Bell *et al.*, 1978; Dinerstein *et al.*, 1979). However, although a number of other structures in the periphery such as the stomach, parathyroids, and carotid bodies are known to be responsive to dopamine, no other dopaminergic neurons have yet been identified.

4. PHARMACOLOGICAL CHARACTERIZATION OF DOPAMINE RECEPTORS

4.1. The "D-1" Dopamine Receptor and the Dopamine-Sensitive Adenylate Cyclase

Cyclic adenosine monophosphate (cAMP) is a second messenger for a number of neurotransmitters in the periphery. Greengard's studies indicated that the inhibitory postsynaptic potential in the bovine superior cervical ganglion was mediated by dopamine and that its effects could be mimicked by exogenous application of cAMP. Biochemical studies soon demonstrated the presence of a dopamine-sensitive adenylate cyclase in this tissue (Greengard, 1976). In comparable studies, Greengard and associates (Kebabian *et al.*, 1972) demonstrated that homogenates of rat corpus striatum would accumulate cAMP when exposed to dopamine. In contrast to the well-studied effects of catecholamines on beta-adrenergic receptors where dopamine is quite weak and isoproterenol is extremely potent, the dopamine-sensitive adenylate cyclase in rat striatum was stimulated greatly by dopamine, less by norepinephrine, and little at all by isoproterenol. Dopamine elicited maximal stimulation of cAMP accumulation at 100 μM concentrations with half maximal effects at about 2 μM. The regional distribution of the enzyme in brain tissue also suggested an association with dopamine transmission. Thus high enzymatic activity was observed in the corpus striatum, olfactory tubercle, and nucleus accumbens, the three brain regions richest in dopamine innervation, while no enzymatic activity could be demonstrated in other brain areas.

Greengard's group and later Iversen and colleagues (Iversen, 1975) evaluated the effects of neuroleptic drugs on the dopamine-sensitive adenylate cyclase. The phenothiazines were effective competitive inhibitors of the enzyme (Clement-Cormier *et al.*, 1974, 1975; Miller *et al.*, 1974; Iversen *et al.*, 1976). In studies of an extensive series of phenothiazines there as a general parallel between their pharmacological potencies as dopamine antagonists in animals and man and their influences on the cyclase. However, there were marked discrepancies for butyrophenones and other neuroleptics (Iversen, 1975; Snyder *et al.*, 1975). For example, haloperidol, which clinically and pharmacologically is about 10–100 times more potent than chlorpromazine, appeared weaker than, or at best equal to, chlorpromazine in its influences on the cyclase. Furthermore, the most potent butyrophenone, spiroperidol (or spiperone), which is about five times more potent than haloperidol in intact animals and in controlling schizophrenia, was weaker than both haloperidol and chlorpromazine in inhibiting the dopamine-sensitive adenylate cyclase. Surprisingly, the potent antipsychotic sulpiride is almost devoid of inhibitory potency.

TABLE 2
Functional Classification of Dopamine Receptor Subtypes[a]

Characteristic	D-1	D-2
Prototype receptor location	Parathyroid gland	Anterior and intermediate pituitary glands
Adenylate cyclase linkage	Stimulatory	Inhibitory or unlinked
Agonists		
Dopamine	Full agonist (micromolar potency)	Full agonist (nanomolar potency)
Apomorphine	Partial agonist (micromolar potency)	Full agonist (nanomolar potency)
Antagonists		
Phenothiazines	Nanomolar potency	Nanomolar potency
Thioxanthenes	Nanomolar potency	Nanomolar potency
Butyrophenones	Micromolar potency	Nanomolar potency
Substituted benzamides	Inactive	Micromolar potency
Dopaminergic ergots	Antagonists or partial agonists (micromolar potency)	Full agonists (nanomolar potency)

[a] Modified from Kebabian and Calne (1979).

Domperidone is also extremely weak with an IC_{50} approaching the millimolar range.

These discrepancies, initially overlooked, raised the possibility that butyrophenones might not block dopamine receptors at all, but rather act in some other system and influence dopaminergic activity indirectly. This would account for the marked difference in chemical structure between phenothiazines and butyrophenones despite their pharmacological similarities. This hypothesis was reinforced by computer modeling studies which demonstrated that the phenothiazine molecule could easily take on a conformation which mimicked the extended or trans-conformation of dopamine (Horn and Snyder, 1971). Similar calculations for butyrophenones demonstrated that they were no more likely to take up the dopamine-mimicking conformation than any other (Tollenaere *et al.*, 1977). An alternative hypothesis, not considered initially, was that there existed more than one type of dopamine receptor. Thus, butyrophenones would exhibit weak affinity for the receptor responsible for eliciting an increase in cAMP whereas they would exhibit higher potencies at those receptors responsible for behavioral and clinical effects. It now seems that this hypothesis is more tenable. Indeed, Kebabian and Calne (1979) have recently written a seminal review of the pharmacological classification of dopamine receptors. They divided dopamine receptors into two general categories. D-1 receptors are responsible for stimulating dopamine-sensitive adenylate cyclase activity upon agonist activation (Table 2). The location for the prototype D-1 receptor is the parathyroid gland where dopamine agonists stimulate cAMP synthesis concomitantly with parathyroid hormone release (Brown *et al.*, 1977; Brown *et al.*, 1980; Attie *et al.*, 1980). For a more detailed discussion of the dopamine-sensitive adenylate cyclase, see reviews by Miller and McDermed (1979) and Schmidt (1979).

4.2. "D-2" Dopamine Receptors

In contrast to D-1 receptors, D-2 receptors are functionally classified as not enhancing adenylate cyclase activity upon agonist occupation. Instead, the consequences of D-2 receptor stimulation are to either decrease or to have no effect on the formation of cAMP (Table 2). Prototype D-2 receptors exist in the anterior and intermediate pituitary glands. It has been clearly established that in both of these tissues dopamine does not elicit its physiological effects through the stimulation of cAMP synthesis. Indeed, Kebabian and colleagues have elegantly shown that in the intermediate pituitary dopamine inhibits the beta-adrenergic agonist stimulated synthesis of cAMP leading to a diminution of hormone release (*vide infra*).

The pharmacological profile of D-2 receptors is clearly distinct from that of D-1 receptors (Table 2). Agonists consistently demonstrate higher affinities in eliciting a biochemical or physiological response at D-2 receptors than at D-1 receptors. Apomorphine is a potent agonist with full intrinsic activity at D-2 receptors in contrast to its partial agonist activity at D-1 receptors. Similarly, various dopaminergic ergots (e.g., bromocryptine, lisuride, lergotrile) are full, potent (nM) agonists at D-2 receptors but only weak, partial agonists or antagonists at D-1 receptors. It should be noted in passing that SKF38393, a drug which had been hypothesized to be a selective D-1 agonist (Settler *et al.*, 1978), has recently been shown to exhibit agonist activity at D-2 receptors as well (Munemura *et al.*, 1980*a*). With respect to antagonists, phenothiazines and thioxanthenes are potent antagonists of D-2 receptors; however, they exhibit equally high affinity for D-1 receptors and thus they are not useful for discriminating between these subtypes. In contrast, butyrophenones and related drugs (e.g., domperidone) are very potent antagonists of D-2 receptors but exhibit only weak affinity for D-1 receptors. Similarly, substituted benzamides such as sulpiride which are inactive at D-1 receptors exhibit potent behavioral dopamine antagonism and moderate affinity at D-2 receptors.

The preceding pharmacological profiles should allow one to theoretically predict drugs or classes of drugs would be suitable to use in radioligand binding studies of dopamine receptor subtypes. High affinity ($K_d < 10$ nM) is an important constraint in radioligand binding experiments which use filtration to separate bound from free ligand (for a discussion see Bennett, 1978). Therefore, phenothiazines and thioxanthenes are the only drugs which could be expected to label D-1 receptors. Indeed, [^{3}H]-flupenthixol and [^{3}H]piflutixol, two of the most potent thioxanthenes, appear to label D-1 as well as D-2 receptors in the striatum (Hyttel, 1978*a,b*; Cross and Owen, 1980; Hyttel, 1981). Butyrophenone or butyrophenone-like antagonists (e.g., [^{3}H]spiroperidol, [^{3}H]haloperidol, [^{3}H]-domperidone) have been found to preferentially label D-2 receptors. At first glance, one would predict that only the D-2 receptor would be labeled with [^{3}H]agonists (Table 2); however, the situation is more complex as one must consider the possibility of "desensitized" or other agonist-specific states of D-1 receptors which may have higher affinity for agonists *in membrane preparations*. The pros and cons for [^{3}H]agonists labeling the D-1, the D-2, and possibly additional dopamine receptor binding sites will be discussed extensively throughout this review.

5. DOPAMINE RECEPTORS IN THE PITUITARY

Our studies indicate that the interpretation of radioligand binding to dopamine receptors in the pituitary may be more straightforward than binding to receptors in the brain. This apparently results from the presence of only a singular dopamine receptor subtype (D-2) in the pituitary, in contrast to multiple receptor types in the brain. Thus, as an introduction to CNS studies, the pituitary provides a good and readily interpretable starting point.

The release of a variety of pituitary hormones is regulated by dopamine originating from the tuberohypophyseal neuron system. The cell bodies of this system are located in the hypothalamic arcuate and periventricular nuclei, and they project axons ventromedially to the median eminence (reviewed in Moore and Bloom, 1978). Some axons continue beyond the median eminence and traverse the pituitary stalk to directly innervate the posterior and intermediate pituitary. The physiological significance of this innervation has heretofore been unclear, but recent evidence (*vide infra*) suggests that dopamine regulates alpha-MSH and beta-endorphin release from the intermediate lobe and possibly oxytocin release from the posterior pituitary. Other axons terminate within the median eminence and pituitary stalk in close approximation to the capillaries that form the hypophyseal portal vessels. Dopamine released from these terminals is transported in the portal blood to the anterior pituitary where it inhibits the release of prolactin (PRL).

Indeed, the release of PRL from the anterior pituitary appears to be under tonic inhibitory hypothalamic control. Convincing evidence suggests that dopamine might be the only inhibitory hypothalamic factor controlling the secretion of PRL (reviewed in Weiner and Ganong, 1978, and MacLeod *et al.*, 1980). Briefly, dopamine and dopamine agonists suppress PRL secretion *in vivo*, from the isolated pituitary gland *in vitro*, and from dispersed pituitary cells in culture; correspondingly, dopamine antagonists stimulate PRL secretion *in vivo* and block the inhibiting action of dopamine agonists *in vitro*. Moreover, the stereoselectivity and rank order of potency of catecholamines, phenothiazines, and related drugs in regulating PRL release *in vitro* directly implicates the presence of specific dopamine receptor sites in the anterior pituitary.

Accordingly, several groups (Creese *et al.*, 1977*a*; Caron *et al.*, 1978; Cronin *et al.*, 1978; Calabro and MacLeod, 1978) have used radioactive dopamine agonists and antagonists to identify a high-affinity, stereoselective and saturable dopamine receptor in anterior pituitary membrane preparations. The rank order of agonists and antagonists for competing with radioligand binding to the dopamine receptor agrees closely with their rank order in inhibiting or disinhibiting PRL release. In addition, one group has provided immunocytochemical evidence that these dopamine receptors are largely confined to the mammotroph cells (R. I. Weiner *et al.*, 1979; Goldsmith *et al.*, 1979). Since the anterior pituitary contains a potentially homogeneous population of dopamine receptors, we have been investigating the radioligand-receptor binding characteristics in this tissue in detail.

The radiolabeled dopamine antagonist, [^{3}H]spiroperidol, has previously been shown to bind exclusively to dopamine receptors in the anterior pituitary of cattle (Creese *et al.*, 1977*a*), sheep (Cronin and Weiner, 1979), and rats (Stefanini *et al.*, 1980). In bovine anterior pituitary membranes, the specific binding of [^{3}H]spiroperidol is saturable and of high affinity. Scatchard analysis of the saturation data indicates a homogeneous population of binding sites with a dissociation constant (K_d) of approximately 0.3 nM. The maximum number of binding sites (B_{max}) is about 4 pmol/g tissue—only 20% of the number of sites detected in bovine caudate. Using [^{3}H]spiroperidol as the radioligand, it can be demonstrated that antagonist competition curves exhibit monophasic, mass-action characteristics with pseudo-Hill coefficients equal to 1. For example, Figure 1 shows the experimental data and the resulting computer-modeled competition curve for the antagonist (+)-butaclamol. The computer analysis employed is a nonlinear least-squares curve-fitting program which can analyze the data in terms of one or more classes of binding sites (De Lean *et al.*, 1980; Munson and Rodbard, 1980). The (+)-butaclamol curve models best to a single homogeneous receptor state with a K_d of 1.1 nM.

In contrast, agonist/[^{3}H]spiroperidol competition curves exhibit heterogenous characteristics with pseudo-Hill coefficients less than unity. As shown in Fig. 2, in the absence of guanine nucleotides, the (−)-apomorphine/[^{3}H]spiroperidol curve is shallow (pseudo-Hill coefficient equal to 0.58), with computer analysis indicating that the data are best explained

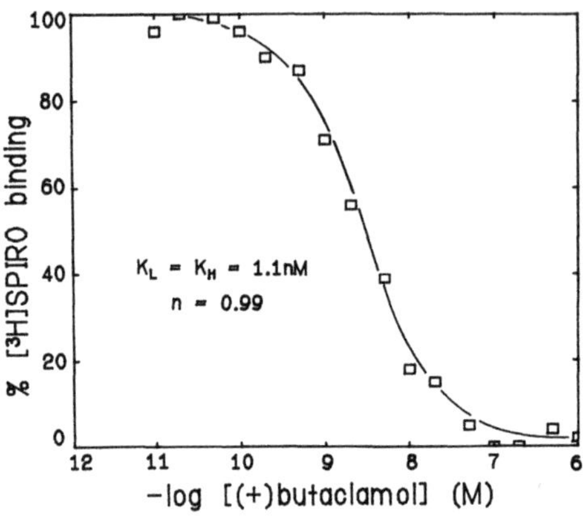

FIG. 1. Computer-fitted curve for a (+)-butaclamol/[^{3}H]spiroperidol competition experiment in bovine anterior pituitary membranes. The data points are shown by open circles and are from a single representative experiment. The computer-drawn curve represents the best fit to the data assuming a single homogenous binding site. The assumption of a two-site model does not improve the fit. The pseudo-Hill coefficient n = 0.99.

by a two site/state binding model. The K_d for the high- and the low-affinity binding sites/states (R_H and R_L) have been designated K_H and K_L, respectively. Interestingly, the two sites/states are present in approximately equal proportions in the membranes. In the presence of a saturating concentration of Gpp(NH)p, a nonmetabolizable analog of GTP, the (−)-apomorphine curve is shifted to the right and is steepened (pseudo-Hill coefficient equal to 0.94). Moreover, computer analysis of the data now indicates a single homogeneous population of binding sites whose affinity for (−)-apomorphine is not significantly different from the K_L value of the control curve (Fig. 2). Three additional agonists, (±)-ADTN, (−)-NPA, and dopamine, have been investigated and give qualitatively identical results.

Recently, we have characterized the binding of the radiolabeled agonist [^{3}H]-NPA to dopamine receptors in bovine anterior pituitary membranes (Sibley and Creese, 1979, 1980*a*). The identification of high-

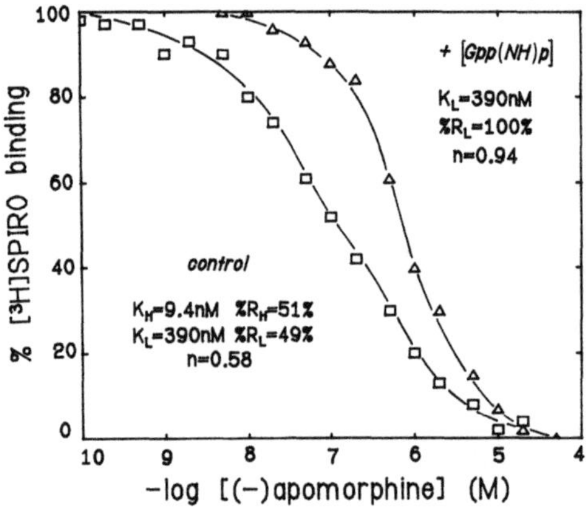

FIG. 2. Computer-fitted curves for a (−)-apomorphine/[^{3}H]spiroperidol competition experiment in bovine anterior pituitary membranes. The (−)-apomorphine control curve is best fitted by assumming a two-site model whereas in the presence of 10^{-4} M guanyl-5′-ylimidodiphosphate [Gpp(NH)p] a one-site model is sufficient to explain the data. When the two curves are analyzed simultaneously and constrained to share the same K_L value, there was no worsening of the fit. R_H and R_L represent the high- and low-affinity binding sites, respectively.

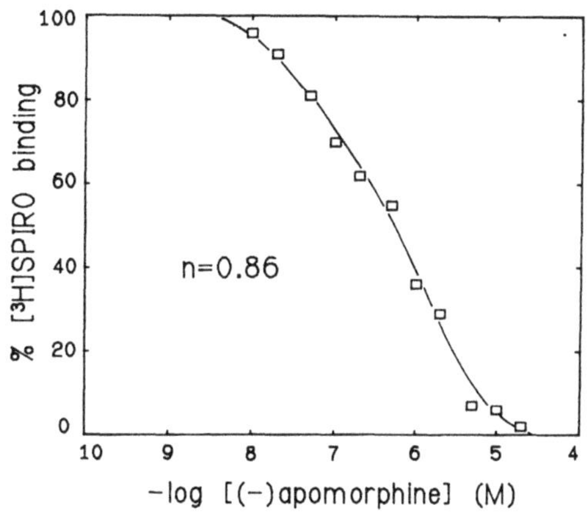

FIG. 3. Competition curve for [^{3}H]spiroperidol binding by (−)-apomorphine in intact bovine anterior pituitary cells which were dispersed by collagenase (0.4%) treatment. The data points represent the means ± S.E.M. of four individual experiments. The pseudo-Hill coefficient n = 0.86.

affinity [^{3}H]agonist binding in a tissue with no direct dopaminergic innervation reinforces our hypothesis that under our assay conditions [^{3}H]-agonists can label "postsynaptic" receptors (Creese and Sibley, 1979). One of the more striking findings with this radioligand is that its B_{max} is approximately 50% of that of [^{3}H]spiroperidol's, suggesting that it labels the high-affinity agonist site/state (R_H) seen in agonist/[^{3}H]spiroperidol curves. This is further suggested by the finding that agonist/[^{3}H]-NPA competition curves are homogeneous with single affinities that are not significantly different from the K_H values obtained from the corresponding agonist/[^{3}H]spiroperidol curve. Furthermore, saturating concentrations of guanine nucleotides completely abolish the specific [^{3}H]-NPA binding to pituitary membranes.

Two major explanations for the data are available. One is that the R_H and R_L sites represent two discrete dopamine receptors, i.e., two separate protein molecules. The two receptors would have identical affinity for all antagonists but differential affinity for all agonists. In addition, guanine nucleotides would inhibit agonist binding to the R_H receptor in some "allosteric" fashion. The second possibility is that the R_H and R_L sites actually represent high- and low-affinity agonist binding states of a single receptor molecule. In this model, guanine nucleotides regulate an interconversion between the high- and the low-affinity states. Evidence supporting this latter possibility can be seen in Fig. 3. In this experiment, bovine anterior pituitaries were first dispersed into single whole cells via collagenase treatment and then used directly in the binding experiment. Strikingly, the (−)-apomorphine/[^{3}H]spiroperidol curve is now steep (pseudo-Hill coefficient equal to 0.86) and comparable to the (−)-apomorphine/[^{3}H]spiroperidol + Gpp(NH)p curve in Fig. 2. Additionally, exogenously added guanine nucleotides no longer affect the (−)-apomorphine/[^{3}H]spiroperidol curve. The finding that the (−)-apomorphine competition curve does not appear to be maximally shifted and steepened in intact cells may be attributable to a nonsaturating intracellular GTP concentration at the receptor. Thus, in whole cells endogenous GTP regulates agonist binding in a fashion identical to that of exogenously added GTP in membrane preparations. Importantly, it should also be noted that specific[^{3}H]-NPA binding is not detectable in intact cells, directly confirming the absence of a detectable R_H state in these cells. However, membranes prepared from these cells exhibit identical binding properties as membranes directly prepared from the whole gland, indicating that the lack of high-affinity agonist binding is not the result of receptor degradation occurring during the collagenase-mediated dispersion. Thus, the R_H and R_L sites are presumably not functionally discrete receptor molecules, since if they were they would *both* be demonstrable in whole cells as well as in membranes.

Recently, Lefkowitz and co-workers have examined in detail the radioligand-receptor binding characteristics of the frog erythrocyte beta-adrenergic receptor system (reviewed in Lefkowitz, 1980, and Hoffman and Lefkowitz, 1980). Their data with the frog erythrocyte beta receptor is qualitatively identical to our anterior pituitary dopamine receptor data. That is, agonist/[^{3}H]antagonist competition curves model to two affinity states in membranes, with the high-affinity state being dispelled with exogenous guanine nucleotides and being undetectable in intact cells (Kent *et al.*, 1980). De Lean *et al.* (1980) have proposed a ternary complex model to explain the binding data in the frog erythrocyte system. This model is similar to the floating receptor (Jacobs and Cuatrecasas, 1976) or two-step (Boeynaems and Dumont, 1977) models previously described. Briefly, agonists or antagonists can bind to the receptor to form an initial drug–receptor complex. The binding of agonists, however, induces a conformational change in the receptor so that it can now couple to a second membrane component. It is this ternary complex which is responsible for the high-affinity agonist binding state. Limbird *et al.* (1980*a*) have provided evidence that the second component is the guanine nucleotide-binding protein of the adenylate cyclase complex. Presumably, it is the ternary complex of agonist, receptor, and nucleotide-binding protein which is responsible for activating adenylate cyclase in the presence of GTP. This complex is formed only transiently, however, since the endogenous GTP rapidly induces its dispersal in intact cells. Although high-affinity [^{3}H]-agonist binding is not demonstrable under equilibrium conditions in intact cells, it is likely that it is present upon initial exposure to the ligand.

The application of this model to the anterior pituitary dopamine receptor system is extremely attractive. However, dopamine does not appear to elicit an increase in anterior pituitary adenylate cyclase activity (Schmidt and Hill, 1977; Clement-Cormier *et al.*, 1977; Mowles *et al.*, 1978;

MacLeod *et al.*, 1980); but see Ahn *et al.* (1979). On the contrary, recent evidence suggests that dopamine may actually decrease cAMP formation in the anterior pituitary (DeCamilli *et al.*, 1979; Pawlikowski *et al.*, 1979, 1981; Labrie *et al.*, 1980; Giannattasio *et al.*, 1981; Ray and Wallis, 1980) and can reverse the activation of adenylate cyclase by vasoactive intestinal peptide (E. Costa, personal communication). Thus, the consequences of agonist–receptor complexation may be to decrease mammotroph cAMP content and thus to decrease PRL release. This hypothesis is additionally supported by recent work which suggests that increased mammotroph cAMP leads to an enhancement of PRL release (Dannies *et al.*, 1976; Naor *et al.*, 1980).

Some of the biochemical mechanisms involved in the dopaminergic regulation of hormone release have been better elucidated in the intermediate pituitary. It is known that the intermediate pituitary is predominantly composed of corticotrophic cells which synthesize and secrete a variety of peptides related to beta-lipotropin and ACTH, including beta-endorphin and alpha-MSH. Interestingly, Vale *et al.* (1979) showed that dopamine agonists can inhibit the release of beta-endorphin from rat neurointermediate pituitary cell cultures whereas cAMP analogs and phosphodiesterase inhibitors could stimulate this release. This latter stimulation was blocked by dopamine agonists, suggesting that dopamine may regulate beta-endorphin secretion by decreasing cAMP levels. More detailed studies of intermediate pituitary corticotroph regulation have been performed by Kebabian and colleagues (Cote *et al.*, 1980, 1981; Munemura *et al.*, 1980*a,b*). Using dispersed cells from rat intermediate pituitaries, they demonstrated that beta-adrenergic agonists, cAMP analogs, and phosphodiesterase inhibitors enhanced the secretion of alpha-MSH. Activation of the beta-receptor was accompanied by an increase in corticotrophic cAMP. Strikingly, dopamine inhibited the basal and isoproterenol (ISO)-enhanced release of alpha-MSH as well as the (ISO)-induced accumulation of cAMP. When homogenates of the intermediate pituitary were prepared and adenylate cyclase activity was measured directly, dopamine agonists inhibited the basal as well as the ISO-stimulated cyclase activity. The inhibition of the ISO response was noncompetitive in nature as dopamine inhibited the maximum ISO-stimulated increase in cyclase activity without affecting the EC_{50} for ISO. Evidence suggesting that dopamine's mechanism of action is distal to the beta-receptor came from radioligand binding experiments with the beta-antagonist [^{125}I]hydroxybenzylpindolol ([^{125}I]-HYP). Dopamine or dopamine-agonists had no direct effect on [^{125}I]-HYP binding nor did they interfere with the ability of beta-agonists to compete for [^{125}I]-HYP binding.

Recently we have directly labeled the dopamine receptor in bovine intermediate pituitary membranes using the radioligands [^{3}H]spiroperidol and [^{3}H]-NPA (Sibley and Creese, 1980*b*). The dopamine receptor binding characteristics in this tissue are remarkably similar to those seen in the anterior pituitary. For example, agonist/[^{3}H]spiroperidol curves are shallow (pseudo-Hill coefficients less than 1) but shift and steepen in the presence of GTP. Additionally, there are approximately twice as many [^{3}H]spiroperidol sites as there are [^{3}H]-NPA sites. These observations suggest the presence of identical dopamine receptors in the anterior and intermediate pituitaries. We have also detected specific [^{3}H]spiroperidol binding in bovine posterior pituitary membranes although this binding has not been extensively characterized (unpublished observations). It is interesting to note that recent evidence suggests a role for dopamine in regulating oxytocin release from the posterior pituitary (Moos and Richard, 1979).

Few studies have been performed on the regulation of pituitary dopamine receptors, but recent work (reviewed in Cronin *et al.*, 1980) suggests some similarities to central dopamine receptors. For instance, destruction of the tuberohypophyseal dopamine neuron system results in enhanced mammotroph sensitivity to dopamine agonists (Cronin *et al.*, 1980). Similarly, chronic pharmacological blockade with dopamine antagonists results in an enhanced effectiveness of dopamine agonists in inhibiting PRL release (Lal *et al.*, 1977; Annunziato *et al.*, 1980). Paradoxically, following either pharmacological blockade (Friend *et al.*, 1978) or medial basal hypothalamic lesions (Cronin *et al.*, 1980) there is a decrease in radioligand binding to anterior pituitary dopamine receptors. In the lesioned animals, this occurs despite the fact that there is an increase in mammotroph density. In contrast to the anterior pituitary, such lesions produce an increase in [^{3}H]spiroperidol binding to the neural and intermediate lobes (Cronin *et al.*, 1980).

Recently, sex steroids have been shown to exert antidopaminergic regulation at the pituitary level (Labrie *et al.*, 1979, 1980). Briefly, when rat anterior pituitary cells in primary culture are preincubated with estradiol, there is a stimulation of basal as well as TRH-induced PRL release. Whereas in control cultures dopamine agonists can maximally suppress 70–95% of PRL release, in cultures preincubated with estradiol the maximal inhibition is reduced to 20–45%. As yet, the effects of estradiol preincubation of radioligand binding to anterior pituitary dopamine receptors has not been successfully examined.

The biochemical and pharmacological criteria outlined above can be used to classify dopamine receptors into two general categories (Kebabian and Calne, 1979) (Table 2). D-1 dopamine receptors are defined by their ability to elicit an increase in adenylate cyclase activity whereas D-2 receptors do not activate this enzyme. Utilizing this as well as other criteria (*vide supra*) the dopamine receptors in the anterior and intermediate pituitary can be placed into the D-2 classification. Thus, [^{3}H]spiroperidol selectively labels the D-2 dopamine receptor while [^{3}H]-NPA labels a guanine nucleotide-sensitive agonist-specific binding state of the D-2 receptor. The guanine nucleotide sensitivity of agonist binding to this receptor may be a reflection of its linkage to adenylate cyclase; but in this case, however, agonist binding to the D-2 receptor may lead to a decrease rather than an increase in hormone-stimulated adenylate cyclase activity.

Recent evidence has suggested the existence of D-2 dopamine receptors which are unassociated with adenylate cyclase activity (Creese *et al.*, 1979*b*) (*vide infra*). Thus, D-2 dopamine receptors may be subdivided into two separate subclasses: those that inhibit hormone-stimulated adenylate cyclase activity and those that are unassociated with this enzyme. This subclassification would be similar to that seen with alpha-adrenergic receptors where alpha-1 receptors are unassociated with adenylate cyclase whereas alpha-2 receptors are generally found to exhibit negative modulation of this enzyme.

6. DOPAMINE RECEPTORS IN THE STRIATUM

The very first dopamine receptor binding studies utilized [^{3}H]dopamine and [^{3}H]haloperidol as ligands (Creese *et al.*, 1975; Burt *et al.*, 1975; Seeman *et al.*, 1975) in the examination of receptors in mammalian striatum. [^{3}H]Haloperidol bound to a site with high affinity very much like the D-2 receptor since described in anterior pituitary (Seeman *et al.*, 1975; Creese *et al.*, 1975). Bovine straitum also possessed high-affinity sites of [^{3}H]dopamine and other agonist ligands that, unlike the R_H state of the pituitary D-2 receptor, had very low (approximately micromolar) affinity but butyrophenones (Table 3) (Creese *et al.*, 1975; Burt *et al.*, 1976; Seeman *et al.*, 1976*a*). This led to the suggestion that mammalian striatum contained two distinct dopaminergic binding sites (Furchgott, 1978). Much of the controversy of the last few years within this area of research has centered around the neuronal localization of these two sites, and their relationship to the dopamine-stimulated adenylate cyclase. Further disputes have involved whether or not these subclasses can be further subdivided, and whether or not there exists yet more dopamine receptor subtypes detectable under different assay conditions. Important experimental approaches to these controversies have included the characterization of both competitive and irreversible inhibition of [^{3}H]ligand binding, solubilization and physical separation of the different binding sites, lesion studies, and studies comparing striatal binding with that of other dopamine-sensitive tissues.

TABLE 3

Characteristics of Dopaminergic Binding Sites in Membrane Preparations

Criteria differentiating receptor subtypes	D-1	D-2 R_H	D-2 R_L	D-3
Usable radioligands				
[^{3}H]Thioxanthenes	+	+	+	?
[^{3}H]Butyrophenones	—	+	+	—
[^{3}H]Agonists	?	+	—	+
Agonist affinity	Micromolar	Nanomolar	Micromolar	Nanomolar
Butyrophenone affinity	Micromolar	Nanomolar	Micromolar	Micromolar
Adenylate cyclase association	Stimulatory	Inhibitory or unassociated		?
Guanine nucleotide sensitivity	+	+	—	+
Striatal location	Intrinsic neurons	Intrinsic neurons/corticostriate afferents?		Nigrostriatal terminals?
Phenoxybenzamine sensitivity	+	++		+

6.1. Studies with Competitive Ligands

6.1.1. [^{3}H]Butyrophenone Binding: Labeling the D-2 Sites

Several lines of evidence suggest that at least the majority of high-affinity binding sites for [^{3}H]butyrophenones in the striatum are identical to the D-2 pituitary receptor. The K_d for [^{3}H]spiroperidol binding to dopamine receptors in striatum determined under a variety of conditions in rat, bovine, and human striatal membranes has been reported as 0.1–0.3 nM (Fields *et al.*, 1977; Creese *et al.*, 1977*a*; Howlett and Nahorski, 1978; Leysen *et al.*, 1978*a*; Quik and Iversen, 1979), in excellent agreement with the value obtained in bovine anterior pituitary. Early equilibrium studies produced linear Scatchard plots for [^{3}H]haloperidol (Burt *et al.*, 1976) and [^{3}H]spiroperidol (Creese *et al.*, 1977*a*), and kinetic analysis yielded association and dissociation rates consistent with the existence of homogeneous binding sites. This evidence indicated that, as in the pituitary, there existed one D-2 receptor. As in pituitary, [^{3}H]agonist ligands can, under appropriate conditions, label these same sites with high affinity (*vida infra*); and the affinity of agonists is reduced by guanine nucleotides with a specificity similar to that of pituitary (Zahniser and Molinoff, 1978; Creese *et al.*, 1979*c*). These D-2 sites are present in considerably higher numbers in striatum than pituitary, as mentioned above, with reported B_{max} values for [^{3}H]butyrophenones typically from 25 to 50 pmol/g tissue (Creese *et al.*, 1977*a*; Leysen *et al.*, 1978*a*) or 250 to 600 fmol/mg protein (Fields *et al.*, 1977; Howlett and Nahorski, 1978; Quik and Iversen, 1979).

Biochemically, the function of the striatal D-2 receptor is not known, although it now seems certain that it is not positively linked to a dopamine-sensitive adenylate cyclase. This D-2 site displays a much different pharmacological specificity (Creese *et al.*, 1975; Hyttel, 1978*b*), ontogenetic time course (Pardo *et al.*, 1977), and regional (Quik and Iversen, 1979) and cellular (*vida infra*) distribution than the dopamine-stimulated adenylate cyclase. This contention has further been supported by irreversible inhibition studies with phenoxybenzamine (*vida infra*). That the striatal D-2 receptor mediates the inhibition of a hormone-stimulated adenylate cyclase (as suggested for the intermediate pituitary lobe D-2 receptor) is purely conjectural at this time. It should be borne in mind, however, that guanine nucleotide sensitivity is at least consistent with such a hypothesis.

On the behavioral level, by contrast, the functional relevance of the striatal D-2 receptors is extremely well documented. The affinities of a number of structurally diverse dopamine antagonists for butyrophenone binding sites correlate well with their molar potencies in antagonism of apomorphine- and amphetamine-induced stereotyped behavior in rat

(Creese *et al.*, 1978*a*; Ogren *et al.*, 1978). Blockade of apomorphine-induced emesis in dog also correlates closely with D-2 binding site affinities. This latter test may avoid the complicating factor of differential drug distribution, since it is presumed to involve dopamine receptors in the area postrema of the brainstem, an area where the blood–brain barrier is less effective. Of greatest clinical importance is the correlation between the potency of these drugs as antipsychotic agents in man and their potency in competition for [^{3}H]butyrophenone binding (Creese *et al.*, 1976; Seeman *et al.*, 1976*b*). The affinity of an antagonist for [^{3}H]-butyrophenone binding is thus a powerful predicter of *in vivo* dopamine receptor antagonism and antipsychotic activity. The nanomolar affinity of the antipsychotic drugs for dopamine receptor binding sites is also commensurate with the plasma concentrations of these drugs at therapeutic dose levels as measured by the neuroleptic radioreceptor assay and by other methods (Creese and Snyder, 1977). A similar analysis has indicated that the anti-Parkinsonian effects of dopamine agonists are also mediated through the butyrophenone-labeled D-2 receptors (Titeler and Seeman, 1978; Schachter *et al.*, 1980).

It should be noted that a number of studies have shown that the most commonly used butyrophenone ligand, [^{3}H]spiroperidol, also labels serotonergic 5-HT_2 receptors in both striatum and other brain areas (Leysen *et al.*, 1978*b*; Peroutka and Snyder, 1979). Care must therefore be taken in studies utilizing butyrophenone ligands to direct binding to the desired receptors so as to avoid spurious identification of multiple "dopaminergic" butyrophenone sites, sites which could be either dopamine or serotonin receptors. This may be accomplished either by using an appropriate dopamine receptor selective "blank" to determine specific binding, or by including a competing drug as a "mask" of the undesired site. ADTN appears to offer a dopaminergic receptor selective blank when used in appropriate concentration (Quik *et al.*, 1978). Several serotonergic antagonists such as mianserin (Withy *et al.*, 1980) and R41468 (Leysen *et al.*, 1981) appear suitable for use as masks.

Recently, Schwartz and co-workers (Martres *et al.*, 1980; Sokoloff *et al.*, 1980*b*) have proposed the existence of another butyrophenone binding site ("D-4" receptor) in striatum, characterized by high affinity for butyrophenones and other dopamine antagonists and low affinity for dopamine agonists. As these authors note, however, D-2 receptors appear to convert to the D-4 type with the addition of GTP. This strongly suggests that, rather than being a separate receptor subtype, the "D-4" site is merely the R_L state of the D-2 receptor. The use of the term "D-4" receptor should therefore be abandoned in the interest of clarity and consistency with the nomenclature of other receptor systems.

Some evidence exists, however, that [^{3}H]butyrophenone binding in pituitary and striatum do differ. As noted above, guanine nucleotides shift the psuedo-Hill coefficient of agonist/[^{3}H]butyrophenone competition curves in pitutary membranes to approximately 1, indicating the existence of only one [^{3}H]butyrophenone binding site in this tissue. This guanine nucleotide change in pseudo-Hill slope, however, is incomplete for some agonists (Zahniser and Molinoff, 1978) in the striatum, despite the fact that these agonist displacement curves are shifted to the right and steepened by guanine nucleotides. Furthermore, the guanine nucleotide shift is of a lower magnitude in the striatum compared to the pituitary. In recent studies using methods to eliminate the confounding presence of 5-HT_2 receptor binding, we have confirmed that, even under these more rigorous conditions, some agonist displacements of [^{3}H]spiroperidol binding to bovine striatal membranes in the presence of maximal GTP have a pseudo-Hill slope less than 1. Thus, in striatum there may well be more than one D-2 receptor, with nearly equal affinities for butyrophenones, but with differing affinities for some agonists. One of these D-2 subtypes may be identical to that found in pituitary, itself interconverting between two agonist affinity states under the influence of guanine nucleotides; a separate D-2 receptor subtype may be present which is insensitive to guanine nucleotides. Kainic acid lesion studies have lent some support for this suggestion (*vide infra*): it has proved possible to selectively remove guanine nucleotide-sensitive [^{3}H]butyrophenone binding sites, leaving a population of nucleotide-insensitive sites intact (Creese *et al.*, 1979*b*). Thus there is evidence, albeit incomplete at this time, for two distinct subtypes of butyrophenone-binding D-2 receptors in both rat and bovine brain.

6.1.2. [^{3}H]*Agonist Binding Sites: Labeling the D-2 and D-3 Sites*

Putative dopamine receptors in striatum have also been identified by the binding of the tritiated dopamine agonists apomorphine, ADTN, and NPA as well as [^{3}H]dopamine itself. Unlike the binding of the [^{3}H]-butyrophenone ligands, that of the [^{3}H]agonist ligands is markedly dependent upon assay conditions. Under some conditions [^{3}H]agonist ligands can also bind to D-2 receptors with high affinity, as they do in anterior pituitary. A subset of the [^{3}H]agonist binding sites, however, differ from the butyrophenone labeled D-2 binding sites in that butyrophenones have micromolar affinities for these sites. Thus, it has been proposed that these agonist binding sites represent yet another distinct dopamine receptor, the "D-3" receptor (Titeler *et al.*, 1979). It should be noted that it remains a possibility that this D-3 agonist binding is actually to a high-affinity-agonist ("desensitized") state of the D-1 receptor. A comparison of antagonist affinities for this [^{3}H]agonist binding versus blockade of dopamine-stimulated cyclase activity is needed to clarify this point. For the purposes of this review, we will, for now, continue to use the D-3 nomenclature.

The D-3 receptors are operationally defined as binding sites with high affinity for [^{3}H]dopamine, [^{3}H]apomorphine, [^{3}H]-NPA, or [^{3}H]-ADTN and low affinity for butyrophenones. While D-3 receptors are apparently absent in pituitary, they are present in mammalian straitum at 10–40 pmol/g wet weight tissue (Burt *et al.*, 1976; Thal *et al.*, 1978; Creese *et al.*, 1979*a*; Komiskey *et al.*, 1978; Creese and Synder, 1978) or about 50–700 fmol/mg membrane protein (Seeman *et al.*, 1975; Cronin *et al.*, 1978; Titeler and Seeman, 1979; List *et al.*, 1980) depending on the conditions employed. In bovine striatal membranes at 37°C in the presence of "physiological" (extracellular) concentrations of ions, [^{3}H]dopamine specifically labels only D-3 sites, with a K_d of about 10–20 nM (Creese *et al.*, 1975; Burt *et al.*, 1976). Such sites have also been labeled in both calf and rat striatum under various conditions with [^{3}H]apomorphine (Thal *et al.*, 1978; Seeman *et al.*, 1979), [^{3}H]-NPA (Creese *et al.*, 1979*a*; Titeler and Seeman, 1979), and [^{3}H]-ADTN (Creese and Snyder, 1978; Seeman *et al.*, 1979) with affinities (K_d) in the nanomolar range. Oddly, under these same roughly physiological conditions, high-affinity [^{3}H]dopamine binding to rat striatal membranes is not reproducible (Creese *et al.*, 1979*d*), although Seeman and co-workers have been able to obtain such binding under other conditions (Titeler *et al.*, 1979; List *et al.*, 1980). We have recently explained these divergent results by characterizing the effects of temperature and ionic conditions on [^{3}H]dopamine binding in rat caudate membranes. In the absence of metal cations and chelating agents, following a preincubation in buffer at 37°C, specific [^{3}H]dopamine binding is entirely to D-3 sites. Addition of millimolar Ca^{2+}, Mg^{2+}, Mn^{2+}, or Co^{2+} allows [^{3}H]dopamine labeling of both the D-2 and D-3 sites with nearly equal affinity. This reflects prevention by these cations of an irreversible degradation of D-2 sites as previously described using [^{3}H]spiroperidol to define D-2 receptors (Usdin *et al.*, 1980). EDTA and EGTA (0.1 μM to 10 mM) paradoxically have a similar effect, although the maximal enhancement in D-2 binding seen with chelators is less than that seen with divalent cations. Chelators and divalent cations have a further effect in greatly decreasing nonspecific binding of [^{3}H]dopamine. Na^{+} (10–150 mM), on the other hand, decreases [^{3}H]dopamine binding to both D-2 and D-3 sites by decreasing agonist but not antagonist affinity. This Na^{+}-mediated decrease in agonist but not antagonist affinity is similar to that observed for the opiate receptor (Pert *et al.*, 1973), the alpha-1 (Glossman and Hornung, 1980) and the alpha-2 (Tsai and Lefkowitz, 1978) adrenergic receptors, and the histamine-1 receptor (Chang and Snyder, 1980). [^{3}H]-Dopamine binding to both D-2 and D-3 sites is also reduced by increasing incubation temperature, although [^{3}H]butyrophenone binding to D-2 sites it not. The combined effect of sodium and temperature is sufficient to place [^{3}H]dopamine affinity for D-3 sites in rat membranes when assayed at 37°C in the presence of Na^{+} outside the range detectable in filtration assays. Under these conditions, the K_i for unlabeled dopamine in displacement of the more potent [^{3}H]agonists is about 200–300 nM (Creese *et al.*, 1979*d*). At 22–25°C in the absence of sodium, however, [^{3}H] dopamine displays a K_d of about 2 nM for D-3 sites. [^{3}H]-NPA (Titeler and Seeman, 1979) and [^{3}H]apomorphine (Thal *et al.*, 1978, Titeler *et al.*, 1978) also label D-2 receptors with high affinity as they do in anterior pituitary, as well as D-3 sites. This dual labeling of D-2 and D-3 receptors by these agonists leads to their biphasic displacement by butyrophenone antagonists with both high (nanomolar) and low (micromolar) affinity components (Burt *et al.*, 1976). Thus, depending on tissue preparation and incubation conditions, dopamine [^{3}H]agonists can label D-2 or D-3 receptors either selectively or together.

The function of the D-3 sites is unclear. Antagonist affinities at these sites do not correlate with their antipsychotic (Creese *et al.*, 1976) or anti-Parkinsonian (Titeler and Seeman, 1978) potencies nor do they correspond well with their ability to block stimulation of dopamine-sensitive adenylate cyclase. Lesion studies suggest that the D-3 site may represent autoreceptors or nigrostriatal terminals (*vide infra*).

6.1.3. Thioxanthene Binding: Labeling the D-1 Site

A high-affinity (K_D = 4 nM) striatal binding site for [^{3}H]flupentixol (Hyttel, 1978*a,b*; Hyttel, 1980; Cross and Owen, 1980) has been identified which appears from competition studies to be the D-1 (adenylate cyclase-linked) receptor, a receptor not apparently labeled with high affinity by either [^{3}H]butyrophenones or [^{3}H]agonists. The potencies of a number of dopaminergic antagonists from a variety of structural classes in inhibiting dopamine-sensitive adenylate cyclase activity correlates well with their potencies in displacing [^{3}H]flupentixol binding. For example, thioxanthenes, which possess very high affinity for [^{3}H]flupentixol binding sites, also have nanomolar potency in the inhibition of the dopamine-stimulated adenylate cyclase. Butyrophenone affinities for both the cyclase and [^{3}H]-flupentixol binding sites are one to two orders of magnitude lower. Agonists are active in both displacing [^{3}H]flupentixol and in stimulating cyclase in the micromolar range. These sites are present in approximately three times the numbers seen for D-2 or D-3 binding sites. Detailed displacement studies have revealed that a minor portion, about 20%, of [^{3}H]flupentixol binding is to the D-2 receptors (Cross and Owen, 1980). [^{3}H]Flupentixol binding can be directed to label exclusively the putative D-1 receptor by the inclusion of an appropriate "masking" drug, i.e., low concentration of unlabeled butyrophenones, in the assay to saturate the D-2 receptors. Recently, a new thioxanthene ligand, [^{3}H]piflutixol, has been characterized. It possesses the same receptor specificity as [^{3}H]-flupentixol but has somewhat higher affinity for D-1 sites (approximately

0.4 nM) (Hyttel, 1981). One of the chief obstacles to the further characterization of the D-1 [^{3}H]thioxanthene binding sites has been the high level of nonspecific binding (generally about 60% of the total) seen with both [^{3}H]flupentixol and [^{3}H]piflutixol. Characteristics such as guanine nucleotide sensitivity require a better D-1 ligand before their detailed study becomes feasible.

6.1.4. *Substituted Benzamide Binding Sites*

Three groups have described saturable binding for the substituted benzamide, [^{3}H]sulpiride, in rat striatum that is quite different from that of other dopaminergic ligands. The affinity of sulpiride for these sites, K_d = 7 nM (Woodruff and Freedman, 1981), K_d = 17 nM (Memo *et al.*, 1980), or K_d = 27 nM (Theodorou *et al.*, 1979), is much greater than its affinity for D-1 sites labeled by [^{3}H]flupentixol (K_i > 10,000 nM) (Hyttel, 1978*b*; Cross and Owen, 1980), D-2 sites labeled by [^{3}H]butyrophenones (IC_{50} = 100–1000 nM) (Leysen *et al.*, 1978*a*; Creese *et al.*, 1979*d*; Seeman *et al.*, 1978), or D-3 sites labeled by [^{3}H]apomorphine (K_i = 100,000 nM) (Sokoloff *et al.*, 1980*b*). There is disagreement, however, as to whether unlabeled *cis*-flupentixol has high (IC_{50} = 1.8 nM), (Freedman and Woodruff, 1980) or low (IC_{50} > 5000 nM) (Theodorou *et al.*, 1979) affinity for these sites. [^{3}H]Sulpiride binding is dependent on the presence of Na^+ and inhibited by Ca^{2+} (Theodorou *et al.*, 1980), a feature not seen with the binding of other dopaminergic ligands. The high-affinity [^{3}H]-sulpiride binding site is found predominately in dopaminergically innervated tissues (Woodruff and Freedman, 1981; Memo *et al.*, 1980) and has the highest affinity for dopamine among the neurotransmitters screened. This binding site may thus represent yet another receptor for dopamine, although a much more extensive characterization is needed before this conclusion is warranted.

7. IRREVERSIBLE MODIFICATION OF DOPAMINE RECEPTORS

Identification of dopamine receptor mechanisms has been aided by the use of several agents to irreversibly inactive receptors. Selective inactivation of specific dopamine receptor subtypes allows relatively unencumbered characterization of the remaining subtypes. Phenoxybenzamine appears to irreversibly alkylate D-2 sites, while the reactive aporphine, (−)-*N*-chloronorapomorphine (NCA), has promise of being D-1 and D-3 selective. NCA can also be used *in vivo* for correlation of receptor and behavioral changes. Heat treatment has an inhibitory effect on the binding of [^{3}H]agonist ligands similar to that of NCA, but may exert its effects indirectly through the denaturation of a guanine nucleotide-binding regulatory protein.

7.1. Phenoxybenzamine: Selective Alkylation of [^{3}H]Butyrophenone Binding Sites

Recently, we have shown that phenoxybenzamine selectively and irreversibly eliminates ^{3}H-butyrophenone-labeled D-2 binding sites while producing little effect on ^{3}H-dopamine-labeled D-3 binding sites (Hamblin and Creese, 1980, 1982*a*). Preincubation of bovine caudate homogenates with phenoxybenzamine rapidly results in a time- and concentration-dependent decrease in subsequent [^{3}H]spiroperidol binding, with maximum effect observed by 10 min and with a pseudo-IC_{50} of 1 μM (Fig. 4). The binding of the antagonists [^{3}H]domperidone and [^{3}H]haloperidol is affected similarly. Binding of [^{3}H]dopamine, however, when assayed under conditions selective for D-3 sites, is inhibited, with an IC_{50} 100 times higher than that for [^{3}H]-spiroperidol binding—a homogenate treatment with 10 μM phenoxybenzamine almost completely eliminates [^{3}H]spiroperidol binding while leaving [^{3}H]dopamine binding nearly unaffected. Thus, as suggested earlier by displacement studies, binding sites for [^{3}H]-dopamine and [^{3}H]butyrophenones appear to be physically distinct and do not interconvert under the conditions of the assay. This decrease in [^{3}H]spiroperidol binding is mediated by a decline in the number of binding sites with little change in their affinity, consistent with a covalent attachment of phenoxybenzamine at this site, as suggested for its action at the alpha-adrenergic receptor. This is further supported by the resist-

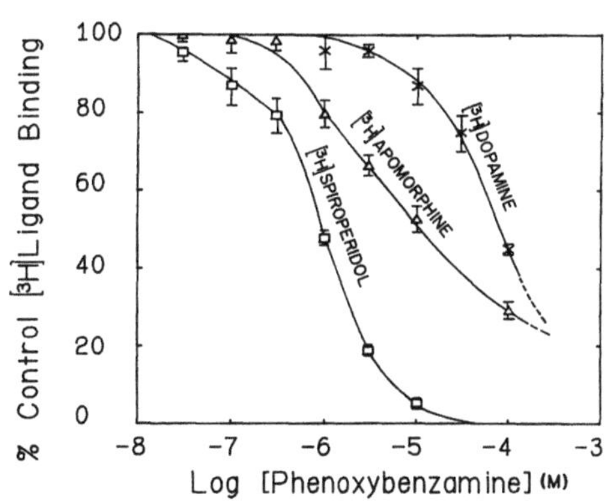

FIG. 4. Phenoxybenzamine inhibition of [^{3}H]spiroperidol, [^{3}H]apomorphine, and [^{3}H]dopamine specific binding to bovine caudate membranes. Phenoxybenzamine is most potent in eliminating the D-2 specific binding of [^{3}H]spiroperidol and least potent in decreasing the D-3 specific binding of [^{3}H]dopamine. Potency at [^{3}H]apomorphine binding is intermediate to that for [^{3}H]spiroperidol and [^{3}H]dopamine. Results are expressed as percent control specific [^{3}H]ligand binding remaining after exposure of homogenates to various concentrations of phenoxybenzamine for 10 min followed by thorough washing. Concentrations of ligands used were 0.5 nM [^{3}H]spiroperidol, 0.8 nM [^{3}H]apomorphine, and 3 nM [^{3}H]dopamine. Each point represents the mean ± S.E.M. of three to five independent determinations.

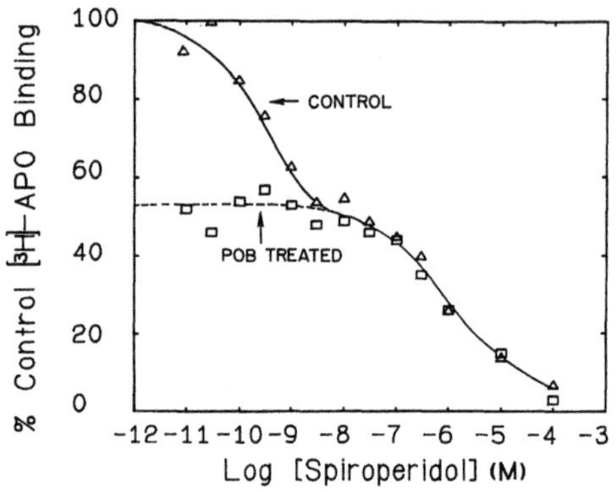

FIG. 5. Displacement of [^{3}H]apomorphine binding by spiroperidol in control and phenoxybenzamine-treated bovine caudate membranes. Displacement of [^{3}H]-apomorphine by spiroperidol from control homogenates is biphasic (solid lines). Pretreatment of homogenates with phenoxybenzamine eliminates the high-affinity displacement phase, leaving the low-affinity phase unaffected (dotted lines). Caudate homogenates are pretreated with 10 μM phenoxybenzamine for 10 min and then thoroughly washed. Various concentrations of unlabeled spiroperidol were added to tubes containing 0.8 nM [^{3}H]-apomorphine, tissue sample, and, for nonspecific binding determinations, 10 μM (+)-butaclamol. Results are expressed as the percentage of specific binding to control membranes without displacing drug. Points represent the mean of two separate experiments with S.E.M. less than 10%.

ance of this inhibition to reversal by repeated washings. [^{3}H]Spiroperidol binding sites are protected from phenoxybenzamine attack by occupancy both by agonists such as dopamine or apomorphine and by antagonists such as domperidone, indicating that the phenoxybenzamine effect is mediated through site-directed attack and not merely through a nonspecific membrane effect.

Binding of the agonist ligand [^{3}H]apomorphine assayed under identical conditions is affected to a degree intermediate to that of [^{3}H]-spiroperidol and [^{3}H]dopamine. The decrease in [^{3}H]apomorphine binding that is seen, as with that for [^{3}H]spiroperidol binding, occurs in a site-

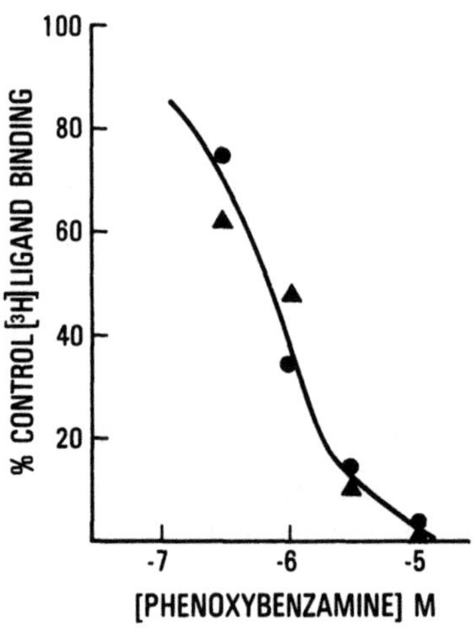

FIG. 6. Phenoxybenzamine inhibition of [^{3}H]spiroperidol, ●; and [^{3}H]apomorphine, ▲ specific binding to bovine anterior pituitary membranes. In contrast to what is observed in caudate membranes, in pituitary the sensitivity of [^{3}H]spiroperidol and [^{3}H]apomorphine binding to phenoxybenzamine is equal. This reflects the absence of D-3 sites in pituitary. The experimental procedure was identical to that employed in Fig. 4.

directed manner, with a decrease in B_{max} unaccompanied by a major change in K_d. The increased sensitivity of [^{3}H]apomorphine high-affinity binding sites in comparison with those for [^{3}H]dopamine suggested that even under conditions where [^{3}H]dopamine binding is D-3 selective, [^{3}H]-apomorphine labels *both* the relatively phenoxybenzamine-resistant D-3 site and the relatively phenoxybenzamine-sensitive D-2 site. Displacement of total [^{3}H]apomorphine specific binding from control membranes by spiroperidol (Creese *et al.*, 1978*b*) or domperidone (Sokoloff *et al.*, 1980*b*) is clearly biphasic with an overall pseudo-Hill slope of about 0.5, consistent with the presence of more than one type of [^{3}H]apomorphine binding site (Fig. 5). Treatment with 10 μM phenoxybenzamine for 10 min, which eliminates 95% of all [^{3}H]spiroperidol high-affinity (D-2) binding sites, eliminates only that [^{3}H]apomorphine binding displaceable with high affinity by unlabeled spiroperidol, i.e., that to D-2 sites. Such treatment has no significant effect on those [^{3}H]apomorphine sites with low affinity for spiroperidol (D-3). Guanine nucleotides decrease both the binding of [^{3}H]apomorphine and the potency of apomorphine in displacement of [^{3}H]spiroperidol at this site. Since the apparent K_d of [^{3}H]apomorphine for calf striatal membranes is unchanged by removal of the phenoxybenzamine labile sites, [^{3}H]apomorphine affinity must be nearly identical at the D-2 and D-3 sites.

Phenoxybenzamine at higher concentrations also inactivates the dopamine-stimulated adenylate cyclase (Walton *et al.*, 1978). Marchais and Bockaert (1980) demonstrated that homogenate pretreatment with 10 μM phenoxybenzamine, which completely eliminates [^{3}H]spiroperidol binding, leaves 35% of the dopamine-stimulated adenylate cyclase, supporting the hypothesis that [^{3}H]spiroperidol sites are not linked to the cyclase in a stimulatory fashion. [^{3}H]Flupentixol binding, like the cyclase, is more resistant to phenoxybenzamine attack than other [^{3}H]antagonists, with about 20% of specific binding remaining after 10 μM phenoxybenzamine treatment. This is consistent with the suggestion that [^{3}H]flupentixol labels, in part, the D-1 receptor.

As anterior pituitary is believed to contain D-2 but not D-3 binding sites, it would be anticipated that [^{3}H]apomorphine binding would show identical POB sensitivity to [^{3}H]spiroperidol binding in this tissue. This is indeed the case (Fig. 6), reinforcing the hypothesis that both agonist and antagonist [^{3}H]ligands label the same singular receptor in the anterior pituitary.

7.2. *N*-Chloroethylnorapomorphine: Effects on [^{3}H]-NPA Binding and Dopamine-Receptor-Mediated Behavior

Neumeyer *et al.* (1980) have synthesized a nitrogen mustard derivative of apomorphine, (−)-*N*-chloroethylnorapomorphine (NCA), that not only irreversibly prevents [^{3}H]agonist binding, but also possesses useful *in vivo* activity that allows behavioral investigations and the potential study of receptor turnover.

Treatment of rat striatal homogenates with NCA results in a dose-dependent decrease in subsequent [^{3}H]-NPA binding with an "IC_{50}" of 1.8 μM (Costall *et al.*, 1980*a*). The NCA-mediated decrease is a result of a decline in the number of binding sites, with no change in affinity, suggesting that the NCA–receptor interaction may be due to covalent binding. This is further supported by the observation that repeated washes after NCA exposure do not reverse the inhibition. The inactivation appears to be site directed as 10 μM ADTN is able to protect against NCA attack. NCA displays a selectivity quite different from that of phenoxybenzamine. Although active in decreasing binding to ^{3}H-NPA-labeled D-3 sites, up to 25 μM NCA has no effect on [^{3}H]spiroperidol (D-2) binding. However, this selectivity has yet to be reported in detail. A similar elimination of [^{3}H]-NPA sites is observed after *in vivo* administration of NCA intrastriatally (Costall *et al.*, 1980*a*).

The dopamine-stimulated adenylate cyclase is also irreversibly inhibited by *in vitro* NCA treatment (Neumeyer *et al.*, 1980; Baldessarini *et al.*, 1980), presumably by attack on the D-1 receptor. The "IC_{50}" for this inhibition, although not determined under identical conditions as that employed for the binding studies, is 25–30 μM, over 10 times higher than that for inhibition of [^{3}H]-NPA and [^{3}H]apomorphine binding (D-3). This provides additional evidence that the D-3 sites labeled by [^{3}H]apomorphine ligands are not linked to adenylate cyclase; a finding similar to that which was shown for D-2 sites through discrimination by phenoxybenzamine (Marchais and Bochaert, 1980).

NCA causes many of the behavioral effects expected of dopaminergic antagonists when administered *in vivo*. Thus, low doses of NCA injected s.c. reduce apomorphine-induced ipsilateral circling behavior in mice with unilateral striatal electrolytic lesions (Costall *et al.*, 1980*a*), and reduce apomorphine-induced climbing behavior in unlesioned mice (Costall *et al.*, 1980*b*). Intrastriatal administration of NCA, in addition to reducing [^{3}H]-NPA binding sites, also rapidly induces an ipsilateral circling that is exacerbated by peripheral administration of apomorphine. The one exception to this dopamine antagonist pattern is the failure of NCA to antagonize apomorphine-induced stereotypy in mice (Costall *et al.*, 1980*b*). This suggests that apomorphine may act at different receptor subtypes, distinguishable on the basis of their NCA labilities, to cause climbing and circling behavior on the one hand, and stereotypy on the other. The effects on circling and climbing behavior, as well as the concomitant changes in [^{3}H]-NPA binding seen after NCA administration, are long lasting, with significant effects on all three of these parameters still apparent 2–4 days after a single dose (Costall *et al.*, 1980*a,b*). By 7 days postadministration, however, the NCA-induced changes are not detectable, possibly reflecting the availability of new receptors by synthesis or redistribution.

7.3. Heat Treatment: Multiple Effects Mimicking GTP

Lew and Goldstein (1979) first reported that briefly raising the temperature of striatal homogenates to 53°C results in a large decrease in [^{3}H]dopamine binding, while leaving [^{3}H]spiroperidol binding largely unchanged. This was interpreted as a heat-induced denaturation of [^{3}H]-dopamine binding sites, but not the separate [^{3}H]spiroperidol binding site. Additional evidence now suggests that it is not the [^{3}H]dopamine binding site, the D-3 site, itself which is denatured. Heat treatment produces changes in dopaminergic ligand binding analogous to those produced for beta-adrenergic ligands by disruption of adrenergic receptor interaction with a guanine nucleotide binding protein (Hamblin and Creese, 1982*b*). This suggests that such a protein may also be involved in regulation of dopaminergic ligand binding, and that it is this moiety that is inactivated by heat.

Exposure of caudate homogenates to 53°C causes a rapid decrease in specific binding of the agonist ligands [^{3}H]apomorphine and [^{3}H]dopamine, with more than one-half eliminated within 30 sec (Hamblin and Creese, 1982*b*). The binding of [^{3}H]spiroperidol is nearly unaffected. Unlike treatment with phenoxybenzamine, heat treatment equally affects the binding of [^{3}H]dopamine, which under the conditions employed here labels only the D-3 site, and that of [^{3}H]apomorphine, which labels both D-2 and D-3 sites. Thus, heat treatment not only eliminates [^{3}H]dopamine and [^{3}H]apomorphine binding to the D-3 site, but also the binding of [^{3}H]-apomorphine to the D-2 site, despite a lack of any alteration in [^{3}H]-spiroperidol binding to the D-2 site. Thus this effect cannot be explained merely as a loss of the binding sites *per se*, and is highly reminiscent of the effects of guanine nucleotides on agonist binding.

In addition to causing an apparent reduction in the number of high-affinity [^{3}H]agonist binding sites, however, heat treatment has a second effect in causing a reduction if potency of unlabeled agonists in displacement of [^{3}H]spiroperidol—the IC_{50}'s for dopamine and apomorphine are shifted 10–15 times higher after exposure of homogenates to 53°C for 4 min. (Fig. 7). Micromolar concentrations of GDP, GTP, or Gpp(NH)p also cause a similar decrease in agonist potency. The effects of 4-min heat treatment and maximal GTP included in the assay are not additive, consistent with a common site of action.

Finally, heat treatment has a third effect: it causes an increase in the pseudo-Hill slope of dopamine/[^{3}H]spiroperidol displacements from 0.4 for control homogenates to 0.8 in those exposed to 53°C for 4 min. This

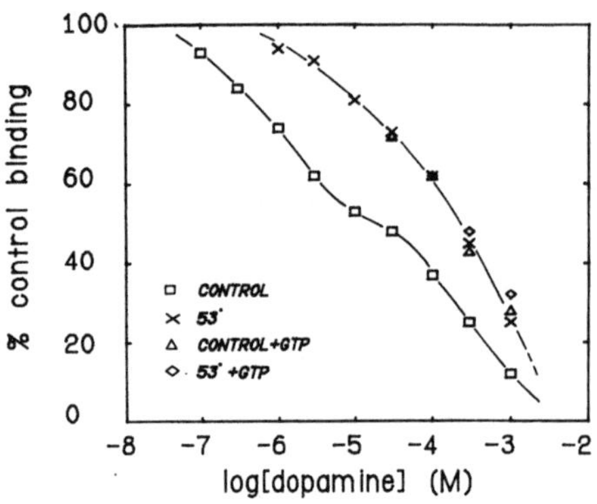

FIG. 7. Displacement by dopamine of [^{3}H]-spiroperidol binding in heat-treated caudate homogenates ± GTP. Aliquots of homogenate and 0.5 nM [^{3}H]spiroperidol, with or without 1 μM (+)-butaclamol blank, were incubated with various concentrations of unlabeled dopamine, with or without 300 μM GTP. "53°C" homogenates were treated at 50°C for 4 min. "Control" homogenates were treated identically except for the heat treatment. Each point represents the mean of two independent determinations with S.E.M. < 15%. Similar results were obtained when Gpp(NH)p was used rather than GTP.

steepening of the displacement curve (Fig. 7) is once again an effect on D-2 binding also seen with addition of guanine nucleotides (Zahniser and Molinoff, 1978; Creese and Sibley, 1979).

A single explanation of these three common, nonadditive effects of heat treatment and guanine nucleotides is suggested by the characterization of GTP-binding regulatory protein ("N") that modulates beta-adrenergic receptor function (Ross and Gilman, 1980). This protein, when coupled with the beta receptor—and possibly many other neurotransmitter and hormone receptors (Rodbell, 1980)—enables high-affinity binding of [^{3}H]-agonist ligands and potent displacement of [^{3}H]antagonists by agonists. When N/receptor association is prevented, either by the addition of GTP (Limbird *et al.*, 1980*a*) or manipulations eliminating N directly (Pike and Lefkowitz, 1980; Ross *et al.*, 1977; Howlett *et al.*, 1978; Limbird *et al.*, 1980*b*), high-affinity [^{3}H]agonist binding is lost, and agonist/[^{3}H]antagonist displacements are right-shifted and steepened. Antagonist binding remains unaffected. Thus, heat denaturation of such a regulatory moiety, rather than the receptor itself, would explain the observed binding changes (Hamblin and Creese, 1982*b*). Additional evidence for this hypothesis comes from the recent observation that high-affinity [^{3}H]-GTP binding sites in brain homogenates, which may correspond to the N protein, are eliminated by pretreatment of the homogenates at 53°C for 15 min (Rosenblatt *et al.*, 1980). These studies suggest that the same, or at least a very similar, heat-labile factor regulates binding at both the ^{3}H-butyrophenone-labeled D-2 site and the ^{3}H-dopamine-labeled D-3 site.

Another agent that decreases binding of [^{3}H]dopamine but not of the antagonist [^{3}H]spiroperidol is the sulfhydryl alkylating agent *N*-ethylamaleimide (NEM) (Suen *et al.*, 1980). It has not yet been reported, however, whether this represents a decrease in receptor number or affinity, or if the inhibition is irreversible with respect to repeated washings. As mentioned above, NEM also decreases beta-adrenergic agonist, but not antagonist, binding to the beta receptor (Pike and Lefkowitz, 1980; Howlett *et al.*, 1978; Williams and Lefkowitz, 1977) through the mechanism now proposed to mediate heat effects on dopaminergic binding—that is, by inactivation of a GTP-binding regulatory protein rather than the receptor itself. It thus remains possible that the effect of NEM on D-3 binding is also mediated by alteration of N.

8. SOLUBILIZATION AND ISOLATION OF DOPAMINE RECEPTORS

Complete characterization of the various dopamine receptors will ultimately require the isolation of the purified protein and other membrane components involved, followed by successful reconstitution. Substantial steps in this direction have already been taken.

Several groups have not reported the solubilization of [^{3}H]butyrophenone binding sites. Gorissen and Laduron (1979) employed 1% digitonin treatment of dog striatal membranes followed by ultracentrifugation. This results in a supernatant containing binding sites for [^{3}H]spiroperidol, assayable using gel filtration to separate bound from free [^{3}H]ligand. These solubilized sites possess affinities for a large number of dopaminergic and nondopaminergic compounds very close to those observed for membrane-bound [^{3}H]spiroperidol sites, and displacement by the isomers of butaclamol is stereospecific. These affinities also correlate well with the potencies of these compounds in antagonizing apomorphine-induced emesis in dogs. The solubilized binding sites show a regional distribution in brain consistent with a dopaminergic nature. Similar results have now been reported using rat (Gorrisen *et al.*, 1980) and human (Madras *et al.*, 1980) striatum. We have successfully followed a similar solubilization procedure and found that the guanine nucleotide sensitivity of agonist displacement of [^{3}H]spiroperidol binding is lost following solubilization (Leff and Creese, 1982). Agonist displacement is steep (pseudo-Hill slope equal to 1) and of low affinity. This suggests that either the guanine nucleotide binding protein is not solubilized along with the [^{3}H]butyrophenone site or that it can no longer couple functionally with the receptor after solubilization.

Salt extraction with potassium chloride has also been employed to solubilize [^{3}H]butyrophenone binding sites from calf caudate (Clement-Cormier and Kendrick, 1980; Clement-Cormier *et al.*, 1980). Saturation and displacement studies, as well as gel filtration elution patterns, suggest the presence of several subtypes of [^{3}H]spiroperidol binding sites using this technique. At least one of these sites demonstrate affinity, specificity, and regional distribution comparable to D-2 membrane binding sites. However, in light of the previously reported solubilization of a high-affinity "spirodecanone" (i.e., nondopaminergic) [^{3}H]spiroperidol binding site (Gorissen *et al.*, 1980), the relationship of these additional sites to multiple membrane-bound [^{3}H]spiroperidol binding sites must be considered unknown pending a more detailed characterization. Similar multiple [^{3}H]spiroperidol binding sites have been subfractionated from chloroform–methanol extracts of calf striatum (Boyan-Salyers and Clement-Cormier, 1980), leading to the suggestion that functional D-2 receptors are proteolipids. Again, these results must be considered preliminary, as a complete characterization of binding has not been reported. Clement-Cormier and collaborators have also reported preliminary studies using KCl and the detergent deoxycholate and octyl-beta-glycosyl pyranoside to solubilize high-affinity binding sites for the agonist ligands [^{3}H]-NPA, [^{3}H]apomorphine, and [^{3}H]-ADTN (Clement-Cormier *et al.*, 1980).

Nishikori *et al.* (1980) have reported that [^{3}H]dopamine can be used as a photoaffinity label for dopamine receptors. Subsequent solubilization yields two distinct binding sites for [^{3}H]dopamine. Using a conventional filtration assay to characterize reversible [^{3}H]dopamine binding to the two sites in membrane homogenates prior to covalent labeling, they report that one site possesses high (K_d = 12 nM) affinity for [^{3}H]dopamine, consistent with it being the D-3 site, while another site possesses much lower affinity (K_d = 3.6 μM), suggesting that it may be the D-1 receptor. [^{3}H]Dopamine binding to these two sites prior to ultraviolet illumination is saturable, reversible, and cannot be demonstrated in canine cerebellum. After photoaffinity labeling, the two presumably covalently labeled receptors may be separated by gel filtration, supporting the evidence from membrane-binding studies that the high-affinity D-3 [^{3}H]dopamine binding site is not merely a high agonist affinity state of the D-1 receptor. Some caution may be advisable in the interpretation of these studies, however. In the absence of kinetic data, it is difficult to understand how the conventional filtration binding assays employed in these studies could detect reversible [^{3}H]dopamine membrane binding with such low ($K_d \geq$ μM) affinity, unprecedented in receptor literature. Other laboratories, while not using exactly the same protocol as employed above, have been unable to achieve [^{3}H]dopamine/receptor photoaffinity labeling (Davies *et al.*, 1980; our unpublished observations).

9. NEUROANATOMICAL LOCALIZATION OF DOPAMINE RECEPTORS IN THE CNS

While a good correlation between the localization of dopamine receptor binding and those regions in the CNS receiving dopaminergic innervation was demonstrated in early studies (Creese *et al.*, 1975; Burt *et al.*, 1975; Seeman *et al.*, 1975, 1976*a*), the specific anatomical localization of these binding sites awaited the advent of lesion techniques. Although some questions remain unanswered, these techniques have been used to determine which dopamine receptor subtypes are associated with postsynaptic neurons, dopamine neuron terminals, terminals from nondopaminergic afferents, glial cells, and cerebral vasculature.

The lesion techniques exploit the differential pre- vs. postsynaptic distribution of the receptor subtypes in striatum and substantial nigra (Fig. 8). Briefly, the dopaminergic input to the striatum can be disrupted—thus removing the presynaptic terminals—by hemisection of the medial forebrain bundle, electrolytic destruction of tissue in the substantia nigra pars compacta, or local injection of 6-hydroxydopamine (6-OHDA), a neurotoxin specifically taken up by catecholaminergic cells (Ungerstedt, 1968). The latter technique not only allows selective study of postsynaptic

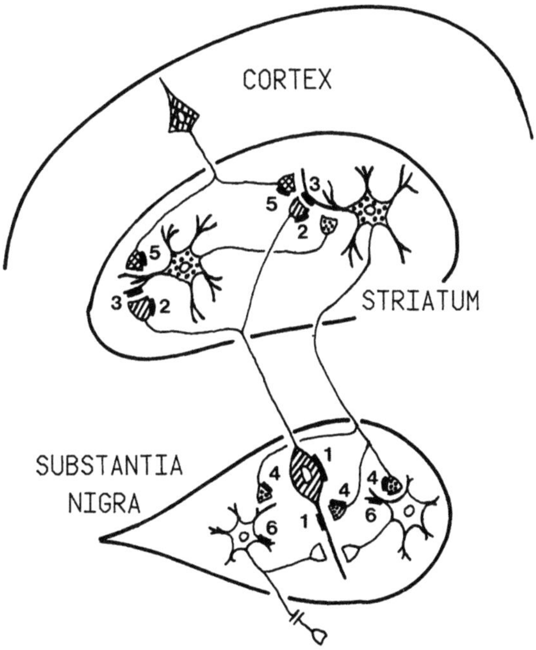

FIG. 8. Schematic diagram of putative sites of dopamine receptors in the nigrostriatal axis. (1) Dopamine autoreceptors on soma and dendrites of dopaminergic neurons of the SN pars compacta; (2) the same, on their terminals in striatum; (3) postsynaptic dopamine receptors on striatal neurons; (4) dopaminergic receptors on terminals of striatonigral afferents; (5) dopamine receptors on terminals of cortico-striate afferents; (6) dopamine receptors regulating activity of some interneurons and substantia nigra efferents of the pars reticulata. Hatched cells and terminals (dopaminergic cell of the substantia nigra pars compacta) degenerate after 6-hydroxydopamine lesion of substantia nigra; stippled cells and terminals degenerate after striatal kainate injection; cross-hatched cortico-striate neuron and terminals are lost after cortical ablation.

receptors in the striatum deafferented of its dopaminergic innervation, but it also allows the study of dopamine receptors on nondopaminergic cells or afferents in the substantia nigra. Kainic acid, a neurotoxin that selectively produces degeneration of neuronal perikarya around the injection site while sparing afferent terminals and passing axons (Coyle and Schwarcz, 1976; McGeer and McGeer, 1976), has been used to cause a selective degeneration of intrinsic striatal neurons. Degeneration is specific to the postsynaptic cells of the striatum and thus all striatal efferent fibers such as the striatal innervation of the substantia nigra (Coyle and Schwarcz, 1976). Cortical ablation has been used as a technique to remove cortico-striate terminals, another putative target of dopaminergic innervation in the striatum.

Cellular and subcellular fractionation techniques have attempted to identify the cell types associated with dopamine receptors (Henn *et al.*, 1977; 1978). An enrichment in glial cell fractions from calf caudate of [^{3}H]haloperidol, [^{3}H]dopamine, and [^{3}H]apomorphine binding, and of dopamine-stimulated adenylate cyclase activity, was reported in these two studies. However, the fractionation procedure cannot prevent contamination of the glial fraction by unidentified membrane fragments, and the characterization of [^{3}H]ligand binding and adenylate cyclase activity was incomplete. Furthermore, increases in dopamine receptor sites are not seen after kainate lesion-induced gliosis in the striatum while a complete loss of the dopamine-sensitive adenylate cyclase is found. Thus the degree to which dopamine receptors are associated with astroglia of the caudate is uncertain.

9.1. Neostriatum

Studies in rat striatum have provided evidence for a differential localization of D-1, D-2, and D-3 receptor subtypes. Kainic acid-induced lesions of intrinsic striatal neurons almost completely eliminates striatal dopamine-sensitive adenylate cyclase activity, indicating that almost all of this enzyme, and thus presumably D-1 receptors, is present on these cells (Schwarcz *et al.*, 1978; Govoni *et al.*, 1978). We have observed that 75% of the binding of [^{3}H]flupentixol, which labels D-1 receptors, is eliminated after kainic acid lesion as opposed to the 55% decrease observed for the D-2 specific ligand [^{3}H]spiroperidol (Leff *et al.*, 1981). This decrease in [^{3}H]flupentixol binding provides further support for the hypothesis that striatal D-1 receptors are confined to intrinsic striatal neurons. The residual 25% of [^{3}H]flupentixol binding seen after lesion is consistent with the ability of this ligand to also label D-2 receptors, which constitutes about 20% of the total control binding (Cross and Owen, 1980).

As mentioned, striatal kainate lesions indicate that about 50% of [^{3}H]-butyrophenone binding sites are localized on intrinsic neurons (Govoni *et al.*, 1978; Schwarcz *et al.*, 1978; Fields *et al.*, 1979). Most of the remaining butyrophenone binding sites in the striatum appear to be localized on cortico-striate terminals and are removable by cortical ablation (Schwarcz *et al.*, 1978; Garau *et al.*, 1978). These later sites appear to have low affinity for agonists and are guanine nucleotide insensitive (Creese *et al.*, 1979*b*). They are also not associated with adenylate cyclase activity which is localized on the intrinsic neurons. Thus it would appear that the cortico-striate terminal "D-2" sites are not linked to a regulatory guanine nucleotide binding protein ("N"). In the absence of additional supporting data, however, the presence of a separate nonguanine nucleotide sensitive D-2 receptor on cortical afferents must remain tentative. It remains a possibility, for example, that kainic acid removes GTP regulation of cortical afferent D-2 receptors by reducing the level of active N protein.

These studies demonstrate that a portion of [^{3}H]butyrophenone binding in the striatum is anatomically dissociated from the dopamine-stimulated adenylate cyclase; they do not demonstrate, however, that the remaining [^{3}H]butyrophenone binding on intrinsic striatal neurons is associated with the dopamine-stimulated adenylate cyclase. As discussed earlier, butyrophenone affinities in inhibiting dopamine-stimulated cyclase activity in striatal preparations are far weaker ($K_i \sim$ micromolar) than their binding affinities ($K_d \sim$ nanomolar). Thus, [^{3}H]butyrophenone binding sites on intrinsic striatal neurons are probably D-2 and not D-1 receptors. [^{3}H]Domperidone, a butyrophenone-like dopamine antagonist which is almost inactive in inhibiting dopamine-stimulated adenylate cyclase and whose pharmacological specificity closely matches that of [^{3}H]spiroperidol (Baudry *et al.*, 1979), exhibits a loss of binding after kainic acid lesions comparable to that of [^{3}H]spiroperidol (Leff *et al.*, 1981). We have found that agonist displacements of both [^{3}H]antagonist ligands show an identical sensitivity to guanine nucleotides. This GTP-regulated D-2 receptor may be linked in an inhibitory manner to a cyclase activated by another modulator as mentioned earlier.

The localization of D-3 sites identified with [^{3}H]agonists has been more elusive. Several dopaminergic agonists have been used in such studies. Differing assay conditions have led to a lack of consensus concerning the pharmacological specificity, neuronal localizations, and functional roles of these sites labeled by [^{3}H]agonists (Creese and Sibley, 1979; Titeler *et al.*, 1979; Leysen, 1979; Sokoloff *et al.*, 1980*a,b*). Some studies suggest that D-3 sites are localized on nigrostriatal terminals and may represent autoreceptors (Nagy *et al.*, 1978; Titeler *et al.*, 1979; Bannon *et al.*, 1980*a*; Sokoloff *et al.*, 1980*a*).

Dopamine autoreceptors have previously been physiologically identified by their modulation of dopamine synthesis and release, and by their inhibitory effects on dopamine neuronal activity (Walters and Roth, 1976; Roth, 1979). They appear to exist on both the cell body and the terminal regions of the dopamine neurons. *In vivo* studies indicate that agonists are more potent on autoreceptors than on postsynaptic dopamine receptors, suggesting that they might be labeled with [^{3}H]agonists. Furthermore autoreceptors do not appear to be linked to adenylate cyclase since presynaptic lesions do not decrease striatal dopamine-stimulated adenylate cyclase activity. Insufficient data are available to satisfactorily classify them in terms of their quantitative sensitivities to antagonists.

Initial studies comparing striatal kainate lesion-induced losses in the binding of [^{3}H]butyrophenones and [^{3}H]agonists, such as [^{3}H]apomorphine and [^{3}H]-ADTN (Creese *et al.*, 1979*b*; Fujita *et al.*, 1980; Fuxe *et al.*, 1979), reported that high-affinity agonist sites were decreased to a greater extent than [^{3}H]butyrophenone binding. Since this greater loss of agonist binding (70% or greater) correlated more closely with the loss seen in dopamine-stimulated adenylate cyclase activity, and since both measures show guanine nucleotide sensitivity, it was hypothesized that [^{3}H]agonists may label a "desensitized" state of the D-1 receptor in striatal membrane preparations. This conjecture has yet to be disproved. Further studies using [^{3}H]-NPA as an agonist ligand confirmed the original findings of a 70–80% loss of agonist binding in kainate-lesioned striata in the same samples used to demonstrate the 75% decrease in [^{3}H]flupentixol binding (Leff *et al.*, 1981).

Using different assay conditions, some groups have reported no change in [^{3}H]apomorphine binding after unilateral striatal kainate lesions (Weinreich and Seeman, 1980; Bannon *et al.*, 1980*a*). Since the observed control binding in these studies appears to show a lower density of [^{3}H]-apomorphine sites than observed by other groups (Creese *et al.*, 1979*d*; Leysen, 1979), the differing assay conditions may select for the labeling of different classes of [^{3}H]agonist sites. This possibility is supported by the observation that some differences exist in the pharmacological specificities of the [^{3}H]agonist binding seen in our studies and in those by Seeman and colleagues (Titeler *et al.*, 1979).

It is possible that differing assay conditions might also contribute to the different changes in [^{3}H]agonist binding seen in striatum after 6-OHDA lesions of dopaminergic cells projecting from the substantia nigra. Such lesions when made unilaterally were found by Ungerstedt (1968) to produce asymmetric body posture and rotational turning ipsilaterally (Ungerstedt and Arbuthnott, 1970). He later found that turning could be produced ipsilaterally or contralaterally in response to systemic injection of amphetamine or apomorphine, respectively. Such behavioral phenomena were hypothesized to be due to denervation hypersensitivity of dopamine receptors (Ungerstedt, 1971*a*,*b*). Thus, 6-OHDA lesions would result in an elimination of the binding to presynaptic receptors and also an increase in binding to postsynaptic receptors. Indeed, increased numbers of putative dopamine receptors in striatum were identified by both butyrophenone binding (Creese *et al.*, 1977*b*; Nagy *et al.*, 1978) and with [^{3}H]apomorphine (Creese and Snyder, 1979). Contrary to Creese's finding, Nagy *et al.* (1978), using different assay conditions, reported a decrease in [^{3}H]apomorphine binding in striata ipsilateral to the 6-OHDA lesion which led them to hypothesize that [^{3}H]apomorphine specifically labels presynaptic autoreceptors. Some groups failed to replicate these findings using similar assay conditions (Leysen, 1979).

A recent study by Sokoloff *et al.* (1980*a*) found a change in the distribution of pharmacologically differentiable [^{3}H]apomorphine sites in striatum after 6-OHDA lesions of the nigrostriatal tract. [^{3}H]Apomorphine sites having nanomolar affinity for butyrophenones and domperidone (D-2 sites—termed by them as Class I) increased 25–30% in density while sites having lower or micromolar affinity for butyrophenones and domperidone (D-3 sites—termed by them as Class II) showed 30–50% decreases in density. In this same study, striatal kainate lesions produced a 57% decrease in Class I sites and no change in Class II sites. These data suggest that a portion of [^{3}H]apomorphine binding, i.e., a portion of Class II sites, could be presynaptic autoreceptors. In these studies the modest increase in Class I (putatively postsynaptic) sites, as opposed to the large increases seen by others (Creese and Snyder, 1979), is possibly due to the degree of lesion. While Sokoloff *et al.* (1980*a*) reported 79–89% depletion of striatal tyrosine hydroxylase activity (from nigrostriatal terminals), some studies indicate that greater than 90% depletions of tyrosine hydroxylase must be produced before significant postsynaptic receptor hypersensitivity results (Creese and Snyder, 1979). Nevertheless using [^{3}H]-NPA and unlabeled domperidone to discriminate between these classes of agonist sites, we too have found a decrease in Class II (D-3) type sites after unilateral 6-OHDA lesions while total [^{3}H]-NPA binding and thus Class I (D-2) sites increased 10–20% (unpublished observation). However, it should be pointed out that the changes in relative amounts of such pharmacologically defined sites may not represent changes in receptor numbers *per se* but rather denervation-induced changes in their pharmacological characteristics.

While the above studies are consistent with the hypothesis that [^{3}H]-agonists can label both pre- and postsynaptic receptor sites, they are only correlative. Future studies will be required to pursue the pharmacological and morphological identity of these putative subclasses of dopamine receptors in order to further comprehend their role in the physiological and behavioral function of the neostriatum.

Recently Bacopoulos (1981) has demonstrated that [^{3}H]dopamine binding is increased by preincubation of homogenates with dopamine. We have confirmed with finding (Hamblin and Creese, 1982*c*). Since normal membrane preparations contain endogenous dopamine this suggests that during membrane preparation of such tissue, [^{3}H]dopamine binding is in fact enhanced. Bacopoulos (1982) has demonstrated that this is indeed true as reserpinization leads to a reduction in both endogenous dopamine and subsequent [^{3}H]dopamine binding. This has obvious implication for studies of [^{3}H]agonist binding following 6-OHDA lesions which also deplete endogenous dopamine as well as remove presynaptic terminals. We have recently found (Leff and Creese, 1982) that the effects of reserpinization and 6-OHDA lesion on [^{3}H]dopamine are identical and nonadditive, demonstrating that the lesion effects leading to apparent decreases in [^{3}H]dopamine binding are not the results of a loss of autoreceptors *per se* but rather represent the loss of endogenous dopamine and the subsequent enhancement of [^{3}H]dopamine binding during membrane preparation. Whether or not some [^{3}H]agonist binding does label autoreceptors is still unclear. However at this point in time it would seem that the majority of [^{3}H]agonist binding is to receptors postsynaptic to dopamine terminals.

9.2. Substantia Nigra

The use of specific lesion techniques has enabled a partial characterization and localization of [^{3}H]butyrophenone and [^{3}H]agonist binding sites and of the dopamine-stimulated adenylate cyclase in the substantia nigra. Several groups have reported the presence of dopamine-stimulated adenylate cyclase activity in rat substantia nigra (Kebabian and Saavedra, 1976; Phillipson and Horn, 1976; Spano *et al.*, 1976; Traficante *et al.*, 1976). Succeeding studies employing 6-OHDA injections into substantia nigra or median forebrain bundle, hemisections, and deafferentation of substantia nigra from striatum by such techniques as kainate lesions were able to determine that the dopamine-stimulated adenylate cyclase is localized largely on presynaptic terminals of descending afferents to the substantia nigra (Gale *et al.*, 1977; Phillipson *et al.*, 1977; Premont *et al.*, 1976; Schwarcz and Coyle, 1977; Spano *et al.*, 1977).

Quik *et al.* (1979) have further characterized subtypes of putative dopamine receptors in substantia nigra through the use of 6-OHDA nigral lesions and striatal kainate lesions. While supporting previous findings that the dopamine-stimulated adenylate cyclase appears localized on descending afferent terminals to the substantia nigra, Quik *et al.* (1979) found that a large portion (40%) of nigral [^{3}H]spiroperidol binding was lost after 6-OHDA lesions of dopaminergic nigral neurons. Striatal kainate lesions which produced large losses in nigral dopamine-stimulated adenylate cyclase activity did not affect levels of [^{3}H]spiroperidol binding. Murrin *et al.* (1979) found similar localizations of [^{3}H]spiroperidol sites in substantia nigra using light-microscopic autoradiography. Lesions of the nigrostriatal dopaminergic pathway produced a 48% decrease in [^{3}H]-spiroperidol sites in the pars compacta of substantia nigra, while disruptions of the striatonigral pathway produced no significant changes in [^{3}H]-spiroperidol labeling. These data thus provide evidence for a localization of D-2 receptors on dopaminergic cell bodies in the nigra.

D-Amphetamine and dopamine agonists administered peripherally, or applied centrally by microinjections into substantia nigra, have been found to decrease firing rates of nigrostriatal dopaminergic neurons (Bunney *et al.*, 1973*a*; Aghajanian and Bunney, 1973; Groves *et al.*, 1975). This regulation of the activity of dopaminergic cells of the substantia nigra has been hypothesized to involve both neuronal feedback mechanisms (Bunney *et al.*, 1973*b*; Bunney and Aghajanian, 1973) and dopamine agonist-mediated inhibition at the level of the substantia nigra (Aghajanian and Bunney, 1973; Groves *et al.*, 1975). Dopamine agonist inhibition of nigral cell firing rates could be mediated through dopamine autoreceptors associated with either dopamine cell bodies or dendrites (Groves *et al.*, 1975; Aghajanian and Bunney, 1977; Cheramy *et al.*, 1981). [^{3}H]Spiperone could in part be labeling these autoreceptors (Reisine *et al.*, 1979). Alternatively, dopamine agonist inhibition of dopaminergic nigral cell firing rates could be mediated through dopamine-sensitive adenylate cyclase-linked receptors on striatonigral terminals. Activation of this adenylate cyclase could be involved in the regulation of nigral GABA release and thus, indirectly, the activity of nigral dopamine neurons (Gale *et al.*, 1977; Reubi *et al.*, 1977; Cheramy *et al.*, 1981). Since the dendrites of the dopaminergic neurons of the substantia nigra have the capacity to synthesize (Pickel *et al.*, 1975), store (Bjorklund and Lindvall, 1975; Geffen *et al.*, 1976), and release (Geffen *et al.*, 1976; Cuello and Iversen, 1978; Nieoullon *et al.*, 1977) dopamine, it is not surprising that dopamine receptors have been found in substantia nigra. Though a major function of dendritically released dopamine appears to be the regulation of nigrostriatal cell activity through self or lateral inhibition, other functions of dopamine in the substantia nigra may be elucidated in future studies (for a review, see Cheramy *et al.*, 1981).

9.3. Retina

Evidence indicates that dopamine is a neurotransmitter in the vertebrate retina. Retinal neurons have been identified which store, synthesize, and metabolize dopamine and which show specific uptake and (upon light stimulation) release of dopamine (Kramer, 1976; Ehinger, 1976; Sarthy and Lam, 1979). Dowling and colleagues have described two types of neurons in retina which contain dopamine: an amacrine cell which has connections with other amacrine cells in the inner-plexiform layer (Dowling and Ehinger, 1978*a*); and in some species, interplexiform cells which have connections in the inner and outer plexiform layer to discrete cell types and their processes (Dowling and Ehinger, 1978*b*). Since well-defined synapses between dopamine-containing cells and target neurons can be found in the retina, it is expected that similar classes of dopamine receptors may be found in retina as in the rest of the CNS. Though characterization of dopamine receptors in retina have been limited to date, studies indicate that a number of striking similarities do exist between the profile of dopamine receptors in retina and striatum.

A "receptor" for dopamine in retina was first identified by the demonstration of a dopamine-stimulated adenylate cyclase in both homogenate and intact preparations (Brown and Makman, 1972; Schorderet, 1977). These findings indicated that a D-1 (adenylate cyclase "linked") receptor is present in retina. Succeeding studies characterized the dopamine-sensitive adenylate cyclase in retina of a number of species and showed it to share a similar pharmacological specificity to the dopamine-stimulated adenylate cyclase in rat striatum (Watling *et al.*, 1979; Redburn *et al.*, 1980; Watling and Dowling, 1981; Dowling and Watling, 1981). Characterizations of [^{3}H]spiperone binding in bovine (Magistretti and Schorderet, 1979; Makman *et al.*, 1980 *a,b*) and goldfish retina Redburn *et al.*, (1980) indicated that [^{3}H]spiperone specific binding sites in retina are similar to those found in mammalian striatum except that retina contains fewer serotonin receptor-associated [^{3}H]spiperone sites (5-HT_2) (see Peroutka and Snyder, 1979). Attempts to show high-affinity binding of [^{3}H]domperidone to retinal membranes were initially negative (Watling *et al.*, 1979; Redburn *et al.*, 1980). Since domperidone has been shown to be impotent (IC_{50} = 0.2 mM) in inhibiting dopamine-stimulated adenylate cyclase, it was postulated that all or nearly all dopamine receptors in retina were of the D-1 type (Watling *et al.*, 1979; Magistretti and Schorderet, 1979; Redburn *et al.*, 1980). However, it is unlikely that [^{3}H]spiperone could be labeling a D-1 receptor with nanomolar affinity while showing micromolar affinity in inhibiting dopamine-stimulated cyclase activity itself. [^{3}H]Spiperone binding is thus probably associated with D-2 receptors. The failure to label retinal membrane D-2 receptors with [^{3}H]-domperidone is probably the result of higher levels of nonspecific binding and the slightly lowered affinity that this compound shows for retinal receptors (K_i = 15 nM in inhibiting [^{3}H]spiroperidol binding in bovine retinal membranes) (Creese and Sibley, 1979) when compared to the striatum.

High-affinity [^{3}H]-ADTN binding to rat and bovine retinal membranes has also been reported (Makman *et al.*, 1980*b*). The identity of these binding sites is unclear. The affinity of [^{3}H]-ADTN binding (K_d =

9 nM) did not correlate well with its ability to inhibit [^{3}H]spiperone binding (K_i = 580 nM) in unwashed bovine retinal membrane homogenates. However, the qualitative agonist and antagonist specificities for high-affinity [^{3}H]-ADTN binding and dopamine agonist-stimulated adenylate cyclase were similar (Makman *et al.*, 1980*a*), although dopamine agonists showed about 100 times greater affinity for [^{3}H]-ADTN binding than for stimulating retinal adenylate cyclase activity. Although a portion of [^{3}H]-ADTN binding sites in retinal membrane preparations may be associated with D-1 receptors, further studies (perhaps with a labeled D-1 selective antagonist) will be required. These data indicate marked similarities in the binding characteristics of dopaminergic drugs in retina and striatum. The retina should prove to be an advantageous system in which to study the role of receptors in dopamine transmission and function due to its relative accessibility to anatomical, physiological, and pharmacological study.

10. REGULATION OF DOPAMINE RECEPTORS

Rather than being static components of the plasma membrane, dopamine receptors are dynamic macromolecules that are under the influence of a large variety of factors. Dopamine receptors appear to be subject to regulatory mechanisms similar to those that modulate other neurotransmitter and hormone receptors. The pharmacological manipulation of these regulatory mechanisms may prove to be a sensitive means of therapy in the numerous psychiatric and neurological diseases in which dopamine system dysfunction has been implicated.

10.1. Up Regulation

10.1.1. Denervation

In Parkinson's disease the dopamine cells in the substantia nigra degenerate, progressively denervating the striatum. This results in the characteristic syndrome of akinesia, rigidity, and tremor. This syndrome can be mimicked in animals by lesioning the nigrostriatal pathway with 6-OHDA, a toxin selective for catecholamine neurons. As discussed earlier, supersensitive behavioral responsiveness exhibited by these lesioned rats to dopamine agonists is accompanied by an increase in the number (with no change in the affinity) of D-2 receptors in the striatum labeled by [^{3}H]-butyrophenones and [^{3}H]apomorphine (Creese *et al.*, 1977*b*). Investigations of the dopamine-stimulated adenylate cyclase are controversial, with studies reporting both increases and no change in activity after denervation

(Creese and Sibley, 1980). Seeman and colleagues investigated dopamine receptor binding in patients who died with Parkinson's disease (Lee *et al.*, 1978*a*). As might be expected, they found increases in dopamine receptor number on the order of 50% in these patients. These patients were not treated with L-DOPA prior to their death, a crucial variable to be discussed below.

10.1.2. Chronic Receptor Blockade

It would be anticipated that pharmacological blockade might produce postsynaptic supersensitivity similar to that produced by denervation, and indeed this occurs in both man and animals (Creese and Sibley, 1980). Schizophrenic patients are often treated for many years with antipsychotic medication. Tardive dyskinesia, a disabling syndrome which is characterized by abnormal repetitive movements of the face and extremities, develops in a significant proportion of these schizophrenic patients (Baldessarini and Tarsy, 1979). Klawans and colleagues (Klawans, 1973) suggested that tardive dyskinesia might be directly caused by an up regulation of dopamine receptors in the extrapyramidal motor system since tardive dyskinesia can be temporarily inhibited by increasing the dose of neuroleptic drug, and it is exacerbated by reducing the antipsychotic medication or treatment with L-DOPA. Chronic neuroleptic treatments of animals for periods as short as 1 day to as long as many months also result in behavioral changes suggestive of a dopamine receptor supersensitivity. In support of this hypothesis, receptor-binding studies utilizing both [^{3}H]antagonists and [^{3}H]agonists have demonstrated that following 1 or more weeks of neuroleptic treatment, followed by a brief withdrawal period, there is a 20 to 35% increase in striatal antagonist binding sites and a smaller increase in agonist binding sites (Creese and Sibley, 1980; Muller and Seeman, 1978). In very long-term treatment regimens of 6 months or more, more closely simulating the dosage of human schizophrenic patients, [^{3}H]butyrophenone binding sites increase by as much as 65% (Owen *et al.*, 1980). On the other hand, pituitary dopamine receptors labeled with [^{3}H]haloperidol are reported to be decreased by chronic neuroleptic treatment, while the striatal dopamine cyclase may or may not be changed (Muller and Seeman, 1978). Such findings again suggest that these measures identify distinct dopamine receptors.

It is of interest that lithium appears to exert a "stabilizing" effect on dopamine receptors, preventing not only the behavioral supersensitivity seen following chronic neuroleptics but also the increase in dopamine receptors that accompanies it (Klawans *et al.*, 1977; Pert *et al.*, 1978). These findings may have profound implications for the treatment of tardive dyskinesia as well as for that of affective disorders.

10.1.3. Dopamine Receptor Increase in Schizophrenia

Schizophrenia has been hypothesized to be associated with increased dopaminergic neurotransmission. Supporting evidence includes the worsening of schizophrenic symptoms by the administration of amphetamine (a drug which enhances dopaminergic neurotransmission) and the induction of psychosis following chronic amphetamine abuse. On the other hand, the inhibition of dopamine synthesis or its storage ameliorates schizophrenic symptoms. However, the dopamine hypothesis of schizophrenia is not supported by the measurement of CSF dopamine metabolite concentrations or serum prolactin levels in drug-free schizophrenic patients. Although raised dopamine concentrations have been reported in the caudate nucleus and nucleus accumbens from postmortem schizophrenic brains, they have not been consistently associated with increased homovanillic acid concentrations, which suggests that no increase in synaptic turnover of dopamine occurs. This lack of support for increased dopamine turnover in schizophrenia has led to the suggestion that dopamine receptor supersensitivity could lead to increased synaptic activity without increased dopamine release.

Four studies support this hypothesis reporting significant (50–200%) increases in postmortem [^{3}H]butyrophenone binding (Crow *et al.*, 1978; Lee and Seeman, 1980; Lee *et al.*, 1978*b*; Mackay *et al.*, 1980*a*) in the brains of schizophrenic patients. However, most of these schizophrenics had been previously treated with antipsychotic medication. Thus it is unclear whether the observed receptor increase is a primary cause of the disease process or simply iatrogenic, e.g., the result of chronic neuroleptic treatment. Three studies addressing this question have yielded contradictory results. A recent study (MacKay *et al.*, 1980*b*) suggests that the elevation in dopamine receptors observed in schizophrenic brain is the result of neuroleptic treatment. In this study brain tissue was obtained from patients diagnosed as psychotic. The case histories of all patients showed that previously they had received at least three years of neuroleptic treatments. However, a few of these patients had received no neuroleptic medication for a month or more before death. The density of [^{3}H]-spiroperidol binding sites was significantly raised in the psychotic group in both the caudate nucleus and nucleus accumbens. However, this increase was limited to the patients who were receiving neuroleptic medication up until death. Patients who had been free of neuroleptic drugs for at least one month before death did not show increases in [^{3}H]spiroperidol binding sites. However, such a rapid reversal of a neuroleptic-induced up regulation of dopamine receptors would be surprising, especially in light of the fact that tardive dyskinesia is often thought to be irreversible. In contrast, in the postmortem series of Crow *et al.* (1978) in which an increased B_{max} for [^{3}H]spiroperidol was observed in the caudate nucleus, five of twenty cases apparently had not received neuroleptic drugs for at least one year before death. Though still present, the increase in B_{max} was not as great in this group. This was also true for a number of "drug-free" cases reported by Lee and Seeman (1980). These findings suggest that dopamine receptor supersensitivity might be associated with schizophrenic illness rather than with the drugs used to treat it. These results are both exciting and tantalizing. The demonstration that a psychiatric illness is the result of a deficit in receptor regulation would markedly change current psychiatric concepts and have a major clinical impact. Obviously more studies of drug-naive schizophrenics are needed to settle this important question.

10.2. Down Regulation

10.2.1. Chronic Receptor Stimulation

Acute or chronic treatment with agonists might well be expected to lead to both a behavioral and receptor subsensitivity as is found in many other neurotransmitter/hormone systems. However, the chronic treatment of rats and guinea pigs with amphetamine leads to a paradoxical increase in behavioral sensitivity to subsequent amphetamine or apomorphine treatments (Segal *et al.*, 1980; W. J. Weiner *et al.*, 1979). The behavioral supersensitivity may indicate that low doses of agonists preferentially activate presynaptic autoreceptors and decrease dopamine release. The resulting decreased dopamine release would then be compensated for by a postsynaptic receptor supersensitivity. This may represent one mechanism by which amphetamine can induce psychosis in man.

Too few receptor binding studies have been conducted to allow any firm conclusions to be drawn at the present time about agonist-induced receptor changes. We have found that amphetamine treatment for 5 days (2.5 mg/kg four times per day) results in a paradoxical 20% decrease in the maximum number of [^{3}H]spiroperidol binding sites in rat striatum with a significant decrease in [^{3}H]agonist (NPA) binding as well. Similar findings have been reported by Howlett and Nahorski (1979). In contrast, Muller and Seeman (1979) have not been able to demonstrate changes in [^{3}H]haloperidol binding in response to apomorphine or amphetamine administration (10 mg/kg per day) for 14 days, although they did find a 25% decrease in [^{3}H]apomorphine binding. All of these studies are inconclusive because chronic amphetamine causes marked changes in dopamine, norepinephrine, and serotonin levels. It is thus unclear whether the apparent behavioral supersensitivity that these animals demonstrate is due solely to dopaminergic mechanisms or whether it is the result of interactions between multiple neuronal systems.

Quik and Iversen (1978) have shown that chronic treatment with the dopaminergic ergot bromocryptine decreases the maximum amount of

[^{3}H]spiroperidol binding by 25 to 50% and also decreases dopamine-sensitive adenylate cyclase activity in striatal slices. However, bromocriptine is an antagonist at the striatal dopamine-stimulated adenylate cyclase, so it is unclear why apparent desensitization should occur. Also, bromocriptine has recently been suggested to be an irreversible ligand of dopamine receptors (Bannon *et al.*, 1980*b*), so that the loss of receptors found here may not provide information about dopamine receptor regulatory mechanisms *per se*. However, Mishra *et al.* (1978) have also reported that chronic L-DOPA, as well as bromocriptine treatment, can abolish dopamine stimulation of dopamine-sensitive adenylate cyclase activity and decrease [^{3}H]haloperidol binding.

10.2.2. Reversal of Supersensitivity

In contrast to the contradictory picture concerning the induction of subsensitivity in drug-naive animals, there have been clear demonstrations of agonist-induced "down regulation" from an established supersensitive state. Friedhoff *et al.* (1977) and List and Seeman (1979) have demonstrated that the behavioral and receptor supersensitivity that develops following chronic neuroleptic treatment can be reversed by subsequent treatment of the animals with L-DOPA. They have suggested that clinical treatment with L-DOPA might have therapeutic potential in the treatment of tardive dyskinesia. That such agonist-induced down regulation may occur in man was demonstrated by the study of Lee *et al.* (1978*a*) of brains of patients with Parkinson's disease. In patients who were not receiving L-DOPA treatment prior to death, there was a significant increase in the number of dopamine receptors. However, patients who were on L-DOPA therapy demonstrated no difference from controls in their dopamine receptor numbers.

11. CONCLUDING COMMENTS

The interpretation of early dopamine receptor binding studies with striatal membranes was difficult—the data did not describe a system containing a single set of homogeneous receptors. A number of approaches have now allowed the clear division of CNS dopamine receptors into subtypes. Among the most important advances have been the examination of binding characteristics in other tissues such as anterior pituitary, the use of discrete CNS lesions to remove particular presynaptic or postsynaptic cellular elements, and the development of methods to irreversibly modify dopamine receptors. Such studies have now characterized three dopamine receptor subtypes known as D-1, D-2 and D-3, which have the characteristics shown in Table 3.

This classification is yet preliminary, as a number of questions remain. For instance, the possibility that a portion of the D-3 binding sites is merely a high-affinity agonist binding state of the D-1 receptor has not yet been rigorously excluded. Also, other classes of dopamine receptor sites, such as those having high affinity for substituted benzamides, may be further elucidated in future studies. Additionally, the degree to which D-2 receptors on cortico-striate terminals and intrinsic striatal neurons differ is as yet uncertain.

Uncertainties also exist with respect to the function of each dopamine receptor subtype. It is still not clear, for example, that D-2 receptors increase in schizophrenic patients, as has been reported, independent of the effects of chronic neuroleptic treatment. Thus, the involvement of dopamine receptors and their regulation in this disorder remains obscure. In addition, although there is a clear role for D-1 receptors in stimulating, and for some D-2 receptors in inhibiting, an adenylate cyclase, the biochemical function of D-3 receptors is unknown. The availability of selective agonists and antagonists has been central to the past therapeutic and experimental advances in the field of dopaminergic transmission. The advent of even more selective D-1, D-2, and D-3 agents may not only allow the resolution of the above questions, but also allow better pharmacological treatment of disorders involving these receptor subtypes.

12. REFERENCES

Aghajanian, G. K., and Bunney, B. S., 1973, Central dopaminergic neurons: neurophysiological identification and responses to drugs, in: *Frontiers in Catecholamine Research* (S. H. Snyder and E. Usdin, eds.), pp. 643–648, Pergamon Press, New York.

Aghajanian, G. K., and Bunney, B. S., 1977, Dopamine "autoreceptors": pharmacological characterization by microiontophoretic single cell recording studies, *Naunyn-Schmiedeberg's Arch. Pharmacol.* **297:**1–7.

Ahn, H. M., Gardner, E., and Makman, M. H., 1979, Anterior pituitary adenylate cyclase: stimulation by dopamine and other monoamines, *Eur. J. Pharmacol.* **53:**313–317.

Annunziato, L., Quattrone, A., Schettini, G., and DiRenzo, G., 1980, Supersensitivity of pituitary dopamine receptors involved in the inhibition of prolactin secretion, *Adv. Biochem. Psychopharmacol.* **24:**379–385.

Attie M. F., Brown, E. M., Gardner, D. G., Spiegel, A. M., and Aurbach, G. D., 1980, Characterization of dopamine-responsive adenylate eyclase of bovine parathyroid cells and its relationship to parathyroid hormone secretion, *Endocrinology* **107:**1776–1781.

Bacopoulos, N. B., 1981, Acute changes in the state of dopamine receptors: *in vitro* monitoring with [^{3}H]dopamine, *Life Sci.* **29:**2407–2414.

Baldessarini, R. J., and Tarsy, D., 1979, Relationship of the actions of neuroleptic drugs to the pathophysiology of tardive dyskinesia, *Int. Rev. Neurobiol.* **21:**1–45.

Baldessarini, R. J., Kula, N. S., Arana, G. W., Neumeyer, J. L., and Law, S. J., 1980, Chloroethylnorapomorphine, a proposed long-acting dopamine antagonist: interactions with dopamine receptors of mammalian forebrain *in vitro*, *Eur. J. Pharmacol.* **67:**105–110.

Bannon, M. J., Bunney, E. B., Zigun, J. R., Skirboll, L. R., and Roth, R. H., 1980*a*, Presynaptic dopamine receptors: insensitivity to kainic acid and the development of supersensitivity following chronic haloperidol, *Naunyn-Schmiedeberg's Arch. Pharmacol.* **312**:161–165.

BANNON, M. J., GRACE, A. A., BUNNEY, B. S., and ROTH, R. H., 1980*b*, Evidence for an irreversible interaction of bromocryptine with central dopamine receptors, *Naunyn-Schmiedeberg's Arch. Pharmacol.* **312**:37–42.

BAUDRY, M., MARTRES, M. P., and SCHWARTZ, J. C., 1979, [^{3}H]Domperidone: a selective ligand for dopamine receptors, *Naunyn-Schmiedeberg's Arch. Pharmacol.* **308**:231–237.

BECKSTEAD, R. M., DOMESICK, V. B., and NAUTA, W. J. H., 1979, Efferent connections of the substantia nigra and ventral tegmental area in the rat, *Brain Res.* **175**:191–217.

BELL, C., LANG, W. J., and LASKA, F., 1978, Dopamine-containing vasomotor nerves in the dog kidney, *J. Neurochem.* **31**:77–83.

BENNETT, J. P., JR., 1978, Methods in binding studies, in: *Neurotransmitter Receptor Binding* (H. I. Yamamura, S. J. Enna, and M. J. Kuhar, eds.), pp. 57–90, Raven Press, New York.

BERGER, B., TASSIN, J. P., BLANC, G., MOYNE, M. A., and THIERRY, A. M., 1974, Histochemical confirmation for dopaminergic innervation of the rat cerebral cortex after destruction of the noradrenergic ascending pathways, *Brain Res.* **81**:332–337.

BJORKLUND, A., and LINDVALL, O., 1975, Dopamine in dendrites of substantia nigra neurons: suggestions for a role in dendritic terminals, *Brain Res.* **83**:531–537.

BJORKLUND, A., LINDVALL, O., and NOBIN, A., 1975, Evidence of an incerto-hypothalamic dopamine neurone system in the rat, *Brain Res.* **89**:29–42.

BLESSING, W. W., and CHALMERS, J. P., 1979, Direct projection of catecholamine (presumably dopamine)-containing neurons from hypothalamus to spinal cord, *Neurosci. Lett.* **11**:35–40.

BOEYNAEMS, J. M., and DUMONT, J. E., 1977, The two-step model of ligand–receptor interaction, *Mol. Cell. Endocrinol.* **7**:33–47.

BOYAN-SALYERS, B. D., and CLEMENT-CORMIER, Y. C., 1980, Identification and partial purification of a hydrophobic protein component associated with [^{3}H]spiroperidol-binding activity, *Biochim. Biophys. Acta* **617**:274–281.

BROWN, E. M., CARROL, R. J., AURBACH, G. D., 1977, Dopaminergic stimulation of cyclic AMP accumulation and parathyroid hormone release from dispersed bovine parathyroid cells, *Proc. Natl. Acad. Sci. USA* **74**:4210–4213.

BROWN, E. M., ATTIE, M. F., REEN, S., GARDNER, D. G., KEBABIAN, J., and AURBACH, G. D., 1980, Characterization of dopaminergic receptors in dispersed bovine parathyroid cells, *Mol. Pharmacol.* **18**:335–340.

BROWN, J. H., and MAKMAN, M. H., 1972, Stimulation by dopamine of adenylate cyclase in retinal homogenates and of adenosine 3′-5′-cyclic monophosphate formation in intact retina, *Proc. Natl. Acad. Sci. USA* **69**:539–543.

BUNNEY, B. S., and AGHAJANIAN, G. K., 1973, Electrophysiological effects of amphetamine on dopaminergic neurons, in: *Frontiers in Catecholamine Research* (S. H. Snyder and E. Usdin, eds.), pp. 957–962, Pergamon Press, New York.

BUNNEY, B. S., WALTERS, J. R., ROTH, R. H., and AGHAJANIAN, G. K., 1973*a*, Dopaminergic neurons: effect of antipsychotic drugs and amphetamine on single cell activity, *J. Pharmacol. Exp. Ther.* **185**:560–571.

BUNNEY, B. S., AGHAJANIAN, G. F., and ROTH, R. H., 1973*b*, Comparison of effects of L-Dopa, amphetamine, and apomorphine on firing rate of rat dopaminergic neurons, *Nature New Biol.* **245**:123–125.

BURT, D. R., ENNA, S. J., CREESE, I., and SNYDER, S. H., 1975, Dopamine receptor binding in the corpus striatum of mammalian brain, *Proc. Natl. Acad. Sci. USA* **72**:4655–4659.

BURT, D. R., CREESE, I., and SNYDER, S. H., 1976, Properties of [^{3}H]haloperidol and [^{3}H]-dopamine binding associated with dopamine receptors in calf brain membranes, *Mol. Pharmacol.* **12**:800–812.

CALABRO, M. A., and MACLEOD, R. M., 1978, Binding of dopamine to bovine anterior pituitary gland membranes, *Neuroendocrinology* **25**:32–46.

CARON, M. C., BEAULIEU, M., RAYMOND, V., GAGNE, B., DROUIN, J., LEFKOWITZ, R. J., and

LABRIE, F., 1978, Dopaminergic receptors in the anterior pituitary gland, *J. Biol. Chem.* **253**:2244–2253.

CHANG, R. S. L., and SNYDER, S. H., 1980, Histamine H_1-receptor binding sites in guinea pig brain membranes: regulation of agonist interactions by guanine nucleotides and cations, *J. Neurochem.* **34**:916–922.

CHERAMY, A., LEVIEL, V., and GLOWINSKI, J., 1981, Dendritic release of dopamine in the substantia nigra, *Nature (London)* **289**:537–542.

CLEMENT-CORMIER, Y. C., and KENDRICK, P. E., 1980, Solubilization and characterization of [^{3}H]spiroperidol binding sites from subcellular fractions of the calf striatum, *Biochem. Pharmacol.* **29**:897–903.

CLEMENT-CORMIER, Y. C., KEBABIAN, J. W., PETZOLD, G. L., and GREENGARD, P., 1974, Dopamine-sensitive adenylate cyclase in mammalian brain: a possible site of action of antipsychotic drugs, *Proc. Natl. Acad. Sci. USA* **71**:1113–1117.

CLEMENT-CORMIER, Y. C., PARRISH, R. A., PETZOLD, G. L., KEBABIAN, J. W., and GREENGARD, P., 1975, Characterization of a dopamine-sensitive adenylate cyclase in the rat caudate nucleus, *J. Neurochem.* **25**:143–149.

CLEMENT-CORMIER, Y. C., HEINDEL, J. J., and ROBISON, G. A., 1977, Adenylyl cyclase from a prolactin-producing tumour cell: the effect of phenothiazines, *Life Sci.* **21**:1357–1364.

CLEMENT-CORMIER, Y. C., MEYERSON, L. R., and MCISAAC, A., 1980, Solubilization of multiple binding sites for the dopamine receptor from calf striatal membranes, *Biochem. Pharmacol.* **29**:2009–2016.

COMMISSIONG, J. W., and NEFF, N. H., 1979, Current status of dopamine in the mammalian spinal cord, *Biochem. Pharmacol.* **28**:1569–1573.

COMMISSIONG, J. W., GENTLEMAN, S., and NEFF, N. H., 1979, Spinal cord dopaminergic neurons: evidence for an uncrossed nigrospinal pathway, *Neuropharmacology* **18**:565–568.

COSTALL, B., FORTUNE, D. H., LAW, S.-J., NAYLOR, R. J., NEUMEYER, J. L., and NOHRIA, V., 1980*a*, (–)-*N*-(Chloroethyl)norapomorphine inhibits striatal dopamine function via irreversible receptor binding, *Nature (London)* **285**:571–573.

COSTALL, B., FORTUNE, D. H., GRANCHELLI, F. E., LAW, S.-J., NAYLOR, R. J., NEUMEYER, J. L., and NOHRIA, V., 1980*b*, On the ability of *N*-chloroethylapomorphine derivatives to cause irreversible inhibition of dopamine receptor mechanisms, *J. Pharm. Pharmacol.* **32**:571–576.

COTE, T., MUNEMURA, M., ESKAY, R. L., and KEBABIAN, J. W., 1980, Biochemical identification of the beta-adrenoceptor and evidence for the involvement of a cyclic AMP system in the beta-adrenergic-induced release of alpha-melanocyte stimulating hormone in the intermediate lobe of the rat pituitary gland, *Endocrinology* **107**:108–116.

COTE, T. E., GREWE, C. W., and KEBABIAN, J. W., 1981, Stimulation of the D-2 dopamine receptor in the intermediate lobe of the rat pituitary gland decreases the responsiveness of the beta-adrenoceptor: biochemical mechanism, *Endocrinology* **108**:420–426.

COYLE, J. T., and SCHWARCZ, R., 1976, Lesion of the striatal neurones with kainic acid provides a model for Huntington's chorea, *Nature (London)* **263**:244–246.

CREESE, I., and SIBLEY, D. R., 1979, Radioligand binding studies: evidence for multiple dopamine receptors, *Commun. Psychopharmacol.* **3**:385–395.

CREESE, I., and SIBLEY, D. R., 1980, Regulation of dopamine receptors, in: *Psychopharmacology and Biochemistry of Neurotransmitter Receptors* (H. I. Yamamura, R. W. Olsen, and E. Usdin, eds.), pp. 387–410, Raven Press, New York.

CREESE, I., and SNYDER, S. H., 1977, A novel, simple and sensitive radioreceptor assay for antischizophrenic drugs in blood, *Nature (London)* **270**:180–182.

CREESE, I., and SNYDER, S. H., 1978, Dopamine receptor binding of [^{3}H]-ADTN (2-amino-6,7-dihydroxy-1,2,3,4-tetrahydronaphthalene) regulated by guanyl nucleotides, *Eur. J. Pharmacol.* **50**:459–461.

CREESE, I., and SNYDER, S. H., 1979, Nigrostriatal lesions enhance striatal [^{3}H]apomorphine and [^{3}H]spiroperidol binding, *Eur. J. Pharmacol.* **56**:277–281.

Creese, I., Burt, D. R., and Snyder, S. H., 1975, Dopamine receptor binding: differentiation of agonist and antagonist states with [^{3}H]dopamine and [^{3}H]haloperidol, *Life Sci.* **17**:993–1002.

Creese, I., Burt, D. R., and Snyder, S. H., 1976, Dopamine receptor binding predicts clinical and pharmacological potencies of antischizophrenic drugs, *Science* **192**:481–483.

Creese, I., Schneider, R., and Snyder, S. H., 1977*a*, [^{3}H]Spiroperidol labels dopamine receptors in pituitary and brain, *Eur. J. Pharmacol.* **46**:377–381.

Creese, I., Burt, D. R., and Snyder, S. H., 1977*b*, Dopamine receptor binding enhancement accompanies lesion-induced behavioral supersensitivity, *Science* **197**:596–598.

Creese, I., Burt, D. R., and Snyder, S. H., 1978*a*, Biochemical actions of neuroleptic drugs: focus on the dopamine receptor, in: *Handbook of Psychopharmacology*, Vol. 10 (L. L. Iversen, S. D. Iversen, and S. H. Snyder, eds.), pp. 37–89, Plenum Press, New York.

Creese, I., Prosser, T., and Snyder, S. H., 1978*b*, Dopamine receptor binding: specificity, localization and regulation by ions and guanyl nucleotides, *Life Sci.* **23**:495–500.

Creese, I., Padgett, L., Fazzini, E., and Lopez, F., 1979*a*, [^{3}H]-*N*-propylnorapomorphine: a novel agonist ligand for central dopamine receptors, *Eur. J. Pharmacol.* **56**:411–412.

Creese, I., Usdin, T., and Snyder, S. H., 1979*b*, Guanine nucleotides distinguish between two dopamine receptors, *Nature (London)* **278**:577–578.

Creese, I., Usdin, T. B., and Snyder, S. H., 1979*c*, Dopamine receptor binding regulated by guanine nucleotides, *Mol. Pharmacol.* **16**:69–76.

Creese, I., Stewart, K., and Snyder, S. H., 1979*d*, Species variations in dopamine receptor binding, *Eur. J. Pharmacol.* **60**:55–66.

Cronin, M. J., and Weiner, R. I., 1979, [^{3}H]Spiroperidol (spiperone) binding to a putative dopamine receptor in sheep and steer pituitary and stalk median eminence, *Endocrinology* **104**:307–312.

Cronin, M. J., Roberts, J. M., and Weiner, R. I., 1978, Dopamine and dihydroergocryptine binding to the anterior pituitary and other brain areas of the rat and sheep, *Endocrinology* **103**:302–309.

Cronin, M. J., Cheung, C. Y., Beach, J. E., Faure, N., Goldsmith, P. C., and Weiner, R. I., 1980, Dopamine receptors on prolactin-secreting cells, in: *Central and Peripheral Regulation of Prolactin Function* (R. M. MacLeod and U. Scapagnini, eds.), pp. 43–58, Raven Press, New York.

Cross, A. J., and Owen, F., 1980, Characteristics of [^{3}H]-*cis*-flupenthixol binding to calf brain membranes, *Eur. J. Pharmacol.* **65**:341–347.

Crow, T. J., Owen, F., Cross, A. J., Lofthouse, R., and Longden, A., 1978, Letters to the Editor, *Lancet* **i**:36–37.

Cuello, A. C., and Iverson, L. L., 1978, Interactions of dopamine with other neurotransmitters in the rat substantia nigra: a possible functional role of dendritic dopamine, in: *Interactions between Putative Neurotransmitters in the Brain* (S. Garattini, J. F. Pujol, and R. Samanin, eds.), pp. 127–149, Raven Press, New York.

Dahlstrom, A., and Fuxe, K., 1964, Evidence for the existence of monoamine-containing neurons in the central nervous system. I. Demonstration of monoamines in the cell bodies of brainstem neurons, *Acta Physiol. Scand. Suppl. 62*, 232:1–55.

Dannies, P. S., Gautvik, K. M., and Tashjian, A. H., 1976, A possible role of cyclic AMP in mediating the effects of thyrotropin-releasing hormone on prolactin release and on prolactin and growth-hormone synthesis in pituitary cells in culture, *Endocrinology* **98**:1147–1159.

Davies, B., Abood, L., and Tometsko, A. M., 1980, Utilization of [^{3}H]dopamine as photoaffinity label of brain synaptosomes, *Life Sci.* **26**:85–88.

De Camilli, P., Macconi, D., and Sdada, A., 1979, Dopamine inhibits adenylate cyclase in human prolactin-secreting pituitary adenomas, *Nature (London)* **278**:252–254.

De Lean, A., Stadel, J. M., and Lefkowitz, R. J., 1980, A ternary complex model explains the agonist-specific binding properties of the adenylate cyclase-coupled beta-adrenergic receptor, *J. Biol. Chem.* **255:**7108–7117.

Dinerstein, R. J., Vannice, J., Henderson, R. C., Roth, L. J., Goldberg, L. I., and Hoffmann, 1979, Histofluorescence techniques provide evidence for dopamine-containing neuronal elements in canine kidney, *Science* **205:**497–499.

Dowling, J. E., and Ehinger, B., 1978*a*, Synaptic organization of the dopaminergic neurones in the rabbit retina, *J. Comp. Neurol.* **180:**203–220.

Dowling, J. E., and Ehinger, B., 1978*b*, The interplexiform cell system. I. Synapses of dopaminergic neurones of the goldfish retina, *Proc. R. Soc. London Ser. B* **201:**7–26.

Dowling, J. E., and Watling, K. J., 1981, Dopaminergic mechanisms in the Teleost retina. II. Factors affecting the accumulation of cyclic AMP in pieces of intact carp retina, *J. Neurochem.* **36:**569–579.

Ehinger, B., 1976, Biogenic amines as transmitters in the retina, in: *Transmitters in the Visual Process* (S. L. Bonting, ed.), pp. 145–163, Pergamon Press, Oxford.

Fielding, S., and Lal, H., 1978, Behavioral actions of neuroleptics, in: *Handbook of Psychopharmacology*, Vol. 10 (L. L. Iversen, S. D. Iversen, and S. H. Snyder, eds.), pp. 91–128, Plenum Press, New York.

Fields, J. Z., Reisine, T. D., and Yamamura, H. I., 1977, Biochemical demonstration of dopaminergic receptors in rat and human brain using [^{3}H]spiroperidol, *Brain Res.* **136:**578–584.

Fields, J. Z., Reisine, T. D., and Yamamura, H. I., 1979, Loss of striatal dopaminergic receptors after intrastriatal kainic acid injection, *Life Sci.* **23:**569–574.

Freedman, S. B., and Woodruff, G. N., 1981, Effect of drugs on [^{3}H]sulpiride binding in rat striatal synaptic membranes, *Br. J. Pharmacol.* **72:**129P–130P.

Friedhoff, A. J., Bonnet, K., and Rosengarten, H., 1977, Reversal of two manifestations of dopamine receptor supersensitivity by administration of L-dopa, *Chem. Pathol. Pharmacol.* **16:**411–423.

Friend, W. C., Brown, G. M., Jawahir, T. L., and Seeman, P., 1978, Effect of haloperidol and apomorphine treatment on dopamine receptors in pituitary and striatum, *Am. J. Psychiatry* **135:**839–841.

Fujita, N., Saito, K., Iwatsubo, K., Hirata, A., Noguchi, Y., and Yoshida, H., 1980, Binding of [^{3}H]apomorphine to striatal membranes prepared from rat brain after 6-hydroxydopamine and kainic acid lesions, *Brain Res.* **190:**593–596.

Furchgott, R. F., 1978, Pharmacological characterization of receptors: its relation to radioligand-binding studies, *Fed. Proc.* **37:**115–120.

Fuxe, K., Hall, H., and Kohler, C., 1979, Evidence for an exclusive localization of ^{3}H-ADTN binding sites to postsynaptic nerve cells in the striatum of the rat, *Eur. J. Pharmacol.* **58:**515–517.

Gale, K., Guidotti, A., and Costa, E., 1977, Dopamine-sensitive adenylate cyclase: location in substantia nigra, *Science* **195:**503–505.

Garau, L., Govoni, S., Stefanini, E., Trabucchi, M., and Spano, P. F., 1978, Dopamine receptors: pharmacological and anatomical evidences indicate that two distinct dopamine receptor populations are present in rat striatum, *Life Sci.* **23:**1745–1750.

Geffen, L. B., Jessel, T. M., Cuello, A. C., and Iversen, L. L., 1976, Release of dopamine from dendrites in rat substantia nigra, *Nature (London)* **260:**258–260.

Giannattasio, G., DeFerrari, M. E., and Spada, A., 1981, Dopamine-inhibited adenylate cyclase in female rat adenohypophysis, *Life Sci.* **28:**1605–1612.

Glossmann, H., and Hornung, R., 1980, Alpha-adrenoreceptors in rat brain: sodium changes the affinity of agonists for prazosin sites, *Eur. J. Pharmacol.* **61:**407–408.

Goldsmith, P. C., Cronin, M. J., and Weiner, R. I., 1979, Dopamine receptor sites in the anterior pituitary, *J. Histochem. Cytochem.* **27:**1205–1207.

Gorissen, H., and Laduron, P., 1979, Solubilisation of high-affinity dopamine receptors, *Nature (London)* **279:**72–74.

GORISSEN, H., ILIEN, B., AERTS, G., and LADURON, P., 1980, Differentiation of solubilized dopamine receptors from spirodecanone binding sites in rat striatum, *FEBS Lett.* **121**:133–138.
GOVONI, S., OLGIATI, V. R., TRABUCCHI, M., GARAU, L., STEFANINI, E., and SPANO, P. F., 1978, [^{3}H]Haloperidol and [^{3}H]spiroperidol receptor binding after striatal injection of kainic acid, *Neurosci. Lett.* **8**:207–210.
GREENGARD, P., 1976, Possible role for cyclic nucleotides and phosphorylated membrane proteins in postsynaptic actions of neurotransmitters, *Nature (London)* **260**:101–108.
GROVES, P. M., WILSON, C. J., YOUNG, S. J., and REBEC, G. V., 1975, Self-inhibition by dopaminergic neurons, *Science* **190**:522–528.
HAMBLIN, M., and CREESE, I., 1980, Phenoxybenzamine discriminates multiple dopamine receptors, *Eur. J. Pharmacol.* **65**:119–121.
HAMBLIN, M. W., and CREESE, I., 1982*a*, Phenoxybenzamine treatment differentiates dopaminergic [^{3}H]ligand binding sites in bovine caudate membranes, *Mol. Pharmacol.* **21**:41–51.
HAMBLIN, M. W., and CREESE, I., 1982*b*, Heat treatment mimics guanosine-5-triphosphate effects on dopaminergic [^{3}H]ligand binding to bovine caudate membranes, *Mol. Pharmacol.* **21**:52–56.
HAMBLIN, M. W., and CREESE, I., 1982*c*, [^{3}H]Dopamine binding to rat striatal D-2 and D-3 sites: enhancement by magnesium and inhibition by sodium, *Life Sci.* **30**:1587–1595.
HENN, F. A., ANDERSON, D. J., and SELLSTROM, A., 1977, Possible relationship between glial cells, dopamine and the effects of antipsychotic drugs, *Nature (London)* **266**:637–638.
HENN, F. A., TITELER, M., ANDERSON, D. J., and MAY, K., 1978, Investigations concerning the cellular origin of dopamine receptors, *Life Sci.* **23**:617–622.
HOFFMAN, B. B., and LEFKOWITZ, R. J., 1980, Radioligand binding studies of adrenergic receptors: new insights into molecular and physiological regulation, *Annu. Rev. Pharmacol. Toxicol.* **20**:581–608.
HÖKFELT, T., HALASZ, N., LJUNGDAHL, A., JOHANSSON, O., GOLDSTEIN, M., and PARK, D., 1975, Histochemical support for a dopaminergic mechanism in the dendrites of certain periglomerular cells in the rat olfactory bulb, *Neurosci. Lett.* **1**:85–90.
HORN, A. S., and SNYDER, S. H., 1971, Chlorpromazine and dopamine: conformational similarities that correlate with the antischizophrenic activity of phenothiazine drugs, *Proc. Natl. Acad. Sci. USA* **68**:2325–2328.
HORNYKIEWICZ, O., 1966, Dopamine and brain function, *Pharmacol. Res.* **18**:925–964.
HOWLETT, A. C., VAN ARSDALE, P. M., and GILMAN, A. G., 1978, Efficiency of coupling between the beta-adrenergic receptor and adenylate cyclase, *Mol. Pharmacol.* **14**:531–539.
HOWLETT, D. R., and NAHORSKI, S. R., 1978, A comparative study of [^{3}H]haloperidol and [^{3}H]spiroperidol binding to receptors on rat cerebral membranes, *FEBS Lett.* **87**:152–156.
HOWLETT, D. R., and NAHORSKI, S. R., 1979, Acute and chronic amphetamine treatments modulate striatal dopamine receptor binding sites, *Brain Res.* **161**:173–178.
HYTTEL, J., 1978*a*, A comparison of the effect of neuroleptic drugs on the binding of [^{3}H]-haloperidol and [^{3}H]-*cis*(*Z*)-flupenthixol and on adenylate cyclase activity in rat striatal tissue *in vitro, Prog. Neuro-Psychopharmacol.* **2**:329–335.
HYTTEL, J., 1978*b*, Effects of neuroleptics on [^{3}H]haloperidol and [^{3}H]-*cis*(*Z*)-flupenthixol binding and on adenylate cyclase activity *in vitro, Life Sci.* **23**:551–556.
HYTTEL, J., 1980, Further evidence that [^{3}H]-*cis*(*Z*)flupenthixol binds to the adenylate cyclase-associated dopamine receptor (D-1) in rat corpus striatum, *Psychopharmacology* **67**:107–109.
HYTTEL, J., 1981, Similarities between the binding of [^{3}H]piflutixol and [^{3}H]flupentixol to rat striatal dopamine receptors *in vitro, Life Sci.* **28**:563–569.
IVERSEN, L. L., 1975, Dopamine receptors in the brain, *Science* **188**:1084–1089.
IVERSEN, L. L., ROGAWSKI, M. A., and MILLER, R. J., 1976, Comparison of the effects of neuroleptic drugs on pre- and postsynaptic dopaminergic mechanisms in the rat striatum, *Mol. Pharmacol.* **12**:251–262.
IVERSEN, S. D., 1977, Brain dopamine systems and behavior, in: *Handbook of Psychopharmacology*,

Vol. 8 (L. L. Iversen, S. D. Iversen, and S. H. Snyder, eds.), pp. 333–384, Plenum Press, New York.

JACOBS, S., and CUATRECASAS, P., 1976, The mobile receptor hypothesis and "cooperativity" of hormone binding, *Biochim. Biophys. Acta* **433**:482–495.

JANSSEN, P. A. J., and VANBEVER, W. F. M., 1978, Structure–activity relationships of the butyrophenones and biphenylbutylpiperidines, in: *Handbook of Psychopharmacology*, Vol. 10 (L. L. Iversen, S. D. Iversen, and S. H. Snyder, eds.), pp. 1–36, Plenum Press, New York.

JENNER, P., and MARSDEN, C. D., 1979, The substituted benzamides—a novel class of dopamine antagonists, *Life Sci.* **25**:479–486.

KEBABIAN, J. W., and SAAVEDRA, J. M., 1976, Dopamine-sensitive adenylate cyclase occurs in a region of substantia nigra containing dopaminergic dendrites, *Science* **193**:683–685.

KEBABIAN, J. W., and CALNE, D. B., 1979, Multiple receptors for dopamine, *Nature (London)* **277**:93–96.

KEBABIAN, J. W., PETZOLD, G. L., and GREENGARD, P., 1972, Dopamine-sensitive adenylate cyclase in caudate nucleus of rat brain and its similarity to the "dopamine receptor," *Proc. Natl. Acad. Sci. USA* **79**:2145–2149.

KENT, R. S., DE LEAN, A., and LEFKOWITZ, R. J., 1980, A quantitative analysis of beta-adrenergic receptor interactions: resolution of high- and low-affinity states of the receptor by computer modeling of ligand binding data, *Mol. Pharmacol.* **17**:14–23.

KLAWANS, H. L., 1973, The pharmacology of tardive dyskinesias, *Am. J. Psychiatry* **130**:82–86.

KLAWANS, H. L., WEINER, W. J., and NAUSIEDA, P. A., 1977, The effect of lithium on an animal model of tardive dyskinesia, *Prog. Neuro-Psychopharmacol.* **1**:53–60.

KOMISKEY, H. L., BOSSART, J. F., MILLER, D. D., and PATIL, P. N., 1978, Conformation of dopamine at the dopamine receptor, *Proc. Natl. Acad. Sci. USA* **75**:2641–2643.

KRAMER, S. G., 1976, Dopamine in retinal neurotransmission, in: *Transmitters in the Visual Process* (S. L. Bontig, ed.), pp. 165–198, Pergamon Press, Oxford.

LABRIE, F., BEAULIEU, M., FERLAND, L., RAYMOND, V., DIPAOLO, T., CARON, M. G., VEILLEUX, R., DENIZEAU, F., EUVRAD, C., RAYNAUD, J. P., and BOISSIER, J. R., 1979, Control of prolactin secretion at the pituitary level: a model for postsynaptic dopaminergic systems, in: *Central Nervous System Effects of Hypothalamic Hormones and Other Peptides* (R. Collu, A. Barbeau, and J. Rochefort, eds.), pp. 207–234, Raven Press, New York.

LABRIE, F., FERLAND, L., DIPAOLO, T., and VEILLEUX, R., 1980, Modulation of prolactin secretion by sex steroids and thyroid hormones, in: *Central and Peripheral Regulation of Prolactin Function* (R. M. MacLeod and U. Scapagnini, eds.), pp. 97–113, Raven Press, New York.

LADURON, P. M., and LEYSEN, J. E., 1979, Domperidone, a specific *in vitro* dopamine antagonist, devoid of *in vivo* central dopaminergic activity, *Biochem. Pharmacol.* **28**:2161–2165.

LAL, H., BROWN, W., DRAWBAUGH, R., HYNES, M., and BROWN, G., 1977, Enhanced prolactin inhibition following chronic treatment with haloperidol and morphine, *Life Sci.* **20**:101–106.

LEE, T., and SEEMAN, P., 1980, Elevation of brain neuroleptic/dopamine receptors in schizophrenia, *Am. J. Psychiatry* **137**:191–197.

LEE, T., SEEMAN, P., RAJPUT, A., FARLEY, I., and HORNYKIEWICZ, O., 1978*a*, Receptor basis for dopaminergic supersensitivity in Parkinson's disease, *Nature (London)* **273**:59–61.

LEE, T., SEEMAN, P., TOURTELLOTTE, W., FARLEY, W. W., and HORNYKIEWICZ, O., 1978*b*, Binding of [^{3}H]neuroleptics and [^{3}H]apomorphine in schizophrenic brains, *Nature (London)* **274**:897–900.

LEFF, S., and CREESE, I., 1982, Solubilization of a Quanine nucleotide sensitive form of the D-2 dopamine receptor from brain requires agonist occupancy, *Fed. Proc.* **41**:1633.

LEFF, S., and CREESE, I., 1982, Acute reserpine mimics the effects of nigrostriatal 6-hydroxydopamine lesions on "D-3" specific [^{3}H]dopamine binding in rat striatum, *Soc. Neurosci. Abst.* **8**:654.

LEFF, S., ADAMS, L., HYTTEL, J., and CREESE, I., 1981, Kainate lesion dissociates striatal dopamine receptor radioligand binding sites, *Eur. J. Pharmacol.* **70**:71–75.

LEFKOWITZ, R. J., 1980, Modification of adenylate cyclase activity by alpha- and beta-adrenergic receptors: insights from radioligand binding studies, in: *Psychopharmacology and Biochemistry of Neurotransmitter Receptors* (H. I. Yamamura, R. W. Olsen, and E. Usdin, eds.), pp. 155–170, Elsevier Press, New York.

LEW, J. Y., and GOLDSTEIN, M., 1979, Dopamine receptor binding for agonists and antagonists in thermal exposed membranes, *Eur. J. Pharmacol.* **55**:429–430.

LEYSEN, J. E., 1979, Unitary dopaminergic receptor composed of cooperatively linked agonist and antagonist sub-unit binding sites, *Commun. Psychopharmacol.* **3**:397–410.

LEYSEN, J. E., GOMMEREN, W., and LADURON, P. M., 1978*a*, Spiperone: a ligand of choice for neuroleptic receptors. 1. Kinetics and characteristics of *in vitro* binding, *Biochem. Pharmacol.* **27**:307–316.

LEYSEN, J. E., NIEMEGEERS, C. J. E., TOLLENAERE, J. P., and LADURON, P. M., 1978*b*, Serotonergic component of neuroleptic receptors, *Nature (London)* **272**:168–171.

LEYSEN, J. E., AWOUTERS, F., KENNIS, L., LADURON, P. M., VANDENBERK, J., and JANSSEN, P. A. J., 1981, Receptor binding profile of R 41 468, a novel antagonist at 5-HT_2 receptors, *Life Sci.* **28**:1015–1022.

LIBET, B., 1976, The SIF cell as a functional dopamine-releasing interneuron in the rabbit superior cervical ganglion, in: *SIF Cells. Structure and Function of the Small, Intensely Fluorescent Sympathetic Cells*, Fogarty Int. Ctr. Proc. 30, pp. 163–177, DHEW-NIH 76-942.

LIMBIRD, L. E., GILL, D. M., and LEFKOWITZ, R. J., 1980*a*, Agonist-promoted coupling of the beta-adrenergic receptor with the guanine nucleotide regulatory protein of the adenylate cyclase·system, *Proc. Natl. Acad. Sci. USA* **77**:775–779.

LIMBIRD, L. E., GILL, D. M., STADEL, J. M., HICKEY, A. R., and LEFKOWITZ, R. J., 1980*b*, Loss of beta-adrenergic receptor–guanine nucleotide regulatory protein interactions accompanies decline in catecholamine responsiveness of adenylate cyclase in maturing rat erythrocytes, *J. Biol. Chem.* **255**:1854–1861.

LINDVALL, O., and BJORKLUND, A., 1974, The glyoxylic acid fluorescence histochemical method: a detailed account of the methodology for the visualization of central catecholamine neurons, *Histochemistry* **39**:97–127.

LINDVALL, O., and BJORKLUND, A., 1977, Organization of catecholamine neurons in the rat central nervous system, in: *Handbook of Psychopharmacology* Vol. 9 (L. Iversen, S. Iversen, and S. H. Snyder, eds.), pp. 139–231, Plenum Press, New York.

LINDVALL, O., BJORKLUND, A., and DIVAC, I., 1978, Organization of catecholamine neurons projecting to the frontal cortex in the rat, *Brain Res.* **142**:1–24.

LIST, S. J., and SEEMAN, P., 1979, Dopamine agonists reverse the elevated [^{3}H]neuroleptic binding in neuroleptic-pretreated rats, *Life Sci.* **24**:1447–1452.

LIST, S., TITELER, M., and SEEMAN, P., 1980, High-affinity [^{3}H]dopamine receptors (D_3 sites) in human and rat brain, *Biochem. Pharmacol.* **29**:1621–1622.

MACKAY, A. V. P., BIRD, E. D., IVERSEN, L. L., SPOKES, E. G., CREESE, I., and SNYDER, S. H., 1980*a*, Dopaminergic abnormalities in postmortem schizophrenic brain, in: *Long-Term Effects of Neuroleptics* (F. Cattabeni, G. Racagni, P. F. Spano, and E. Costa, eds.), pp. 325–333, Raven Press, New York.

MACKAY, A. V. P., BIRD, E. D., SPOKES, E. G., ROSSOR, M., IVERSEN, L. L., CREESE, I., and SNYDER, S. H., 1980*b*, Dopamine receptors and schizophrenia: drug effect or illness? *Lancet* **ii**:915–916.

MACLEOD, R. M., NAGY, I., LOGIN, I. S., KIMURA, H., VALDENEGRO, C. A., and THORNER, M. D., 1980, The role of dopamine, cAMP, and calcium in prolactin secretion, in: *Central and Peripheral Regulation of Prolactin Function* (R. M. MacLeod and U. Scapagnini, eds.), pp. 27–41, Raven Press, New York.

MADRAS, B. K., DAVIS, A., KUNASHKO, P., and SEEMAN, P., 1980, Solubilization of dopamine receptors from dog and human brains, in: *Psychopharmacology and Biochemistry of Neurotransmitter Receptors* (H. I. Yamamura, R. W. Olsen, and E. Usdin, eds.), pp. 411–419, Elsevier/North-Holland, New York.

MAGISTRETTI, P. J., and SCHORDERET, M., 1979, Dopamine receptors in bovine retina: characterization of the [^{3}H]spiroperidol binding and its use for screening dopamine receptor affinity of drugs, *Life Sci.* **25:**1675–1686.

MAKMAN, M. H., DVORKIN, B., HOROWITZ, S. G., and THAL, L. J., 1980*a*, Properties of dopamine agonist and antagonist binding sites in mammalian retina, *Brain Res.* **194:**403–418.

MAKMAN, M. H., DVORKIN, B., HOROWITZ, S. G., and THAL, L. J., 1980*b*, Retina contains guanine nucleotide sensitive and insensitive classes of dopamine receptors, *Eur. J. Pharmacol.* **63:**217–222.

MARCHAIS, D., and BOCKAERT, J., 1980, Is there a connection between high-affinity [^{3}H]-spiperone binding sites and DA-sensitive adenylate cyclase in corpus striatum? *Biochem. Pharmacol.* **29:**1331–1336.

MARTRES, M.-P., SOKOLOFF, P., and SCHWARTZ, J. C., 1980, Three classes of dopaminergic receptors evidenced by two radioligands: [^{3}H]apomorphine and [^{3}H]domperidone, in: *Psychopharmacology and Biochemistry of Neurotransmitter Receptors* (H. I. Yamamura, R. W. Olsen, and E. Usdin, eds.), pp. 421–434, Elsevier/North-Holland, New York.

MCGEER, E. G., and MCGEER, P. L., 1976, Duplication of biochemical changes of Huntington's chorea by intrastriatal injection of glutamic and kainic acid, *Nature (London)* **263:**517–519.

MEMO, M., SPANO, P. F., and TRABUCCHI, M., 1981, Characterization of dopamine-D_2 central receptors, *Br. J. Pharmacol.* **72:**124P–125P.

MILLER, R. J., and MCDERMED, J., 1979, Dopamine-sensitive adenylate cyclase, in: *The Neürobiology of Dopamine* (A. S. Horn, J. Korf, and B. H. C. Westerink, eds.), pp. 159–177, Academic Press, New York.

MILLER, R. J., HORN, A. S., and IVERSEN, L. L., 1974, The action of neuroleptic drugs on dopamine-stimulated adenosine cyclic 3′,5′-monophosphate production in rat neostriatum and limbic forebrain, *Mol. Pharmacol.* **10:**759–766.

MISHRA, R. K., WONG, Y-W., VARMUZA, S. L., and TUFF, L., 1978, Chemical lesion and drug-induced supersensitivity and subsensitivity of caudate dopamine receptors, *Life Sci.* **23:**443–446.

MOORE, R. Y., and BLOOM, F. E., 1978, Central catecholamine neuron systems: anatomy and physiology of the dopamine systems, *Annu. Rev. Neurosci.* **1:**129–169.

MOOS, F., and RICHARD, P., 1979, Effects of dopaminergic antagonist and agonist on oxytocin release induced by various stimuli, *Neuroendocrinology* **28:**138–144.

MOWLES, T. F., BURGHARDT, B., BURGHARDT, C., CHARNEKI, A., and SHEPPARD, H., 1978, The dopamine receptor of the rat mammotroph in cell culture as a model for drug action, *Life Sci.* **22:**2103–2108.

MULLER, P., and SEEMAN, P., 1978, Dopaminergic supersensitivity after neuroleptics: time course and specificity, *Psychopharmacology* **60:**1–11.

MULLER, P., and SEEMAN, P., 1979, Presynaptic subsensitivity as a possible basis for sensitization by long-term dopamine mimetics, *Eur. J. Pharmacol.* **55:**149–157.

MUNEMURA, M., COTE, T. E., TSURUTA, K., ESKAY, R. L., and KEBABIAN, J. W., 1980*a*, The dopamine receptor in the intermediate lobe of the rat pituitary: pharmacological characterization, *Endocrinology* **107:**1683–1686.

MUNEMURA, M., ESKAY, R. L., and KEBABIAN, J. W., 1980*b*, Release of alpha-melanocyte-stimulating hormone from dispersed cells of the intermediate lobe of the rat pituitary gland: involvement of catecholamines and adenosine 3′,5′-monophosphate, *Endocrinology* **106:**1795–1803.

MUNSON, P. J., and RODBARD, D., 1980, LIGAND: a versatile computerized approach for characterization of ligand-binding systems, *Anal. Biochem.* **107:**220–239.

MURRIN, L. C., GALE, K., and KUHAR, M. J., 1979, Autoradiographic localization of neuroleptic and dopamine receptors in the caudate-putamen and substantia nigra: effects of lesions, *Eur. J. Pharmacol.* **60:**229–235.

NAGY, J. I., LEE, T., SEEMAN, P., and FIBIGER, H. C., 1978, Direct evidence for presynaptic and postsynaptic dopamine receptors in brain, *Nature (London)* **274:**278–281.

NAOR, Z., SNYDER, G., FAWCETT, C. P., and MCCANN, S. M., 1980, Pituitary cyclic nucleotides and thyrotropin-releasing hormone action: the relationship of adenosine 3′,5′-monophosphate and guanosine 3′,5′-monophosphate to the release of thyrotropin and prolactin, *Endocrinology* **106:**1304–1310.

NEUMEYER, J. L., LAW, S. J., BALDESSARINI, R. J., and KULA, N. S., 1980, (−)-*N*-(2-Chloroethyl)-10,11-dihydroxynoraporphine (Chloroethylnorapomorphine), a novel irreversible dopamine receptor antagonist, *J. Med. Chem.* **23:**595–599.

NIEOULLON, A., CHERAMY, A., and GLOWINSKI, J., 1977, Release of dopamine *in vivo* from cat substantia nigra, *Nature (London)* **266:**375–377.

NISHIKORI, K., OSAMU, N., SANO, K., and MAENO, H., 1980, Characterization, solubilization, and separation of two distinct dopamine receptors in canine caudate nucleus, *J. Biol. Chem.* **255:**10909–10915.

OGREN, S. O., HALL, H., and KOHLER, C., 1978, Studies on the stereoselective dopamine receptor blockade in the rat brain by rigid spiro amines, *Life Sci.* **23:**1769–1774.

OWEN, F., CROSS, A. J., WADDINGTON, J. L., POULTER, M., GAMBLE, S. J., and CROW, T. J., 1980, Dopamine-mediated behaviour and [^{3}H]spiperone binding to striatal membranes in rats after nine months haloperidol administration, *Life Sci.* **26:**55–59.

PARDO, J. S., CREESE, I., BURT, D. R., and SNYDER, S. H., 1977, Ontogenesis of dopamine receptor binding in the corpus striatum of the rat, *Brain Res.* **125:**376–382.

PAWLIKOWSKI, M., KARASEK, E., KUNERT-RADEK, J., and STEPIEN, H., 1979, Dopamine blockade of the thyroliberin-induced cyclic AMP accumulation in rat anterior pituitary, *J. Neural Transm.* **45:**75–79.

PAWLIKOWSKI, M., KARASEK, E., KUNERT-RADEK, J., and JARANOWSKA, M., 1981, Effects of dopamine on cyclic AMP concentration in the anterior pituitary gland *in vitro, J. Neural Transm.* **50:**179–184.

PEROUTKA, S. J., and SNYDER, S. H., 1979, Multiple serotonin receptors: differential binding of [^{3}H]5-hydroxytryptamine, [^{3}H]lysergic acid diethylamide, and [^{3}H]spiroperidol, *Mol. Pharmacol.* **16:**687–699.

PERT, A., ROSENBLATT, J., SWIT, C., PERT, C., and BUNNEY, W. E., 1978, Long-term treatment with lithium prevents the development of dopamine receptor supersensitivity, *Science* **201:**171–173.

PERT, C. B., PASTERNAK, G., and SNYDER, S. H., 1973, Opiate agonists and antagonists discriminated by receptor binding in brain, *Science* **182:**1359–1361.

PHILLIPSON, O. T., and HORN, A. S., 1976, Substantia nigra of the rat contains a dopamine-sensitive adenylate cyclase, *Nature (London)* **261:**418–420.

PHILLIPSON, O. T., EMSON, P. C., HORN, A. S., and JESSELL, T., 1977, Evidence concerning the anatomical location of the dopamine-stimulated adenylate cyclase in the substantia nigra, *Brain Res.* **136:**45–58.

PICKEL, V. M., JOH, T. H., FIELD, P. M., BECKER, C. G., and REIS, D. J., 1975, Cellular localization of tyrosine hydroxylase by immunohistochemistry, *J. Histochem. Cytochem.* **23:**1–12.

PIKE, L. J., and LEFKOWITZ, R. J., 1980, Activation and desensitization of beta-adrenergic receptor-coupled GTPase and adenylate cyclase of frog and turkey erythrocyte membranes, *J. Biol. Chem.* **255:**6860–6867.

PREMONT, J., THIERRY, A. M., TAASSIN, J. P., GLOWINSKI, J. G., and BOCKAERT, J., 1976, Is the dopamine-sensitive adenylate cyclase in the rat substantia nigra coupled with autoreceptors? *FEBS Lett.* **68:**99–104.

QUIK, M., and IVERSEN, L. L., 1978, Subsensitivity of the rat striatal dopaminergic system after treatment with bromocriptine: effects of [^{3}H]spiperone binding and dopamine-stimulated cyclic AMP formation, *Naunyn-Schmiedeberg's Arch. Pharmacol.* **304:**141–145.

QUIK, M., and IVERSEN, L. L., 1979, Regional study of [^{3}H]spiperone binding and the dopamine-sensitive adenylate cyclase in rat brain, *Eur. J. Pharmacol.* **56:**323–330.

QUIK, M., IVERSEN, L. L., LARDER, A., and MACKAY, A. V. P., 1978, Use of ADTN to define specific [^{3}H]spiperone binding to receptors in brain, *Nature (London)* **274:**513–514.

QUIK, M., EMSON, P. C., and JOYCE, E., 1979, Dissociation between the presynaptic dopamine-sensitive adenylate cyclase and [^{3}H]spiperone binding sites in rat substantia nigra, *Brain Res.* **167:**355–375.

Ray, K. P., and Wallis, M., 1980, Is cyclic adenosine 3′5′-monophosphate involved in the dopamine-mediated inhibition of prolactin secretion? *J. Endocrinol.* **85:**59p.

Redburn, D. A., Clement-Cormier, Y., and Lam, D. M. K., 1980, Dopamine receptors in the goldfish retina: [^{3}H]spiroperidol and [^{3}H]domperidone binding; and dopamine-stimulated adenylate cyclase activity, *Life Sci.* **27:**23–31.

Reisine, T. D., Nagy, J. I., Fibiger, H. C., and Yamamura, H. I., 1979, Localization of dopamine receptors in rat brain, *Brain Res.* **169:**209–214.

Reubi, J.-C., Iversen, L. L., and Jessell, T. M., 1977, Dopamine selectively increases [^{3}H]-GABA release from slices of rat substantia nigra *in vitro, Nature (London)* **268:**652–654.

Rodbell, M., 1980, The role of hormone receptors and GTP-regulatory proteins in membrane transduction, *Nature (London)* **284:**17–22.

Rosenblatt, J. E., Del Carmen, R., and Wyatt, R., 1980, A high-affinity GTP binding site in rat brain, *Eur. J. Pharmacol.* **64:**365–366.

Ross, E. M., and Gilman, A. G., 1980, Biochemical properties of hormone-sensitive adenylate cyclase, *Annu. Rev. Biochem.* **49:**533–564.

Ross, E. M., Maguire, M. E., Sturgill, T. W., Biltonen, R. L., and Gilman, A. G., 1977, Relationship between the beta-adrenergic receptor and adenylate cyclase. Studies of ligand binding and enzyme activity in purified membranes of S49 lymphoma cells, *J. Biol. Chem.* **252:**5761–5775.

Roth, R. H., 1979, Dopamine autoreceptors: pharmacology, function, and comparison with postsynaptic dopamine receptors, *Commun. Psychopharmacol.* **3:**429–445.

Sarthy, P. J., and Lam, D. M. K., 1979, The uptake and release of [^{3}H]dopamine in the goldfish retina, *J. Neurochem.* **32:**1269–1277.

Schachter, M., Bedard, P., Debono, A. G., Jenner, P., Marsden, C. D., Price, P., Parkes, J. D., Keenan, J., Smith, B., Rosenthaler, J., Horowski, R., and Dorow, R., 1980, The role of D-1 and D-2 receptors, *Nature (London)* **286:**157–159.

Schmidt, M. J., 1979, Perspectives on dopamine-sensitive adenylate cyclase in the brain, in: *Neuropharmacology of Cyclic Nucleotides* (G. C. Palmer, ed.), pp. 1–52, Urban & Schwartzenberg, Baltimore.

Schmidt, M. J., and Hill, L. E., 1977, Effects of ergots on adenylate cyclase activity in the corpus striatum and pituitary, *Life Sci.* **20:**789–798.

Schorderet, M., 1977, Pharmacological characterization of the dopamine-mediated accumulation of cyclic AMP in intact retina of rabbit, *Life Sci.* **20:**1741–1748.

Schwarcz, R., and Coyle, J. T., 1977, Neurochemical sequelae of kainate injections in corpus striatum and substantia nigra of the rat, *Life Sci.* **20:**431–436.

Schwarcz, R., Creese, I., Coyle, J. T., and Snyder, S. H., 1978, Dopamine receptors localised on cerebral cortical afferents to rat corpus striatum, *Nature (London)* **271:**766–768.

Seeman, P., Chau-Wong, M., Tedesco, J., and Wong, K., 1975, Brain receptors for antipsychotic drugs and dopamine: direct binding assays, *Proc. Natl. Acad. Sci. USA* **72:**4376–4380.

Seeman, P., Lee, T., Chau-Wong, M., Tedesco, J., and Wong, K., 1976*a*, Dopamine receptors in human and calf brains, using [^{3}H]apomorphine and an antipsychotic drug, *Proc. Natl. Acad. Sci. USA* **73:**4354–4358.

Seeman, P., Lee, T., Chau-Wong, M., and Wong, K., 1976*b*, Antipsychotic drug doses and neuroleptic/dopamine receptors, *Nature (London)* **261:**717–719.

Seeman, P., Tedesco, J. L., Lee, M., Chau-Wong, M., Muller, P., Bowles, J., Whitaker, P. M., McManus, C., Tittler, M., Weinreich, P., Friend, W. C., and Brown, G. M., 1978, Dopamine receptors in the central nervous system, *Fed. Proc.* **37:**130–136.

Seeman, P., Woodruff, G. N., and Poat, J. A., 1979, Similar binding of [^{3}H]-ADTN and [^{3}H]apomorphine to calf brain dopamine receptors, *Eur. J. Pharmacol.* **55:**137–142.

Segal, D. S., Weinberger, S. B., Cahill, J., and McCunney, S. J., 1980, Multiple daily amphetamine administration: behavioral and neurochemical alterations, *Science* **207**:904–907.

SETTLER, P. E., SARAU, H. M., ZIRCLE, C. L., SAUNDERS, H. L., 1978, The central effects of a novel dopamine agonist, *Eur. J. Pharmacol.* **50**:419–430.

SIBLEY, D. R., and CREESE, I., 1979, Multiple pituitary dopamine receptors: effects of guanine nucleotides, *Soc. Neurosci. Abstr.* **5**:352.

SIBLEY, D. R., and CREESE, I., 1980*a*, Anterior pituitary dopamine receptors: heterogeneity of agonist binding, *Fed. Proc.* **39**:1098.

SIBLEY, D. R., and CREESE, I., 1980*b*, Dopamine receptor binding in bovine intermediate lobe pituitary membranes, *Endocrinology* **107**:1405–1409.

SNYDER, S. H., CREESE, I., and BURT, D. R., 1975, The brain's dopamine receptor: labeling with [^{3}H]dopamine and [^{3}H]haloperidol, *Psychopharmacol. Commun.* **1**:663–673.

SOKOLOFF, P., MARTRES, M.-P., and SCHWARTZ, J.-C., 1980*a*, [^{3}H]Apomorphine labels both dopamine postsynaptic receptors and autoreceptors, *Nature (London)* **288**:283–286.

SOKOLOFF, P., MARTRES, M.-P., and SCHWARTZ, J.-C., 1980*b*, Three classes of dopamine receptor (D-2, D-3, D-4) identified by binding studies with [^{3}H]apomorphine and [^{3}H] domperidone, *Naunyn-Schmiedeberg's Arch. Pharmacol.* **315**:89–102.

SPANO, P. F., DICHIARA, G., TONON, G. C., and TRABUCCHI, M., 1976, A dopamine-stimulated adenylate cyclase in rat substantia nigra, *J. Neurochem.* **27**:1565–1568.

SPANO, P. F., TRABUCCHI, M., and DICHIARA, G., 1977, Localization of nigral dopamine-sensitive adenylate cyclase on neurons originating from the corpus striatum, *Science* **196**:1343–1345.

STEFANINI, E., DEJOTO, P., MARCHISIO, A., VERNALEONE, F., and COLLU, R., 1980, [^{3}H] Spiroperidol binding to a putative dopaminergic receptor in rat pituitary gland, *Life Sci.* **26**:583–587.

SUEN, E. T., STEFANINI, E., and CLEMENT-CORMIER, Y. C., 1980, Evidence for essential thiol groups and disulfide bonds in agonist and antagonist binding to the dopamine receptor, *Biochem. Biophys. Res. Commun.* **96**:953–960.

THAL, L., CREESE, I., and SNYDER, S. H., 1978, [^{3}H]Apomorphine interactions with dopamine receptors in calf brain, *Eur. J. Pharmacol.* **49**:295–299.

THEODOROU, A., CROCKETT, M., JENNER, P., and MARSDEN, C. D., 1979, Specific binding of [^{3}H]sulpiride to rat striatal preparations, *J. Pharm. Pharmacol.* **31**:424–426.

THEODOROU, A. E., HALL, M. D., JENNER, P., and MARSDEN, C. D., 1980, Cation regulation differentiates specific binding of [^{3}H]sulpiride and [^{3}H]spiperone to rat striatal preparations, *J. Pharm. Pharmacol.* **32**:441–444.

TITELER, M., and SEEMAN, P., 1978, Antiparkinsonian drug doses and neuroleptic receptors, *Experientia* **34**:1490–1492.

TITELER, M., and SEEMAN, P., 1979, Selective labeling of different dopamine receptors by a new agonist [^{3}H]ligand: [^{3}H]-*N*-propylnorapomorphine, *Eur. J. Pharmacol.* **56**:291–292.

TITELER, M., WEINREICH, P., SINCLAIR, D., and SEEMAN, P., 1978, Multiple receptors for brain dopamine, *Proc. Natl. Acad. Sci. USA* **75**:1153–1156.

TITELER, M., LIST, S., and SEEMAN, P., 1979, High-affinity dopamine receptors (D_3) in rat brain, *Commun. Psychopharmacol.* **3**:411–420.

TOLLENAERE, J. P., MOEREELS, H., and KOCH, M. H. J., 1977, On the conformation of neuroleptic drugs in the three aggregation states and their conformational resemblance to dopamine, *Eur. J. Med. Chem.* **12**:199–211.

TRAFICANTE, L. J., FRIEDMAN, E., OLESHANSKY, M. A., and GERSHON, S., 1976, Dopamine-sensitive adenylate cyclase and cAMP phosphodiesterase in substantia nigra and corpus striatum of rat brain, *Life Sci.* **19**:1061–1066.

TSAI, B. S., and LEFKOWITZ, R. J., 1978, Agonist-specific effects of monovalent and divalent cations on adenylate cyclase-coupled alpha adrenergic receptors in rabbit platelets, *Mol. Pharmacol.* **14**:540–548.

UNGERSTEDT, U., 1968, 6-Hydroxydopamine-induced degeneration of central monoamine neurons, *Eur. J. Pharmacol.* **5:**107–110.

UNGERSTEDT, U., 1971*a*, Striatal dopamine release after amphetamine or nerve degeneration revealed by rotational behavior, *Acta Physiol. Scand.* 82 (Suppl. 367):49–68.

UNGERSTEDT, U., 1971*b*, Postsynaptic supersensitivity after 6-hydroxydopamine-induced degeneration of the nigrostriatal dopamine system of the rat brain, *Acta Physiol. Scand.* 82 (Suppl. 367):69–93.

UNGERSTEDT, U., and ARBUTHNOTT, G. W., 1970, Quantitative recording of rotational behavior in rats after 6-hydroxydopamine lesions of the nigrostriatal dopamine system, *Brain Res.* **24:**485–493.

USDIN, T. B., CREESE, I., and SNYDER, S. H., 1980, Regulation by cations of [^{3}H]spiroperidol binding associated with dopamine receptors of rat brain, *J. Neurochem.* **34:**669–676.

VALE, W., RIVIER, J., GUILLEMIN, R., and RIVIER, C., 1979, Effects of purified CRF and other substances on the secretion of ACTH and beta-endorphin-like immunoactivities by cultured anterior or neurointermediate pituitary cells, in: *Central Nervous System Effects of Hypothalamic Hormones and Other Peptides* (R. Collu, A. Barbeau, J. Ducharne, and J. Rochefort, eds.), pp. 163–176, Raven Press, New York.

WALTERS, J. R., and ROTH, R. H., 1976, Dopaminergic neurons: an *in vivo* system for measuring drug interactions with presynaptic receptors, *Naunyn-Schmiedeberg's Arch. Pharmacol.* **296:**5–14.

WALTON, K. G., LIEPMANN, P., and BALDESSARINI, R. J., 1978, Inhibition of dopamine-stimulated adenylate cyclase activity by phenoxybenzamine, *Eur. J. Pharmacol.* **52:**231–234.

WATLING, K. J., DOWLING, J. E., 1981, Dopaminergic mechanisms in the Teleost retina. I. Dopamine-sensitive adenylate cyclase in homogenates of carp retina; effects of agonists, antagonists, and ergots, *J. Neurochem.* **36:**559–568.

WATLING, K. J., DOWLING, J. E., and IVERSEN, L. L., 1979, Dopamine receptors in the retina may all be linked to adenylate cyclase, *Nature* (*London*) **281:**578–580.

WEINER, R. I., and GANONG, W. F., 1978, Role of brain monoamines and histamine in regulation of anterior pituitary secretion, *Physiol. Rev.* **58:**905–976.

WEINER, R. I., CRONIN, M. J., CHEUNG, C. Y., FAURE, N., CLARK, B. R., and GOLDSMITH, P. C., 1979, Anterior pituitary dopamine receptors and prolactin, in: *Catecholamines: Basic and Clinical Frontiers* (E. Usdin, I. J. Kopin, and J. Barchas, eds.), pp. 1218–1220, Pergamon Press, New York.

WEINER, W. J., GOETZ, C. G., NAUSIEDA, P. A., and KLAWANS, H. L., 1979, Amphetamine-induced hypersensitivity in guinea pigs, *Neurology* **29:**1054–1057.

WEINRICH, P., and SEEMAN, P., 1980, Effect of kainic acid on striatal dopamine receptors, *Brain Res.* **198:**491–496.

WILLIAMS, L. T., and LEFKOWITZ, R. J., 1977, Slowly reversible binding of catecholamine to a nucleotide-sensitive state of the beta-adrenergic receptor, *J. Biol. Chem.* **252:**7207–7212.

WITHY, R. M., MAYER, R. J., and STRANGE, P. G., 1980, [^{3}H]Spiroperidol binding to brain neurotransmitter receptors, *FEBS Lett.* **112:**293–295.

WOODRUFF, G. N., and FREEDMAN, S. B., 1981, Binding of [^{3}H]sulpiride to purified rat striatal synaptic membranes, *Neuroscience* **6:**407–410.

YORK, D. H., 1975, Amine receptors in CNS. II. Dopamine, in: *Handbook of Psychopharmacology*, Vol. 6 (L. L. Iversen, S. D. Iversen, and S. H. Snyder, eds.), pp. 23–61, Plenum Press, New York.

ZAHNISER, N. R., and MOLINOFF, P. B., 1978, Effect of guanine nucleotides on striatal dopamine receptors, *Nature* (*London*) **275:**453–455.

4

5-HYDROXYTRYPTAMINE RECEPTORS IN BRAIN

G. Fillion

1. INTRODUCTION

The biogenic amine, 5-hydroxytryptamine (5-HT), was first described in extracts of natural materials almost 30 years ago (see review by Collier, 1957). Twarog and Page (1953) and Amin *et al.* (1954) were the first to demonstrate, on pharmacological evidence, the occurence of 5-HT in the central nervous system of the dog which was confirmed by Bogdanski *et al.* (1956) and Correale (1956) by physicochemical criteria.

The question of the possible function of 5-HT as a neurotransmitter has often been raised. To support such a hypothesis, several criteria have to be fulfilled, as with other putative neurotransmitter substances. Most of the criteria are related to the distribution, the metabolism, and the physiological function of the substance. In the case of 5-HT, several such criteria have been met: 5-HT is distributed heterogeneously within the brain (Dahlström and Fuxe, 1964; Fuxe, 1965; Descarries *et al.*, 1975; Gaudin-Chazal *et al.*, 1979); the effect of the iontophoretic injections of 5-HT mimics the electrical stimulation of neurons containing 5-HT (see review by Aghajanian *et al.*, 1975). The amine is synthesized in the nervous tissue, and the main enzymes involved in its metabolism have been identified (Grahame-Smith, 1964; Gàl, 1965; Nakamuru *et al.*, 1965; Sims and Bloom, 1973; Hamon *et al.*, 1973; Green and Grahame-Smith, 1975).

G. Fillion • Pharmacology Department, Pasteur Institute, 75015 Paris, France.

The molecule is also inactivated by a neuronal uptake mechanism, as observed originally by Shaskan and Snyder (1970). Serotoninergic pathways have been described in the brain; they correspond to an organized neuronal system originating principally from neurons in the raphe area and projecting to several brain regions (Fuxe *et al.*, 1968; Kuhar *et al.*, 1972; Fuxe and Jonsson, 1973; Aghajanian *et al.*, 1973; Descarries *et al.*, 1975; Calas *et al.*, 1976; Chan-Palay, 1976; Steinbusch *et al.*, 1978; Felten and Cummings, 1979; Möllård and Wicklund, 1979; Azmitia, 1979). These criteria and several others favor the hypothesis of 5-HT as a neurotransmitter. However, an important criterion to be met is the existence of receptors specific for this amine and involved in mediating the physiological effects of 5-HT. In this article, we will review some of the results which have been obtained from studies of serotoninergic receptors in the central nervous system.

Our knowledge of 5-HT receptors is not as advanced as for the nicotinic cholinergic receptor. Acetylcholine has been studied as a neurotransmitter for almost a century, and its receptor is one of the most thoroughly understood (see Changeux, 1975). In the case of the acetylcholine receptor, experimental conditions could be found which greatly favored the study of the receptor: first, the use of a tissue (electric organ of *Torpedo marmorata*) extremely rich in nicotinic receptors offering also the possibility of *in vivo* electrophysiological and pharmacological experimentation as well as *in vitro* determinations of binding or other parallel biochemical measurements; second, the existence of snake venom toxins which are highly specific and quasi-irreversible ligands for the receptor. The study of the 5-HT receptor does not benefit from such advantages, either with respect to the concentration of the receptors in tissues, or the characteristics of the available ligands. Serotoninergic systems are found in nonmammalian brains (Osborne, 1980; Cottrel and Cobett, 1978; Gerschenfeld and Paupardin-Tritsch, 1974; Beddok and Mansour, 1979; Nathanson and Greengard, 1974); however, although they present certain advantages, these preparations are still relatively poor in 5-HT receptors. The serotoninergic ligands currently used are generally not found to possess advantages comparable to those of the α-toxins for the nicotinic receptor. Most of the serotoninergic ligands bind reversibly to the serotoninergic site with a relatively high rate of dissociation, and many of them do not possess a very high degree of specificity toward the serotoninergic receptor.

However, recent work on 5-HT receptors does allow us to answer some basic questions. The first question concerns the existence of specific sites able to recognize 5-HT. Do they constitute single or multiple populations of sites? What are their biochemical characteristics? The second question is directed toward the first physiological events following the binding of an agonist to this specific receptor. As far as is known at the present time, an adenylate cyclase system is the only one involved at this step. However, this does not exclude the possibility that 5-HT may act also on other types of effectors. The third question is the molecular mechanisms involved in the function of this receptor system. The model proposed below allows an exploration of the mechanism of action of some drugs, in particular, antidepressants.

2. SEROTONINERGIC RECOGNITION SITES IN THE CNS

Initial attempts to observe serotoninergic receptors directly by binding studies were performed by Marchbanks (1966, 1967). Several classes of low-affinity binding sites were shown to correspond to association constants of approximately 2×10^6 and 5×10^5. These sites did not appear to be receptors involved in neurotransmission.

Later Farrow and van Vunakis (1972, 1973) used tritiated lysergic acid diethylamide (LSD) which, at the time, was available with a radioactive specific activity higher than that for [^{3}H]-5-HT. Synaptosomal preparations isolated from rat brain contained two populations of sites which were found to bind the hallucinogenic drug, one with a high affinity constant ($K_{D1} = 9 \times 10^{-9}$ M) and the other with a lower affinity ($K_{D2} = 1.2 \times 10^{-6}$ M). Tritiated LSD was displaced from these sites by other hallucinogens and also by 5-HT.

The discovery of a specific ligand to label 5-HT receptors would be of great importance since the labeling of receptors by some of the available ligands can lead to complex situations. For example, [^{3}H]-LSD labels not only serotonin receptors but also dopamine receptors (Burt *et al.*, 1976; Whittaker and Seeman, 1979). The choice of the ligand in the case of the serotoninergic receptor is rather difficult since none of those currently available is totally satisfactory. A toxin or other natural substance which binds specifically and irreversibly to serotoninergic receptors has not been reported. In our own laboratory we have tried to find such a substance in various venoms of amphibian origin. Some of them contained indoleamines or other substances which bind to serotoninergic receptors; however, the dissociation rates of the complexes formed with the receptor were too high to make such substances suitable as serotoninergic markers. Among other candidates as a specific ligand, 5-HT itself offers some advantages: it is commercially available as a tritiated substance with a high specific activity (up to 30 Ci/mmol) and is the obvious ligand to label the sites corresponding to the physiological effects induced by the amine. However, it should be remembered that *in vitro* conditions are not necessarily identical to those occurring *in vivo*.

Antagonist ligands have also been used; the results obtained will be discussed later. In any event, these drugs are in general not entirely specific for serotoninergic receptors, and may be pharmacologically active at other types of sites as well. Their use as specific markers of serotoninergic receptors is thus complex. However, interesting results have been presented for some of them which appear to label, at least partly, serotoninergic receptors. Among radioactivity labeled antagonists which have been synthesized are [^{3}H]methergoline (Hamon *et al.*, 1981) and [^{3}H]methiothepine (Nelson *et al.*, 1979). They do not have any decisive advantages for routine experiments, although they do appear to label serotoninergic sites.

2.1. Neuronal Binding Sites

Fractions of synaptic membranes, purified on successive discontinuous density gradients according to Cotman and Matthews (1971), contain a single population of sites which bind 5-HT with a high affinity (K_D = 1–3 $\times$ 10^{-9} M) (Fillion *et al.*, 1976). These sites are present in striatum and hippocampus of rat, horse, ox (Fillion *et al.*, 1978), and rabbit brain (Fillion, G., Bahers, J. M., Fillion, M. P., unpublished experiments). The density of sites varies somewhat among these various species, for example, 0.430 $\pm$ 0.05 pmol/mg protein of purified membranes were found in rat, whereas the value was slightly lower in the horse (0.220 $\pm$ 0.02 pmol/mg protein).

Lysed crude mitochondrial fractions isolated from the same species contain, in addition to this high-affinity binding, a second class of sites which bind 5-HT with a slightly lower affinity (K_D close to 10 nM) (Fillion *et al.*, 1978). These findings were confirmed by Shih and Young (1978) in human brain and Segawa *et al.* (1979) in rat brain.

The existence of high-affinity binding sites for 5-HT has also been observed by other authors using crude mitochondrial fractions isolated from rat brain (Snyder and Bennett, 1975; Bennett and Snyder, 1976; Middlemiss *et al.*, 1977; Whittaker and Seeman, 1978*a*; Nelson *et al.*, 1978; Gripenberg *et al.*, 1978; Peroutka and Snyder, 1979). Surprisingly the second class of sites of lower affinity was not detected by these authors. Several explanations might account for this discrepancy. All of these authors used the same (or similar) techniques as described by Bennett and Snyder (1976) in which homogenization of cerebral tissue was done differently. Therefore, the membrane preparations used might contain a smaller proportion of low-affinity binding, making it difficult to identify. Moreoever, the binding of [^{3}H]-5-HT was often tested over relatively narrow ranges of concentrations (1–7, 8, or 10 nM), which might not be suitable to detect the presence of a population of sites binding 5-HT with an affinity constant close to 10 nM.

The importance of the experimental conditions used for these studies is illustrated by one example regarding the washing of the membrane preparations before use. Crude mitochondrial fractions used directly after their preparation without any washing still contain significant amount of endogenous ligand (either 5-HT or another unknown ligand) (Nelson *et al.*, 1978). This endogenous material competes with [^{3}H]-5-HT for the occupation of the binding sites during the assay and modifies the observed K_D; for example, Snyder and collaborators reported in unwashed preparations an affinity constant K_D = 30 nM (Snyder and Bennett, 1975) or K_D = 8 to 12 nM (Bennett and Snyder, 1976), whereas in a subsequent study using washed preparations a K_D close to 2 nM was observed (Peroutka and Snyder, 1979).

Table 1 summarizes the affinity constants reported by various laboratories for 5-HT binding sites, assumed to correspond to the serotoninergic receptor. The observed K_D values in washed membranes are in good agreement with values ranging from 1 to 6 nM. Thus, these studies apparently deal with the same type of high-affinity binding sites. The binding is saturable, reversible, and specific for 5-HT. Serotoninergic agonists are able to displace [^{3}H]-5-HT from these sites, whereas molecules having a chemical structure related to 5-HT but devoid of its pharmacological properties (e.g., 5-hydroxyindoleacetic acid) do not modify [^{3}H]-5-HT binding (Bennett and Snyder, 1976; Fillion *et al.*, 1976, 1978; Whittaker and Seeman, 1978*a*).

Serotoninergic antagonists (cyproheptadine, cinanserin, methysergide, methiothepine, methergoline) are also able to inhibit the binding of [^{3}H]-5-HT, but a variety of pharmacologically unrelated molecules do not displace [^{3}H]-5-HT from its binding sites (Bennett and Snyder, 1976; Fillion *et al.*, 1976–1978; Nelson *et al.*, 1978). The pre- or postsynaptic nature of the sites which bind 5-HT with high affinity has been examined by several laboratories. Degeneration of the presynaptic serotoninergic pathways produced by 5,6- or 5,7-dihydroxytryptamine lead to either no change (Bennett and Snyder, 1976; Fillion *et al.*, 1976) or even a moderate increase (20 to 30%) in [^{3}H]-5-HT binding (Nelson *et al.*, 1978). This indicated that the binding sites were not located presynaptically on 5-HT neurons, but could be either postsynaptic or nonneuronal (although the latter possibility appeared unlikely since the sites were present in purified synaptic membrane preparations).

Schwarcz *et al.* (1977) and Fillion *et al.* (1979*a*) observed a marked decrease (> 80%) in the high-affinity binding sites for [^{3}H]-5-HT after stereotaxic injections of kainic acid, a procedure which produces a degeneration of neurons postsynaptic to the serotoninergic synapse. These results clearly indicated that high-affinity 5-HT binding sites were postsynaptically located. Nelson *et al.* (1980) showed a moderate decrease of this binding when injections of kainic acid in the striatum were performed in 6- to 10-day-old rats. The fact that synaptogenesis is probably not complete at this age in the rat, however, makes the interpretation of these results more difficult.

TABLE 1

Affinity Constants of [^{3}H]-5-HT Binding in Various Tissue Preparations

References	Preparation	Brain tissue species	K_D
Bennett and Snyder (1976)	Homogenate, 50.000-g pellet	Rat cerebral cortex	8–12 nM
Fillion *et al.* (1976)	Purified synaptosomal membranes	Rat striatum	1 nM
Fillion *et al.* (1978)	P_2 lysed	Bovine or horse striatum	2.5 nM
	Synaptic membranes		2.6 ± 0.96 nM
Heltzel *et al.* (1981)	Homogenate 50.000-g pellet	Human different brain areas	1.2–4.9 nM (depending on the area)
Middlemiss *et al.* (1977)	P_2 lysed	Whole rat brain	11.7 ± 2.5 nM
Nelson *et al.* (1978)	P_2 lysed	Rat whole forebrain	1.6 nM
Nelson *et al.* (1980)	P_2 lysed	Rat striatum	1.61 ± 0.08 nM
		Hippocampus	1.82 ± 0.1 nM
Ögren *et al.* (1979)	P_2	Rat cerebral cortex	5.6 nM
Peroutka *et al.* (1979)	P_2	Rat, guinea pig, or calf frontal cerebral cortex	3 nM
Peroutka and Snyder (1979)	P_2	Rat frontal cerebral cortex	4.6 nM
Quayle *et al.* (1977)	Synaptic membranes	Whole rat brain	0.5 nM
			5 nM
			10 nM
			70 nM
Samanin *et al.* (1980)	P_2 lysed	Rat cortex	13.0 ± 1.9 nM
Savage *et al.* (1980)	Homogenate	Rat cerebral cortex	1.85 ± 0.14
		Hippocampus	1.85 ± 0.11
Segal (1980)	P_2	Rat hippocampus	6.8 nM
Segawa *et al.* (1979)	P_2	Rat whole brain	0.96 ± 0.06 nM
Shih and Young (1978)	P_2	Human cerebral cortex	7.3 nM
Snyder and Bennet (1975)	Microsomal fraction (P_3)	Rat cerebral cortex	≃ 30 nM
Trulson and Jacobs (1979)	Homogenate, 50.000-g pellet	Rat forebrain	30 ± 1.1 nM
		Rat brainstem plus spinal cord	29.4 ± 1 nM

2.2. Glial Binding Sites

Several laboratories (Fillion *et al.*, 1976; Shih and Young, 1978; Segawa *et al.*, 1979) have shown that crude mitochondrial fractions or crude homogenates of mammalian brain possess two different populations of [^{3}H]-5-HT binding sites: one corresponding to a high-affinity site post-synaptically located on neurons (see neuronal binding sites), and a second population of sites characterized by a lower affinity (K_D about 10 nM). Attempts to determine the nature of this latter population of sites (Fillion *et al.*, 1978) suggested that they were not located on neuronal membranes, since they were not present in preparations enriched in neuronal membranes, but were present in fractions enriched in microsomal membranes. The same authors (Fillion *et al.*, 1979*a*) suggested that the low-affinity binding sites might be located on glial cells. This hypothesis was supported by the finding that a fraction enriched in glial membranes prepared from equine striatum contained only a single population of "low"-affinity binding sites (Fillion *et al.*, 1980*a*). The affinity constant for [^{3}H]-5-HT binding at these sites was 10 $\pm$ 1 nM. The binding appeared to be specifically serotoninergic as [^{3}H]-5-HT was displaced by serotoninergic agonists or antagonists and not (or only poorly) by nonserotonin-related drugs. These binding sites were clearly distinct from those involved in the uptake of 5-HT by glial cells as previously described by Henn and Hamberger (1971) and Suddith *et al.* (1978).

The glial cell fractions prepared from various brain areas by centrifugation techniques are invariably contaminated to some extent by neuronal elements (Blomstrand and Hamberger, 1970). Therefore, the low-affinity binding might be neuronal rather than glial. However, if the low-affinity binding sites were neuronal, one should observe low-affinity binding in fractions enriched in neuronal membranes, but this is not the case. Furthermore, if the neuronal contamination were important in the glial preparations, the high-affinity [^{3}H]-5-HT binding sites, characteristic of the purified synaptic membranes, should also be detected in the glial preparations, and again this was not the case. Thus, the low-affinity binding seems to occur to nonneuronal (presumably glial) membranes.

Further support for the glial location of these sites is provided by the observation of a single population of binding sites of low affinity (K_D = 0.2 μM) in membrane prepared from C_6 glioma cell lines (Beaudoin *et al.*, in preparation).

2.3. Multiple Serotonin Receptors

The results reviewed so far suggest the existence of different classes of sites binding [^{3}H]-5-HT. Most reports in the literature agree on the existence of a population of high-affinity 5-HT binding sites (K_D in the nanomolar range) located postsynaptically (see Table 1). In addition, as described above, a second population of sites binding 5-HT with a lower affinity constant (K_D in the 10 nM range) may be localized on glial cells (Fillion *et al.*, 1980*a*; Beaudoin *et al.*, 1980).

Moreover, Nelson *et al.*, (1980, 1981) found that the inhibition of [^{3}H]-5-HT binding by various drugs involved biphasic curves, suggesting a heterogeneity of binding sites; they suggested the existence of multiple sites labeled by [^{3}H]-5-HT.

This concept of multiple serotonin receptor sites has also been studied using ligands other than 5-HT. In particular [^{3}H]-D-lysergic acid diethylamide ([^{3}H]-LSD) has been widely used. The potent hallucinogen LSD has long been thought to act on serotonin receptors, following the discovery by Gaddum and his colleagues of its potent antagonism of 5-HT actions on peripheral smooth muscle (Gaddum and Hameed, 1954). LSD also has structural similarities to 5-HT and affects 5-HT metabolism in CNS (Anden *et al.*, 1968; Boakes *et al.*, 1970; Freedman *et al.*, 1970; Schubert *et al.*, 1970). Moreover, LSD was shown to inhibit the firing of serotoninergic cells in the raphe nuclei (Aghajanian *et al.*, 1968, 1972, 1973, 1975). Initial attempts to demonstrate serotonin receptors by radioligand binding were carried out with this molecule: Farrow and van Vunakis (1972) observed two populations of binding sites in brain with affinity constant K_{D_1} = 9 nM and K_{D_2} = 1.2μM for the high- and low-affinity binding, respectively. Bennett and Aghajanian (1974*a,b*) and Bennett and Snyder (1975) studied the binding of [^{3}H]-D-LSD to various membrane preparations from rat brain. It was shown that LSD bound stereospecifically and with high affinity to rat brain membranes (K_D = 10 nM) and could be displaced by hallucinogenic drugs with potencies parallelling their psychotropic activities. 5-HT was also active as a displacer of [^{3}H]-LSD. On the basis of their biochemical and pharmacological properties, the authors concluded that the majority of the sites labeled by [^{3}H]-LSD were serotonin receptors.

A comparison between the binding properties of [^{3}H]-5-HT and [^{3}H]-LSD was made by Bennett and Snyder (1976) and by Fillion *et al.* (1978). Both studies showed that agonists preferentially displaced [^{3}H]-5-HT, whereas antagonists acted more potently on [^{3}H]-LSD binding. These results showed that the sites with a high affinity for binding [^{3}H]-5-HT were not identical to those binding [^{3}H]-LSD. It was suggested that they might represent two independent categories of sites or possibly two different conformations of the same site.

It is important to note that [^{3}H]-LSD labels not only serotonin receptors but also other categories of sites; in particular, Burt *et al.* (1976) have shown that in calf caudate, [^{3}H]-LSD binds to dopamine receptors, and this has been confirmed by others (Whittaker and Seeman, 1978*b*, 1979; Duchemin *et al.*, 1979).

In summary, these findings indicate that [^{3}H]-LSD labels serotonin receptors that are different from those labeled by [^{3}H]-5-HT, and can also label nonserotoninergic sites to various degrees, depending on the brain area in question.

Another ligand whose use has led to the concept of a multiplicity of serotonin-related sites is [^{3}H]spiperone. This potent neuroleptic drug was originally used by Fields *et al.* (1977) to label dopamine receptors. Leysen and Laduron (1977), Leysen *et al.* (1978*a,b*), and Creese and Snyder (1978) subsequently presented evidence indicating that this ligand also labels serotonin receptor sites in frontal cortex. Recently, Peroutka and Snyder (1979) proposed that [^{3}H]-5-HT and [^{3}H]-spiperone label distinct populations of serotonin receptors designated as 5-HT_1 and 5-HT_2, by analogy with β_1- and β_2-adrenoceptors (U'Prichard *et al.*, 1978; Barnett *et al.*, 1978; Minneman *et al.*, 1979) and α_1- and α_2-adrenoceptors (Peroutka *et al.*, 1978). They showed that [^{3}H]-LSD bound with high affinity to both 5-HT_1 and 5-HT_2 sites, whereas [^{3}H]-5-HT and [^{3}H]spiperone showed high affinity for only 5-HT_1, or 5-HT_2 sites, respectively (see Table 2).

On the other hand, [^{3}H]spiperone itself labels at least two different classes of sites in certain brain regions (Briley and Langer, 1978; Pedigo *et al.*, 1978; Hewlett and Nahorski, 1978; Andorn and Maguire, 1980). These results support the hypothesis that one of the classes of sites labeled by [^{3}H]spiperone is related to dopamine receptors, with the other possibly to serotonin receptors or α-adrenoceptors.

These results suggest the existence of a multiplicity of sites able to recognize 5-HT. However, it is difficult to be sure of the relationship between the sites labeled with different radioligands and to determine to what extent these sites correspond to different conformations of a single site or to different populations of sites. It is also difficult to reconcile the recent biochemical evidence on the existence of multiple 5-HT receptors with earlier pharmacological data. Thus, Gaddum and colleagues (Gaddum and Picarelli, 1957) suggested the existence of two types of serotonin receptor in peripheral tissues. The "D" receptors were located on smooth muscle, and were potently blocked by LSD and dibenzyline, while a second receptor type, the "M" receptor, was located on neural cells in the myenteric plexus of the gut mediating indirect actions of 5-HT. The M receptors were relatively insensitive to blockade by LSD, but could be blocked by morphine. The M and D receptors clearly cannot correspond to the 5-HT_1 and 5-HT_2 categories, since both of the latter categories exhibit a high affinity for LSD. One may also question the nature of "serotoninergic" sites such as the postulated "5-HT_2" receptors which have a certain specificity for 5-HT-related drugs but exhibit only a low affinity for their natural ligand, 5-HT itself. One may speculate on the possible existence of endogenous ligands having a slightly different structure from 5-HT, but still acting at serotonin receptors. The existence of such

TABLE 2

Effects of Various Drugs on Binding of Several Serotoninergic [^{3}H]Ligands to Cerebral Membrane Preparations[a]

Drugs	[^{3}H]-5-HT	[^{3}H]-LSD	[^{3}H]Spiperone	[^{3}H]Methergoline	[^{3}H]Methiothepine
Tryptamines					
5-HT	2(4)	100(1)	3,000(3)	1,650(9)	
	1.6(5)	950(2)	2,700(7)		
	3.8(7)	300(3)	12,000(8)		
	1.8(10)	100(4)	37,200 and 12.5(10)		
		110(7)			
5-Methoxytryptamine	60(4)	150(4)	2,700(7)		
	11(7)	210(7)	9,000(8)		
5-Methoxy-*N*,*N*-dimethyltryptamine		480(2)		16,000(9)	
5-Hydroxy-*N*,*N*-dimethyltryptamine	17(4)	450(2)	1,000(3)		
	37(7)	400(3)	840(7)		
		250(4)			
		220(7)			
N,*N*-Dimethyltryptamine	425(4)	150(1)	2,000(8)		
		1,100(2)			
		1,700(4)			
5-6-Dihydroxytryptamine	300(7)	3,700(7)	22,000(7)		
LSD analogs					
D-LSD	4(4)	4(4)	13(7)		
	10(7)	7.3(7)	5(9)		
2-Bromo-LSD	75(4)	3(1)	2.5(7)		
	89(7)	15(4)			
		13(7)			
Methysergide	500(4)	50(1)	2(3)	28(9)	
	88(7)	5(2)	2.6(7)		
		4(3)			
		80(4)			
		84(7)			

Antagonists					
Cyproheptadine	1,000(4) 1,500(7)	2,000(1) 100(2) 500(4) 90(7)	20(7)	7.5(9)	
Mianserin	390(5) 860(7)	33(7)	5.7(7)		
Cianserin	4,700(4) 1,800(7)	350(0) 50(3) 1,000(4) 130(7)	50(3) 18(7) 30(8)	440(9)	
Methergoline	3.06(5)			6(9)	
Methiothepine	308(5) 150(6)	2.5(2)		2.1(9)	
Neuroleptics					
Spiperone	730(7)	3(3) 18(7)	2(3) 0.5(7)	25(9)	170(6)
Haloperidol	13,000(4) 16,000(7)	150(2) 300(3) 15,000(4) 960(7)	50(3) 42(7)	63(9)	
Promethazine	10,000(7)	50(1) 2,000(7)	170(7)		
Clozapine	180(5) 1,000(7)	75(2) 79(7)	24(7)		
Other drugs					
DA	20,000(7)	150,000(1) 21,000(2) 51,000(7)	75,000(7)	10,700(9)	
ADTN	70,000(7)	110,000(7)	83,000(7)		

[a] The apparent affinity constants (K_I) for these drugs were obtained directly from the literature or calculated from the available EC_{50} following the equation $K_I = EC_{50}/(1 + L/K_D)$), where L is the concentration of the radioactive ligand used in the assay and K_D is the affinity constant of the corresponding binding. The K_D values used in these calculations were 2–4 nM for [^{3}H]-5-HT, 4 nM for [^{3}H]-LSD, 0.5–1 nM for [^{3}H]spiperone, 1 nM for methergoline, and 1 nM for methiothepine. Numbers in brackets refer to data reported by the following authors: (1) Bennett and Aghajanian (1974*b*); (2) Lovell and Freedman (1976); (3) Leysen and Laduron (1977); (4) Fillion *et al.* (1978); (5) Nelson *et al.* (1978); (6) Nelson *et al.* (1979); (7) Peroutka and Snyder (1979); (8) Quick and Iversen (1979); (9) Hamon *et al.* (1981); (10) Pedigo *et al.* (1981).

endogenous molecules has been suggested by the results of Mehl and collaborators (Mehl *et al.*, 1977; Mehl and Guiard, 1978; Stolzki *et al.*, 1978), although this possibility has not yet been clarified.

There is also little information on the functional role of these various sites. Progress on this particular point would greatly advance our knowledge of the serotoninergic system. The majority of the binding sites described as serotoninergic cannot yet be related to any physiological function. MacCall and Aghajanian (1979) have recently described distinct electrophysiological effects of 5-HT in different parts of the brain: inhibitory effects in forebrain and midbrain, and facilitatory ones in facial nucleus and reticular formation. Methysergide blocked the facilitatory effects of 5-HT but not the inhibitory ones. Thus, these results also suggested the existence of different types of receptors; however, at present it is not possible to relate these electrophysiological findings to particular subcategories of serotoninergic binding sites.

3. 5-HT-SENSITIVE ADENYLATE CYCLASE

The binding of a neurotransmitter to a specific site represents the first event in mediating its physiological effect. However, it is certainly insufficient to describe a binding site as a receptor simply on the basis of its specificity toward a particular chemical structure. Several examples of specific binding sites have been described which do not correspond to actual receptors (see review by Burt, 1978). One of the best criteria indicating that a binding site constitutes part of a receptor unit is the existence of some relationship between the binding of a ligand to a site and the occurrence of a physiological signal. Two particular events are often triggered by the occupation of receptors by neurotransmitters: the opening of an ionophore and the activation of an enzyme system.

A classical example of the first case is that of the nicotinic cholinergic receptors (see review by Changeux, 1975). No direct evidence has yet been presented in favor of the existence of a receptor-coupled ionophore mechanism for 5-HT. However, this hypothesis cannot be excluded, and has even received some indirect support (Segal, 1981). The second possibility is more likely, at least for one type of serotonin receptor. The modification of the activity of an enzymatic system induced by the binding of a neurotransmitter was originally discovered by Sutherland and collaborators, who showed that a great number of hormonal effects were mediated by cAMP. They developed a theory of hormone action involving cAMP as a universal intracellular "second messenger" (Sutherland *et al.*, 1965). The first results indicating that 5-HT might modify an adenylate cyclase system were reported in the liver fluke by Mansour *et al.* (1960).

A 5-HT-sensitive adenylate cyclase was described in cockroach ganglia by Nathanson and Greengard (1974) and more recently in other invertebrate tissues by Abrahams *et al.* (1976); Gentleman and Mansour (1977), and Drummond *et al.* (1980).

In mammalian brain the activation of an adenylate cyclase by 5-HT was described by Von Hungen *et al.* (1974) who reported an apparent affinity constant (K_A) for 5-HT close to 1 μM. Other authors have since confirmed and extended these findings (Enjalbert *et al.*, 1978*a,b*; Daszuta *et al.*, 1978). These authors, using brain slices or whole homogenates, measured the activation of adenylate cyclase by determining the formation of [^{32}P]-cAMP from [^{32}P]-ATP according to the technique described by Salomon *et al.* (1974). They found that 5-HT caused a marked activation (75 to 144%) of adenylate cyclase activity in the brains of newborn rats, whereas the relative increase in activity was much lower (less than 20%) in adult rat brain. The apparent potency of 5-HT corresponded to a K_A in the micromolar range, and the response was inhibited by serotonin antagonist drugs and also by various neuroleptics. The authors suggested that these low-affinity adenylate cyclase-coupled sites represented postsynaptic 5-HT receptors.

In our own laboratory we have studied the 5-HT-sensitive adenylate cyclase, using a radioimmunoassay which allowed the detection of femtomole quantities of cAMP. The tissue studied was either brain homogenates or various subfractions, including ones enriched in synaptosomal membranes. Homogenates of adult mammalian brain or crude mitochondrial fractions were found to contain not only an adenylate cyclase sensitive to 5-HT with an affinity constant of about 1 μM, but also another class of adenylate cyclase-coupled sites stimulated by very low concentrations of 5-HT (K_A approximately 1 nM) (Fillion *et al.*, 1977–1979*b*). The basal level of adenylate cyclase activity corresponded to 30–50 pmol cAMP/mg protein/min and the maximum stimulation by 5-HT represented an increase of 80 to 100% for the low-affinity response, and 40 to 50% for the high-affinity one.

3.1. Neuronal 5-HT-Sensitive Adenylate Cyclase

The two types of 5-HT-sensitive adenylate cyclases described above (Fillion *et al.*, 1977, 1979*b*) proved to be physically separable and thus completely distinct: the high-affinity adenylate cyclase response was present as a single type in purified synaptic membranes, whereas the low-affinity system was found in the microsomal fraction known to contain glial membranes (Fillion *et al.*, 1979*b*).

The adenylate cyclase activity present in synaptic membranes was inhibited by serotonin antagonist drugs and synergistically activated by

GTP, and the 5-HT response was clearly additive to that induced by dopamine in the same preparation. Its postsynaptic nature was indicated by the fact that degeneration of the postsynaptic neurons following stereotaxic injections of kainic acid abolished the 5-HT-induced increase in cAMP production (Fillion *et al.*, 1979*a*). Other authors have presented data indicating that 5-HT stimulates adenylate cyclase activity with a high affinity; several peaks of activation were observed by Pagel *et al.* (1976) in a concentration range of 5-HT varying from 0.5 to 20 nM.

3.2. Glial 5-HT-Sensitive Adenylate Cyclase

As described above, mitochondrial fractions from various mammalian brain areas contain two types of adenylate cyclase. One of them corresponds to a low affinity constant ($K_{A\,app}$ = 1 μM) and is present in microsomal fractions (Fillion *et al.*, 1979*b*). Further studies were performed to determine the nature of this second class of 5-HT-coupled adenylate cyclase activation. Glial cell-enriched fractions were prepared from equine striatum according to Blomstrand and Hamberger (1970), and glial membrane preparations isolated from these fractions were found to contain a single class of 5-HT-sensitive adenylate cyclase having an activation constant corresponding to a low apparent affinity (K_A = 1 μM) (Fillion *et al.*, 1980*a*). Antiserotonin drugs inhibited the adenylate cyclase response, whereas neuroleptics and α- and β-adrenoceptor blockers were only weakly inhibitory.

These results confirmed that the high- and low-affinity types of 5-HT adenylate cyclase observed in brain homogenates were clearly distinct; moreover, the results suggested that the low-affinity adenylate cyclase was located on glial cell membranes.

The degree of neuronal contamination of the glial membrane fraction used in these experiments, however, is not known, so the glial location of the adenylate cyclase cannot be established with certainty. However, several other results firmly support this proposal. The low-affinity type of adenylate cyclase activation by 5-HT was presumably not neuronal since it was absent in fractions enriched in neuronal membranes; conversely, the activation observed in the neuronal membranes (high-affinity type) was not present in detectable amounts in glial preparations. Micromolar concentrations of 5-HT were also found to activate cAMP production in a C_6 glial cell line, and the activation corresponded to a single low affinity constant ($K_{A\,app}$ = 4.7 ± 1.65 μM). The maximum effect represented an increase of 30% over the basal level (Beaudoin *et al.*, in preparation). Moreover, primary cultures of glial cells, prepared according to Berwald-Netter (1981) (in collaboration), also contained a 5-HT-sensitive adenylate cyclase with a low affinity ($K_{A\,app}$ = 2.5 μM). The maximum increase in cAMP production was modest but clearly measurable (+ 25%) (Beaudoin *et al.*, in preparation).

4. MECHANISM OF REGULATION OF SEROTONIN RECEPTORS

Binding studies with [^{3}H]-5-HT and observations on the 5-HT sensitive adenylate cyclase led to the following conclusions.

1. Neuronal membranes contain a single class of high-affinity 5-HT binding sites and a single type of adenylate cyclase activated by 5-HT with a high apparent affinity constant. The presence of these two elements in close association with one another is suggested by the fact that they were observed as single populations of sites in purified postsynaptic membranes (Fillion *et al.*, 1976–1979*b*), and that they disappeared in parallel after neuronal postsynaptic degeneration, indicating that they were presumably postsynaptically located (Schwarcz *et al.*, 1977; Fillion *et al.*, 1979*a*).
2. Glial membranes are capable of binding [^{3}H]-5-HT with low affinity and also contain an adenylate cyclase sensitive to the amine with low apparent affinity constant. The experimental data of Nelson *et al.* (1980*b*) are in agreement with these conclusions, since they demonstrated that high-affinity [^{3}H]-5-HT binding sites were not related to the low-affinity adenylate cyclase activation by 5-HT.

This naturally raises the question whether the binding sites and the 5-HT-sensitive adenylate cyclase might be considered as components of a single 5-HT receptor system. To attempt to answer the question, we will consider in more detail the [^{3}H]-5-HT binding and the 5-HT-induced activation of adenylate cyclase activity observed in purified synaptic membranes.

4.1. Regulation at the Binding Site

As shown in Table 1, for most of the sites recognizing 5-HT with high affinity, the dissociation constants which have been reported are in agreement with each other, with values varying from 1 to 6 nM. In most cases the binding assays were performed under experimental conditions which allowed binding to reach equilibrium; in particular, the incubation periods were relatively long (7–10 min at 37°C, or 15–30 min at 22°C).

Fillion and Fillion (1980), using purified synaptic membrane fractions prepared from rat or horse striatum, found that a shorter incubation time (5–8 min at 22°C) resulted in the binding of [^{3}H]-5-HT with a lower apparent affinity constant (K_D = 12 nM) compared to control membranes incubated in parallel for longer times (15–30 min at 22°C) (K_D = 4 nM). Interestingly, the maximum number of binding sites (B_{max}) observed was the same after long or short incubations. This indicates that equilibrium was reached under these experimental conditions. Intermediate incubation times suggested the binding of 5-HT to two different populations of sites, one corresponding to the high-affinity ones already described, and the other to lower-affinity ones. The addition of the maximum numbers of sites for these two sites was equal to the B_{max} of the high-affinity sites, when observed as a single population. The fact that the total number of sites was constant suggested the existence of a single population of sites able to bind 5-HT with two different affinities, the high-affinity sites arising from the low-affinity ones.

To explain these results, Fillion and Fillion (1980) proposed the hypothesis of a conformational change in the serotonin receptor following occupation of the site by an agonist ligand. A similar hypothesis had already been suggested for the cholinergic nicotinic receptor by Katz and Thesleff (1957) and further developed by Changeux *et al.* (1973). This group (see reviews by Changeux, 1975, and Heidmann and Changeux, 1978) has presented much evidence in favor of the hypothesis of the existence of an allosterically controlled conformational change in the receptor site leading to structural modifications of the ionophore that is coupled to it. According to the three-state model proposed by these authors, the cholinergic binding site may exist in different conformations with different affinities for agonists; conformational changes would be induced by the binding of the agonist and would correspond to the transformation of the site from a resting state of low affinity to an active state with a higher affinity and finally to a desensitized state with the highest affinity for agonists (Heidman and Changeux, 1979*a,b*).

The serotonin system can be examined in the light of a similar three-state model. According to this hypothesis, the binding of the agonistic ligand induces a structural change in the receptor protein. This phenomenon has indeed been observed for the 5-HT site, assuming that the change in affinity described above reflects a structural modification of the receptor protein. Indeed, pre-exposure of purified synaptic membranes to 5-HT (10 nM) followed by a rapid wash resulted in an increase in the affinity from a K_D of 12 nM in the control to a K_D of 3–4 nM in 5-HT pre-exposed membranes. The phenomenon was dependent on the concentration of 5-HT and the time of exposure. The agonist molecules (5-HT, 5-hydroxy-*N,N*-dimethyltryptamine, 5-methoxy-*N,N*-dimethyltryptamine) were able to cause this increase in affinity for the [^{3}H]-5-HT; in contrast, antagonist drugs did not change the affinity of [^{3}H]-5-HT binding or even lowered it.

As it is known that disulfide bonds play a crucial role in allosteric changes in proteins (Changeux, 1975; Moore and Raftery, 1979), the effects of sulfhydryl reagents on the binding of 5-HT have been studied. The covalent blockade of sulfhydryl groups by *N*-ethylmaleimide totally inhibited the increase in affinity normally induced by exposure to 5-HT (Fillion and Fillion, 1980). These results tended to support the hypothesis that an allosteric conformational change might underlie the observed increase in affinity.

The affinity of [^{3}H]-5-HT binding is also affected by guanosine triphosphate (GTP). The nucleotide rapidly dissociates the high-affinity binding of the amine (Peroutka *et al.*, 1979; G. Fillion, unpublished experiments). This phenomenon occurs even in 5-HT pre-exposed membranes. Thus, GTP appears to stabilize a low-affinity conformation of the receptor site. Peroutka *et al.* (1979) described a decrease of the binding of [^{3}H]-5-HT to serotonin sites by GTP, and their results also indicate a decrease in the binding affinity. These results might be compared to those reported by others for the effects of guanyl nucleotides on various different receptors: α-adrenoceptors (U'Prichard and Snyder, 1978), β-adrenoceptors (Williams and Lefkowitz, 1977), dopamine (Creese *et al.*, 1978), or opiate (Blume, 1978) receptors, where guanine nucleotides also selectively affect agonist binding, usually decreasing it.

Gripenberg and Jansson (1978), using synaptic plasma membranes, reported changes in the time course of binding of 5-HT that depended on the ionic composition of the incubation medium, the duration of the incubation, and the conditions of the membrane preparation.

The authors concluded that "these factors should be taken into consideration when studying the pharmacological and binding characteristics of the receptor." Their results were presumably related to the various conformations of the 5-HT binding site favored by the different experimental conditions utilized.

4.2. Regulation of 5-HT-Sensitive Adenylate Cyclase Activity

Adenylate cyclase activity was measured by monitoring the increase in production of cAMP following the incubation of synaptosomal membranes in the presence of 5-HT. Membranes exposed to 5-HT for a short time (2 to 5 min at 30°C) in the presence of ATP responded by increasing their cAMP production by 50 to 80% over the basal level. However, membranes pre-exposed to 5-HT for 10 min and then submitted to the same incubation in the presence of ATP showed a 60 to 90% reduction in the 5-HT response. This phenomenon was clearly dependent on the time of pre-exposure and the concentration of 5-HT used. This desensitization process was suppressed or at least markedly retarded in the presence of GTP (Fillion *et al.*, 1980*b*).

4.3. Proposed Mechanism of Regulation of the 5-HT Receptor

Several points emerge from these studies on the binding of 5-HT and the 5-HT-sensitive adenylate cyclase in synaptic membranes. Membranes washed in buffer are initially able to bind [^{3}H]-5-HT with low affinity (K_D close to 10 nM) and respond to the amine by producing cAMP; the same membranes pre-exposed to 5-HT and then rapidly washed are able to bind 5-HT with a high affinity (K_D close to 1 nM) but are desensitized to the effect of the amine since 5-HT is no longer capable of increasing cAMP production. GTP stabilizes a structural conformation of the protein which binds [^{3}H]-5-HT with low affinity, and it allows the activation of the adenylate cyclase by the amine.

These findings and also kinetic considerations led us to propose a scheme (Fig. 1) based partly on a similar model already put forward for cholinergic receptors (Changeux, 1975) and partly on previous findings on adenylate cyclases involved in other types of receptors (Cassel and

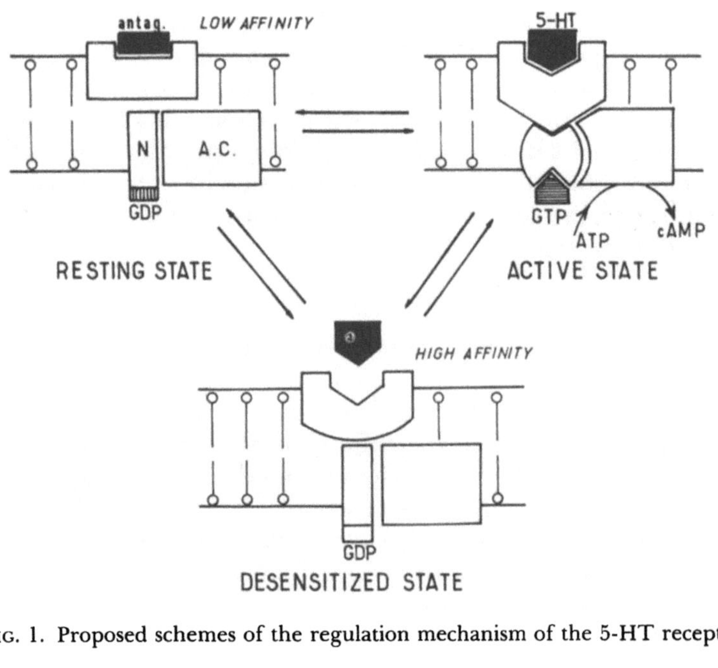

FIG. 1. Proposed schemes of the regulation mechanism of the 5-HT receptor.

Selinger, 1978). We assume that the receptor system for 5-HT consists of at least three main subunits: the recognition site which binds 5-HT, the adenylate cyclase unit, and a protein binding the nucleotides, similar (or identical) to that described by Pfeuffer (1977) for the adenylate cyclase stimulated by the β-adrenoceptor agonists. In the resting state, the recognition site would be able to bind agonists or antagonists, with antagonists simply blocking the system in its initial state. Agonists would bind with relatively low affinity to the site and induce a conformational change which, in the absence of GTP, would give rise to a state in which there was a high affinity for the amine and in parallel an inability to activate the adenylate cyclase. In the presence of GTP, the site would remain in a "low"-affinity state capable of causing activation of the enzyme.

Several points in this scheme may be discussed. First, it may appear paradoxical that the active state of the receptor is related to the low-affinity conformation of the protein, and that its inactive state is linked to the high-affinity state. The phenomenon, however, is simply an expression of the second law of thermodynamics, which predicts that a given system goes spontaneously toward a less organized state; accordingly, the ligand–receptor complex is moving toward a more stable state corresponding to one of higher-affinity binding.

Childers and Synder (1980) have reported that 30% of the sites binding GTP in brain membranes were associated with synaptic components. These results indicate that GTP may play a role at the synapse. The fact that GTP stabilizes the low-affinity conformation of the recognition site and appears necessary for the activation process of the adenylate cyclase by 5-HT favors the hypothesis that a nucleotide binding protein similar or identical to that described by Pfeuffer (1977), Cassell and Pfeuffer (1978), and Cassel and Selinger (1978) is also involved in the case of the 5-HT receptor. An allosteric process may take place between the recognition site for 5-HT and the nucleotide binding protein. This is also suggested by the fact that the cooperativity of the [^{3}H]-5-HT binding appears to be modified by GTP (Fillion *et al.*, in preparation). At this stage it is difficult to determine whether the 5-HT binding site is coupled to the adenylate cyclase unit solely through a nucleotide binding protein, or whether there is also some direct link between the binding site and the adenylate cyclase. In any event, the role of the nucleotide binding protein appears to be important in the regulation mechanism of the receptor. Agonists induce a conversion of the site to a state of high affinity in the presence or absence of GDP and GMP, whereas, in contrast, GTP stabilizes the low-affinity state. Similarly the presence of GDP or GMP does not modify or reduce adenylate cyclase activity, whereas GTP is necessary for the activation induced by 5-HT. The rapid conversion of GTP to GDP reported by Cassel and Selinger (1978) and Childers and Snyder (1980) probably plays a crucial role in this regulatory step.

In summary, the serotonin receptor appears to be regulated through a basic process related to structural interconversions of the binding site having different affinities for the ligand; these various conformations correspond to resting, active, and desensitized states of the receptor. The role of the nucleotide binding protein appears rather important; it is able to stabilize, presumably by an allosteric process, the structural conformation of the binding site leading to the activation of the effector system.

4.4. Effects of Antidepressant Drugs on the Postsynaptic Serotonin Receptor

The mechanism of action of tricyclic antidepressants is still controversial. These substances were known to act as inhibitors of 5-HT and (or) catecholamine uptake (Glowinski and Axelrod, 1964; Iversen, 1965;

TABLE 3

Affinity Constants for [^{3}H]-5-HT Binding after Preexposure of the Purified Synaptic Membranes to Various Drugs[a]

Drugs tested	K_D (nM)
None	15.3 ± 1.26(14)
5-HT agonists	
5-HT	1.8 ± 0.4(8)
Bufotenine	3.2(2)
5-MeOH-*N,N*-DMT	2.1(2)
Tricyclic antidepressants	
Imipramine	2.4 ± 0.2(6)
Chlorimipramine	1.7 ± 1(3)
Desipramine	4.9 ± 0.69(3)
Amitryptiline	4.2 ± 0.35(3)
Trimeprimine	4.0 ± 0.3(3)
Zimelidine	3.0 (2)
Atypical antidepressants	
Iprindol	3.9 ± 1.0(3)
Mianserin	2.9 ± 0.88(3)
Other drugs	
Chlorpromazine	14.24 ± 1.9(5)
Fluphenazine	10.7 ± 1.9(3)
Haloperidol	13.5(2)

[a] The synaptic membranes prepared according to Cotman and Matthews (1971) were washed in a Tris-HCl buffered medium (50 mM, pH 7.4) at 37°C for 10 min, centrifuged, and resuspended in the same medium. They were pre-exposed to 10 nM of the various drugs for 10 min at 37°C. The drugs were washed out by centrifugation and the membranes rapidly resuspended in the medium in the presence of [^{3}H]-5-HT for 8 min at 22°C; the K_D values were obtained from the Scatchard plot of the binding curves using eight to twelve concentrations of [^{3}H]-5-HT. The numbers in brackets represent the number of assays performed for each drugs.

Carlsson *et al.*, 1966). However, some clinically effective antidepressants have no effect or are only weak inhibitors of the uptake of these amines (mianserin, iprindole) (Leonard, 1974). On the other hand, Aprison *et al.* (1978) suggested a possible postsynaptic effect of these drugs. Thus, it was interesting to study the possible effect of these substances on the function of the 5-HT receptor.

Imipramine displaces [^{3}H]-5-HT from its postsynaptic binding sites with an apparent affinity in the micromolar range. However, pre-exposure of synaptic membranes to low concentrations of imipramine (1 to 10 nM) or other tricyclic antidepressants under the same conditions as those used for 5-HT (see above) led to an increase in affinity for [^{3}H]-5-HT very similar to that observed after 5-HT pre-exposure (see Table 3). This phenomenon was observed with a large series of tricyclic antidepressants, but did not occur with nonantidepressant drugs (i.e., neuroleptics) (Fillion *et al.*, 1980*b*; Fillion and Fillion, 1981). The interconversion of the 5-HT binding site by imipramine leads presumably to an inactivation of the receptor by conversion to a desensitized state. This effect might be related to the reduction in the numbers of serotonin receptors in brain observed after long-term administration of antidepressants (Segawa *et al.*, 1979; Ögren *et al.*, 1979; Childers and Snyder, 1980).

5. CONCLUSIONS

Several types of receptor sites in the mammalian central nervous system appear to be able to recognize serotonin. Moreover, the various radioligands currently used appear to label sites which are markedly different from each other. Thus, it is quite difficult at the present time to describe clearly the relationships which may or may not exist between the various sites described. A better understanding of the functional roles of these sites would be of a great help in clarifying their identities.

Use of [^{3}H]-5-HT in various regions of the brain shows two main populations of sites; one, directly related to an adenylate cyclase, is thought to represent the postsynaptic 5-HT receptor; the other, with lower affinity for 5-HT, possibly linked also to an adenylate cyclase, is different from the first one and appears to be located on glial cell membranes, with an unknown functional role. Among the [^{3}H]-5-HT binding sites which have been described, presynaptic sites have not been demonstrated directly, although their existence is strongly suggested by experimental evidence (Cerrito and Raiteri, 1979; Göthert and Weinheimer, 1979; Feniuk *et al.*, 1979). This might be due to the relative paucity of such sites compared to other classes of binding sites.

The regulation of the postsynaptic 5-HT receptor takes place through a system involving various conformational states of the recognition site, which may modulate the "coupling" with the effector adenylate cyclase. It is quite interesting to note that this phenomenon appears to be a widespread one. Evidence for a similar phenomenon is quite strong in the case of the nicotinic cholinergic receptor, and it may also occur for other types of receptors (α-, β-adrenoceptors, glucagon, and possible other peptide receptors). This is of interest from both the theoretical and practical point of view in guiding future research on neurotransmitter receptors.

6. REFERENCES

Abrahams, S. L., Northup, J. K., and Mansour, T., 1976, Adenosine cyclic 3′,5′ monophosphate in the liver fluke *fasciola hepatica*, *Mol. Pharmacol.* **12**:49–58.

Aghajanian, G. K., Foote, W. E., and Sheard, M. H., 1968, Lysergic acid diethylamide: sensitive neuronal units in the midbrain raphe, *Science* **161**:706–708.

Aghajanian, G. K., Haigler, H. J., and Bloom, F. E., 1972, Lysergic acid diethylamide and serotonin: direct actions on serotonin-containing neurons, *Life Sci.* **11**:615–622.

Aghajanian, G. K., Kuhar, M. J., and Roth, R. H., 1973, Serotonin-containing neuronal perikarya and terminals: differential effects of *p*-chlorophenylalanine, *Brain Res.* **54**:85–101.

Aghajanian, G., Haigler, H., and Bennett, J. L., 1975, Amine receptors in CNS. III. 5-Hydroxytryptamine in brain, in: *Handbook of Psychopharmacology* (L. L. Iversen, S. D. Iversen, and S. H. Snyder, eds.), Vol 6, pp. 63–96, Plenum Press, New York–London.

Amin, A. H., Crawford, T. B., and Gaddum, J. H., 1954, The distribution of substance P and 5-hydroxytryptamine in the central nervous system of the dog, *J. Physiol.* (*London*) **126**:596–618.

Anden, N. E., Corrodi, H., Fuxe, K., and Höckfelt, T., 1968, Evidence for a central 5-hydroxytryptamine receptor stimulation by lysergic acid diethylamide, *Br. J. Pharmacol.* **34**:1–7.

Andorn, A. C., and Magurie, M. E., 1980, [^{3}H]spiroperidol binding in rat striatum: two high-affinity sites of differing selectivities, *J. Neurochem.* **35**:1105–1113.

Aprison, M. H., Takahashi, R., and Tachiki, K., 1978, Hypersensitive serotonergic receptors involved in clinical depression—a theory, in: *Neuropharmacology and Behavior* (B. Haber and M. M. Aprison, eds.), pp. 23–53, Plenum Press, New York.

Azmitia, E. C., 1981, The serotonin-producing neurons in the midbrain median and dorsal raphe nuclei, in: *Handbook of Psychopharmacology*, (L. L. Iversen, S. D. Iversen, and S. H. Snyder), vol. 9, pp. 233–314, Plenum Press, New York.

Barnett, O. B., Rugg, E. L., and Nahorski, S., 1978, Direct evidence of β-adrenoceptor binding sites in lung tissue, *Nature* (*London*) **273**:166–168.

Beaudoin, D., Netter, Y., and Fillion, G., 1980, Interaction of 5-hydroxytryptamine with cultured glia and neurones, 1st International Meeting of ISDN, Strasbourg, 1980, Abstracts 151.

Beddok, R., and Mansour, T., 1979, Antagonism of serotonin-activated adenylate cyclase in the liver fluke *fasciola hepatica* by levorphanol and dextrorphan, *Biochem. Pharmacol.* **28**:3689–3692.

Bennett, J. L., and Aghajanian, G. K., 1974*a*, Stereospecific binding of *d*-LSD on physiological response of serotoninergic neurones, *Fed. Proc.* **33**:256.

Bennett, J. L., and Aghajanian, G. K., 1974*b*, D-LSD binding to brain homogenates: possible relationship to serotonin receptors, *Life Sci.* **15**:1935–1944.

Bennett, J. P., and Snyder, S., 1975, Stereospecific binding of *d*-lysergic acid diethylamide (LSD) to brain membranes: relationship to serotonin receptors, *Brain Res.* **94**:523–544.

BENNETT, J. P., and SNYDER, S. H., 1976, Serotonin and lysergic acid diethylamide binding in rat brain membranes: relationship to postsynaptic serotonin receptors, *Mol. Pharmacol.* **12:**373–389.

BERWALD-NETTER, Y., 1981, Na^+ Channel-associated scorpion toxin receptor sites as probes for neuronal evolution *in vivo* and *in vitro, Proc. Natl. Acad. Sci. USA* **78:**1245–1249.

BLOMSTRAND, C., and HAMBERGER, A., 1970, Amino acid incorporation *in vitro* in proteins of neuronal and glial cell enriched fractions, *J. Neurochem.* **17:**1187–1195.

BLUME, A. J., 1978, Interaction of ligands with the opiate receptors of brain membranes: regulation by ions and nucleotides, *Proc. Natl. Acad. Sci. USA* **75:**1713–1717.

BOAKES, R. J., BRADLEY, P. B., BRIGGS, I., and DRAY, A., 1970, Antagonism of 5-hydroxytryptamine by LSD 25 in the central nervous system. A possible neuronal basis for the actions of LSD 25, *Br. J. Pharmacol.* **40:**202–218.

BOGDANSKI, D. F., PLETSCHER, A., BRODIES, B. B., and UDENFRIEND, S., 1956, Identification and assay of serotonin in brain, *J. Pharm. Exp. Ther.* **117:**82–88.

BRILEY, M., and LANGER, S. Z., 1978, Two binding sites for [^{3}H]spiroderidol on rat striatal membranes, *Eur. J. Pharmacol.* **50:**283–284.

BURT, D. R., 1978, Criteria for receptor identification in: *Neurotransmitter Receptor Binding* (H. I. Yamamura, S. J. Enna, and M. J. Kuhar, eds.), Raven Press, New York.

BURT, D. R., CREESE, I., and SNYDER, S. H., 1976, Binding interactions of lysergic acid diethylamide and related agents with dopamine receptors in the brain, *Mol. Pharmacol.* **12:**631–638.

CALAS, A., BESSON, J. M., GAUCH, C., ALONSO, G., GLOWINSKI, J., and CHERAMY, A., 1976, Radioautographic study of *in vivo* incorporation of [^{3}H]monoamines in the cat caudate nucleus: identification of serotoninergic fibers, *Brain Res.* **118:**1–3.

CARLSSON, A., FUXE, K., HAMBERGER, B., and LINDQVIST, M., 1966, Biochemical and histochemical studies on the effects of imipramine-like drugs and (+)-amphetamine on central and peripheral catecholamine neurons, *Acta Physiol. Scand.* **67:**481–497.

CASSEL, D., and PFEUFFER, T., 1978, Mechanism of cholera toxin action: covalent modification of the guanyl nucleotide binding protein of the adenylate cyclase system, *Proc. Natl. Acad. Sci. USA* **75:**2669–2673.

CASSEL, D., and SELINGER, Z., 1978, Mechanism of adenylate cyclase activation through the β-adrenergic receptor: catecholamine-induced displacement of bound GDP by GTP, *Proc. Natl. Acad. Sci. USA* **75:**4155–4159.

CERRITO, F., and RAITERI, M., 1979, Serotonin release is modulated by presynaptic autoreceptors, *Eur. J. Pharmacol.* **57:**427–430.

CHAN-PALAY, V., 1976, Serotonin axons in the supra and subependymal plexuses and the leptomeninges; their roles in local alterations of cerebrospinal fluid and vasomotor activity, *Brain Res.* **102:**103–130.

CHANGEUX, J. P., 1975, The cholinergic receptor protein from fish electric organ, in: *Handbook of Psychopharmacology* (L. L. Iversen, S. D. Iversen, and S. H. Snyder, eds.), Vol. 6, pp. 235–301, Plenum Press, New York.

CHANGEUX, J. P., MEUNIER, J. C., OLSEN, R. W., WEBER, M., BOURGEOIS, J. P., POPOT, J. L., COHEN, J. B., HAZELBAUER, G. L., and LESTER, H. A., 1973, Studies on the mode of action of cholinergic agonists at the molecular level, in: *Drug Receptors* (H. P. Rang, ed.), pp. 273–294, Macmillan, London.

CHILDERS, S. R., and SNYDER, S. H., 1980, Characterization of [^{3}H]guanine nucleotide binding sites in brain membranes, *J. Neurochem.* **35:**183–192.

COLLIER, H. O., 1957, The occurrence of 5-hydroxytryptamine in *Nature.*, in: *5-Hydroxytryptamine* (G. P., Lewis, ed.), pp. 5–19, Pergamon Press, New York.

CORREALE, P., 1956, The occurrence and distribution of 5-hydroxytryptamine (enteramine) in the central nervous system of vertebrates, *J. Neurochem.* **1:**2–31.

COTMAN, C. W., and MATTHEWS, D. A., 1971, Synaptic plasma membranes from rat brain synaptosomes: isolation and partial characterization, *Biochim. Biophys. Acta* **249:**380–394.

Cottrell, G. A., and Cobbett, A., 1978, Actions of propranolol on 5-HT receptors of snail neurons, *J. Pharm. Pharmacol.* **30**:820.

Creese, I., and Snyder, S. H., 1978, [^{3}H]spiroperidol labels serotonin receptors in rat cerebral cortex and hippocampus, *Eur. J. Pharmacol.* **49**:201–202.

Creese, I., Usdin, T. B., and Snyder, S. H., 1978, Dopamine receptor binding regulated by guanine nucleotides, *Mol. Pharmacol.* **16**:69–76.

Dahlström, A., and Fuxe, K., 1964, Evidence for the existence of monoamine-containing neurones in the central nervous system, *Acta Physiol. Scand.* **62**(232):1–55.

Daszuta, A., Pons, F., and Cadilhac, J., 1978, Effect of serotonin on cyclic AMP level in rat hypothalamus slices during development, *Eur. J. Pharmacol.* **56**:397–401.

Descarries, L., Beaudet, A., and Watkins, K. C., 1975, Serotonin nerve terminal in adult rat neocortex, *Brain Res.* **100**:563–588.

Duchemin, A. M., Quach, T. T., Rose, C., and Schwartz, J. C., 1979, Pharmacological characterization of [^{3}H]-LSD binding sites in mouse brain *in vivo*, *Life Sci.* **24**:401–410.

Enjalbert, A., Bourgoin, S., Hamon, M., Adrien, J., and Bockaert, J., 1978*a*, Postsynaptic serotonin-sensitive adenylate cyclase in the central nervous system. I. Development and distribution of serotonin- and dopamine-sensitive adenylate cyclases in rat and guinea pig brain, *Mol. Pharmacol.* **14**:2–10.

Enjalbert, A., Hamon, M., Bourgoin, S., and Bockaert, J., 1978*b*,Postsynaptic serotonin-sensitive adenylate cyclase in the central nervous system. II. Comparison with dopamine- and isoproterenol-sensitive adenylate cyclases in rat brain, *Mol. Pharmacol.* **14**:11–23.

Farrow, J. T., and van Vunakis, H., 1972, Binding of *d*-lysergic acid diethylamide to subcellular fractions from rat brain, *Nature (London)* **237**:164–166.

Farrow, J. T., and van Vunakis, H., 1973, Characteristics of *d*-lysergic acid diethylamide binding to subcellular fractions derived from rat brain, *Biochem. Pharmacol.* **22**:1103–1113.

Felten, D., and Cummings, J., 1979, The raphe nuclei of the rabbit brain stem, *J. Comp. Neurol.* **187**:199–244.

Feniuk, W., Humphrey, P. P. A., and Watts, A. D., 1979, Preliminary characterization of the presynaptic receptor for 5-hydroxytryptamine in dog isolated saphenous vein, *Br. J. Pharmaol.* **67**:423P–424P.

Fields, J. Z., Reisine, T. D., and Yamamura, H. I., 1977, Biochemical demonstration of dopaminergic receptors in rat and human brain using [^{3}H]spiroperidol, *Brain Res.* **136**:578–582.

Fillion, G., and Fillion, M. P., 1980, Transitional states of the neuronal serotoninergic site, *Eur. J. Pharmacol.* **65**:109–112.

Fillion, G., and Fillion, M. P., 1981, Modulation of affinity of postsynaptic serotonin receptors by antidepressant drugs, *Nature (London)* **292**:349–351.

Fillion, G., Fillion, M. P., Spirakis, C., Bahers, J. M., and Jacob, J., 1976, 5-Hydroxytryptamine binding to synaptic membranes from rat brain, *Life Sci.* **18**:65–74.

Fillion, G., Rousselle, J. C., Goiny, M., Pradelles, P., Dray, F., and Jacob, J., 1977, Activation adenylatecyclasique sensible à la 5-hydroxytryptamine sur des préparations membranaires synaptosomales de cerveau, *C. R. Acad. Sci. Paris* **285**:265–268.

Fillion, G., Rousselle, J. C., Fillion, M. P., Beaudoin, D., Goiny, M., Deniau, J. M., and Jacob, J., 1978, High-affinity binding of [^{3}H]-5-hydroxytryptamine to brain synaptosomal membranes: comparison with [^{3}H]lysergic acid diethylamide binding, *Mol. Pharmacol.* **14**:50–59.

Fillion, G., Beaudoin, D., Rousselle, J. C., Deniau, J. M., Fillion, M. P., Dray, F., and Jacob, J., 1979*a*, Decrease of [^{3}H]-5-HT high-affinity binding and 5-HT adenylate cyclase activation after kainic acid lesion in rat brain striatum, *J. Neurochem.* **33**:567–570.

Fillion, G., Rousselle, J. C., Beaudoin, D., Pradelles, P., Goiny, M., Dray, F., and Jacob, J., 1979*b*, Serotonin-sensitive adenylate cyclase in horse brain synaptosomal membranes, *Life Sci.* **24**:1813–1822.

Fillion, G. M. B., Beaudoin, D., Rousselle, J. C., and Jacob, J., 1980*a*, [^{3}H]-5-HT binding sites and 5-HT-sensitive adenylate cyclase in glial cell membrane fraction, *Brain Res.* **198:**361–375.

FILLION, G., FILLION, M. P., and ROUSSELLE, J. C., 1980*b*, Regulation of the serotoninergic receptor activity in synaptosomal membranes, *Neurosci. Lett. Suppl.* **5:**69.

FREEDMAN, D. X., GOTTLIEB, R., and LOVELL, R. A., 1970, Psychotomimetic drugs and brain 5-hydroxyindole metabolism, *Biochem. Pharmacol.* **19:**1181–1188.

FUXE, K., 1965, The distribution of monoamine terminals in the central nervous system, *Acta Phys. Scand. Suppl. 64* **247:**39–85.

FUXE, K., and JONSSON, G., 1973, Further mapping of central 5-hydroxytryptamine neuron studies with neurotoxic dihydroxytryptamine, *Adv. Biochem. Psychopharmacol.* **10:**1–12.

FUXE, K., HÖCKFELT, T., and UNGERSTEDT, U., 1968, Localization of indolalkylamines in CNS, *Adv. Pharmacol.* **6A:**235–251.

GADDUM, J. H., and HAMEED, K. A., 1954, Drugs which antagonize 5-hydroxytryptamine, *Br. J. Pharmacol.* **9:**240–248.

GADDUM, J. H., and PICARELLI, Z. P., 1957, Two kinds of tryptamine receptor, *Br. J. Pharmacol.* **12:**323–328.

GÀL, E. M., 1965, *In vitro* hydroxylation of tryptophan by brain tissue, *Fed. Proc.* **24:**580.

GAUDIN-CHAZAL, G., DASZUTA, A., FAUDON, M., and TERNAUX, J. P., 1979, 5-HT concentration in cat's brain, *Brain Res.* **160:**281–293.

GENTLEMAN, S., and MANSOUR, T. E., 1977, Control of Ca^{2+} efflux and cyclic AMP by 5-hydroxytryptamine and dopamine in abalone gill, *Life Sci.* **20:**687–694.

GERSCHENFELD, H. M., and PAUPARDIN-TRITSCH, D., 1974, Ionic mechanisms and receptor properties underlying the responses of molluscan neurons to 5-hydroxytryptamine, *J. Physiol.* **243:**427–432.

GLOWINSKI, J., and AXELROD, J., 1964, Inhibition of uptake of tritiated noradrenaline in the intact rat brain by imipramine and structurally related compounds, *Nature (London)* **204:**1318–1319.

GÖTHERT, M., and WEINHEIMER, G., 1979, Extracellular 5-hydroxytryptamine inhibits 5-HT release from rat brain cortex slices, *Arch. Pharmacol.* **310:**93–96.

GRAHAME-SMITH, D. G., 1964, Tryptophan-hydroxylation in brain, *Biochem. Biophys. Res. Commun* **16:**586–592.

GREEN, A. R., and GRAHAME-SMITH, D. G., 1975, 5-hydroxytryptamine and other indoles in the central nervous system, in: *Handbook of Psychopharmacology* (L. L. Iversen, S. D., Iversen, and S. H. Snyder, eds.), Vol. 3, pp. 169–245, Plenum Press, New York.

GRIPENBERG, J., and JANSSON, S. E., 1978, Binding of [^{3}H]-5-hydroxytryptamine to synaptic plasma membranes of rat brain, *Acta Phys. Scand.* **102:**123–125.

HAMON, M., BOURGOIN, S., and GLOWINSKI, J., 1973, Feedback regulation of 5-HT synthesis in rat striatal slices, *J. Neurochem.* **20:**1727–1745.

HAMON, M., MALLAT, M., HERBET, A., NELSON, D., AUDINOT, M., PICHAT, L., and GLOWINSKI, J., 1981, [^{3}H]methergoline: a new ligand of 5-HT receptors in the rat brain, *J. Neurochem.* **36:**613–626.

HEIDMANN, T., and CHANGEUX, J. P., 1978, The acetylcholine nicotinic receptor, *Annu. Rev. Biochem.* **47:**317–357.

HEIDMANN, T., and CHANGEUX, J. P., 1979*a*, Fast kinetic studies on the interaction of a fluorescent agonist with the membrane-bound acetylcholine receptor from *Torpedo marmorata*, *Eur. J. Biochem.* **94:**255–279.

HEIDMANN, T., and CHANGEUX, J. P., 1979*b*, Fast kinetic studies on the allosteric interactions between acetylcholine receptor and local anesthetic binding sites, *Eur. J. Biochem.* **94:**281–296.

HELTZEL, J. A., BOEHME, D. H., and VOGEL, W., 1981, Serotonin binding to different regions of human brain, *Brain Res.* **204:**451–454.

HENN, F. A., and HAMBERGER, A., 1971, Glial cell function: uptake of transmitter substances, *Proc. Natl. Acad. Sci. USA* **68:**2686–2690.

HEWLETT, D. R., and NAHORSKI, S. R., 1978, A comparative study of [^{3}H]haloperidol and [^{3}H]spiroperidol binding to receptors on rat cerebral membranes, *FEBS Lett.* **87:**152–155.
IVERSEN, L., 1965, Inhibition of noradrenaline uptake by drugs, *J. Pharm. Pharmacol.* **17:**62–64.
KATZ, B., and THESLEFF, S., 1957, A study of the "desensitization" produced by acetylcholine at the motor end plate, *J. Physiol. (London)* **138:**63–80.
KUHAR, M. J., AGHAJANIAN, G. K., and ROTH, R. H., 1972, Tryptophan hydroxylase activity and synaptosomal uptake of serotonin in discrete brain regions after midbrain raphe lesions: correlations with serotonin levels and histochemical fluorescence, *Brain Res.* **44:**165–176.
LEONARD, B. E., 1974, Some effects of a new tetracyclic antidepressant compound mianserine on the metabolism of monoamines in the rat brain, *Psychopharmacologia* **36:**221–236.
LEYSEN, J. E., and LADURON, P. M., 1977, A serotonergic component of neuroleptic receptors, *Arch. Int. Pharmacodyn.* **230:**337–339.
LEYSEN, J. E., GOMMEREN, W., and LADURON, P., 1978, Distinction between dopaminergic and serotonergic components of neuropeptic binding sites in limbic brain areas, *Biochem. Pharmacol.* **28:**447–448.
LEYSEN, J. E., NIEMEGERS, C. J. E., TOLLENAERE, J. P., and LADURON, P. M., 1978*b*, Serotonergic component of neuroleptic receptors, *Nature (London)* **272:**163–166.
LOVELL, R. A., and FREEDMAN, D. X., 1976, Stereospecific receptor sites for *d*-lysergic acid diethylamide in rat brain: effect of neurotransmitters, amine antagonists and other psychotrophic drugs, *Mol. Pharmacol.* **12:**620–630.
MACCALL, R. B., and AGHAJANIAN, G. K., 1979, Serotonergic facilitation of facial motoneuron excitation, *Brain Res.* **169:**11–28.
MANSOUR, T. E., SUTHERLAND, E. W., RALL, T. W., and BUEDING, E., 1960, The effect of serotonin (5-hydroxytryptamine) on the formation of adenosine 3′,5′-phosphate by tissue particles from the liver fluke (*Fasciola hepatica*), *J. Biol. Chem.* **235:**466–470.
MARCHBANKS, R. M., 1966, Serotonin binding to nerve-ending particles and other preparations from rat brain, *J. Neurochem.* **13:**1481–1493.
MARCHBANKS, R. M., 1967, Inhibitory effects of lysergic acid derivatives and reserpine on 5-HT binding to nerve-ending particles, *Biochem. Pharmacol.* **16:**1971–1979.
MEHL, E., and GUIARD, L., 1978, Physiological ligands of putative LSD–serotonin receptors: heterogeneity of LSD–Serotonin receptors: heterogeneity of LSD-displacing factors in human body fluids and nervous tissue, *Hoppe Seyler's Z. Physiol. Chem.* **359:**539–542.
MEHL, E., RÜTHER, E., and REDEMANN, J., 1977, Endogenous ligands of a putative LSD–serotonin receptor in the cerebrospinal fluid: higher level of LSD-displacing factor (LDF) in unmedicated psychotic patients. *Psychopharmacology* **54:**9–16.
MIDDLEMISS, D., BLAKEBOROUGH, L., and LEATHER, S. R., 1977, Direct evidence for an interaction of β-adrenergic blockers with the 5-HT receptor. *Nature (London)* **267:**289–290.
MINNEMAN, K. P., HEGSTRAND, L. R., and MOLINOFF, P. B., 1979, Simultaneous determination of β_1- and β_2-adrenergic receptors in tissues containing both receptor subtypes, *Mol. Pharmacol.* **16:**34–46.
MÖLLGÅRD, K., and WIKLUND, L., 1979, Serotoninergic synapses on ependymal and hypendymal cells of the rat subcommissural organ, *J. Neurocytol.* **8:**445–467.
MOORE, H. P. H., and RAFTERY, M., 1979, Ligand-induced interconversion of affinity states in membrane-bound acetylcholine receptor from *Torpedo californica*. Effects of sulfhydryl and disulfide reagents, *Biochemistry* **10:**1907–1911.
NAKAMURA, S., ICHIYAMA, A., and HAYAISHI, O., 1965, Purification and properties of tryptophan hydroxylase in brain, *Fed. Proc.* **24:**604.
NATHANSON, J. A., and GREENGARD, P., 1974, Serotonin-sensitive adenylate cyclase in neural tissue and its similarity to the serotonin receptor: a possible site of action of lysergic acid diethylamide, *Proc. Natl. Acad. Sci. USA* **71:**797–801.
NELSON, D. L., HERBET, A., BOURGOIN, S., GLOWINSKI, J., and HAMON, M., 1978, Character-

istics of central 5-HT receptors and their adaptative changes following intracerebral 5,7-dihydroxytryptamine administration in the rat, *Mol. Pharmacol.* **14**:983–995.
NELSON, D., HERBET, A., PICHAT, L., and GLOWINSKI, J., 1979, *In vitro* and *in vivo* disposition of [^{3}H]methiothepin in brain tissue, *Arch. Pharmacol.* **310**:25–33.
NELSON, D., HERBET, A., ENJALBERT, A., BOCKAERT, J., and HAMON, M., 1980, Serotonin-sensitive adenylate cyclase and [^{3}H]serotonin binding sites in the CNS of the rat. I and II, *Biochem. Pharmacol.* **29**:2445–2453.
NELSON, D. L., PEDIGO, N. W., and YAMAMURA, H., 1981, Multiple [^{3}H]-5-HT binding sites in rat brain. The serotoninergic neuron, *J. Physiol.* (*Paris*) **77**:369–372.
ÖGREN, S. O., FUXE, K., AGNATI, L. F., GUSTAFSSON, J. A., JÖNSSON, G., and HOLM, A. C., 1979, Reevaluation of the indoleamine hypothesis of depression. Evidence for a reduction of functional activity of central 5-HT systems by antidepressant drugs, *J. Neural Transm.* **46**:85–103.
OSBORNE, N. N., 1980, Reasons for using the snail brain in pharmacological research, *Trends in Pharmacol. Sci.* **1**:290–292.
PAGEL, J., CHRISTIAN, S. T., QUAYLE, E. S., and MONTI, J. A., 1976, A serotonin-sensitive adenylate cyclase in mature rat brain synaptic membranes, *Life Sci.* **19**:819–824.
PEDIGO, N. W., REISINE, T. D., FIELDS, J. Z., and YAMAMURA, H. I., 1978, [^{3}H]Spiroperidol binding to two receptor sites in both the corpus striatum and frontal cortex of rat brain, *Eur. J. Pharmacol.* **50**:451–453.
PEDIGO, N. W., YAMAMURA, H. I., and NELSON, D. L., 1981, Discrimination of multiple [^{3}H]-5-hydroxytryptamine binding sites by the neuroleptic spiperone in rat brain, *J. Neurochem.* **36**:220–226.
PEROUTKA, S., and SNYDER, S., 1979, Multiple serotonin receptors: differential binding of [^{3}H]-5-hydroxytryptamine, [^{3}H]lysergic acid diethylamide, and [^{3}H]-spiroperidol, *Mol. Pharmacol.* **16**:687–699.
PEROUTKA, S. J., GREENBERG, D. A., U'PRICHARD, D. C., and SNYDER, S. H., 1978, Regional variations in alpha-adrenergic receptor interactions of [^{3}H]dihydroergotryptine in calf brain; implication for a two-site model of α-receptor function, *Mol. Pharmacol.* **14**:403–412.
PEROUTKA, S. J., LEBOVITZ, R. M., and SNYDER, S. H., 1979, Serotonin receptor binding sites affected differentially by guanine nucleotides, *Mol. Pharmacol.* **16**:700–708.
PFEUFFER, T., 1977, GTP-binding proteins in membranes and the control of adenylate cyclase activity, *J. Biol. Chem.* **252**:7224–7234.
QUAYLE, E. S., PAGEL, J., MONTI, S. A., and CHRISTIAN, S., 1977, Specific serotonin binding and adenylate cyclase stimulation: a correlative study using isolated synaptosomal membranes from mature rat brain, *Alabama J. Med. Sci.* **14**:259–263.
QUIK, M., and IVERSEN, L. L., 1979, Regional study of [^{3}H]spiperone binding and the dopamine-sensitive adenylate cyclase in the rat brain, *Eur. J. Pharmacol.* **56**:323–330.
SALOMON, Y., LONDOS, C., and RODBELL, M., 1974, A highly sensitive adenylate cyclase assay, *Anal. Biochem.* **58**:541–548.
SAMANIN, R., MENNINE, T., FERRARIS, A., BENDOTTI, C., and BORSINI, F., 1980, Hyper- and hyposensitivity of central serotonin receptors: [^{3}H]serotonin binding and functional studies in the rat, *Brain Res.* **189**:449–457.
SAVAGE, D., MENDELS, J., and FRAZER, A., 1980, Monoamine oxidase inhibitors and serotonin uptake inhibitors: differential effects on [^{3}H]serotonin binding sites in rat brain, *J. Pharmacol. Exp. Ther.* **212**:259–263.
SCHUBERT, J., NYBÄCK, H., and SEDVALL, G., 1970, Accumulation and disappearance of [^{3}H]-5-hydroxytryptamine formed from [^{3}H]tryptophan in mouse brain: effect of LSD 25, *Eur. J. Pharmacol.* **10**:215–224.
SCHWARCZ, R., BENNET, J. P., and COYLE, J. T., 1977, Loss of striatal serotonin synaptic receptor binding induced by kainic scid lesions: correlation with Huntington's disease, *J. Neurochem.* **28**:867–869.

SEGAL, M., 1981, Regional differences in neuronal responses to 5-HT: Intracellular studies in hippocampal slices, *J. Physiol. (Paris)* **77**:373–375.

SEGAWA, T., MIDZUTA, T., and NOMURA, Y., 1979, Modifications of central 5-hydroxytryptamine binding sites in synaptic membranes from rat brain after long-term administration of tricyclic antidepressants, *Eur. J. Pharmacol.* **58**:75–83.

SHASKAN, E. G., and SNYDER, S. H., 1970, Kinetics of serotonin accumulation into slices from rat brain; relationships to catecholamine uptake, *J. Pharmacol. Exp. Ther.* **175**:404–418.

SHIH, J. C., and YOUNG, H., 1978, The alteration of serotonin binding sites in aged human brain, *Life Sci.* **23**:1441–1448.

SIMS, K. L., and BLOOM, F. E., 1973, Rat brain L-3,4-dihydroxyphenylalanine and L-5-hydroxytryptophan decarboxylase activities: differential effect of 6-hydroxydopamine, *Brain Res.* **49**:165–175.

SNYDER, S. H., and BENNETT, J., 1975, Biochemical identification of the postsynaptic serotonin receptor in mammalian brain, in: *Pre- and Postsynaptic Receptors* (E. Usdin and W. Bunney, Jr., eds.), Vol. 3 pp. 191–206, Marcel Dekker, Inc., New York.

STEINBUSCH, H. W., VERHOSFSTAD, A. A., and JOOSTEN, H. W., 1978, Localization of serotonin in the central nervous system by immunohistochemistry: description of a specific and sensitive technique and some applications, *Neuroscience* **3**:811–819.

STOLZKI, B., KAISER, H., and MEHL, E., 1978, Heterogeneity of LSD-displacing factors and multiple types of high-affinity LSD-binding sites, *Life Sci.* **23**:593–598.

SUDDITH, R. L., HUTCHISON, H. T., and HABER, B., 1978, Uptake of biogenic amines by glial cells in culture. 1. Neuronal-like transport system for serotonin, *Life Sci.* **22**:2179–2188.

SUTHERLAND, E. W., OYE, I., and BUTCHER, R. W., 1965, The action of epinephrine and the role of the adenylate cyclase system in hormone action, *Recent Prog. Horm. Res.* **21**:623–647.

TRULSON, M. E., and JACOBS, B. L., 1979, Alterations of serotonin and LSD receptor binding following repeated administration of LSD, *Life Sci.* **24**:2053–2062.

TWAROG, B. M., and PAGE, I. H., 1953, Serotonin content of some mammalian tissues and urine, *Am. J. Physiol.* **175**:157–161.

U'PRICHARD, D. C., BYLUND, O. B., and SNYDER, S. H., 1978, (±)-[^{3}H]epinephrine and (−)-[^{3}H]dihydroalprenolol binding to β_1 and β_2 noradrenergic receptors in brain, heart, and lung membranes, *J. Biol. Chem.* **253**:5909–5102.

U'PRICHARD, D. C., and SNYDER, S. H., 1978, Guanyl nucleotide influences on [^{3}H]ligand binding to α adrenergic receptors in calf brain membranes, *J. Biol. Chem.* **253**:3444–3452.

VON HUNGEN, K., ROBERTS, S., and HILL, D. F., 1974, Development and regional variations in neurotransmitter-sensitive adenylate cyclase systems in cell-free preparations from rat brain, *J. Neurochem.* **22**:811–817.

WHITTAKER, P. M., and SEEMAN, P., 1978*a*, High-affinity [^{3}H]serotonin binding to caudate: inhibition by hallucinogens and serotonin drugs, *Psychopharmacology* **59**:1–5.

WHITTAKER, P. M., and SEEMAN, P., 1978*b*, Selective labelling of serotonin receptors by delta [^{3}H]lysergic acid diethylamide in calf caudate, *Proc. Natl. Acad. Sci. USA* **75**:5783–5787.

WHITTAKER, P. M., and SEEMAN, P., 1979, Selective labeling of apomorphine receptors by [^{3}H]-LSD, *Eur. J. Pharmacol.* **65**:269–271.

WILLIAMS, L. T., and LEFKOWITZ, R. J., 1977, Slowly reversible binding of catecholamines to a nucleotide-sensitivie state of the β-adrenergic receptor, *J. Biol. Chem.* **252**:1207–1213.

5

RECEPTORS FOR AMINO ACID TRANSMITTERS

S. *Robert Snodgrass*

1. INTRODUCTION

This chapter will review plasma membrane receptors for amino acids which are or may be neurotransmitters. Our focus will be on the mammalian brain, with brief consideration of simpler animals and peripheral nervous tissue. Although the number of amino acid transmitters is not known at present, it is convenient to divide neuroactive amino acids into excitatory and inhibitory transmitters. This traditional approach must be viewed with some skepticism. Invertebrate experience clearly indicates that the same molecule may mediate excitation, inhibition, or mixtures of both, depending upon the nature of the postsynaptic receptor which it encounters (Blankenship *et al.*, 1971; Gerschenfeld, 1973; Swann and Carpenter, 1975; Walker *et al.*, 1975; Carpenter *et al.*, 1977). The notion that GABA is always inhibitory and glutamate is always excitatory in the mammalian CNS derives from the earliest studies, which utilized extracellular recording. It will soon require modification. Evidence is accumulating that GABA can produce either depolarizing or hyperpolarizing responses, when studied by intracellular recording (Brown *et al.*, 1980; Assaf *et al.*, 1981). GABA has been known to produce depolarizing responses in sympathetic ganglia for some years (Bowery and Brown, 1974), but depolarizing GABA responses remain of uncertain importance

S. Robert Snodgrass • Department of Neurology, University of Southern California, Children's Hospital of Los Angeles, Los Angeles, California 90027.

in the mammalian CNS. Conversely, glutamate responses are not always excitatory (Yamamoto *et al.*, 1976). We shall pay particular attention to GABA receptors. The relationship between GABA receptors and other receptors, such as those for benzodiazepines and picrotoxinin, will also be reviewed briefly (Section 2.10). This chapter will stress review articles, and the reader is referred to previous general reviews, which contain much valuable information (Kelly and Beart, 1975; McLennan, 1975; Enna and Maggi, 1979; Nistri and Constanti, 1979; Watkins *et al.*, 1981; Enna, 1981; Krogsgaard-Larsen and Falch, 1981).

1.1. Historical Review

The term receptive substance was first used in 1905 by Langley (Langley, 1905), and Ehrlich in the following year announced his concept that drugs do not act without binding; yet these modern-sounding ideas did not have a great impact for some decades. The notion of receptors remained a hypothetical construct used mainly by pharmacologists until the advent of radioligand binding studies in the early 1970s. It is instructive to take two well-known historical reference points: the first in 1963, at the time of Curtis' review in *Pharmacological Reviews* (Curtis, 1963); the second, the well-known review by Curtis and Johnston in 1974 (Curtis and Johnston, 1974). These are not the only important review articles of the last generation; they were chosen because of their broad scope and influence on teaching and research. Intracellular recording in the vertebrate CNS was necessary to establish the chemical nature of synaptic transmission, and to show that both inhibitory and excitatory postsynaptic potentials were seen. By 1963, inhibitory postsynaptic potentials were well known, but the principal success of transmitter neuropharmacology had been in the area of inhibitory neurotransmission, where the role of GABA as a transmitter was beginning to take shape. However, understanding of even inhibitory amino acid receptors was confused by present standards. Strychnine actions were not explicable, although it was known that strychnine had a special effect on spinal reflexes, and the pharmacology of pre- and postsynaptic inhibition was just beginning to be understood (Eccles *et al.*, 1963). In 1974 there was much improvement in the understanding of the inhibitory amino acid responses, and glycine was well established as a second transmitter for the production of so-called "fast ipsp's," but excitatory receptors and excitatory transmitters remained a problem. The first studies of kainic acid were just beginning, and these appear to have provided the impetus to understanding of the complex and overlapping excitatory amino acid responses. Kainic acid and its cytotoxic effects on nerve cells became the focus of many investigations, stimulated by its utility for anatomical studies and its possible relevance to diseases such as Huntington's chorea and epilepsy. It was soon apparent that kainate and glutamate responses differed in a number of ways and that kainate could not be regarded as a glutamate analog.

The author believes that one barrier to understanding of amino acid receptors has been excessive "localism." For example, many of the pioneering physiological studies were done in the cat or frog spinal cord. Although they provided a useful framework of understanding for other tissues, the assumption that agonist and antagonist specificity will be invariant in all CNS regions is not supported by increased experimental sophistication. It has been customary to state that strychnine-sensitive inhibition is due to glycine, and bicuculline-sensitive inhibition to GABA, but there is increasing evidence that there are significant overlaps in the antagonist specificity of GABA and glycine receptors, as will be discussed below. We now have evidence for the existence of GABA receptors which are insensitive to either bicuculline or strychnine, thus further complicating matters.

Because of the success, relative ease, and widespread adoption of radioligand binding studies, it is important to remember that such techniques provide only partial information about receptors and must always be supplemented by other approaches. The history of amino acid transmitters is intimately linked with that of iontophoresis, and this technique has been more central to the study of these transmitters than is the case for monoamines or opiate peptides, for example. This review will contrast the information available from different approaches, and it will be apparent that for some receptors, we have very one-sided pictures, with only physiological or only radioligand binding information. For other problems, the different experimental approaches may give conflicting answers.

1.2. Definitions

A plasma membrane receptor is a macromolecule with a recognition site which binds the agonist and results in a change in the internal milieu of the cell, in most cases by a change in membrane conductance or permeability to small ions. In general, receptors show stereospecificity, and receptor binding does not chemically change a ligand; however, some receptors produce an internalization of the ligand which may be associated with metabolic degradation. Receptors may be studied by neurophysiological techniques, radioligand binding techniques, or by isolated organ techniques. Each approach has its limitations, and this review will contrast the differing contributions made by each to the understanding of various amino acid receptors.

1.3. Neurophysiological Techniques

The iontophoretic studies of Curtis and his colleagues provided the first evidence for the existence of amino acid receptors. These early studies indicated that a variety of dicarboxylic amino acids depolarized neurons, whereas a number of other amino acids had inhibitory actions. Furthermore, a dicarboxylic excitant could generally be converted into an inhibitory substance by decarboxylation (Curtis and Watkins, 1960). Amino acids with direct membrane effects generally contain a free amino group and at least one ionizable acidic group, separated by a distance equivalent to a straight-line carbon chain of one to four carbon atoms (Buu *et al.*, 1976). Detailed information about the membrane effects of amino acids and other transmitters generally requires intracellular recording, which is technically difficult in the intact CNS. Therefore, much of the best neurophysiological data has come from studies of neurons grown in tissue culture (Barker and Ransom, 1978; Dichter, 1980; Barker *et al.*, 1981) or from *in vitro* slice preparations such as the hippocampal slice preparation (Andersen *et al.*, 1980; Segal, 1980). Significant information continues to accumulate from iontophoretic studies utilizing extracellular recording, but exact delineation of the membrane effects of drugs is not possible in such systems. An obvious advantage of the physiological approach is that it deals with (relatively) intact cells and that sequential measurements over time are possible. The destructive sampling involved in ligand binding studies generally rules out sequential studies of the same tissue, although parallel samples may be taken at different time points. The use of sequential sampling is especially valuable in the study of phenomena such as desensitization (Katz and Thesleff, 1957) or rapid changes in synaptic efficacy (Lynch *et al.*, 1976).

A major limitation of neurophysiological studies is that only a few cells can be studied and they are generally not representative of the tissue as a whole. For example, the smallest neurons cannot be successfully impaled for intracellular recording. Because the CNS is so heterogeneous, neurophysiological data are of limited usefulness unless the neurons studied are morphologically identified (Bloom, 1974; Kelly, 1975). Iontophoresis is at best a semiquantitative technique, and the number of ions ejected per unit of current passed changes during an experiment. This serious limitation can be overcome in tissue culture and slice studies by the use of microperfusion or "puffer" techniques, in which air pressure is used to eject solutions from small pipettes less than 5 μm in tip diameter (Choi, *et al.*, 1977; Dichter, 1980). This technique can be quantitative if precautions are taken (Choi and Fischbach, 1981), but it has not been successfully applied to the intact CNS. Because precise control of drug concentrations delivered to receptors is essential to quantitative studies of altered receptor sensitivity, it is likely that this field will be dominated by

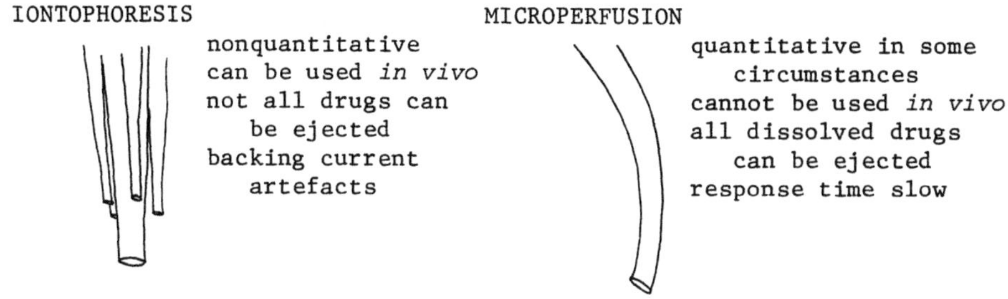

FIG. 1. A comparison of iontophoresis and microperfusion.

tissue culture and slice techniques. To the extent that tissue culture and slice preparations differ from *in vivo* conditions, the precise data which microperfusion studies offer may be misleading. Figure 1 compares the strengths and weaknesses of iontophoresis and microperfusion. It should be clear that iontophoresis, although less quantitative than the ideal technique, will continue to be important because of its application to *in vivo* studies.

1.4. Radioligand Binding Studies

Modern ligand binding studies depend upon the availability of high-specific-activity tritiated or iodinated ligands. Small amounts of radioactive impurities can produce serious artefacts and problems in data interpretation (Builder and Segel, 1978). The first successful ligand binding studies involved α-bungarotoxin, a particularly favorable drug for study because its binding to the nicotinic receptor is irreversible or nearly so (Heidmann and Changeux, 1978). No analogous long-acting agent is available for any of the amino acid receptors, and in the case of excitatory amino acid receptors, the lack of good antagonists has been a major obstacle to progress. Antagonists typically have higher affinity for binding sites than do agonists and they are less likely to be substrates for uptake systems (Snyder and Bennett, 1976).

The usual ligand binding study involves broken cell preparations, and suffers from the mixing together of membranes from a variety of different cell types. Receptor autoradiography can offset many of these limitations by demonstrating the location of receptors on intact cells (Anderson and Greenwald, 1969). However, receptor autoradiography is difficult when the ligand is diffusible and binds transiently to the receptor. In the case of amino acid receptors, we are limited by the ephemeral nature of agonist binding, by the ubiquity of amino acid uptake systems, and often by the presence of endogenous molecules which compete with the radioactive ligand for binding sites (Kuhn *et al.*, 1981). It is possible in some cases to study radioligand binding in intact cells, and the results of such studies may provide useful comparisons to those obtained in membrane preparations. In the study of broken cell preparations, endogenous transmitters may be liberated or produced from larger molecules by catabolic enzymes. The removal of endogenous ligands is a major problem for studies of amino acid receptors, one which has not been fully solved. If the membranes are extensively washed during preparation of the sample, we may find changes in the nature of the membranes whose binding properties are studied. If multiple washes are not employed, variable amounts of endogenous ligands may confound the results. On occasion, apparent receptor super- or subsensitivity has been due to different amounts of endogenous ligand when compared to control conditions (Creese and Sibley, 1981). We call attention to two examples of the importance of endogenous ligands and modulators: the effects of Triton X-100 and other detergents on the binding of GABA and GABA agonists to CNS membranes (Wong and Horng, 1977), and the usefulness of the enzyme adenosine deaminase in the study of adenosine receptors. Even with repeated washing, some adenosine appears to adhere to the membranes; when the enzyme is added during preincubation, better binding (greater specific binding) of adenosine ligands can be shown (Bruns *et al.*, 1980; Williams and Risley, 1980). Perhaps the use of analogous enzymes would facilitate the *in vitro* binding studies of excitatory amino acid receptors. Although subject to a host of methodological problems, *in vitro* ligand binding assays have brought great progress to the receptor field. One of their advantages is that we can compare animal results with those obtained in postmortem human brain tissue, affording a rare chance to validate concepts derived from animals studies to human disease. Additional important benefits are that we can predict *in vitro* (and usually without great expense) whether a drug acts at a given receptor, and we can study the mechanism of action of various toxins and their utility in purification of receptors. Purified receptors may be used to raise antibodies for the purposes of receptor localization by immunohistochemistry. They are especially useful for structure–activity studies, which utilize a large number of related compounds, and are central to the development of better drugs (Barlow, 1980).

When one compares the concentrations of amino acids (or their antagonists) needed to produce a conductance change with that required to displace the transmitter in radioligand assays, a consistent difference emerges. More agonist is needed to produce the physiological response than would be expected from the binding assay. This quantitative problem cannot be studied well by iontophoresis. Homma and Rovainen (1978) felt that uptake might account for part of this discrepancy. Uptake systems may account for some of the decline in drug potency over time, but they cannot explain the basic discrepancy between binding and "functional" data. For example, strychnine, which is not a substrate for uptake, has an affinity in the nanomolar range, and yet a concentration near to or above 1 μM is typically required before one sees physiological effects. The role of uptake in controlling agonist responses is poorly understood in the whole area of amino acid transmitters. See J. Cohen (1978) and Strickland and Loeb (1981) for helpful further discussion about the discrepancy between K_d seen in binding studies and concentrations needed for physiological responses, and Crnic *et al.* (1973) for a theory about a confounding role of uptake processes in effects of amino acids such as GABA.

In receptor studies, the correlation between animal data and human data has been very good. Further, receptors have fortuitously turned out to be generally stable postmortem, unlike some substrates and enzymes. It is preferable to use a compound different from the labeled compound as a displacer, for example, to determine the amount of nonspecific binding (Maayani and Weinstein, 1980). When the same compound is used as radioligand and displacer, a variety of methodological problems may occur, especially if the drug is a hydrophobic compound (Mendel and Almon, 1979). This problem is exemplified by the case of phencyclidine or PCP, an important drug for which saturable binding has been reported, but whose receptor status remains uncertain (Zukin *et al.*, 1979; Maayani and Weinstein, 1980; Vincent *et al.*, 1980). In the case of the amino acids, until very recently there was no alternative to using the same drug as ligand and displacer, which contributed to confusing data. Other problems arise from the need to exclude the effects of ubiquitous uptake systems: binding studies of amino acids and their derivatives are typically done in sodium-free media. If receptor behavior is different in sodium-free media, as reported for frog spinal cord by Nistri (1981*b*), the results should be validated by studies which include sodium ions and block uptake in some other way. Several reports indicate that Na^+ ions alter GABA receptor binding in a way that is unlikely to be uptake-related (Kurioka *et al.*, 1981).

Kinetic studies of radioligand assays have been extensively performed. Such studies can provide useful hints of complex relationships between linked receptors, as is true of benzodiazepine and GABA receptors (see Section 2.10). A drug which alters receptor binding in a noncompetitive manner may act at an allosteric site (Barlow, 1980). However, the use of kinetic studies to demonstrate multiplicity of receptor sites has not always been subject to careful scrutiny. It has been customary to take nonlinear Scatchard plots as evidence of multiple binding sites. In fact, such nonlinearity has many possible causes, and most experiments have not presented sufficient data to choose between them (Boeynaems and Dumont, 1975). Furthermore, careful studies of ligand binding in a variety of different buffers have often shown linear or biphasic Scatchard plots depending upon the medium chosen (Hill and Bowery, 1981, for example). In other cases, as with classical GABA receptors, it seems that a variety of treatments may be used to convert one form of receptor to another and give a variety of different binding isotherms which may be very far removed from what existed in the intact neuron (Lloyd and Beaumont, 1980; Browner *et al.*, 1981). Our measuring device (the synaptic membrane preparation) may never reproduce the natural state, particularly if extensive manipulations are used, as in the preparation of junctional complexes. See E. Roberts *et al.* (1981) for a discussion of the differences between stereospecificity of the isolated crayfish stretch receptor and mouse brain membranes.

Receptor autoradiography has been studied extensively by Kuhar and his colleagues, whose methods have broad applicability to virtually all receptors on the exterior surface of cells (Young and Kuhar, 1979). This approach can usually demonstrate which cells have the receptors in question. However, electron microscopy is required for best anatomical definition.

1.5. Isolated Tissue or Organ Techniques

Isolated tissue or organ preparations began with the organ bath preparations used to study peripheral tissues such as gut and uterus. Such methods are of limited use in the study of CNS tissues, although they may be useful for the study of nerve-muscle synapses and for the study of autonomic tissues such as adrenal medulla. The greatest value of such preparations for CNS studies has been for the measurement of changes in cyclic nucleotides after stimulation of slices synaptosomes, or other CNS preparations with drugs. Data resulting from such studies have led to the concept that most transmitters act via changes in cyclic nucleotides (Rodbell, 1980). This concept appears valid for catecholamines, but a similar role for cyclic nucleotides remains to be established in the case of any of the amino acid transmitters, in spite of experiments seeking to prove such roles (see Section 2.5). Whether this indicates a fundamental difference between amino acid transmitters, which are much faster acting in general, and other transmitters and hormones, is not yet clear. A variant of this approach uses the incorporation of radioactive phosphorus into brain phospholipids as a signal "downstream from the receptors," as several transmitters have been shown to alter this incorporation (Friedel *et al.*, 1973; Subramanian *et al.*, 1980). Again the cells responsible for the change observed are not identified.

A new variant of this approach employs autoradiography of small molecules which enter cells only after receptor activation. Yoshikami (1981) used tritiated agmatine, a primary amine which enters sympathetic neurons only after activation of nicotinic acetylcholine receptors. This method may be useful for quantitating the sensitivity of cells to a transmitter, and the change in sensitivity which follows denervation appears to be demonstrable by this method. Such an approach would be best used in tissue culture or *in vitro* slice experiments, where the drug can be applied in the medium and there is no restriction of access to the receptor sites. This approach may also serve to map the sensitivity of processes which are too small to be impaled with microelectrodes, if electron microscopy is used. It must, however, be remembered that all isolated tissue techniques, whether they measure cyclic nucleotides or other chemical change, have a "higher-order relationship" to the synaptic receptors (Werman, 1969; Nistri and Constanti, 1979; Strickland and Loeb, 1981). There is no fixed relationship between the number of receptors occupied and the effect seen. Furthermore, the quantitative relationship between receptor activation and ion flow may change with age, innervation, or with changes in the lipid environment of the membrane (Chang and Bock, 1979; Lloyd and Beaumont, 1980). Although this review stresses the deficiencies of the nonphysiological approaches, physiological approaches have their problems as well. In addition to the severe sampling problem, even when conductance can be measured, conductance changes are often found to be voltage-dependent, meaning that the membrane potential should ideally be held at a constant value, most generally the resting value.

2. GABA RECEPTORS

2.1. Historical Review

Although GABA was discovered as recently as 1950 (Awapara *et al.*, 1950; Roberts and Frankel, 1950), GABA receptors are much better understood than those of any other amino acid. This may be related to the apparently large number of GABA synapses (it is estimated from autoradiographic and immunocytochemical studies that perhaps 50% of all forebrain nerve endings are GABA terminals). Early iontophoretic studies by Curtis and his students defined a major role for GABA as a ubiquitous neuronal inhibitor. We will return later to the question why all or virtually all neurons have receptors for GABA and glutamate. It is noteworthy that bicuculline, the prototypical GABA receptor antagonist, was not used until the 1970s. The early parts of this decade saw considerable discussion about the relative merits of bicuculline and picrotoxin as GABA antagonists (Godfraind *et al.*, 1970; Biscoe *et al.*, 1972; Freeman, 1973). Among the first steps toward understanding the function of the neuroactive amino acids was the discovery of active uptake systems for all the neuroactive amino acids. These uptake systems proved to be a problem for subsequent radioligand binding studies, but could generally be controlled with the use of sodium-free media (Snyder and Bennett, 1976).

The major problem for early GABA receptor studies was that of separating uptake sites from receptor sites. An additional problem was that of fleeting binding between GABA and its receptor, which meant that filtration assays were difficult to use with GABA and indeed with almost all the amino acid transmitters. Snyder's laboratory was the first to demonstrate saturable and specific GABA receptor binding, using GABA itself as a displacer (Zukin *et al.*, 1974). This assay involved extensive freezing and thawing of membranes, which was later superceded by the discovery that treatment of the membranes with low concentrations of Triton X-100 markedly increased specific binding (Wong and Horng, 1977; Enna and Snyder, 1977). Extensive washing of membranes is not always desired, as it may eliminate important interactions between GABA receptors and those for drugs such as barbiturates. We have now entered the era of receptor solubilization, which is important for receptor purification and biochemical studies. Many binding properties of receptors, including GABA receptors, are retained in the solubilized preparation (Gavish and Snyder, 1980*a*). It is possible to differentially solubilize GABA receptor sites and GABA uptake sites (Lester *et al.*, 1981). It is possible also to solubilize GABA and benzodiazepine (BDZ) receptors together (Gavish and Snyder, 1980*a*) or separately (Massotti *et al.*, 1981).

2.2. Separation of GABA Receptors from Uptake

Uptake processes appear to be important in the function of GABA synapses (Iversen, 1972), and because GABA uptake sites greatly outnumber GABA receptor sites (Table 1) (Coyle and Enna, 1976), the addition of GABA to membrane preparations would primarily study uptake sites unless the latter were destroyed or inactivated. Some early studies used drugs to suppress GABA uptake systems (for example, see Peck *et al.*, 1976); these were generally not successful, whereas removal of sodium ions from the incubation medium more effectively inhibited uptake sites, which tend to survive freezing and thawing rather well. In studying the phenomenon of receptor desensitization, it was important to be certain that GABA uptake did not account for the desensitization which is commonly seen with prolonged GABA application. Since uptake inhibition

TABLE 1
Comparison of GABA Uptake Sites to Receptor Numbers in Four Species

Species	[^{3}H]-GABA binding sites fmol/mg wet weight	[^{3}H]-GABA uptake sites fmol/mg wet weight
Rat[a]		
Frontal cortex	1.15	9.38
Occipital cortex	1.29	6.33
Caudate-putamen	2.85	12.19
Hypothalamus	1.63	3.85
Substantia nigra	2.65	13.20
Cerebellar cortex	6.84	4.01
Cervical cord	0.66	1.09
Cat[a]		
Frontal cortex	1.49	4.72
Occipital cortex	1.98	6.25
Caudate	3.25	3.90
Hippocampus	2.61	9.78
Hypothalamus	1.94	2.09
Substantia nigra	3.31	5.05
Cerebellar cortex	4.16	3.12
Cervical cord	0.48	1.41
Monkey[b]		
Frontal cortex		
Temporal cortex	1.54	6.07
Cingulate cortex	2.04	7.39
Amygdala	1.51	7.63
Hippocampus	1.41	8.48
Caudate	0.73	3.90
Substantia nigra	2.52	5.86
Cerebellar cortex	1.13	6.75
Cervical cortex	1.71	2.30
Cervical cord	0.09	1.52
Postmortem human brain[a]		
Frontal cortex	2.39	4.75
Temporal cortex	4.05	1.61
Occipital cortex	4.61	3.39
Hippocampus	1.38	5.09
Hypothalamus	0.95	1.27
Caudate	0.79	6.00
Substantia nigra	2.65	4.59
Cerebellar cortex	2.99	6.32
Cervical cord	0.37	0.90

[a] The rat, cat, and human data come from the author's studies and all involve brain membranes treated with Triton X-100. [^{3}H]-GABA binding utilized a ligand concentration of 7.5 nM in sodium-free buffer, while GABA uptake utilized 1 μM [^{3}H]-GABA in a HEPES buffer containing 122 nM NaCl. The human postmortem material was taken from the brains of adult patients believed to have no neurological disorder, by both clinical history and gross inspection of the brain.
[b] The monkey data given here were calculated from Table I of Enna (1981).

fails to alter the time course of GABA densensitization in the dorsal root ganglion (Desarmenien *et al.*, 1981), it is unlikely that the ubiquitous GABA uptake system accounts for desensitization. However, blockade of GABA uptake allows exogenous GABA to penetrate deeper into the dorsal root gangion, with resulting greater net depolarization (Desarmenien *et al.*, 1981); see Section 2.7. In the more complex CNS, studies have been inconclusive. They suggest that uptake has a role in terminating the action of GABA, but not of muscimol or THIP.

2.3. GABA Agonists

The first GABA binding studies used [^{3}H]-GABA itself. GABA is less potent than muscimol, an alkaloid from the mushroom *Amanita muscaria* (see Fig. 2). Muscimol is typically about 10 times as potent as GABA in displacing [^{3}H]-GABA or bicuculline from membrane preparations. Muscimol has a semirigid conformation, with a flexible side chain. Its marked potency at most GABA receptors suggests that they react best with GABA in an extended conformation (Azanza and Walker, 1975). Muscimol is less potent than GABA at the bicuculline-insensitive "GABA receptor" described by Bowery and co-workers, and 3-aminopropanesulfonic acid, another potent GABA agonist, is even weaker at this receptor. Thiomus-

TABLE 2
Comparison of the in Vitro and in Vivo Potency of GABA Drugs

Drug	Turning response	Displacement of [^{3}H]-GABA
	Data from Author's laboratory	
Muscimol	0.05	20.7
THIP	0.96	1150
Homomuscimol	5.9	17000
Isoguvacine	19.5	1420
3-APS	225	325
GHBA	400	>24000
Guvacine	360	>24000
	Data of Arnt *et al.* (1979)	
S(−)-5′-methyl muscimol	5.1	64
Thiomuscimol	28	12
Azamuscimol	36	>100,000
Isomuscimol	276	>100,000
Isonipecotic acid	471	15,000

[a] The table compares the potency of various GABA-related drugs, determined in the author's laboratory or taken from the paper of Arnt *et al.* (1979). In every case, turning response is indicated as ED_{50}, estimated from measuring rotational behavior. The potency of each drug was also studied in binding assays. The concentration of [^{3}H]-GABA employed by the author was 7.5 nM while the data from Arnt *et al.* employed a similar but slightly different binding assay.

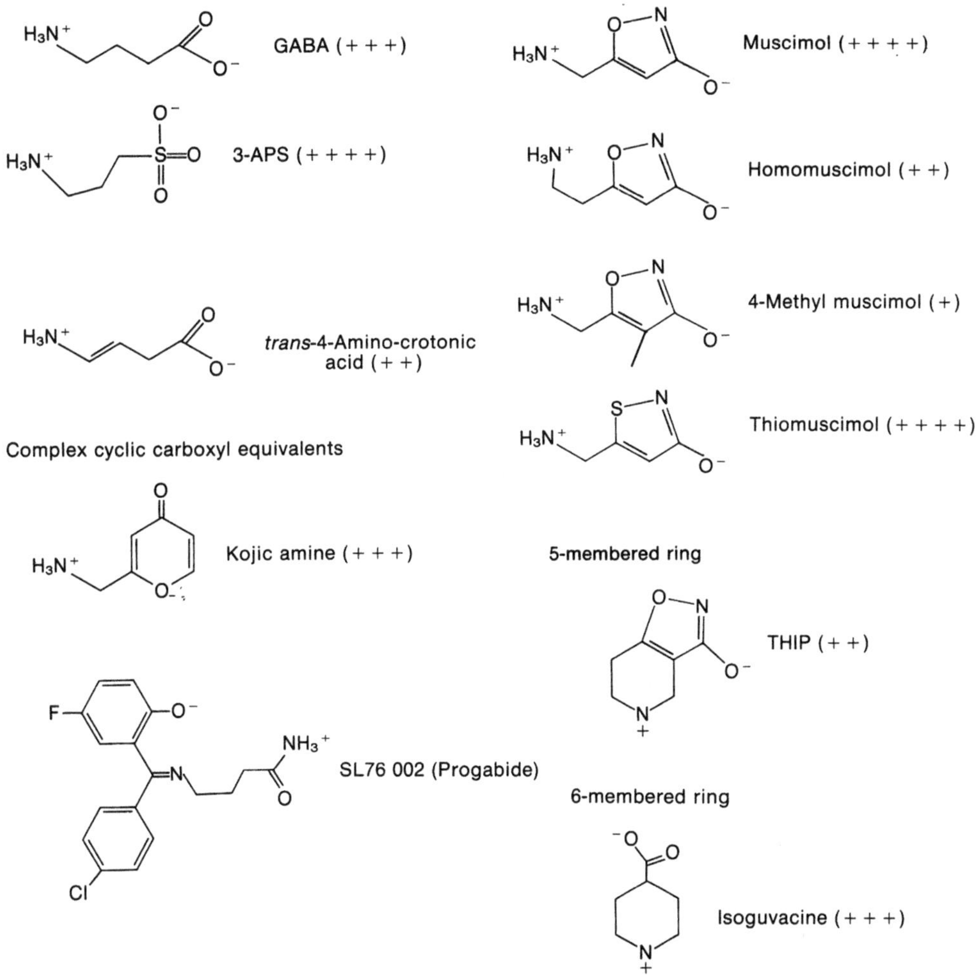

FIG. 2. Potency of some GABA agonists in the mammalian CNS. Key: + + + +, more potent than GABA; + + +, similar in molar potency to GABA; + +, moderately potent; +, weak potency. SL 76002 is not ranked for potency since it acts as a prodrug which liberates GABA in the CNS.

cimol is about as potent as muscimol in the studies of the mammalian CNS, yet is substantially less potent after *in vivo* application (see Table 2) or when applied to cockroach and insect neurons (C. Roberts *et al.*, 1981). In the latter test dihydromuscimol lacks the planar isoxazolol ring of muscimol. Homomuscimol is slightly less potent than muscimol, whereas isomuscimol has hardly any receptor activity. These facts give us some information about the structure–activity relationships of the GABA receptor.

Structure–activity studies suggest that the preferred GABA conformation for its classical receptor is partially folded and nearly planar at the carboxyl end of the molecule, as muscimol is a planar molecule. Substitution on the alpha carbon has a much greater effect than does substitution on the beta or gamma carbons; furthermore substitutions in the cyclic muscimol molecule generally reduced affinity more than the same substitution in the GABA series (Nicholson *et al.*, 1979; Krogsgaard-Larsen *et al.*, 1981). Alterations in the muscimol side chain, such as the addition of methyl or ethyl groups, markedly decrease GABA agonist potency. The GABA receptor shows stereospecificity: the *cis* and *trans* isomers of 3-aminocyclopentane-1-carboxylic acid (Enna and Snyder, 1977) differ in potency.

Muscimol is effective after systemic administration, but is subject to metabolic degradation, mainly by transamination (Baraldi *et al.*, 1979). It has been surmised that an active metabolite might be responsible for entering the brain and producing the effects observed (Maggi and Enna, 1979), but known muscimol metabolites have little or no ability to displace muscimol from its receptor. Furthermore, the study of Matthews *et al.* found a good correlation between the ability of systemically administered muscimol to block bicuculline-induced seizures and the brain content of muscimol, suggesting that muscimol is in fact the active agent (Matthews *et al.*, 1981). Muscimol produces myoclonic jerks in mice (Menon and Vivonia, 1981) and man (Lloyd and Worms, 1981); this has seemed paradoxical to those who assume that GABA is always an inhibitory transmitter, that GABA and GABA agonists always have anticonvulsant properties (Enna, 1981). Presynaptic GABA receptors may explain some of these phenomena (see Sections 2.6 and 2.13).

Because GABA uptake sites are so numerous, a search has gone on for GABA agonists with little or no potency for uptake sites. Muscimol, for example, is a poor substrate for the GABA uptake system (Johnston *et al.*, 1978; Snodgrass, 1978), but there are so many uptake sites that it is bound by some when presented to intact tissue. This can be a problem for autoradiographic localization of GABA receptors. Isoguvacine is less potent than muscimol as an uptake substrate, yet it too can be released from preloaded synaptosomes by high potassium media (White and Snodgrass, in press). The search for amino acid agonists and antagonists with absolutely no affinity for uptake systems is probably doomed to disappointment. Such unnatural compounds as *N*-methyl-D-aspartate (NMDA) (Skerritt and Johnston, 1981) and D-α-aminoadipate (Charles and Chang, 1981) have been found to be taken up by sodium-dependent processes in the CNS. It is surely unwise to conclude from this property that the compound in question is a neurotransmitter, as was done by Charles and Chang.

Tunnicliff and Smith (1981) reported that MOPS, HEPES, and other "good" buffers inhibit the sodium-independent binding of [^{3}H]-GABA to brain membranes, with much less effect on muscimol binding. Although the effect is not large, the author has also seen it, and it may indicate a different mechanism of action for GABA and muscimol. Constanti and Nistri (1981) have reported that sodium-free media result in much greater inhibition of muscimol than GABA responses in lobster muscle, even though the two agonists have parallel log-dose conductance curves in normal media. Rennie and Gliemann (1981) reported an interesting artefact of Tris buffer on isulin receptors, resulting in the artefactual appearance of down regulation of insulin receptors, which was not seen with other buffers. It is well to ask whether unusual effects seen with synthetic buffers are also present in bicarbonate buffers. The complex effects of sodium ions upon membrane binding studies again remind us that sodium-free media can be misleading guides to receptor activity.

Cyclic GABA Analogs. Krogsgaard-Larsen has introduced a series of useful 6-carbon cyclic GABA agonists. Figure 3 gives their structure and potency; note that THIP is the cyclic analog of muscimol, while 4-PSA is the analog of 3-aminopropanesulfonic acid. THIP enters the brain well after systemic administration, is not a good substrate for GABA-T, and is well tolerated in chronic animal studies (Krogsgaard-Larsen *et al.*, 1981). Isoguvacine enters the brain poorly, in contrast to THIP (Krogsgaard-Larsen and Falch, 1981). In these cyclic analogs, the aminomethyl side chain of muscimol is replaced by a ring, which makes the molecule unsuitable for transamination. At the same time, the lack of a flexible side chain appears to explain the inability to stimulate BDZ binding. Most flexible GABA agonists (muscimol, dihydromuscimol) are potent stimulators of BDZ binding. The bulky sulfonate group may account for the low potency of aminopropanesulfonic acid, a very flexible molecule, as a stimulator of BDZ binding. The rigid cyclic analogs such as THIP, isoguvacine, and PSA are weak or inactive as stimulators of BDZ binding, as discussed in Section 2.10. Isoguvacine oxide contains a reactive epoxide group in place of the double bond of isoguvacine and may be useful as a receptor alkylating agent. Structures of some representative GABA analogues are shown in Fig. 3.

GABA agonists have been compared *in vitro* and *in vivo*, with an eye to understanding the contribution of factors which cannot be present in the membrane preparation. Since GABA action depends very much upon the site of application, e.g., intravenously administered GABA has a different effect upon the firing of dopaminergic cells than when GABA is iontophoretically applied to these cells (Grace and Bunney, 1979; Waszczak and Walters, 1980), it is necessary to agree upon a standard *in vivo* approach. The commonest method has been to inject GABA and

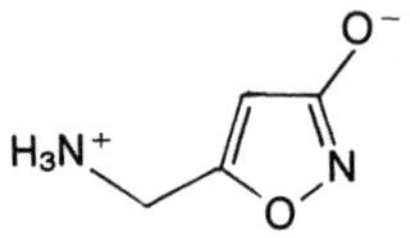

Muscimol

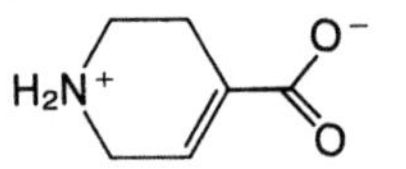

Isoguvacine

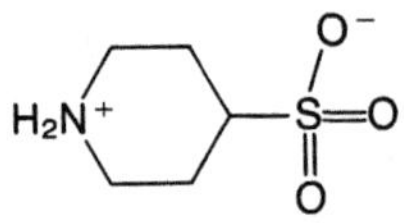

Piperidine-4-sulfonic acid

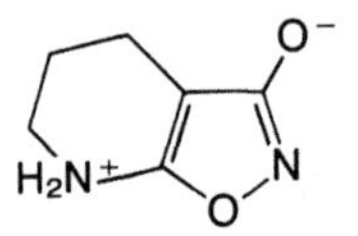

THIP

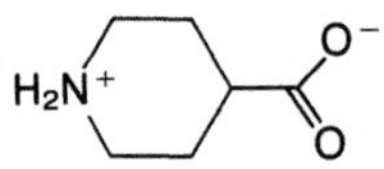

Isonipecotic acid

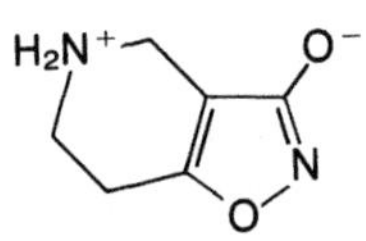

THAZ

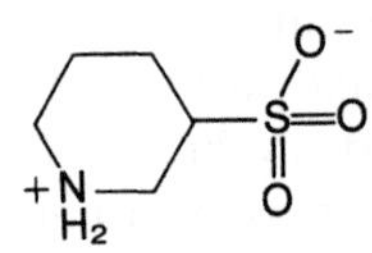

Piperidine-3-sulfonic acid

FIG. 3. Muscimol and its cyclic analogues listed in order of potency. Note that isoguvacine and piperidine-4-sulfonic acid are similar in potency to one another.

GABA agonists into the pars reticulata of the substantia nigra. When injected at this site, GABA and GABA agonists evoke a strong contralateral turning which is readily quantitated (Arnt *et al.*, 1979); however, changes in injection volume or even speed of the injection may markedly alter the results (James and Starr, 1978). Table 2 summarizes the comparison of *in vivo* and *in vitro* potency. Additional GABA agonists which merit further study include progabide or SL76002 (Lloyd and Worms, 1981), isoguvacine oxide, and kojic amine.

The preceding discussion has concerned direct-acting GABA agonists, i.e., those that act directly on GABA receptors. It is also useful to consider the possibility that there may be indirect-acting GABA agonists, analogous to amphetamine, whose action is mediated by discharge of endogenous GABA from nerve endings. As seen in Section 2.8, there is some evidence that baclofen may function as such an indirect agonist.

The activation of GABA receptors classically produces hyperpolarizing ipsp's. This potential has an equilibrium potential similar to the chloride equilibrium potential (Coombs *et al.*, 1955; Araki *et al.*, 1961). The crustacean inhibitory neuromuscular junction has been well studied. GABA is the transmitter and its fast ipsp is mediated almost exclusively by chloride (Takeuchi and Takeuchi, 1966). In some cells, chloride content is high because of a Cl^- pump. This may explain the depolarizing GABA responses seen in sympathetic ganglia and primary afferent terminals (DeGroat, 1972; Nishi *et al.*, 1974). Reversal of the chloride gradient cannot explain all depolarizing GABA responses, however. In *Helix aspersa*, GABA receptors mediate D (depolarizing) and H (hyperpolarizing) responses, but the properties of the GABA binding sites appear to differ (Azanza

TABLE 3
Agonist Classification[a]

	Classification of GABA Agonists: CNS receptors		
	Bicuculline-responsive synaptic inhibition	High-affinity GABA binding (nM)	Activation of BDZ receptors
GABA	+ + +	27	+ +
Type I			
Muscimol	+ + + +		+ + + +
Dihydromuscimol	+ + + +	6	+ + +
Thiomuscimol	+ + + +	9	+ + +
Type 2			
3-APS	+ + + +	21	+
Isoguvacine	+ + +	47	+
Type 3			
THIP	+ + +	95	+ −
Piperidine-4-sulfonate	+	60	0
Type 4			
Baclofen	0	>(1000)	0
Type 5			
GHBA	0	>(1000)	+

[a] This table is modified from that of Meldrum *et al.* (1980). In the first column, the potency at bicuculline-sensitive CNS receptors is given with + + + + as the maximum value. Next the ability of each agonist to displace [^{3}H]-GABA from binding sites on rat forebrain synaptic membranes is given (IC_{50}), and in the last column, the ability of the GABA agonist to activate BDZ receptor binding in rat cerebellar membranes is given.

and Walker, 1975). GABA produces depolarizing responses at a number of extrasynaptic sites on both peripheral and central neurons. Since GABA can produce depolarizing and hyperpolarizing responses from different sites on the same cell, this would require that the Cl^- gradient vary in dendritic processes and soma of the same cell (Barker and Ransom, 1978; Assaf *et al.*, 1981). Note also that while GABA depolarizing actions on sympathetic ganglia are blocked by bicuculline, this antagonist does not prevent GABA's effect on transmitter release, suggesting that sympathetic ganglia possess GABA receptors which are bicuculline insensitive (Kato and Kuba, 1980); see also Section 2.8. GABA agonists differ in their properties. One functional grouping of these drugs is shown in Table 3.

Meldrum *et al.* (1980) classified GABA agonists into four types based on their actions at central bicuculline-sensitive receptors, their effects on peripheral ganglia, and their ability to activate BDZ receptor binding *in vitro*. Note that THIP and P-4-SA, the cyclic analogs of muscimol and APS, not only fail to stimulate [^{3}H]diazepam binding (Maurer, 1979), but block the activation of BDZ binding by muscimol. Meldrum included GHBA because of its effects after intranigral injection and the abnormal EEG patterns it produces, which resemble those produced by systemic muscimol and THIP (Meldrum and Horton, 1980). As noted in Section 2.8, it is likely that GHBA deserves a category separate from baclofen, but Meldrum's schema is a useful start at classification.

2.4. GABA Antagonists

The prototypical GABA antagonist is bicuculline, an alkaloid found in a number of different herbs, including *Dicentra* and *Corydalis* species. This hydrophobic alkaloid was introduced by Curtis' group in Canberra (Curtis *et al.*, 1971); it has a number of other properties, including weak potency as an acetylcholinesterase inhibitor and a blocker of GABA uptake (Olsen *et al.*, 1976). These properties appear unrelated to its ability to block GABA receptors. Some of the early doubts about the usefulness and specificity of this agent may have been related to its chemical instability at pH 6 or above and its slight water solubility (Olsen *et al.*, 1976). At the present time, the first mark of a GABA agonist is that it produces bicuculline-sensitive inhibition. When this is not found, the status of this response as GABA receptor mediated is questionable, at least in the CNS. Bicuculline-insensitive GABA receptors are discussed in Section 2.8. Bicuculline is generally less effective as a physiological GABA antagonist in molluscan and insect GABA receptors than is picrotoxin (Nistri and Constanti, 1979). It is common to speak of bicuculline as a competitive, and picrotoxin as a noncompetitive, antagonist, and this is supported by data from the mammalian CNS (Curtis *et al.*, 1971; Simmonds, 1980), as well as the frog (Nicoll and Wojtowicz, 1980) and lamprey (Homma and Rovainen, 1978) nervous system. The theoretical point is that competitive kinetics imply a competition of agonist and antagonist for the same binding site (Barlow, 1980). In fact, bicuculline antagonism of GABA responses is not always competitive, and sophisticated studies of picrotoxin antagonist mechanisms show that it is mixed in lobster muscle (Constanti, 1978). A later study by the same author showed a similar pattern for bicuculline at the same synapse (Smart and Constanti, 1981). These data suggest that neither bicuculline nor picrotoxinin act at the same site as does GABA in lobster muscle. Bicuculline is largely uncharged at physiological pH, due to the pK_a of 6.48 (Andrews and Johnston, 1973). We do not know which molecular species is the more active at mammalian CNS GABA receptors, although we know that bicuculline specific binding decreases rapidly as pH is decreased below 6.5 (Mohler and Okada, 1977*a*). Even more important is the marked chemical instability of bicuculline at physiological pH, as first pointed out by Olsen *et al.* (1976): In general, one can expect that 50% of bicuculline's effectiveness will be lost after 30 min at physiological pH. In radioligand binding studies, bicuculline and its derivatives compete with GABA at receptor binding sites in the mammalian CNS (Zukin *et al.*, 1974; Collins and Hill, 1974; Mohler and Okada, 1977*a*; Greenlee *et al.*, 1978), whereas picrotoxin does not. However, the binding of bicuculline methiodide bore a nonlinear relationship to the GABA content of the incubation medium when this was specifically studied by Mohler and Okada (1977*a*), suggesting again that pure competitive antagonism of GABA and bicuculline is not common. Furthermore, many of the CNS studies reporting a competitive inhibition employed extracellular recording and may not be definitive.

Receptors are expected to show stereospecificity, even when the natural agonist lacks an asymmetric carbon atom, as GABA does. It is important, therefore, that bicuculline has two stereoisomers which show large differences in potency as a GABA antagonist, whether assayed electrophysiologically or in radioligand binding studies (Collins and Hill, 1974). It should also be noted that Triton X-100 treatment, which markedly increases specific binding of agonists to GABA receptors, does not alter the binding of bicuculline methiodide.

Bicuculline's ability to displace GABA in radioligand studies is enhanced by treatment of membranes with thiocyanate or other chaotropic ions (Enna and Snyder, 1977), whether or not detergent treatment is also used. Furthermore, bicuculline's potency as a GABA displacer varies considerably among the different regions of the rat brain, it being most potent in the midbrain and least potent in the cerebellum (Browner et al., 1981). Thiocyanate appears to eliminate high-affinity GABA binding sites,

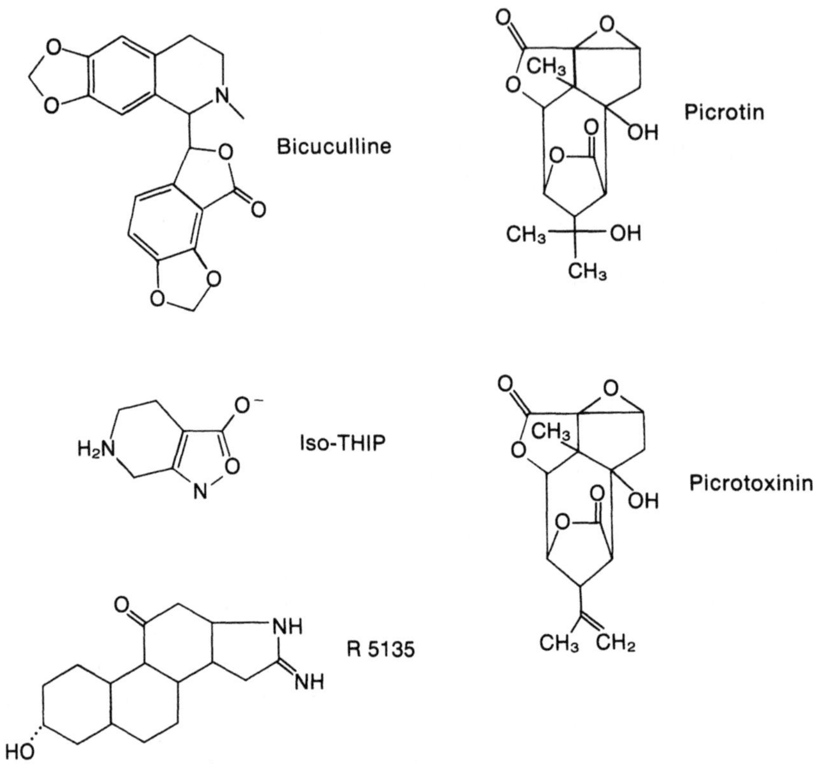

FIG. 4. Some GABA antagonists.

perhaps by solubilizing them (Browner *et al.*, 1981). Since these sites are less sensitive to bicuculline and do not seem to mediate the effects of GABA agonists on BDZ binding, it is reasonable to ask whether they are physiologically important and, if so, just what role they may serve.

Picrotoxin. Picrotoxin another GABA antagonist, occurs naturally in plant seeds and is a mixture of two compounds, picrotoxinin and picrotin (Porter, 1967) (see Fig. 4). Knowledge of picrotoxin goes back at least as far as the 16th century, its early uses including stunning fish. Picrotoxinin, the more active substance, has a double bond which can be reduced. This produces [^{3}H]dihydropicrotoxinin (DHP), which has been used for *in vitro* binding studies, most notably by Olsen, Ticku, and their colleagues (Ticku and Olsen, 1978; Ticku *et al.*, 1978*a,b*). There are two possible DHP isomers, alpha and beta, and the alpha isomer has been used for the receptor binding studies without much study of the different properties of the two isomers in other studies. Studies of picrotoxinin binding have been complicated by two problems: first, the proportion of specific binding is very low, which makes accurate quantitative analysis more difficult; second, the apparent K_d is very high compared to most antagonists, being in the micromolar range (Ticku *et al.*, 1978*a*). Note that picrotoxin contains no nitrogen and is not an alkaloid, although it is of plant origin. Tutin is a similar compound of plant origin which has some ability to antagonize GABA responses and perhaps also glycine (Curtis *et al.*, 1973). Olsen *et al.* (1980) found that tutin was among the most potent competitors for DHP binding sites. Among the others were pyrethroid insecticides, BDZs, and some barbiturates. However, the rank order of potency of the BDZs and barbiturates as DHP antagonists correlated poorly with their potency at classical BDZ receptors or for production of barbiturate responses, as discussed in Section 2.10.

Among other GABA receptor antagonists and related compounds are the bicyclophosporic acid esters (Bowery *et al.*, 1975, 1976*a*), of which the best known is TETS. These so-called "cage convulsants" are noncompetitive GABA blockers and compete for binding sites with dihydropicrotoxinin (Ticku and Olsen, 1979). Their action can be reversed by barbiturates (Nicoll and Wojtowicz, 1980). Barbiturates have been reported to compete for binding with picrotoxinin (Ticku *et al.*, 1978*a*; Leeb-Lundberg *et al.*, 1980), but barbiturates may bind to a different but closely related site (Willow and Johnston, 1981*b*). Of interest is the idea that an endogenous purine may regulate the sensitivity of GABA receptors by binding to picrotoxinin and benzodiazepine receptors (Ticku and Burch, 1980; Skolnick *et al.*, 1980).

Anisatin is a potent GABA antagonist which blocks GABA effects in the frog spinal cord and does not displace GABA from its receptor (Kudo *et al.*, 1981*a*). It appears to be another picrotoxin-like antagonist. Another potent GABA antagonist is the steroid derivative R 5135 (Hunt and Clement-Jewery, 1981). This has not been fully studied but also displaces glycine from its receptor. It has been suggested that picrotoxinin binds to an anion channel in the synaptic membrane (Ticku *et al.*, 1978*b*), but there is no direct evidence for this. If it binds to an ionic channel, this must be a different channel than that which is linked to the glycine receptor, as picrotoxin rarely blocks glycine inhibition.

2.5. Coupling of GABA Receptors to Cyclic Nucleotides

The link between monoamine neurotransmission and changes in cyclic nucleotides has been mentioned. From the first discovery of such a relationship, there have been attempts to find second messengers for neuroactive amino acids. Ferrendelli *et al.* (1974) reported that GABA, glutamate, and glycine all produced increases in cerebellar cyclic GMP when incubated with slices of mouse cerebellum. The effect of glutamate was quickest and quantitatively greatest. The authors suggested that the effect of the inhibitory amino acids might be an indirect one, as it was slow to appear. Subsequent studies have confirmed the ability of glutamate to increase cerebellar cyclic GMP, but have reported opposite effects of GABA, i.e., decreases in cyclic GMP (Costa *et al.*, 1975; Mailman *et al.*, 1978). Muscimol and diazepam both decrease cerebellar cGMP, but the diazepam effect is not seen after blockade of GABA synthesis (Guidotti *et al.*, 1977). The cerebellum is unusual in that cyclic GMP levels are especially high compared to other brain regions, as is the ratio of cGMP to cAMP.

Guidotti *et al.* (1977) reported that GABA agonists could act on pituitary cAMP concentrations. Muscimol reduced pituitary cAMP content and antagonized the effects of isoniazid, which increases cAMP. Muscimol did not prevent reserpine-induced increases in pituitary cAMP. Diazepam and nipecotic acid, which enhance GABA effects by different mechanisms, both blocked the isoniazid effect. The pituitary may be useful for further studies of GABA receptors and "downstream effects." Prolactin secretion is influenced by central and peripheral GABA receptors (Enna, 1981), and GABA agonists unable to enter the brain may have direct pituitary effects on prolactin secretion (Grandison and Guidotti, 1979). Grossman *et al.* (1981) showed that secretion of prolactin by isolated anterior pituitary cells was inhibited by GABA and muscimol. This effect was not potentiated by BDZs, but was blocked by bicuculline and picrotoxin.

2.6. Presynaptic GABA Receptors

Receptors on axons, which are called presynaptic receptors, are very common (Starke, 1981). The first study to suggest the existence of presynaptic receptors involved GABA receptors, as we now understand primary afferent depolarization (Frank and Fuortes, 1957). These investigators studied spinal reflexes in cat motoneurons and observed an apparent reduction in transmitter release, which they ascribed to inhibitory receptors on afferent terminals. By contrast, dorsal root ganglion cells are generally insensitive to acidic amino acids, but both GABA and glutamate appear to be able to depolarize primary afferent terminals. The first biochemical studies of presynaptic receptors were published in the same year, in the sympathetic nervous system (Brown and Gillespie, 1957).

There is much evidence for presynaptic receptors, which decrease transmitter release, in the peripheral adrenergic nervous system, and several studies suggest that similar receptors exist on central GABA neurons (Johnston and Mitchell, 1971; Mitchell and Martin, 1978; Snodgrass, 1978) and are bicuculline sensitive. There is some evidence for tonic inhibition of release due to presynaptic receptor stimulation. Other studies suggest that some but not all GABA neurons have such presynaptic receptors: Arbilla *et al.* (1979) found that muscimol and nonradioactive GABA inhibited the release of [^{3}H]-GABA from slices of substantia nigra but not from occipital cortex.

There is no reason to believe that presynaptic GABA receptors are limited to GABA neurons. Indeed GABA has been reported to alter the release of many different neurotransmitters from brain slices or synaptosomes. Unfortunately, small differences in experimental procedure often drastically change the results obtained. As an example, Ennis and Cox (1981) found that GABA enhanced the potassium-evoked release of [^{3}H]dopamine from 250 μm- but not 100-μm ribbons of rat striatum. The effect was dose dependent and muscimol was less potent than GABA, unlike a standard GABA receptor. These GABA responses were insensitive to picuculline and picrotoxin and resemble the bicuculline-insensitive baclofen receptor of Bowery (Bowery *et al.*, 1980) (Section 2.8) except that Bowery's baclofen responses invariably decreased CNS transmitter release, whereas Ennis and Cox found increased dopamine release. Another interesting and probable presynaptic effect was reported by Gallo *et al.* (1981). These workers found that GABA produced a dose-dependent enhancement of the potassium-evoked release of newly taken up D-[^{3}H]-aspartate or glutamate newly synthesized from [^{14}C]glutamine, but not the release of endogenous nonradioactive glutamate in cerebellar synaptosomes. This effect was sensitive to bicuculline and picrotoxin and was not seen in cortical synaptosomes. A possibly similar GABA receptor was described by Mitchell (1980) in striatal slices. GABA agonists enhanced the potassium-evoked release of [^{3}H]glutamate at this bicuculline-sensitive receptor. An unusual feature of this receptor is the fact that APS and isoguvacine are more potent than muscimol. Baclofen had no effect on glutamate release in Mitchell's experiments. Considering the tremendous number of apparent presynaptic and extrasynaptic GABA receptors in the CNS, it is not difficult to imagine that simultaneous stimulation of all of them might drastically disrupt CNS function and produce seizures.

It has been generally assumed that presynaptic GABA effects were mediated by depolarization, reasoning from the phenomenon of "primary afferent depolarization" (Nishi *et al.*, 1974; Gallagher *et al.*, 1978). Attempts to test this theory in bullfrog sympathetic ganglia produced no evidence for depolarization of nerve endings (Kato and Kuba, 1980). These authors showed that the presynaptic GABA effect was chloride dependent and believe that increased chloride conductance following activation of GABA receptors led to shunting of the action potential. However, the effects of GABA on inward calcium currents (Dunlap and Fischbach, 1978) could also account for presynaptic inhibition and deserve more study.

2.7. Unusual Responses to GABA

It is generally assumed that GABA depolarization of primary afferents is mediated by outward chloride currents, by analogy with studies of dorsal root ganglia. It should be noted, however, that several laboratories have reported this response to be decreased in sodium-free media (Barker and Nicoll, 1973; Kudo, 1978; Nistri, 1981*b*). If the depolarizing GABA response involves calcium channels, as suggested by Padjen and Smith (1981), sodium dependency might be explained by the consequent inhibition of sodium–calcium exchange. However, if a calcium channel is activated, one would not expect suppression of transmitter release. Clearly, the matter of presynaptic inhibition and GABA receptors remains confusing.

GABA receptors have been documented on mammalian sympathetic (DeGroat, 1970) and sensory (DeGroat, 1972; Feltz and Rasminsky, 1974) ganglia, neither of which receive GABA innervation in adult life. The cell bodies of many enteric neurons also possess bicuculline-sensitive GABA receptors, whose activation leads to depolarization. As mentioned in Section 2.3, there is a tendency for extrasynaptic GABA receptors to produce depolarizing responses. The bicuculline-insensitive responses to baclofen will be discussed in Section 2.8. Studies of the depolarizing

TABLE 4
Relative Potencies of GABA Agonists in Sympathetic Ganglia[a]

Compound	Carbons between COOH and NH_2 groups	Uptake blockade increases potency	Potency on dog BP[b]
3-APS		−	
Muscimol	2[c]	+	n.t.
γ-Aminobutyric (GABA)	3[d]	+ +	1.5
GABA-choline	3[d]	n.t.	1
γ-Amino-β-hydroxybutyric acid	3	+	0.1
β-Guanidinopropionic acid	2[d]	n.t.	
Guanidinoacetic acid	1	n.t.	
δ-Aminovaleric acid	4	n.t.	
β-Alanine	2	+ +	0.01
N-Methyl-GABA	3	n.t.	0.007
Taurine	2[c]	n.t.	0.001
ε-Aminocaproic acid	5	n.t.	0
Glycine	1	n.t.	0

[a] The table shows GABA agonists arranged in order of decreasing potency at the rat superior cervical ganglion (3-APS is most potent, glycine and ε-aminocaproic are least potent). The table is modified from tables in Bowery and Brown (1974) and Brown and Galvan (1977).

[b] This column refers to potencies of the drugs in depressing blood pressure of the dog, which are expressed as ratios to that of GABA which is given a value of one. From Stanton and Woodhouse (1960); the values for muscimol and GABA-choline are estimated from unpublished data of the author.

[c] Sulfonic acid rather than carboxylic acid.

[d] More than one nitrogen.

[e] n.t., not tested.

response to GABA in the rat superior cervical ganglion (first described by DeGroat, 1970) led to the sequence of agonist potencies listed in Table 4.

In addition to classical conductance effects, GABa may have effects on the shape and duration of the action potential, as described by Dunlap and Fischbach (1978). These authors reported that inward calcium currents in dorsal root ganglion cells were blocked by GABA, norepinephrine, and 5-HT. The GABA responses were not sensitive to bicuculline or picrotoxin and probably represent a qualitatively different type of receptor, although these responses may be related to baclofen responses in peripheral ganglia (see next section). This is just one example of the voltage-dependent changes in membrane conductance which are beginning to be found by more extensive use of intracellular recording (Pellmar, 1981).

2.8. Other GABA-Related Drugs

2.8.1. Baclofen

Baclofen, or β-chlorophenyl-GABA, was designed to be a GABA agonist. However, by 1974, data began to accumulate suggesting that GABA and baclofen had qualitatively different effects on neuronal membranes (Davidoff and Sears, 1974; Fox *et al.*, 1978). More recent data indicate that baclofen effects show marked stereospecificity (Olpe *et al.*, 1979): only the l-enantiomer has therapeutic effects (actions against spasticity), and more recent studies show that the l-enantiomer is about 100 times as potent as the *d*-isomer (Bowery *et al.*, 1980). This stereospecificity has suggested that baclofen acts at a receptor. Iontophoretic studies in the spinal cord show that small baclofen currents antagonize the results of afferent stimulation without any effect on the postsynaptic responses to a variety of excitatory amino acids (Davies, 1981). This effect was directly opposite to that of D-α-aminoadipate, which blocked the responses to exogenous excitants and had no effect on synaptically evoked excitation. Davies' data are consistent with a mainly presynaptic effect of baclofen, with the possibility of postsynaptic effects at high doses. Is there any relationship between baclofen receptors and GABA receptors (Azanza, 1981)? There is evidence that baclofen can increase BDZ receptor binding, suggesting that baclofen receptors have some relationship to the picrotoxinin–BDZ–barbiturate–purine receptor complex described in Section 2.10. Among its other properties, baclofen has been reported to be a releaser of endogenous GABA (Roberts *et al.*, 1978), but this does not appear to be a general effect. There is evidence that baclofen can inhibit the release of excitatory amino acids (Potashner, 1979; Potashner and Lake, 1981), and this has led to suggestions that it might be useful in the treatment of diseases caused by endogenous "excitotoxic" amino acids. No such diseases have been established, but the resemblance of the lesion of Huntington's chorea to that produced by striatal injections of kainic acid (see Section 6) has occasioned great interest (Coyle and Schwarcz, 1976). Could baclofen treatment halt the progression of HD? Baclofen has been reported to increase BDZ binding (Gallager *et al.*, 1978), consistent with a direct or indirect GABA effect.

Baclofen has also been reported to have direct effects. In the isolated frog spinal cord, baclofen hyperpolarized dorsal roots and motoneurons even in media which should block synaptic transmission (containing tetrodotoxin or free calcium with high Mg^{2+} content). Prolongation of the rise time of the epsp was also seen (Kudo *et al.*, 1981*b*). Baclofen is an interesting drug, and further studies of its receptor and other actions are indicated. It has modest value in the treatment of spasticity and some promise as a possible anticonvulsant. Hill and Bowery have studied the binding of [^{3}H]baclofen to rat brain membranes. It is interesting to note that only very slight specific binding was found in the standard Tris-citrate medium (Hill and Bowery, 1981). However, when calcium or magnesium ions are added to the medium, specific and GABA-displaceable binding emerges. This binding was monophasic in Tris-HCl with added $CaCl_2$, whereas it was biphasic in Krebs–Henseleit medium. Even more convincing was the demonstration that the addition of calcium ions allowed baclofen to displace [^{3}H]-GABA. Most of the standard GABA agonists, cyclic or otherwise, had very little ability to displace [^{3}H]baclofen, and bicuculline and picrotoxin were ineffective. It would be interesting to know the functional effects of a "baclofen antagonist." This baclofen binding site was termed a GABA-b receptor by the authors. It is stereospecific in that (−)-baclofen is much more potent than (+)-baclofen. Although baclofen binding was not inhibited by bicuculline, unpublished work indicates that bicuculline can block the *in vitro* stimulation of BDZ binding which baclofen produces. The relationship between baclofen binding sites, which are bicuculline insensitive, and the bicuculline-sensitive presynaptic receptors demonstrated by a number of workers remains to be clarified.

2.8.2. *Valproic Acid*

This organic acid was originally used as a solvent and was serendipitously found to have anticonvulsant properties. The nature of these anticonvulsant properties remains poorly understood. Valproate (VPA) can produce increases in brain GABA and succinic semialdehyde, but they may not be essential to its anticonvulsant effect. Godin *et al.*, (1969) showed that rat brain GABA content increased 30% one hour after an i.p. injection of VPA, 200 mg/kg, although the specific activity of GABA derived from U-[^{14}C]glucose was not significantly changed. Godin also showed that high concentrations of VPA inhibited GABA-T *in vitro*. This was confirmed by Fowler *et al.* (1975) who showed that VPA inhibition of rabbit brain GABA-T was competitive with a K_i of 42 mM. Harvey *et al.* (1975) showed greater potency of VPA as an inhibitor of succinic semialdehyde dehydrogenase (K_i 1.5 mM). Whittle and Turner (1978) reported VPA to be even more effective as an inhibitor of aldehyde reductase, which is believed responsible for conversion of SSA to α-hydroxybutyrate in the brain. Although valproate inhibits GABA-T only at high concentrations, it is able to increase CSF GABA levels at doses similar to those used for anticonvulsant treatment (Loscher, 1979). If the drug's ability to inhibit aldehyde reductase is important (the low K_i of 50 μM supports this), it is possible that VPA's beneficial actions relate to inhibition of the production of an endogenous convulsant (GHBA). This same aldehyde reductase is known to be inhibited by a number of other anticonvulsants (Whittle and Turner, 1981).

Gale and Iadarola (1980) showed that valproate leads to accumulation of GABA in nerve terminals of the substantia nigra and that this GABA is available for use as a transmitter. The same laboratory has studied the effect of valproate on nigrostriatal dopamine neurons and shown that it produces changes in tyrosine hydroxylase consistent with an increased GABA transmission (Casu and Gale, 1981). However, the time course of anticonvulsant effects appears poorly correlated with GABA accumulation, and it is likely that additional non-GABA mechanisms are involved (Anzelark *et al.*, 1976). Iontophoretic studies suggest that valproate may share some benzodiazepine effects, having no effect on neuronal firing when applied alone, but augmenting the effect of GABA inhibition in a dose-dependent manner (MacDonald and Bergey, 1978; Baldino and Geller, 1981). Other workers have obtained different results, and the nature of valproate effects remain uncertain. If an active metabolite is involved, as seems likely, the *in vitro* studies might be misleading. The recent report of Bruni *et al.* (1981) is quite interesting in that regard, suggesting that 3-hydroxypropylpentanoic acid, a known metabolite of VPA, may have important anticonvulsant properties. The metabolism of VPA shows marked species difference, and neither its toxicity nor its mechanism of anticonvulsant action are well understood at present.

2.8.3. Penicillin

This is a classical convulsant, which has been repeatedly suspected of acting via GABA receptors. Proof of this notion has been difficult to obtain, and multiple different mechanisms appear to underly penicillin epileptogenesis in different species and tissues. Benzylpenicillin has been found to block ipsp's, including GABA-mediated Cl^- conductance changes, in invertebrates (Meyer and Prince, 1973; Hochner *et al.*, 1976). Studies in mammalian spinal cord have given conflicting results, as have studies of neocortical inhibition. In the hippocampal slice preparation, Dingledine and Gjerstad (1981) have found that penicillin blocks recurrent ipsp's which appear to be GABA-mediated. The decrease in postsynaptic GABA responses correlates well with increased responses to orthodromic stimulation. However, the synchronized bursting often reported in studies of penicillin epileptogenesis appears to involve some additional mechanism.

2.8.4. δ-*Aminolevulinic Acid*

Brennan and Cantrill (1979) have reported that δ-ALA reduces the evoked release of exogenous GABA from synaptosomes, and they have suggested that this may be a receptor-dependent action which could explain some of the abnormalities of neurological function seen in acute intermittent porphyria. These interesting data require confirmation in another laboratory. δ-Aminolevulinic acid has a number of effects on amino acid neurotransmission (Brennan and Cantrill, 1981), which all may contribute to physiological derangements in porphyria.

2.8.5. γ-*Hydroxybutyrate*

γ-Hydroxybutyrate (GHBA) is an interesting compound, naturally occurring in the brain and produced from GABA. It has properties of a convulsant anesthetic when given in large doses. GHBA in high doses increases striatal dopamine content (Gessa *et al.*, 1966), and smaller subanesthetic doses decrease the firing rate of nigral dopamine neurons (Roth *et al.*, 1980). Olpe and Koella (1979) studied the effects of iontophoretically applied GHBA on GABA-sensitive nigral cells: large currents produced bicuculline-insensitive inhibition. Kozhechkin (1981) reported that iontophoretic administration of GHBA to rabbit cortical neurons produced bicuculline-sensitive inhibition of firing. However, this study did not exclude a secondary effect due to release of endogenous GABA. Baclofen and GABA have similar effects on forebrain dopamine and 5-HT neurons (Roth and Nowycky, 1977; Waldmeier and Fehr, 1978), but GHBA does not appear to share baclofen's peripheral effects. Baclofen and GHBA can stimulate BDZ binding after *in vivo* administration (Gallager *et al.*, 1978), baclofen but not GHBA can stimulate BDZ binding when added to brain membranes *in vitro*, baclofen and not GHBA inhibited the release of [^{3}H]-GABA from striatal synaptosomes, whereas both drugs inhibited the release of dopamine (S. R. Snodgrass, unpublished data). GHBA, although a sedative drug, has been useful in the treatment of narcolepsy (Broughton and Mamelak, 1980). Baclofen and GHBA seem to belong to different but related categories. Snead, in a series of papers, has reported that GHBA can produce a stuporous state which appears to be caused by seizure discharges. These seizure discharges are antagonized by naloxone and have some resemblance to the group of human seizure disorders loosely referred to as "petit mal epilepsy" (Snead and Bearden, 1980). Since GHBA is present in small amounts in mammalian brain, overproduction might conceivably be involved in seizure disorders (see valproate, Section 2.8.2).

2.8.6. Naloxone

Naloxone, an opiate antagonist, was reported to be a weak antagonist at rat brain GABA receptors (Dingledine *et al.*, 1978). Further studies in mouse spinal neurons in culture, with intracellular techniques, showed that naloxone could produce dose-dependent inhibition of GABA responses. However, the effect was not stereospecific, being shown by (+)-naloxone, which has slight opiate receptor activity, and concentrations of 100 μM or more were needed for GABA antagonist effects when naloxone was added to the medium (Gruol *et al.*, 1980). At these high doses, an increased membrane resistance and potentiation of glutamate responses were often seen. Naloxone competitively antagonizes GABA responses at very high concentrations, concentrations which are unlikely to be seen except in iontophoretic experiments. There have been clinical reports that symptoms of BDZ overdosage in man may respond to naloxone treatment, and that some of the sedative effects of BDZs in rodents are naloxone-reversible (Walz and Davis, 1979), although the doses of naloxone used were very large.

A point of interest is whether GABA receptors are limited to neurons or are also found on glial and other cells, such as blood vessels. The evidence is not definitive, but suggests that glial receptors and endothelial receptors probably do occur (Edvinsson and Krause, 1979; Krause *et al.*, 1980). The existence of GABA receptors on glial tumors and glial cells grown in culture does not mean that this property must be expressed *in vivo*. GABA and glycine have been reported to produce depolarization of cultured glial cells without a change in membrane resistance (Hosli *et al.*, 1978). The drug 4-aminopyridine, which blocks potassium channels of excitable membranes, reversibly alters both the glial depolarization and increased extracellular potassium produced by GABA and glycine, suggesting that glial cells lack coupled receptors for these amino acids, and that all glial effects are indirect (Hosli *et al.*, 1981). Radioligand binding studies by DeFeudis *et al.* (1980), after subcellular fractionation, also led to the conclusion that glial elements have few if any GABA receptors.

2.9. GABA Receptor Autoradiography

Early studies of the localization of GABA receptors involved the direct injection of muscimol into brain or CSF, followed by perfusion fixation and autoradiography (Chan-Palay, 1978*a,b*). The large numbers of GABA uptake sites meant that some muscimol was taken up by the GABA uptake system (Snodgrass, 1978). Young and Kuhar (1979) devised a technique which minimizes this problem. Lightly fixed sections of brain are mounted on microscope slides. The sections are preincubated to deplete endogenous GABA (Palacios *et al.*, 1980). Slides are incubated in the absence of sodium (Snyder's Tris-citrate buffer), in which GABA uptake is minimally active. Other slides are incubated with [^{3}H]muscimol and cold GABA, and the number of autoradiographic grains found on these slides are used to estimate nonspecific binding.

2.10. Complex Interactions between GABA, Benzodiazepines, and Barbiturates

The demonstration of specific neural receptors for benzodiazepines generated much interest (Mohler and Okada, 1977*b*; Squires and Braestrup, 1977). Coming soon after the demonstration of opiate receptors and endogenous opiate ligands in mammalian brain, the possibility of endogenous ligands for BDZ receptors was obvious and appealing. This topic is reviewed in depth in another article. I will touch on a few points of maximal relevance to GABA receptors here. There is no question that BDZs can alter the sensitivity of GABA receptors; however, this property has not been proven to underly the characteristic effects of BDZs, namely restoration of punished behavior and their anticonvulsant properties.

Choi and co-workers were the first to demonstrate clearly that BDZs had no direct membrane effect of their own, but increased the conductance response to GABA (Choi *et al.*, 1977). Similar effects have been demonstrated in mouse neurons grown in culture (MacDonald and Barker, 1978*b*). Of interest is the ability of GABA agonists to stimulate binding of BDZs to their receptors (Tallman, et al., 1978). This is a bicuculline-sensitive effect, and when chloride-deficient media are used in studies of living cells, GABA agonists can no longer enhance BDZ binding (White *et al.*, 1981). It is interesting to note that chloride ions also alter the effects of rigid GABA agonists such as PSA and BDZ binding. Triton X-100 treatment increases the affinity of GABA receptors for GABA agonists. One or more endogenous molecules appear to regulate sensitivity of these receptors (Enna and Snyder, 1977). GABA itself is one of the factors which diminish GABA receptor binding (Napias *et al.*, 1980) but may not be the most important factor. Guidotti *et al.* (1978) have identified a heat-stable, acidic protein which they call GABA-modulin. This protein is reported to alter the binding of either GABA or BDZs to receptors. Its antagonism of GABA binding was noncompetitive, whereas its effect on BDZ binding was competitive. Diazepam did not stimulate GABA binding in membranes which had been freed of the inhibitor protein. GABA-modulin inhibits GABA binding in solubilized receptor preparations where BDZ receptors are not present, and BDZs no longer inhibit heat inactivation of [^{3}H]-GABA binding sites (Massotti *et al.*, 1981). In this circumstance, GABA itself still retards heat inactivation of its receptor. In crude membranes, GABA agonists, including THIP, which does not ordinarily enhance BDZ binding, and bicuculline protect both GABA and BDZ binding sites from heat inactivation (Gavish and Snyder, 1980*b*), but they do not protect other receptors. Since the effects of GABA and diazepam together on heat sensitivity of either receptor are greater than the effects of either agent alone, it is likely that GABA and diazepam act at separate but allosterically linked sites. Furthermore, phylogenetic comparisons show that species which have BDZ receptors appear to show the Triton effect on GABA receptors (Nielsen *et al.*, 1978).

Kinetic studies of BDZ binding produce linear Scatchard plots and suggest a single class of BDZ sites, but triazolopyridazines such as CL 218,872 displace BDZ binding with curvilinear kinetics (Klepner *et al.*, 1979). The effects of β-carbolines (Braestrup *et al.*, 1980) on BDZ binding likewise suggest the existence of multiple receptor types. Table 5 gives two of the groupings of BDZ sites which have been suggested: one based upon triazolopyridazine effects, the other on a supposed neuronal versus glial distinction.

The ability of BDZs to enhance GABA binding is usually lost in washed or Triton-treated membranes (Toffano *et al.*, 1978). Skolnick *et al.* (1980) were able to show that both GABA and 100 μM pentobarbital enhanced the binding of [^{3}H]diazepam to brain membranes. Several endogenous inhibitors of the binding of dihydropicrotoxinin have been reported (Olsen and Leeb-Lundberg, 1980). Several purines were reported by Ticku and Burch (1980) to competitively antagonize diazepam binding with a noncompetitive effect on GABA binding. To further complicate matters, there is at least one purine (EMD 28422) which enhances diazepam binding by a bicuculline-sensitive mechanism (Skolnick *et al.*, 1980). Possible interactions between purines, methylxanthines, BDZs, GABA, and adenosine receptors have been discussed but not clearly demonstrated. The antihelminthic drug, avermectin B1a, can stimulate the binding of BDZs to their receptors after *in vitro* addition to synaptic membranes, but is ineffective in detergent-treated membranes (Supavilai and Karobath, 1981*b*). GABA remains able to stimulate BDZ binding in detergent-treated membranes, and it appears to bind to a different site than avermectin.

Even ethyl alcohol appears to facilitate GABA-mediated inhibition

TABLE 5
Categorization of Benzodiazepine Receptors in the CNS[a]

Categorization by effects of triazolopyridazines				
Brain region	*Kd* 1	B_{max} 1	*Kd* 2	B_{max} 2
Cerebellar vermis	22	83	0	0
Frontal cortex	46	69	181	32
Dorsal hippocampus	30	41	305	57

Neuronal and nonneuronal receptor types

Nonneuronal BDZ receptor[b]			Neuronal BDZ receptor[c]		
Ligand [^{3}H]-RO 5-4864			Ligand: [^{3}H]diazepam (or flunitrazepam)		
No effect on binding of CL 218,272			Binding increases with CL 218,872		
Displaced by	diazepam IC_{50}	23 nM	Displaced by	clonazepam	1.5 nM
	flunitrazepam	25 nM		flunitrazepam	2.8 nM
	lorazepam	8.5 μM		lorazepam	2.7 nM
	clonazepam	9.2 μM		diazepam	6.3 nM
	flurazepam	10 μM		flurazepam	11 nM

[a] The regional study of type 1 and type 2 receptors used the ability of CL 218,872 to stimulate BDZ binding (as described by Klepner *et al.*, 1979) to distinguish type 1 receptors (presumed to be GABA independent) from type 2 receptors, presumed to be GABA-coupled. The data are taken from Lippa *et al.* (1980) and refer to rat brain. Gallager *et al.* (1981) used the ability of clonazepam to displace [^{3}H]diazepam binding to membrane preparations as indicating neuronal type BDZ receptors, while [^{3}H]diazepam binding displaceable with RO 5-4864 was taken as a measure of nonneuronal BDZ receptors. The data above listed under nonneuronal BDZ receptor are taken from Schoemaker *et al.* (1981) and were obtained with a membrane preparation from rat cerebral cortex. The data on neuronal BDZ receptor binding are taken from the paper by Mohler and Okada (1978*b*) and were obtained in rat cortical membranes. However, the same paper presents very similar results in membrane preparations from postmortem human brain. Note that flunitrazepam and diazepam, two potent anticonvulsant BDZs, have significant affinity for the "nonneuronal receptor."

[b] Binding was not affected by β-carbolines, barbiturates, or picrotoxin.

[c] Binding is inhibited by β-carbolines, and under appropriate conditions is sensitive to barbiturates and picrotoxin.

under some circumstances (Nestoros, 1980). A recent study by Davis and Ticku (1981) reported that ethanol increased BDZ binding in a solubilized receptor preparation, that this effect could be blocked by both bicuculline and picrotoxinin, and that the rank order of enhancement of BDZ binding by alcohols (ethanol) > (methanol) > (long-chain alcohols) did not correlate with partition coefficients. There is no question that alcohol and barbiturates have some direct membrane effects (Hill and Bangham, 1975) and that we must remember that their tissue levels are very high, suggesting that a major component of their action is not receptor-mediated. However, it seems probable that some effects of these drugs are mediated by effects on GABA neurotransmission.

Most physiological studies of barbiturate actions have used pentobarbital, which is not a useful anticonvulsant. Because barbiturates have many effects, it has been difficult to know which of the different barbiturate actions might account for their clinical utility as anticonvulsants. Studies by Schulz and MacDonald (1981) on cultured mouse spinal neurons suggest that anticonvulsant actions correlate with an augmentation of postsynaptic GABA responses. Pentobarbital and phenobarbital were similar in their ability to antagonize bicuculline effects or facilitate postsynaptic GABA responses, but pentobarbital produced direct membrane effects at a much lower concentration than phenobarbital. It is important that similar studies be done in cerebral cortex, because this is the site of origin of most human seizure discharges. Barbiturates have effects on GABA uptake systems (Cutler *et al.*, 1974; Brown and Constanti, 1978), but these probably do not account for their anticonvulsant actions (Simmonds, 1981). Simmonds attempted to distinguish between the ability of phenobarbitone to antagonize picrotoxin effects, which he felt correlated best with anticonvulsant properties and the ability to antagonize bicuculline effects or potentiate those of muscimol (in the cuneate nucleus). Benzodiazepines were less potent enhancers of muscimol effect than were barbiturates. Willow and Johnston (1980) have reported that pentobarbitone can enhance GABA binding to crude synaptosomal membranes from rat brain, but most workers have been unable to demonstrate effects of barbiturates on GABA binding, whereas that of BDZs is easily shown. Barbiturates can reverse blockade of GABA responses by bicuculline and picrotoxin (Bowery and Dray, 1976) and inhibit the binding of dihydropicrotoxinin (DHP) to rat brain membranes (Ticku and Olsen, 1978). Interpretation of these interactions shown by ligand binding studies is complex: convulsant and depressant barbiturates both compete, weak barbiturates such as mephobarbital have IC_{50}s which seem unduly low, and the depressant (−)-isomer of 1-methyl-5-phenyl-5-propylbarbiturate is a better displacer of DHP than the convulsant (+)-isomer (Ticku, 1981). The DHP binding data do not predict whether or not a barbiturate has anticonvulsant properties. The same is true of barbiturate effects on GABA binding (Willow and Johnston, 1981*a*). Diphenylhydantoin or phenytoin is a widely used anticonvulsant. It is a moderately potent displacer of DHP binding and can transiently increase the number of BDZ binding sites after systemic injection (Gallager *et al., 1980*). Treatment of brain membranes *in vitro* with the drug has no effect on BDZ or GABA binding. Phenytoin is ineffective as an antagonist of bicuculline seizures, and its relationship to the GABA–BDZ system seems to be indirect and modest.

Barbiturates and BDZs can both potentiate GABA effects, but barbiturates have an additional property not shared by BDZs, namely a suppression of excitation even when GABA is excluded (Nicoll, 1978; Haefely, 1980). The effects of barbiturates and BDZs on dorsal root potentials are dissimilar (Nicoll, 1975, Haefely, 1980): barbiturates depolarize primary afferents and enhance the depolarizing GABA response of sympathetic ganglia, BDZs do not. Neither BDZs nor barbiturates have been shown to increase the effects of GABA on transmitter release, presumed to reflect presynaptic receptor sites. Since BDZs competitively inhibit the effects of "GABA-modulin," it is possible that barbiturates and other GABA modulating drugs all act via this protein, but there is no evidence to support this speculation. Although the sedative properties of barbiturates limit their clinical usefulness as anticonvulsants, they are more effective in general as maintenance anticonvulsants than are BDZs. The ability of BDZs to antagonize the actions of many chemical convulsants does not predict a good response in many clinical settings.

The ability of BDZs to stimulate GABA responses has been shown in several species. Note, however, the suggestion by Stein *et al.* (1975) that 5-HT receptors might be involved in the benzodiazepine behavioral or whole animal responses. Guidotti (1978) showed that the antipunishment effect of BDZs was associated with decreased 5-HT turnover; however, this was unfortunately not localized to specific brain regions. Warbritten *et al.* (1978) showed that intraventricular infusions of 5-HT evoked behavioral inhibition. It is also known that injections of serotonin neurotoxins into the dorsal raphe nucleus (and some other areas) reduce the behavioral consequences of punishment, and that destruction of serotonin neurons removes the antipunishment effects of BDZs (Thiebot *et al.*, 1980). For all these reasons, it is important to consider that 5-HT brainstem systems may play an important role in BDZ antipunishment actions. It is also interesting that Hwang and Van Woert (1979) showed that the characteristic antimyoclonic action of BDZs was abolished by 5-HT receptor blockers and not by bicuculline. Although BDZs are effective against many types of seizures, their clinical efficacy is greatest for myoclonic seizure disorders (Aicardi, 1980). Treatment with serotonin precursors is helpful for some myoclonic disorders in man (Van Woert *et al.*, 1979). Again, there is reason for further studies of 5-HT systems in this drug effect, especially in view of the serious consequences of this type of seizure disorder for some patients. See Sepinwall and Cook (1980), Schenberg (1978), and Tye *et al.* and Graeff (1979) for further discussion of the possible relationships between BDZs and 5-HT neuronal systems.

2.11. Developmental Studies of GABA Receptors

At birth, [^{3}H]-GABA binding in the cortex and cerebellum of rat brain is about 40% of the adult values. The principal developmental change is one of increased V_{max} (Aldinio *et al.*, 1980). In general, GABA receptors appear to develop somewhat in advance of the presynaptic GABA terminal, as judged by GABA uptake (Coyle and Enna, 1976). The ability of GABA to increase specific binding of BDZs decreases with age

(Gallager, 1980). This effect is variable and depends upon the amount of tissue used per assay tube (Regan *et al.*, 1980). The anticonvulsant diphenylhydantoin (DPH) decreases BDZ binding in young rats, while increasing it in older animals. However, the effects of prenatal DPH administration may reflect a simple loss of neurons, in view of the results of Gallager *et al.* (1981*b*).

2.12. Desensitization

With continuing agonist exposure, tissue responsiveness often decreases. First studied by Katz and Thesleff (1957) at the cholinergic neuromuscular junction, this decrease in tissue responsiveness occurs at many but not all receptors. Desensitization of GABA receptors is seen at the crustacean neuromuscular junction (Epstein and Grundfest, 1970); however, it may be absent (Takeuchi and Takeuchi, 1966), weak (Feltz, 1971), or pronounced (Sarne, 1976) in crustacean muscle in response to GABA. It is often seen at mammalian GABA receptors (Curtis *et al.*, 1959; Krnjevic *et al.*, 1977), and occurs at GABA receptors in sympathetic ganglia (Adams and Brown, 1975). There have been two basic theories about the mechanism of desensitization: a change of the receptor to an inactive form (Katz and Thesleff, 1957) and a change in the ion channel or effector mechanism (Nastuk and Parsons, 1970). For receptors which are cyclase coupled, changes in the effector mechanism (adenylate cyclase) usually precede changes in numbers of receptors judged by radioligand assays (Su *et al.*, 1980; Nickols and Brooker, 1979), but both mechanisms may be operative. In the frog erythrocyte, a redistribution or internalization of beta receptor sites occurs during desensitization (Chuang, 1981). When

TABLE 6
Different GABA Receptors in the Crayfish[a]

Type	GABA inhibition[b]	Desensitization	β-GPA[c] inhibition[b]	Desensitization
Postsynaptic				
Opener muscle[d]	+	–	–	0
Closer muscle[d]	++	++	+	+
Presynaptic				
Opener muscle[d]	+	–	+	–
Closer muscle[d]	+	+	+	–

[a] Note that the study of desensitization exposes receptor differences which are otherwise invisible. Adapted from Dudel and Hatt (1976).
[b] The transmitter inhibits the response.
[c] β-Guanidinopropionic acid.
[d] Muscles studied are in the walking legs of the crayfish.

several pulses of agonist are administered, potentiation as well as desensitization can be seen (Feltz and Trautman, 1980). The mechanism of desensitization of GABA receptors is not known, but the process is stimulated by calcium ions as is true of nicotinic cholinergic receptors (Sarne, 1976). GABA receptors are typically more likely to show desensitization than are glycine receptors (Krnjevic *et al.*, 1977). Individual GABA receptors in the same species vary in their propensity to desensitization (see Table 6).

In lamprey spinal cord, Homma and Rovainen found that higher temperatures decreased responses to GABA or glycine, but not when sodium-free media was used. Unstirred layers also reduced sensitivity, which suggests that uptake mechanisms play some role in regulating the responses to these inhibitory amino acids (Homma and Rovainen, 1978). However, mammalian studies suggest that the contributions of uptake in this matter is small (Lodge *et al.*, 1977). Studies to resolve the contribution of uptake to the loss of responsiveness which usually follows agonist exposure are difficult except in slice or tissue culture preparations.

2.13. GABA Agonists and Seizures

Early workers hoped that GABA, as a ubiquitous inhibitory transmitter, might be a unifying concept and therapeutic agent for human epilepsy. Such hopes were reinforced by the discovery that many convulsants were inhibitors of the GABA synthetic enzyme, glutamate decarboxylase, and the discovery that infant formulas lacking in pyridoxine were likely to result in seizures (Tapia, 1975). However, evidence linking common forms of epilepsy to GABA or any other transmitters has been weak and disappointing (Maynaert *et al.*, 1975). The discovery of CNS BDZ receptors rejuvenated the quest for a tie between GABA and human epilepsy, which was heartened by reports that changes in BDZ binding follow chemically or electrically induced seizures (Paul and Skolnick, 1978). Altered BDZ binding has been reported in epileptic mice and baboons (Squires *et al.*, 1979). As noted in Section 2.10 the antimyoclonic actions of BDZs may relate better to 5-HT than to GABA neurons, and it should be noted that apparently specific BDZ antagonists are not convulsants (Hunkeler *et al.*, 1981). However, some relationship between GABA neurons and BDZ anticonvulsant effects is likely, and is made more likely by the recent information of subtle interactions between barbiturates and GABA receptors (Willow and Johnston, 1981*b*).

Why then is muscimol not a useful anticonvulsant? Not only does it produce a number of unwanted side effects, such as hallucinations and myoclonic jerks in man (Shoulson *et al.*, 1978), but it has often failed to prevent convulsions in animal seizure models, including bicuculline-induced seizures, which are the prototype of a GABA-related seizure (Worms and Lloyd, 1980; Patel *et al.*, 1980). Patel *et al.* reported that BDZs, barbiturates, and even ethanol blocked bicuculline seizures, while muscimol and aminooxyacetic acid, a potent antagonist of GABA transaminase, were ineffective. Muscimol is sometimes effective in seizure models, including bicuculline-induced seizures (Matthews *et al.*, 1981), but muscimol and GABA-transaminase inhibitors such as γ-vinyl-GABA are tricky anticonvulsants (Kendall *et al.*, 1981), not always doing what would be expected. Returning to Meldrum's classification, it may be that only certain types of GABA agaonist will prove to be useful anticonvulsants and none is likely to be a universal anticonvulsant. Loscher (1981) studied the anticonvulsant actions of a variety of GABA-related drugs: the ability of the drugs to increase brain GABA content correlated well with their ability to antagonize pentylenetetrazole seizures, but not with changes in the electroconvulsive threshold, which is more relevant to most types of seizures seen in adults. It is possible that GABA replacement therapy cannot work because of the brief duration of GABA responses *in vivo*, in contrast to the slower actions of transmitters such as dopamine, for which replacement therapy is sometimes effective. Current interest centers on THIP and progabide or SL 76002 as possibly useful anticonvulsants which appear relatively nontoxic. They merit further study in spite of the problems alluded to.

There remains the problem of explaining the proconvulsant effects of muscimol, THIP, and other GABA agaonists. Such effects are readily demonstrated in primates, and indeed Meldrum reported them with representatives of three different classes of GABA agonist: muscimol, THIP, and baclofen. If GHBA is to be given status as a GABA agonist of sorts, it would be a fourth type which can promote seizures. It seems that the weapon which stops seizures can easily start them as well and that presynaptic inhibition is the simplest explanation for such paradoxes: the nervous system often has inhibition of inhibitory circuits. Here, activation of the first receptor will excite the final target. As noted by Sepinwall and Cook (1980), under some circumstances BDZs are better anticonvulsants than intrathecally administered GABA and muscimol. Krnjevic (1980) proposed that a primary event in the genesis of hippocampal seizures was loss of postsynaptic responsiveness to GABA, by a desensitization-related process. If he is correct, the prolonged duration of action of muscimol, THIP, and similar GABA agonists, which are poor uptake substrates, may automatically make them convulsants because their prolonged contact with receptors might produce undesirable changes.

2.14. GABA Receptors and Human Disease

GABA receptor studies in human postmortem brain are in close agreement with the results of animal studies. In rat and human, cerebellum has the greatest density of receptors, and Triton X-100 markedly increases GABA binding in each (Lloyd and Dreksler, 1979). Although the magnitude of the Triton effect varies greatly among different brain regions, being greatest in the basal ganglia, the relative potencies of GABA agonists were identical in a variety of brain regions (Enna, 1978).

While dysfunction of GABA neurons is postulated for many human diseases, GABA receptor abnormalities have been found in only Parkinson's disease and Huntington's chorea or HD. Decreased [^{3}H]-GABA binding was found in hippocampus and substantia nigra of Parkinsonian patients (Lloyd and Dreksler, 1978). The latter was interpreted as indicating that most nigral GABA receptors were found on dopamine cell bodies, which are known to be selectively lost in Parkinson's disease. Postmortem binding studies of HD brain tissue showed a large increase in cerebellar binding of [^{3}H]-GABA, with equally large decreases in GABA binding by caudate and putamen (Lloyd and Dreksler, 1978). An interesting additional point was that phospholipase C treatment, which acts like Triton X-100 to convert GABA receptors from low- to high-affinity form, had no further effect on cerebellar membranes from HD patients, as if their phospholipids might differ from the normal (Lloyd and Davidson, 1979; Lloyd and Beaumont, 1980). The decreased GABA receptor content of caudate and putamen may be one reason why GABA agonists have been ineffective in the treatment of HD. Two studies report increased GABA binding in substantia nigra of HD patients, in addition to the increase of cerebellar high-affinity receptors (Enna *et al.*, 1976; Cross and Waddington, 1981). These nigral GABA changes duplicate those seen in the striatal kainic acid lesion, which has been suggested as an animal model for HD (Coyle and Schwarcz, 1976).

Although muscimol and other "GABA replacement" therapies have been disappointing in HD, they appear to have some value in the treatment of tardive dyskinesia (Tamminga *et al.*, 1979). Low doses of neuroleptics have no direct effect on GABA receptors, but they do alter GABA turnover in the basal ganglia (Marco *et al.*, 1976). Chronic neuroleptic treatment has been reported to increase high-affinity GABA binding in the substantia nigra (Gale, 1980). Most authors reporting increased numbers of GABA receptors after various treatments have not included functional studies. Waddington and Cross (1978) reported increased binding and increased rotational response to intranigral muscimol in rats with unilateral lesions of the striatonigral GABA pathway. Waszczak *et al.* (1981) showed that similar lesions led to increased sensitivity of pars reticulata neurons to iontophoretically administered GABA or systemic muscimol. It should be noted that these increases were small in magnitude and not comparable to the large increases in agonist responses produced by denervation at the neuromuscular junction.

2.15. Radioreceptor Assays

The principle of radioreceptor assays was developed by Cuatrecasas in 1969 for insulin receptors. However, the technique seems to have found its greatest usefulness in the GABA radioreceptor assay, developed by Enna and Snyder (1976). This technique permits the estimation of GABA content because nonradioactive GABA quantitatively reduces the binding of labeled GABA or muscimol, and from standard curves one can accurately estimate the GABA content of an unknown sample. The assay can accurately measure 5 pmol/ml of incubation medium, and, fortunately, GABA appears to be the only compound found in brain that can displace labeled GABA when added in small amounts. In the case of other receptors, such as opiates, BDZs, and some peptides, there seem to be many endogenous compounds which can compete for binding, and the technique is less useful, although a discrepancy between radioreceptor results and those found by RIA or chromatographic methods may disclose the presence of a new class of endogenous ligands. The GABA radioreceptor assay should be used cautiously when animals or patients have received psychoactive drugs, but it remains rapid and useful, especially for the study of bodily fluids, such as CSF.

3. GLYCINE RECEPTORS

3.1. Historical Review

The early iontophoretic studies of Curtis and colleagues implicated GABA as a mediator of fast ipsps (Coombs *et al.*, 1955; Curtis *et al.*, 1959; Araki *et al.*, 1961) but did not suggest any special role for glycine, even though spinal cord physiologists knew that strychnine had a special ability to block important inhibitory processes within the cord (Bremer, 1953). One factor was that frog spinal cord appears to lack Renshaw cells (Kudo, 1978) and glycine responses are not prominent there, although strychinine binding sites are clearly present in the frog spinal cord (Muller and Snyder, 1978). Werman and Aprison were primarily responsible for delineating a special role for glycine in the spinal cord (Werman *et al.*, 1967, 1968). Curtis and colleagues promptly reported that the glycine hyperpolarization of spinal motor neurons was antagonized by strychnine and related compounds (Curtis *et al.*, 1967). Reports of strychnine effects on brainstem neurons soon followed (Hosli and Tebecis, 1970). The development of an *in vitro* radioligand binding assay for glycine receptors by Young and Snyder (1973) was a major step forward but was followed by several years of inactivity, with few reports relating to glycine receptor studies. For a time, it seemed that GABA receptors were the only important inhibitory receptors. Reports of glycine studies are now increasing, and attempts are being made to link clinical phenomena to function and malfunction of glycine receptors. Retinal neuropharmacology appears to provide a fertile area for studies of the functional role of glycine receptors, and a number of interesting studies have already appeared (Caldwell and Daw, 1978; Frumkes *et al.*, 1981; Miller *et al.*, 1981). At present, glycine remains best established as a spinal cord transmitter, as stressed by Aprison and Nadi (1978). There is good evidence that Renshaw cells are glycine neurons (Lodge *et al.*, 1977). In the chick spinal cord, there is a good temporal correlation between the onset of physiologically defined inhibition (Stokes and Bignall, 1977) and the maturation of [^{3}H]strychnine binding (Zukin *et al.*, 1975).

3.2. Glycine Effects and Glycine Agonists

Many glycine effects have been inferred from the symptoms of strychnine poisoning: Spinal myoclonus, painful responses to sensory stimulation, and distortion of auditory and visual input all suggest that glycine receptors are important in the function of sensory and nociceptive pathways. Werman and Aprison showed that glycine ipsps were chloride mediated. It is likely that depolarizing glycine responses will ultimately be found. In the radioligand assay of Young and Snyder, β-alanine was the amino acid best able to displace strychnine, after glycine. Although more than 50 GABA agonists have been studied, glycine agonists have received only minimal study. This may change as diseases of glycine receptors are found. Ferrendelli *et al.* (1974) reported that glycine altered cGMP levels in cerebellum, as do GABA agonists. However, there are few glycine receptors in cerebellum, and others have been unable to duplicate this result.

3.3. Glycine Antagonists

We have already alluded to the problem of strychnine specificity. Strychnine could not be categorized as a glycine receptor blocker until glycine was recognized as a neurotransmitter. It was established that "presynaptic inhibition" as defined by Frank and Fuortes (1957) was sensitive to picrotoxin and not to strychnine (Eccles *et al.*, 1963). We now know that a strychnine-sensitive inhibition follows classical presynaptic inhibition, which is most likely GABA-mediated (Bagust *et al.*, 1981), illustrating the principle that pure presynaptic inhibition is rarely seen.

Structure–activity studies of strychnine and related compounds have been done by Mackerer *et al.* (1977). These studies show that there is an excellent correlation between the ability of these compounds to produce convulsions in the mouse and their affinity for the receptor, as judged by membrane binding studies *in vitro.* There was no correlation between receptor binding and the oil–water partition coefficient, suggesting that it is the protonated, hydrophilic form of the strychnine molecule which binds to the receptor. Strychnine appears to label a single population of receptor sites, and glycine interacts with this site in a cooperative fashion, yielding a Hill coefficient of 1.7 (Young and Snyder, 1974). Because protein modification with agents such as acetic anhydride has a differential effect on the ability of glycine and strychnine to compete with [^{3}H]-strychnine binding, it is likely that the agonist and antagonist bind to separate but interacting sites (Young and Snyder, 1974; Muller and Snyder, 1978). It has also been suggested that strychnine may bind to the glycine-activated ion channel. In this regard it is noteworthy that strychnine has relatively little effect on chloride currents induced by GABA, at least in spinal cord, where the overlap between strychnine antagonism of GABA and glycine is slight. Strychnine does, however, antagonize the effects of closely related amino acids such as taurine and β-alanine. If these other amino acids possess true receptors, they may share an ionophore with glycine receptors. This might be most plausible for taurine and glycine in the retina of some vertebrates (see Section 4). Borbe *et al.* (1981) studied the binding of [^{3}H]strychnine in bovine retina and found only a single binding site, antagonized better by glycine than by taurine.

Strychnine neuronography has been used in the past for neuroanatomical investigation. Although we have learned much about strychnine over the past 25 years, we still do not know how to account for its selective actions when applied topically to the cortex (Towe and Mann, 1973). Its potency on topical application to cortex is less than on spinal cord (Bremer, 1953), and its cortical effects may be explicable by effects on GABA receptors (Dichter, 1980). The specificity of strychnine effects for glycine rather than GABA receptors appears to be substantially less in the forebrain, and it should be noted that competition by bicuculline for [^{3}H]-strychnine binding sites has been reported (Goldinger and Muller, 1980).

3.4. Glycine Receptor Autoradiography

Glycine receptor autoradiography has been studied in Kuhar's laboratory (Zarbin *et al.*, 1981). These workers reported that strychnine binding sites were most abundant in spinal gray matter and in laminae believed to be related to nociception, and that brainstem nuclei with cardiovascular and respiratory function were rich in strychnine binding sites, as were auditory structures such as the lateral lemniscus. The inner plexiform layer of the retina was the only part of the visual system with significant numbers of strychnine binding sites. The cerebellum and cerebral cortex lacked significant numbers of strychnine sites, in spite of its known ability to produce seizures when applied to cortex.

3.5. Absence of Desensitization at Glycine Synapses

As a general rule, desensitization of glycine receptors is less readily found than desensitization of GABA receptors, even though both mediate fast ipsp's (Werman *et al.*, 1968; Krnjevic *et al.*, 1977). However, desensitization of glycine receptors is seen in the retina (Frumkes *et al.*, 1981). If there is a general tendency for glycine receptors not to desensitize, this may give important information for the differences between the inhibition mediated by glycine and that mediated by GABA.

4. OTHER "INHIBITORY AMINO ACIDS"

A number of investigators have reported that β-alanine and taurine can inhibit neuronal firing under certain circumstances; hence there has been speculation that these amino acids may be transmitters. The obvious problem is to exclude responses from GABA or glycine receptors. If there are specific receptors for β-alanine or taurine, these amino acids should be more potent than GABA and glycine and it would be preferable to have no blocking by classical GABA and glycine receptor blockers. The limitations of iontophoresis, namely its semiquantitative nature, make it difficult to be certain about relative sensitivity to the amino acids. Early studies by Curtis *et al.* (1959) indicated that both GABA and β-alanine depressed the firing and excitation of motor neurons and other spinal cord cells, that neither amino acid was blocked by strychnine, and that there was no clear indication of greater sensitivity to GABA than β-alanine. These results were consistent with β-alanine action at GABA synapses. Tebecis and Phillis (1969), in the toad spinal cord, reported that β-alanine and glycine always had comparable effects, different from those of GABA in many cases. Strychnine always blocked or reversed β-alanine and glycine effects, while its effect on GABA responses was weak and inconstant. These results were similar to those reported by Curtis and colleagues in the cat spinal cord (Curtis *et al.*, 1968).

In view of the ability of several of these amino acids to release endogenous transmitters (Bowery *et al.*, 1976*b*), it is important to check reports of the receptor actions of these other amino acids to be sure that endogenous transmitter release can be excluded. Studies of Sonnhof *et al.* (1975) on the frog spinal cord appear to make a good case for taurine receptors, but did not in fact exclude indirect effects via the liberation of endogenous transmitters or other mechanisms. These investigators did report that taurine was more potent than either GABA or glycine. If we can believe that the quantitation is valid, this would argue against taurine actions being mediated at GABA or glycine receptors. As in most spinal cord studies, strychnine was a good blocker of taurine effects. The strongest case for true taurine receptors may exist in the retina. Cunningham and Miller have presented data for taurine effects in the rabbit (Cunningham and Miller, 1976) and mudpuppy retina (Cunningham and Miller, 1980). The latter studies were quite sophisticated: release of endogenous transmitter appears to have been prevented by the use of cobalt ions. Taurine and glycine had similar potencies, and each was blocked by strychnine. As noted in Section 3, binding studies of [^{3}H]-strychnine in the retina have shown only a single population of binding sites, at which glycine had greater affinity than did taurine, but this was in the cow, not the mudpuppy. Autoradiographic studies by Ehinger (1976) suggested that Ω-amino acids might act as false transmitters at GABA sites in the retina, and A. Cohen (1978) has suggested that taurine can act as a false transmitter at glycine receptor sites. Studies with [^{3}H]-taurine should ultimately clarify this issue.

Proline has also been reported to have effects on neuronal firing. Felix and Kunzle (1974) reported depressant effects on the firing of cerebellar Purkinje cells, while Bailey and Phillis (1976) found depolarizing effects on dorsal and ventral root in the toad cord. None of the proline studies excluded the release of endogenous transmitters.

5. EXCITATORY AMINO ACIDS

Although potent excitatory effects of glutamic acid were described more than 25 years ago (Hayashi, 1952), the understanding of acidic amino acids and their receptors has been a slow process. Curtis's 1963 review covered only inhibition; this was logical in view of the poor understanding of excitatory synaptic mechanisms. The 1974 review article by Curtis and Johnston (1974) suggested that glutamate and aspartate were probably excitatory transmitters, but raised the question why the receptor showed so little stereospecificity. The list of known excitatory amino acids grows each year; our task is to organize the fragmentary and complex data.

Circumstantial evidence suggests that L-glutamate is the transmitter for cortical pyramidal cells (Divac *et al.*, 1977), cerebellar granule cells (Young *et al.*, 1974) and probably other neurons as well. There was a natural tendency to think of other excitatory amino acids as glutamate analogs. In the last few years it has become increasingly clear that multiple different receptors exist for excitatory amino acids. Watkins and his colleagues have been especially influential in arguing this point (Biscoe *et al.*, 1978; Watkins, 1980; Watkins *et al.*, 1981). Biscoe *et al.* showed that excitatory receptors were similar in frog and rat spinal cord, and that quisqualate, domoate, and kainate (KA) were the strongest excitants in both species, being about two orders of magnitude more potent than L-glutamate, when judged by the magnitude of iontophoretic currents needed to produce similar rates of neuronal firing. However, the number of different receptor types in mammalian brain is not settled. Iontophoresis and related physiological studies have spearheaded our understanding of the excitatory amino acids, whereas radioligand assays have been hampered by technical difficulties and have been only moderately helpful to date. All effective excitatory amino acids contain two separate acidic groups, both charged at body pH. They also possess an amino group. Chemical substitutions which reduce the ionizability of either acidic or amino groups decrease the excitatory potency. Maximal potency results when a 3- or 4-carbon chain separates the acidic groups. Glutamate shows only slight stereospecificity in the vertebrate CNS (Biscoe *et al.*, 1978; Hall *et al.*, 1979). Less flexible glutamate-like compounds, including kainic acid and ibotenic acid, show more stereospecificity. The long-chain amino acid antagonists such as α-aminoadipate show marked stereospecificity: L-α-AA is an excitant, while D-α-AA blocks the actions of some but not all excitatory amino acids.

Table 7 shows the potency of glutamate and related excitatory amino acids in several physiological systems: cerebral cortex, feline spinal cord, frog spinal cord, and in the sodium efflux bioassay of Luini *et al.* (1981) which measures the ability of excitants to displace radioactive sodium form CNS tissue *in vitro*. Kainic acid was studied by MacDonald and Nistri (1978) in the cat spinal cord, but no potency comparison with glutamate was possible because of the unusual shape of the kainate dose-response curve. Its duration of action was much slower, and it showed no tendency to a peak firing rate but continued to increase firing rates until depolarization block occurred (MacDonald and Nistri, 1978). These differences were a clear indication that KA could not be viewed as a "glutamate analog." Invertebrate studies with KA showed it to generally potentiate glutamate responses (Constanti and Nistri, 1976), depress the development of glutamate desensitization (Shinozaki, 1980), and not to exactly parallel

TABLE 7
Glutamate Potency in Various Tissues

Drug	Ratio[a]
Potency ratios in cat cerebral cortex[b]	
N-Methyl-D-aspartate	6.5
DL-Homocysteate	4
L-Glutamate	1
L-Aspartate	1
D-Glutamate	0.7
D-Aspartate	0.8
L-Asparagine	no effect
Potency ratios in feline spinal interneurons[c]	
Quisqualate	8.39 ± 1.20
N-Methyl-D-aspartate	3.94 ± 0.50
Ibotenic acid	1.07 ± 0.31
L-Glutamate	1
Aspartate	0.91 ± 0.07
α-Aminopimelate	0.67 ± 0.06
N-Methyl-DL-aspartate	0.48 ± 0.01
Potency ratios in rat thalamic neurons[d]	
Kainate	10.6 ± 1.7
cis-1-Amino-1,3-dicarboxycyclopentane (ADCP)	9.7 ± 1.6
Ibotenate	7.4 ± 1.3
N-Methyl-DL-aspartate	2.8 ± 0.5
trans-ADCP	2.0 ± 0.1
DL-Homocysteate	1.8 ± 0.4
L-Glutamate	1.0
D-Aspartate	0.9 ± 0.1
L-Aspartate	0.7 ± 0.1
D-Glutamate	0.4 ± 0.1
DL-α-Aminopimelate	0.2 ± 0.1
L-α-Aminoadipate	0.2 ± 0.1
Potency ratios in grog spinal neurons[e]	
Quisqualate	416 ± 68
Domoic acid	280 ± 49
Kainate	101 ± 18
N-Methyl-D-aspartate	12 ± 2
DL-Homocysteate	11 ± 4
2,4,5-Trihydroxyphenylalanine (6-OH-DOPA)	4.3 ± 0.3
D-Glutamate	2.1 ± 0.6
L-Glutamate	1.0
D-Aspartate	0.75 ± 0.1
L-Aspartate	0.49 ± 0.1
L-DOPA	0.26 ± 0.1
α-Allokainic acid	0.25 ± 0.05
Relative potencies of excitatory amino acids as releasers of [$^{22}Na^{+}$][f]	
N-Methyl-*D*-aspartate	50.0
DL-Homocysteate	11.0
Kainate	8.0
Quisqualate	6.4

(continued)

TABLE 7 (*Continued*)
Glutamate Potency in Various Tissues

Drug	Ratio[a]
Kainate-methyl ketone	5.2
D-Glutamate	1.2
Piperidine dicarboxylate	1.2
L-Glutamate	1.0
D-Aspartate	0.6
L-Aspartate	0.6
L-Proline	0.2

[a] Note that the D isomers of glutamate and aspartate may be more potent than the "natural isomers" and that these two amino acids are much less potent than other excitants in some systems.
[b] Ionotophoretic studies of cat cerebral cortex by Crawford and Curtis (1963).
[c] Using barbiturate anesthesia, cat spinal interneurons were studied in urethane–barbiturate animals by MacDonald and Nistri (1978).
[d] Rat thalamic neurons were studied under urethane anesthesia by Hall *et al.* (1979).
[e] Frog spinal neurons were studied in hemisected cord preparations and bath application of drugs by Biscoe *et al.* (1976).
[f] Bath exposure of small striatal slices to amino acids, and measurement of washout of radioactive sodium Luini *et al.* (1981).

glutamate effects. Differences between the effects of KA on the responses to iontophoretic glutamate and stimulation of the excitatory nerves led to the suggestion that KA must act on extrajunctional receptors (Takeuchi and Onedera, 1975).

Among the first indications of differential properties of receptors for excitatory amino acids was the discovery that divalent metal ions could block the effects of some but not all excitants on frog motoneurons (Evans *et al.*, 1977; Watkins *et al.*, 1981). Low magnesium concentrations (as low as 10–25 μM) depressed responses to *N*-methyl-D-aspartate (NMDA) but had little effect on the responses to glutamate, quisqualate, or kainate. Aspartate responses were more sensitive to Mg^{2+} ion than were glutamate, although both were relatively resistant. Similar results were obtained in the cat spinal cord (Davies and Watkins, 1977). These magnesium effects were not calcium reversible (Watkins *et al.*, 1981). A number of organic antagonists were soon found to have effects like that of magnesium, including D-α-aminoadipate (Biscoe *et al.*, 1978; Davies and Watkins, 1977; Collingridge and Davies, 1979), and these were not limited to the spinal cord. The monoamino and diamino dicarboxylic acids are competitive antagonists of NMDA responses, while divalent ions have a noncompetitive effect. L-Homocysteic acid is an excitant which resembles NMDA in producing Mg^{2+}-sensitive responses. DL-α-Aminoadipate is a relatively potent blocker of responses to iontophoretically applied aspartate (Bergey *et al.*, 1980). Recent studies by Watkins and co-workers have produced a number of antagonists which block NMDA responses relatively selectively, and both 2-amino-5-phosphonovalerate (δ-2-APV) and D-α-aminosuberate are quite potent, with K_d estimated at 2.4 and 15.9 μM (Watkins *et al.*,

1981). It is not likely that NMDA or any other D-amino acid is an endogenous ligand in spinal cord, but antagonists selective for these receptors block spinal excitatory transmission (Davies *et al.*, 1979), suggesting that some unknown excitant (probably not glutamate or aspartate) belonging to the NMDA "group" (see Table 8) is a major spinal excitatory transmitter.

Binding studies with [^{3}H]kainic acid indicate that glutamate and aspartate are not good competitors with kainate (Simon *et al.*, 1976; Johnston *et al.*, 1979; London and Coyle, 1979). This kainate binding is localized to brain, correlates with susceptibility to kainate neurotoxicity, and is enriched in synaptosomal fractions. Unlike the usual glutamate receptor in mammalian brain, the kainate receptor shows marked stereospecificity for L-glutamate, and kainate binding and neurotoxicity are not readily blocked by antagonists of other excitatory amino acids. Hence, it seems likely that kainate receptors are a distinct class, different from the primary glutamate and aspartate receptors. Ontogenetic studies also show that KA and glutamate receptors appear at different times during development (Slevin and Coyle, 1981). Domoic acid has a higher affinity than KA itself for KA binding sites. Lactones derived from kainate (Goldberg *et al.*, 1981) appear to be relatively selective blockers of kainate responses, although millimolar concentrations of the lactones are required. Of all the suggested antagonists of excitatory amino acids, only 2-amino-5-phosphonovaleric acid and related "NMDA blockers" are active at lower concentrations (Davies *et al.*, 1981; Luini *et al.*, 1981).

Studies with other antagonists, glutamate diethyl ester (GDEE), and γ-D-glutamylglycine (γ-DGG) suggest that there is a group of excitants, of which the prototype is quisqualate, which are GDEE sensitive, insensitive to 2-APV, and sensitive to *cis*-2,3-piperidine dicarboxylate (PDA), a sub-

TABLE 8
Summary of NMDA Receptors[a]

Agonists	Antagonists
N-Methyl-D-Aspartate	2-Amino-5-phosphonovalerate (APV)
D-Homocysteate	D-α-Aminosuberate
N-Methyl-L-aspartate	α-D-Glutamyglycine
2,3-*trans*-Piperidine dicarboxylate	α-D-Glutamyl-L-proline
D-Aspartate	D-α-Aminoadipate (DAAA)
L-Homocysteate	*cis*-2,3-Piperidine dicarboxylate
D-Glutamate	
L-Aspartate	
L-Glutamate	
Kainate	

[a] The agonists and antagonists are presented in order of potency, considering both physiological and binding studies of spinal cord. A different potency order might apply for NMDA receptors in other CNS regions.

stance which also blocks NMDA and kainate receptors (Watkins, 1981). Unfortunately, most blockers of KA or quisqualate excitation are also NMDA blockers. γ-D- (or L-) glutamylglycine depresses NMDA, KA, and L-aspartate responses but has little effect on glutamate or quisqualate responses. Glutamate fits best into the category of quisqualate receptors, but to some extent appears to activate all categories of excitatory receptors. It is possible to suggest a tentative categorization of excitatory amino acid receptors into three categories (Davies *et al.*, 1980; Watkins, 1981; McLennan *et al.*, 1981) as outlined in Table 9. The ability of GDEE to depress

TABLE 9
Working Classification of Receptors for Excitatory Amino Acids

Receptors	Agonists	Antagonists
Quisqualate-preferring[a]	Quisqualic acid L-Cysteic acid L-Glutamic acid L-Aspartic acid L-Homocysteic acid Ibotenic acid	Glutamic acid diethyl ester *cis*-2,3-Piperidine dicarboxylic acid (PDA)
NMDA-preferring[b]	Ibotenic acid *N*-Methyl-D-aspartic acid (RS)-α-Amino-3-hydroxyl-5-methyl-4-isozazolopropionic acid (AMPA)[c] D-Homocysteic acid L-Homocysteic acid *N*-Methyl-L-aspartic acid D-Aspartic acid L-Aspartic acid	2-APV D-α-Aminosuberate γ-D-Glutamylglycine D-α-Amino adipate (DAA) Co^{2+} Mg^{2+} *cis*-2,3-PDA
Kainate-preferring[d]	Domoic acid Kainic acid L-Glutamic acid Quisqualate NMDA β-*N*-Oxalyl-α β-Diaminopropionate (β-ODAP)	*cis*-PDA Carboxyphenylglycine Kainic acid lactones[e] γ-D-GG

[a] It seems likely that uptake is a significant factor in terminating responses at quisqualic acid-preferring receptors. It is likely that glutamic acid responses in the intact animal often relate to these receptors (because of the blocking effect of GDEE), although all three receptor types do respond to glutamate.
[b] Of the three receptor types, NMDA receptors are the best defined. Antagonist effects suggest their importance in the spinal cord.
[c] Possibly an agonist. From Krogsgaard-Larsen *et al.* (1980).
[d] Kainic acid receptors continue to pose some problems, in part because of the complex interactions between kainic acid and glutamic acid. Although kainic acid inhibits glutamate uptake, the time course of kainic acid effects, in general, suggests either the absence of a role for kainic acid uptake in terminating receptor effects or the requirement for a cotransmitter to be released. Note that blockers of kainic acid excitation usually block other excitants as well.
[e] From Goldberg *et al.* (1981).

excitatory synaptic responses in the mammalian CNS has suggested to many people that quisqualate receptors important in the brain, and that glutamate might be the transmitter at some of these.

A number of workers have studied the binding of [^{3}H]-L-glutamate and aspartate. Roberts (1974*b*) first reported binding sites for glutamate on brain membranes, and a variety of sodium-independent binding sites have since been described (Foster and Roberts, 1981; Biziere *et al.*, 1980). Preincubation with calcium or at increasing temperatures up to 37°C markedly increases the number of glutamate binding sites (Baudry and Lynch, 1979) but not the number of KA sites. These glutamate sites do not appear to be uptake sites, but have not been established as receptor sites. Quisqualate is the most potent competitor for such sites. Two NMDA antagonists, DL-α-aminoadipate and D-α-aminopimelic acid, had significant affinity for glutamate sites, while GDEE had very little affinity. These results are not in agreement with the picture of glutamate receptors that would emerge from electrophysiological studies. Similarly, D-aspartate is a very weak displacer of [^{3}H]-L-aspartate binding in rat cerebellar membranes but is almost equipotent in physiological studies (Sharif and Roberts, 1981). It should be noted that glutamate and aspartate binding sites do not appear to survive freezing and cannot be studied in postmortem brain material at present.

Glutamate binding sites are enriched in synaptic junctions prepared by differential centrifugation (Foster *et al.*, 1981*a,b*). However, these glutamate sites are no longer stimulated by calcium ions. Glutamate sites outnumber aspartate sites at rat brain synaptic junctions. The lengthy isolation procedure may fundamentally distort the characteristics of these binding sites.

Glutamate, KA, and other excitatory amino acids can increase tissue levels of cGMP when added to cerebellar slices (Ferrendelli *et al.*, 1974; Schmidt *et al.*, 1977; Roberts, 1981). This effect has not been obtained in broken-cell preparations. Glutamate's ability to increase cGMP is blocked by GDEE. However, glutamate and aspartate lack stereospecificity for evoking this response, even though they show marked stereospecificity in their binding to synaptic membranes from the same brain region. Furthermore, the usual NMDA blockers such as D-α-aminoadipate fail to block aspartate or NMDA effects on cerebellar cGMP (Roberts, 1981). The effects of KA on cGMP are synergistic with those of glutamate. The mechanism and value of this cGMP effect remain to be demonstrated.

I have not discussed the complex relationship between excitatory amino acids and neurotoxic or "excitotoxic" effects (Olney, 1978). KA appears to require glutamate innervation for expression of its neurotoxic effects (Campochiaro and Coyle, 1978), while destruction of glutamate uptake sites enhances the weak excitotoxic potential of glutamate itself (Kohler and Schwarcz, 1981). Other excitatory amino acids, such as ibotenic acid, appear to have excitotoxic properties which are independent of glutamate innervation (Kohler *et al.*, 1979). It is easy to suggest that KA alters the disposition of glutamate or the response of glutamate receptors. While KA is a weak blocker of glutamate uptake, this does not seem to explain its actions. Recent studies have suggested that folates may act at KA receptors (Ruck *et al.*, 1980; Olney *et al.*, 1981), but this hypothesis has not been proven (P. Roberts *et al.*, 1981). Note that some folates are inhibitors of high-affinity glutamate uptake (Roberts, 1974*a*). This possibility of an endogenous excitotoxic modulator will no doubt be studied extensively in the next few years.

Although the study of a variety or panel of antagonists is beginning to clarify the picture of excitatory amino acid receptors, the membrane events which follow activation of such receptors remain very confused. Glutamate and aspartate, which are surely used as transmitters at some excitatory synapses, produce a variety of effects when studied by intracellular recording: depolarization with or without conductance increase, and hyperpolarization with conductance increase (Marder and Paupardin-Tritsch, 1978; Constanti *et al.*, 1980*b*; Lambert *et al.*, 1981). The old assumption that excitatory amino acids all acted by sodium conductance

TABLE 10
In Vitro Binding Studies with Kainate and NMDA

Compound	Kainate binding, Ki (nM)
Domoic acid	6
Kainic acid	23
α-Ketokainic acid	100
L-Glutamic acid	440
Quisqualic acid	650
Ibotenic acid	3400
DL-Homocysteic acid	10000
N-Methyl-DL-aspartate	10000

Compound	NMDA binding, IC 50 (nM)
Ibotenic acid	8.5
N-Methyl-D-aspartic acid	17
DL-Homocysteic acid	53
N-Methyl-L-aspartic acid	155
D-Aspartic acid	320
D-Glutamic acid	440
Kainic acid	600
L-Aspartic acid	900
L-Glutamic acid	2300

[a] The kainic acid binding data are modified from Coyle *et al.* (1981) and refer to studies with cerebellar membranes. The NMDA binding data are taken from unpublished work done by S. R. Snodgrass using mouse cerebellar membranes.

changes is surely wrong. The first evidence that L-glutamate was a transmitter came from studies of the crayfish neuromuscular junction (Robbins, 1959; Van Harreveld, 1959). Detailed intracellular studies of glutamate effects at this synapse reveal a great variety of mechanisms, and many glutamate receptors appear to be nonsynaptic (Colton and Freeman, 1975; Florey and Rathmayer, 1981). Some drugs act to potentiate excitant effects by interfering with amino acid transport systems—for example, histidine potentiates the effects of quisqualate in rat spinal cord (Lodge, 1981)—and studies of mechanism have much to explain in the realm of excitatory responses.

6. SUMMARY

This review has stressed the variety and complexity of GABA receptors. Just as Dudel and Hatt found four different types within the same muscle of the crayfish, more receptor types are likely to be found in the mammalian CNS. Most but not all GABA receptors mediate changes in chloride permeability. At present, most but not all mammalian GABA receptors mediate hyperpolarization. Radioligand binding studies show that GABA receptors exist in high-affinity and low-affinity forms: the low-affinity forms can be converted to high-affinity sites by a variety of treatments, including detergents, phospholipase C, and freezing and thawing. Denervating brain lesions also produce more high-affinity binding sites, whose functional significance remains uncertain. The interrelationships between BDZs, barbiturates, and GABA receptors are being actively studied and promise to help in the understanding and drug treatment of human disease. The exact relationship between amino acid transmitters and cerebellar cGMP remains to be clarified. A tightly linked "second message" coupled to these transmitters would be helpful in studies of receptor mechanisms. Meldrum's classification of GABA agonists can be used as a first step in understanding the different responses evoked by GABA. Cyclic GABA analogs have been studied extensively by Krogsgaard-Larsen and have been very useful in studies of GABA receptor mechanisms. They are poor substrates for uptake and transamination, and are also poor stimulators of BDZ binding. Antagonists of GABA receptors have always proved to be convulsants, while GABA agonists may produce convulsant, anticonvulsant, or mixed effects, depending upon many factors which are not understood.

Glycine and glycine receptors are beginning to be studied in detail and may be important in some disease states. As we begin to be able to categorize different classes of excitatory amino acids, we are still very uncertain about the fundamental mechanisms of excitatory transmission and why so many different kinds of responses to glutamate exist. After years of study, doubt still exists about the validity of glutamate as the transmitter at the accessible crayfish neuromuscular junction (Shinozaki, 1980). The role of glutamate and aspartate in mammalian CNS is not yet settled, and it seems likely that other additional excitatory transmitters will be found, perhaps small peptides containing these acidic amino acids. Considering that all or nearly all mammalian neurons possess receptors for glutamate and GABA, it can be argued that these receptors carry important developmental information and that they may remain in adult life, even if synaptic connections to GABA and glutamate-releasing neurons have been lost. It is also important that GABA and glutamate carry "fast messages" and that most of the newer peptides and "neuromodulators" produce slow responses, often without conductance changes. Perhaps every neuron uses GABA or glutamate as the vehicle for fast input at some point in its lifetime.

7. REFERENCES

Adams, P. R., and Brown, D. A., 1975, Actions of γ-aminobutyric acid on sympathetic ganglion cells, *J. Physiol.* **250:**85–120.

Aicardi, J., 1980, Seizures and epilepsy in children under two years of age, in: *The Treatment of Epilepsy* (J. H. Tyrer, ed.), pp. 203–250, Lippincott, Philadelphia.

Aldinio, C., Balzano, M. A., and Toffano, G., 1980, Ontogenic development of GABA recognition sites in different brain areas, *Pharmacol. Res. Commun.* **12:**495–500.

Andersen, P., Dingledine, R., Gjerstad, L., Langmoen, I. A., and Mosfeldt-Laursen, A., 1980, Two different responses of hippocampal pyramidal cells to application of gamma-aminobutyric acid (GABA), *J. Physiol.* **305:**279–296.

Anderson, C. H., and Greenwald, G. S., 1969, Autoradiographic analysis of estradiol uptake in the brain and pituitary of the female rat, *Endocrinology* **85:**1160–1165.

Andrews, P., and Johnston, G. A. R., 1973, Molecular orbital and proton magnetic resonance studies of bicuculline, *Nature New Biol.* **243:**29–31.

Anis, N. A., Clark, R. B., Gration, K. A. F., and Usherwood, P. N. R., 1981, Influence of agonists on desensitization of glutamate receptors on locust muscle, *J. Physiol.* **312:**345–364.

Anzelark, G., Horton, R. W., Meldrum, B. S., and Sawaya, M. C. B., 1976, Anticonvulsant action of ethanolamine-*O*-sulfate and di-*n*-propylacetate and the metabolism of γ-aminobutyric acid (GABA) in mice with audiogenic seizures, *Biochem. Pharmacol.* **25:**413–417.

Aprison, M. H., and Nadi, N. S., 1978, Glycine: inhibition from the sacrum to the medulla, in: *Amino Acids as Chemical Transmitters* (F. Fonnum, ed.), pp. 531–570, Plenum Press, New York.

Araki, T., Ito, M., and Oscarsson, O., 1961, Anion permeability of the synaptic and nonsynaptic motoneurone membrane, *J. Physiol.* **159:**410–435.

Arbilla, S., Kamal, L., and Langer, S. Z., 1979, Presynaptic GABA autoreceptors on GABAergic nerve endingts of the rat substantia nigra, *Eur. J. Pharmacol.* **57:**211–217.

Arnt, J, Scheel-Kruger, J., Magelund, G., and Krogsgaard-Larsen, P., 1979, Muscimol and related GABA receptor agonists: the potency of GABAergic drugs *in vivo* determined after intranigral injection, *J. Pharm. Pharmacol.* **31:**306–313.

Assaf, S. Y., Crunelli, V., and Kelly, J. S., 1981, Depolarizing postsynaptic actions of GABA in the rat dentate gyrus, in: *Amino Acid Neurotransmitters* (F. V. DeFeudis and P. Mandel, eds.), pp. 239–248, Raven Press, New York.

Awapara, J., Landina, A. J., Fuerst, R., and Seale, B., 1950, Free γ-aminobutyric acid in brain, *J. Biol. Chem.* **187**:35–39.

Azanza, M. J., 1981, Benzodiazepines, barbiturates, and baclofen interaction within the GABA receptors, *Gen. Pharmacol.* **12**:123–128.

Azanza, M. J., and Walker, R. J., 1975, GABA receptor interactions: models in *Helix aspersa* neurons, *Comp. Biochem. Physiol. C* **50**:155–161.

Bagust, J., Green, K. A., and Kerkut, G. A., 1981, Strychnine-sensitive inhibition in the dorsal horn of mammalian spinal cord, *Brain Res.* **217**:425–429.

Bailey, P. A., and Phillis, J. W., 1976, The interaction of four putative glutamate antagonists with glutamate and their effects on the toad spinal cord, *Gen. Pharmacol.* **7**:283–287.

Baldino, F., and Geller, H. M., 1981, Sodium valproate enhancement of γ-aminobutyric acid (GABA) inhibition: electrophysiological evidence for an anticonvulsant activity, *J. Pharmacol. Exp. Ther.* **217**:445–450.

Baraldi, M., Grandison, L., and Guidotti, A., 1979, Distribution and metabolism of muscimol in the brain and other tissues of the rat, *Neuropharmacology* **18**:57–62.

Barker, J. L., and Mathers, D. A., 1981, GABA analogues activate channels of different duration on cultured mouse spinal neurons, *Science* **212**:358–361.

Barker, J. L., and McBurney, R. N., 1979*a*, GABA and glycine may share the same conductance channel on cultured mammalian neurones, *Nature (London)* **277**:234–236.

Barker, J. L., and McBurney, R. N., 1979*b*, Phenobarbitone modulation of postsynaptic GABA receptor function on cultured mammalian neurons, *Proc. R. Soc. London Ser. B* **206**:319–327.

Barker, J. L., and Nicoll, R. A., 1973, The pharmacology and ionic dependency of amino acid responses in the frog spinal cord, *J. Physiol.* **228**:259–277.

Barker, J. L., and Ransom, B. R., 1978, Amino acid pharmacology of mammalian central neurones grown in tissue culture, *J. Physiol.* **280**:331–354.

Barker, J. L., MacDonald, J. F., Mathers, D. A., McBurney, R. N., and Oertel, W., 1981, GABA receptor functions in cultured mouse spinal neurons, in: *Amino Acid Neurotransmitters* (F. V. DeFeudis and P. Mandel, eds.), pp. 281–293, Raven Press, New York.

Barlow, R. B., 1980, *Quantitative Aspects of Chemical Pharmacology*, University Park Press, Baltimore.

Baudry, A., and Lynch, G., 1979, Regulation of glutamate receptors by cations, *Nature (London)* **282**:748–750.

Baudry, M., and Lynch, G., 1981*a*, Characterization of two [^{3}H]Glutamate binding sites in rat hippocampal membranes, *J. Neurochem.* **36**:811–820.

Baudry, M., and Lynch, G., 1981*b*, Hippocampal glutamate receptors, *Mol. Cell. Biochem.* **38**:5–18.

Baudry, M., Bundman, M. C., Smith, E. K., and Lynch, G. S., 1980, Micromolar calcium stimulates proteolysis and glutamate binding in rat brain synaptic membranes, *Science* **212**:937–938.

Bergey, G. K., Martin, M. R., and Hermes, M., 1980, Effects of D,L-α-aminoadipate on postsynaptic amino acid responses in cultured mouse spinal cord neurons, *Brain Res.* **193**:199–207.

Biggio, G., Costa, E., and Guidotti, A., 1976, Different mechanisms mediating the decrease in cerebellar cGMP elicited by halperidol and diazepam, *Adv. Biochem. Psychopharmacol.* **15**:325–335.

Biggio, G., Corda, M. G., Lamberti, C., and Gessa, G. L., 1979, Changes in benzodiazepine receptors following GABAergic denervation of the substantia nigra, *Eur. J. Pharmacol.* **58**:215–216.

Biscoe, T. J., Duggan, A. W., and Lodge, D., 1972, Antagonism between bicuculline, strychnine, and picrotoxin and depressant amino acids in the rat nervous system, *Comp. Gen. Pharmacol.* **3**:423–433.

Biscoe, T. J., Evans, R. H., Headley, P. M., Martin, M. R., and Watkins, J. C., 1976, Structure-activity relations of excitatory amino acids on frog and rat spinal neurons, *Br. J. Pharmacol.* **58**:373–382.

Biscoe, T. J., Davies, J., Dray, A., Evans, R. H., Martin, M. R., and Watkins, J. C., 1978, D-α-aminoadipate, α-diaminopimelic acid, and HA-966 as antagonists of amino acid-induced and synaptic excitation of mammalian spinal neurones *in vivo*, *Brain Res.* **148**:543–548.

Biziere, K., Thompson, H., and Coyle, J. T., 1980, Characterization of specific high-affinity binding sites for L-[^{3}H]glutamic acid in rat brain membranes, *Brain Res.* **183**:421–433.

Blankenship, J. E., Wachtel, H., and Kandel, E. R., 1971, Ionic mechanisms of excitatory, inhibitory, and dual synaptic actions mediated by an identified interneuron in abdominal ganglion of Aplysia, *J. Neurophysiol.* **34**:76–92.

Bloom, F. E., 1974, To spritz or not to spritz: the doubtful value of aimless iontophoresis, *Life Sci.* **14**:1819–1834.

Boeynaems, J. M., and Dumont, J. E., 1975, Quantitative analysis of the binding of ligands to their receptors, *J. Cycl. Nucleotide Res.* **1**:123–142.

Borbe, H. O., Muller, W. E., and Wollert, U., 1981, Specific [^{3}H]strychnine binding associated with glycine receptors in bovine retina, *Brain Res.* **205**:131–139.

Bowery, N. G., and Brown, D. A., 1974, Depolarizing actions of γ-aminobutyric acid and related compounds on rat superior cervical ganglia *in vitro*, *Br. J. Pharmacol.* **50**:205–218.

Bowery, N. G., and Dray, A., 1976, Barbiturate reversal of amino acid antagonism produced by convulsant agents, *Nature (London)* **264**:276–278.

Bowery, N. G., Brown, D. A., and Collins, J. F., 1975, Tetramethylenedisulphotetramine: an inhibitor of γ-aminobutyric acid-induced depolarization of the isolated superior cervical ganglion of the rat, *Br. J. Pharmacol.* **53**:422–424.

Bowery, N. G., Collins, J. F., and Hill, R. G., 1976*a*, Bicyclic phosphorus esters that are potent convulsants and GABA antagonists, *Nature (London)* **261**:601–604.

Bowery, N. G., Brown, D. A., Collins, G. G. S., Galvan, M., Marsh, S., and Yamini, G., 1976*b*, Indirect effects of amino acids on sympathetic ganglion cells mediated through the release of γ-aminobutyric acid from glial cells, *Br. J. Pharmacol.* **57**:73–91.

Bowery, N. G., Hill, D. R., Hudson, A. L., Doble, A., Middlemis, N., Shaw, J., and Turnbull, M., 1980, (−)-Baclofen decreases neurotransmitter release in the mammalian CNS by an action at a novel GABA receptor, *Nature (London)* **283**:92–94.

Bowery, N. G., Doble, A., Hill, D. R., Hudson, A. L., Turnbull, M. J., and Warrington, R., 1981, Structure/Activity Studies at a baclofen-sensitive, bicuculline-insensitive GABA receptor, in: *Amino Acid Neurotransmitters* (F. V. DeFeudis and P. Mandel, eds.), pp. 333–341, Raven Press, New York.

Braestrup, C., and Squires, R. F., 1978, Pharmacological characterization of benzodiazepine receptors in the brain, *Eur. J. Pharmacol.* **48**:263–270.

Braestrup, C., Nielsen, M., Krogsgaard-Larsen, P., and Falch, E., 1979, Partial agonists for brain GABA/benzodiazepine receptor complex, *Nature (London)* **280**:331–333.

Braestrup, C., Nielsen, M., and Olsen, C. E., 1980, Urinary and brain β-carboline-3-carboxylates as potent inhibitors of brain benzodiazepine receptors, *Proc. Natl. Acad. Sci. USA* **77**:2288–2292.

Bremer, F., 1953, Strychnine tetanus of the spinal cord, in: *The Spinal Cord* (G. W. Wolstenholme and J. S. Freeman, eds.), pp. 78–82, J. and A. Churchill, London.

Brennan, M. J. W., and Cantrill, R. C., 1979, δ-aminolaevulinic acid is a potent agonist for GABA autoreceptors, *Nature (London)* **280**:514–515.

Brennan, M. J. W., and Cantrill, R. C., 1981, δ-aminolevulinic acid and amino acid neurotransmitters, *Mol. Cell. Biochem.* **38**:49–58.

Briley, P. A., Konyoumdjian, J. C., Haidamous, M., and Goumard, P., 1979, Effect of L-Glutamate and kainate on rat cerebellar cGMP levels *in vivo*, *Eur. J. Pharmacol.* **54:**181–184.

Brown, D. A., and Galvan, H. N., 1977, Influence of neuroglial transport on the action of γ-aminobutyric acid on mammalian ganglion cells, *Br. J. Pharmacol.* **59:**373–378.

Broughton, R., and Mamelak, M., 1980, The effects of nocturnal gamma-hydroxybutyrate on sleep/waking patterns in narcolepsy/cataplexy, *Can. J. Neurol. Sci.* **7:**23–31.

Brown, D. A., Galvan, D. A., and Scholfield, C. N., 1980, Depolarization of neurones in slices of the olfactory cortex of the guinea pig by GABA, *Brain Res. Bull. Suppl. 2* **5:**291–293.

Brown, G. L., and Gillespie, J. S., 1957, The output of sympathetic transmitter from the spleen of the cat, *J. Physiol.* **138:**81–102.

Browner, M., Ferkany, J. W., and Enna, S. J., 1981, Biochemical identification of pharmacologically and functionally distinct GABA receptors in rat brain, *J. Neurosci.* **1:**514–518.

Bruni, J., Hammond, E. J., and Wilder, B. J., 1981, Effects of ethyl ester derivatives of valproic acid metabolites on pentylenetetrazol seizures in mice, *Can. J. Neurol. Sci.* **8:**259–264.

Bruns, R. F., Daly, J. W., and Snyder, S. H., 1980, Adenosine receptors in brain membranes: binding of *N*6-cyclohexyl-[^{3}H]adenosine and 1,3-diethyl-8-[^{3}H]phenylxanthine, *Proc. Natl. Acad. Sci. USA* **77:**5547–5551.

Builder, S. E., and Segel, I. H., 1978, Equilibrium ligand binding assays using labeled substrates: nature of the errors introduced by radiochemical impurities, *Anal. Biochem.* **85:**413–424.

Buu, N. T., Puil, E., and Van Gelder, N. M., 1976, Receptors for amino acids in excitable tissues, *Gen. Pharmacol.* **7:**5–14.

Caldwell, J. H., Daw, N. H., and Wyatt, H. J., 1978, Effects of picrotoxin and strychnine on rabbit retinal ganglion cells: lateral interactions for cells with more complex receptive fields, *J. Physiol.* **276:**277–298.

Campochiaro, P., and Coyle, J. T., 1978, Ontogenetic development of kainate neurotoxicity: correlates with glutamatergic innervation, *Proc. Natl. Acad. Sci. USA* **75:**2025–2029.

Carpenter, D. O., Swann, J. W., and Yarowsky, P. J., 1977, Effect of curare on responses to different putative neurotransmitters in Aplysia neurons, *J. Neurobiol.* **8:**119–132.

Casu, M., and Gale, K., 1981, Differential effects of gamma-aminobutyric acid (GABA) elevating agents on the neuroleptic-induced activation of striatal tyrosine hydroxylase: evidence that di-*n*-propylacetate augments GABAergic neurotransmission, *J. Pharmacol. Exp. Ther.* **217:**177–180.

Chang, H. W., and Bock, F., 1979, Structural stabilization of isolated acetylcholine receptor: specific interaction with phospholipids, *Biochemistry* **18:**172–179.

Chan-Palay, V., 1978*a*, Autoradiographic localization of γ-aminobutyric acid receptors in the rat central nervous system by using [^{3}H]muscimol, *Proc. Natl. Acad. Sci USA* **75:**1024–1028.

Chan-Palay, V., 1978*b*, Quantitative visualization of γ-aminobutyric acid receptors in hippocampus and area dentate demonstrated by [^{3}H]muscimol autoradiography, *Proc. Natl. Acad. Sci. USA* **75:**2516–2520.

Charles, A. K., and Chang, Y. F., 1981, Uptake, release and metabolism of D- and L-α-aminoadipate by rat cerebral cortex, *J. Neurochem.* **36:**1127–1136.

Choi, D. W., and Fischbach, G. D., 1981, GABA conductance of chick spinal cord and dorsal root ganglion neurons in cell culture, *J. Neurophysiol.* **45:**605–621.

Choi, D. W., Farb, D. H., and Fischbach, G. D., 1977, Chlordiazepoxide selectively augments GABA action in spinal cord cell cultures, *Nature (London)* **269:**342–344.

Chuang, D. M., 1981, Inhibitors of transglutaminase prevent agonist-mediated internalization of β-adrenergic receptors, *J. Biol. Chem.* **256:**8291–8293.

Cohen, A. I., 1978, Retinal organization and function: possible roles for taurine, in: *Taurine and Neurological Disorders* (A. Barbeau and R. J. Huxtable, eds.), pp. 117–130, Raven Press, New York.

COHEN, J. B., 1978, Ligand binding properties of membrane-bound cholinergic receptors of *Torpedo marmorata*, in: *Molecular Specialization and Symmetry in Membrane Function* (A. K. Solomon and M. Karnovsky, eds.), pp. 99–127, Harvard University Press, Cambridge, Massachusetts.

COLLINS, J. F., and HILL, R. G., 1974, (+)- and (−)-bicuculline methochloride as optical isomers of a GABA antagonist, *Nature (London)* **249:**845–847.

COLTON, C. K., and FREEMAN, A. R., 1975, Dual response of lobster muscle fibers to L-glutamate, *Comp. Biochem. Physiol.* **51:**275–284.

CONNORS, B. W., 1981, A comparison of the effects of pentobarbital and diphenylhydantoin on the GABA sensitivity and excitability of adult sensory ganglion cells, *Brain Res.* **207:**357–369.

CONSTANTI, A., 1978, The mixed effect of picrotoxin on the GABA dose/conductance relation recorded from lobster muscle, *Neuropharmacology* **17:**159–170.

CONSTANTI, A., and NISTRI, A., 1976, A comparative study of the effects of glutamate and kainate on the lobster muscle fibre and the frog spinal cord, *Br. J. Pharmacol.* **57:**359–368.

CONSTANTI, A., and NISTRI, A., 1981, Differential effects of sodium-free media on γ-aminobutyrate and muscimol-evoked conductance increases recorded from lobster muscle fibers, *Neuroscience* **7:**1443–1453.

CONSTANTI, A., KRNJEVIC, K., and NISTRI, A., 1980*a*, Intraneuronal effects of inhibitory amino acids, *Can. J. Physiol. Pharmacol.* **58:**193–204.

CONSTANTI, A., CONNOR, J. D., GALVAN, M., and NISTRI, A., 1980*b*, Intracellularly recorded effects of glutamate and aspartate on neurons in the guinea pig olfactory cortex slice, *Brain Res.* **195:**403–420.

COOMBS, J. S., ECCLES, J. C., and FATT, P., 1955, The specific ionic conductances and ionic movements across the motoneuronal membrane that produce the inhibitory postsynaptic potential, *J. Physiol.* **130:**326–373, 1955.

COSTA, E., GUIDOTTI, A., MAO, C. C., and SURIA, A., 1975, New concepts on the mechanisms of action of the benzodiazepines, *Life Sci.* **17:**167–186.

COSTA, T., RODBARD, D., and PERT, C. B., 1978, Is the benzodiazepine receptor coupled to a chloride anion channel? *Nature (London)* **277:**315–317.

COYLE, J. T., and ENNA, S. J., 1976, Neurochemical aspects of the ontogenesis of GABAergic neurons in the rat brain, *Brain Res.* **111:**119–133.

COYLE, J. T., and SCHWARCZ, R., 1976, Lesion of striatal neurones with kainic acid provides a model for Huntington's chorea, *Nature (London)* **263:**244–246.

CRAWFORD, J. M., and CURTIS, D. R., 1963, The excitation and depression of mammalian cortical neurones by amino acids, *Br. J. Pharmacol.* **23:**313–329.

CREESE, I., and SIBLEY, D. R., 1981, Receptor adaptations to centrally acting drugs, *Annu. Rev. Pharmacol. Toxicol.* **21:**357–391

CREPEL, F., DELHAYE-BOUCHARD, N., and PUMAIN, R., 1981, Amino acids and neurotransmission in rat cerebellar cortex, in: *Amino Acid Neurotransmitters*, (F. V. DeFeudis and P. Mandel, eds.), pp. 295–299, Raven Press, New York.

CRNIC, D. M., HAMMERSTAD, J. P., and CUTLER, R. W. P., 1973, Accelerated efflux of [^{14}C]- and [^{3}H]amino acids from superfused slices of rat brain, *J. Neurochem.* **20:**203–209.

CROSS, A. J., and WADDINGTON, J. L., 1981, Substantia nigra γ-aminobutyric acid receptors in Huntington's Disease, *J. Neurochem.* **37:**321–324.

CUNNINGHAM, R., and MILLER, R. F., 1976, Taurine: its selective action on neuronal pathways in the rabbit retina, *Brain Res.* **117:**341–345.

CUNNINGHAM, R., and MILLER, R. F., 1980, Electrophysiological analysis of taurine and glycine action on neurons of the mudpuppy retina. I. Intracellular recording, *Brain Res.* **197:**123–138.

CURTIS, D. R., 1963, The pharmacology of central and peripheral inhibition, *Pharmacol. Rev.* **15:**333–364.

CURTIS, D. R., and JOHNSTON, G. A. R., 1974, Amino acid transmitters in the mammalian central nervous system, *Ergebn. Physiol.* **69:**97–188.

CURTIS, D. R., and WATKINS, J. C., 1960, The excitation and depression of spinal neurones by structurally related amino acids, *J. Neurochem.* **6:**117–141.

CURTIS, D. R., PHILLIS, J. W., and WATKINS, J. C., 1959, The depression of spinal neurones by γ-aminobutyric acid and β-alanine, *J. Physiol.* **146:**182–203.

CURTIS, D. R., HOSLI, L., and JOHNSTON, G. A. R., 1967, Inhibition of spinal neurones by glycine, *Nature (London)* **215:**1502–1503.

CURTIS, D. R., HOSLI, L., JOHNSTON, G. A. R., and JOHNSTON, I. H., 1968, The hyperpolarization of spinal neurones by glycine and related amino acids, *Exp. Brain Res.* **5:**235–258.

CURTIS, D. R., DUGGAN, A. W., FELIX, D., and JOHNSTON, G. A. R., 1971, Bicuculline, an antagonist of GABA and synaptic inhibition in the spinal cord of the cat, *Brain Res.* **32:**69–96.

CURTIS, D. R., DAVIES, J., GAME, C. J. A., JOHNSTON, G. A. R., and MCCULLOCH, R. M., 1973, Central actions of shikimin and tutin, *Brain Res.* **63:**419–423.

CURTIS, D. R., LODGE, D., and BORNSTEIN, J. C., 1979, Nuciferine and central glutamate receptors, *J. Pharm. Pharmacol.* **31:**795–797.

CUTLER, R. W. P., MARKOWITZ, D., and DUDZINKSI, D. S., 1974, The effects of barbiturates on [^{3}H]GABA transport in rat cerebral cortex slices, *Brain Res.* **81:**189–197.

CZUCZWAR, S. J., TURSKI, L., and KLEINROK, Z., 1981, Diphenylhydantoin potentiates the protective effect of diazepam against pentylenetetrazole but not bicuculline- and isoniazid-induced seizures in mice, *Neuropharmacology* **20:**675–679, 1981.

DAVIDOFF, R. A., and SEARS, E. S., 1974, The effects of lioresal on synaptic activity in the isolated spinal cord, *Neurology* **24:**957–963.

DAVIDSON, N., MACFARLANE, E. I., and MICHIE, D. L., 1977, Comparison of bicyclic phosphorus esters with bicuculline and picrotoxin as antagonists of presynaptic inhibition in the rat cuneate nucleus, *Experientia* **33:**935–936.

DAVIES, J., 1981, Selective depression of synaptic excitation in cat spinal neurones by baclofen: an iontophoretic study, *Br. J. Pharmacol.* **72:**373–384.

DAVIES, J., and POLC, P., 1979, Effects of L-nuciferine on kainate, *N*-methyl-D-aspartate, and acetylcholine excitation of cat spinal neurons, *J. Pharm. Pharmacol.* **31:**178–179.

DAVIES, J., and WATKINS, J. C., 1977, Effect of magnesium ions on the responses of spinal neurones to excitatory amino acids and acetylcholine, *Brain Res.* **130:**364–368.

DAVIES, J. and WATKINS, J. C., 1981, Differentiation of kainate and quisqualate receptors in the cat spinal cord by selective antagonism with γ-D- (and L-) glutamylglycine, *Brain Res.* **206:**172–177.

DAVIES, J., EVANS, R. H., FRANCIS, A. A., and WATKINS, J. C., 1979, Excitatory amino acid receptors and synaptic excitation in the mammalian central nervous system, *J. Physiol. (Paris)* **75:**641–654.

DAVIES, J., EVANS, R. H., FRANCIS, A. A., JONES, A. W., and WATKINS, J. C., 1980, Excitatory amino acid receptors in the vertebrate central nervous system, in: *Neurotransmitters and Their Receptors* (U. Z. Littauer, Y. Dudai, I. Silman, V. I. Teichberg, and Z. Vogel, eds.), Wiley, New York.

DAVIES, J., FRANCIS, A. A., JONES, A. W., and WATKINS, J. C., 1981, 2-Amino-5-phosphonovalerate (2-APV), a potent and selective antagonist of amino acid-induced and synaptic excitation, *Neuroscience Lett.* **21:**77–81.

DAVIS, W. C., and TICKU, M. K., 1981, Ethanol enhances [^{3}H]diazepam binding at the benzodiazepine-γ-aminobutyric acid receptor–ionophore complex, *Mol. Pharmacol.* **20:**287–294.

DeFeudis, F. V., Orensanz-Munoz, L. M., and Fando, J. L., 1978, High-affinity glycine binding sites in rat CNS: regional variation and strychnine sensitivity, *Gen. Pharmacol.* **9**:171–176.

DeFeudis, F. V., Ossola, L., Schmitt, G., and Mandel, P., 1980, Substrate specificity of [^{3}H]-muscimol binding to a particulate fraction of a neuron-enriched culture of embryonic rat brain, *J. Neurochem.* **34**:845–849.

DeGroat, W. C., 1970, The actions of γ-aminobutyric and related amino acids on mammalian autonomic ganglia, *J. Pharmacol. Exp. Ther.* **172**:384–396.

DeGroat, W. C., 1972, GABA depolarization of a sensory ganglion: antagonism by picrotoxin and bicuculline, *Brain Res.* **38**:29–432.

Desarmenien, M., Santangelo, F., Linck, G., Headley, P. M., and Feltz, P., 1981, Physiological study of amino acid uptake and receptor desensitization: the GABA system in dorsal root ganglia, in: *Amino Acid Neurotransmitters* (F. V. DeFeudis and P. Mandel, eds.), pp. 309–319, Raven Press, New York.

Dichter, M. A., 1980, Physiological identification of GABA as the inhibitory transmitter for mammalian cortical neurons in cell culture, *Brain Res.* **190**:111–121.

Dingledine, R., and Gjerstad, L., 1981, Reduced inhibition during epileptiform activity in the *in vitro* hippocampal slice, *J. Physiol.* **305**:297–313.

Dingledine, R., Iversen, L. L., and Brenker, E., 1978, Naloxone as a GABA antagonist: evidence from iontophoretic, receptor binding, and convulsant studies, *Eur. J. Pharmacol.* **47**:19–27.

Divac, I., Fonnum, F., and Storm-Mathisen, J., 1977, High-affinity uptake of glutamate in terminals of corticostriatal axons, *Nature (London)* **266**:377–378.

Doble, A., and Turnbull, M. J., 1981, Lack of effect of benzodiazepines on bicuculline-insensitive GABA receptors in the field-stimulated guinea pig vas deferens preparation, *J. Pharm. Pharmacol.* **33**:267–268.

Dudel, J., and Hatt, H., 1976, Four types of GABA receptors in crayfish leg muscles characterized by desensitization and specific antagonist, *Pflugers Arch.* **364**:217–222.

Dunlap, K., and Fischbach, G. D., 1978, Neurotransmitters decrease the calcium component of sensory neurone action potentials, *Nature (London)* **276**:837–839.

Eccles, J. C., Schmidt, R. F., and Willis, W. D., 1963, Pharmacological studies on presynaptic inhibition, *J. Physiol.* **168**:500–530.

Edvinsson, L., and Krause, D. N., 1979, Pharmacological characterization of GABA receptors mediating vasodilation of cerebral arteries *in vitro*, *Brain Res.* **173**:89–97.

Edwards, C., and Kuffler, S. W., 1959, The blocking effect of γ-aminobutyric acid and the action of related compounds on simple nerve cells, *J. Neurochem.* **4**:19–30.

Ehinger, B., 1976, Selective neuronal accumulation of ω-amino acids in the rabbit retina, *Brain Res.* **107**:541–554.

Engberg, I., Flatman, J. A., and Lambert, J. D. C., 1979, The actions of excitatory amino acids on motoneurones in the feline spinal cord, *J. Physiol.* **288**:227–261.

Enna, S. J., 1978, The GABA receptor assay: focus on human studies, in: *Amino Acids as Chemical Transmitters* (F. Fonnum, ed.), pp. 445–456, Plenum Press, New York.

Enna, S. J., 1981, GABA receptor pharmacology, *Biochem. Pharmacol.* **30**:907–913.

Enna, S. J., and Maggi, A., 1979, Biochemical pharmacology of GABAergic agonists, *Life Sci.* **24**:1727–1738.

Enna, S. J., and Snyder, S. H., 1975, Properties of γ-aminobutyrate (GABA) receptor binding in rat brain synaptic membrane fractions, *Brain Res.* **100**:81–97.

Enna, S. J., and Snyder, S. H., 1976, A simple, sensitive, and specific radioreceptor assay for endogenous GABA in brain tissue, *J. Neurochem.* **26**:221–224.

Enna, S. J., and Snyder, S. H., 1977, Influence of ions, enzymes, and detergents on γ-aminobutyric acid receptor binding in synaptic membranes of rat brain, *Mol. Pharmacol.* **13**:442–453.

Enna, S. J., Bennett, J. P., Bylund, D. B., Snyder, S. H., Bird, E. D., and Iversen, L. L.,

1976, Alterations of brain neurotransmitter receptor binding in Huntington's Chorea, *Brain Res.* **116:**531–537.

Enna, S. J., Collins, J. F., and Snyder, S. H., 1977, Stereospecificity and structure–activity requirements of GABA receptor binding in rat brain, *Brain Res.* **124:**185–190.

Ennis, C., and Cox, B., 1981, GABA enhancement of [^{3}H]dopamine release from slices of rat striatum: dependence on slice size, *Eur. J. Pharmacol.* **70:**417–420.

Epstein, R., and Grundfest, H., 1970, Desensitization of gamma-aminobutyric acid (GABA) receptors in muscle fibers of the crab, *Cancer borealis*, *J. Gen. Physiol.* **56:**33–45.

Evans, R. H., Francis, A. A., and Watkins, J. C., 1977, Selective antagonism by Mg^{2+} of amino acid-induced depolarization of spinal neurons, *Experientia* **33:**489–490.

Evans, R. H., Francis, A. A., and Watkins, J. C., 1978, Mg^{2+}-like selective antagonism of excitatory amino acid-induced responses by γ-diaminopimelic acid, D-γ-aminoadipate, and HA 966 in isolated spinal cord of frog and immature rat, *Brain Res.* **148:**536–542.

Evans, R. H., Francis, A. A., Hunt, K., Oakes, D. J., and Watkins, J. C., 1979, Antagonism of excitatory amino acid-induced responses and of synaptic excitation in the isolated spinal cord of the frog, *Br. J. Pharmacol.* **67:**591–603.

Felix, D., and Kunzle, H., 1974, Iontophoretic and autoradiographic studies on the role of proline in nervous transmission, *Pflugers Arch. Ges. Physiol.* **350:**135–144.

Feltz, A., 1971, Competitive interaction of β-guanidinopropionic acid and γ-aminobutyric acid on the muscle fibre of the crayfish, *J. Physiol.* **216:**391–401.

Feltz A., and Trautman, A., 1980, Interaction between nerve-released acetylcholine and bath-applied agonists at the frog end plate, *J. Physiol.* **299:**533–552.

Feltz, P., and Rasminsky, M., 1974, A model for the mode of action of GABA on primary afferent terminals: depolarizing effects of GABA applied iontophoretically to neurones of mammalian dorsal root ganglia, *Neuropharmacology* **13:**553–563.

Ferrendelli, J. A., Chang, M. M., and Kinscherf, D. A., 1974, Elevation of cyclic GMP levels in central nervous system by excitatory and inhibitory amino acids, *J. Neurochem.* **22:**535–540.

Florey, E., and Rathmayer, M., 1981, Glutamate actions on crustacean muscle in: *Amino Acid Neurotransmitters* (F. V. DeFeudis and P. Mandel, eds.), pp. 351–358, Raven Press, New York.

Foster, A. C., Mena, E. E., Monaghan, D. T., and Cotman, C. W., 1981*a*, Synaptic localization of kainic acid-binding sites, *Nature (London)* **289:**73–75.

Foster, A. C., Mena, E. E., Fagg, G. E., and Cotman, C. W., 1981*b*, Glutamate and aspartate binding sites are enriched in synaptic junctions isolated from rat brain, *J. Neurosci.* **1:**620–625.

Foster, G. A., and Roberts, P. J., 1981, Kainic acid stimulation of cerebellar cyclic GMP level: potentiation by glutamate and related amino acids, *Neurosci. Lett.* **23:**67–70.

Fowler, L. J., Beckford, J., and John, R. A., 1975, An analysis of the kinetics of the inhibition of rabbit brain gamma-aminobutyrate transaminase by sodium di-*n*-propylacetate and some other simple carboxylic acids, *Biochem. Pharmacol.* **24:**1267–1270.

Fox, S., Krnjevic, K., Morris, M. E., Puil, E., and Werman, R., 1978, Actions of baclofen on mammalian synaptic transmission, *Neuroscience* **3:**495–505.

Frank, K., and Fuortes, M. G. F., 1957, Presynaptic and postsynaptic inhibition of monosynaptic reflexes, *J. Physiol.* **16:**39–40.

Frederickson, R. C. A., Neuss, M., Morzorati, S. L., and McBride, W. J., 1978, A comparison of the inhibitory effects of taurine and GABA on identified Purkinje cells and other neurons in the cerebellar cortex of the rat, *Brain Res.* **145:**117–126, 1978.

Freeman, A. R., 1973, Electrophysiological analysis of the actions of strychnine, bicuculline, and picrotoxin on the axonal membrane, *J. Neurobiol.* **4:**567–582.

Friedel, R. O., Johnson, J. R., and Schanberg, S. M., 1973, Effects of sympathomimetic drugs on incorporation *in vivo* of intracisternally injected 33-Pi into phospholipids of rat brain, *J. Pharmacol. Exp. Ther.* **184:**583–589.

FRUMKES, T. E., MILLER, R. F., SLAUGHTER, M., and DACHEUX, R. F., 1981, Physiological and pharmacological basis of GABA and glycine action of neurons of mudpuppy retina. III. Amacrine-mediated inhibitory influences on ganglion cell receptor field organization: a model, *J. Neurophysiol.* **45**:783–804.

GALE, K., 1980, Chronic blockade of dopamine receptors by antischizophrenic drugs enhances GABA binding in substantia nigra, *Nature (London)* **283**:569–579.

GALE, K., and IADAROLA, M. J., 1980, Seizure protection and increased nerve terminal GABA: delayed effects of GABA transaminase inhibition, *Science* **208**:288–291.

GALLAGER, D. W., 1980, Functional importance of benzodiazepine binding in brain, *Brain Res. Bull. Suppl. 2* **5**:833–838.

GALLAGER, D. W., THOMAS, J. W., and TALLMAN, J. F., 1978, Effect of GABAergic drugs on benzodiazepine binding site sensitivity in rat cerebral cortex, *Biochem. Pharmacol.* **27**:2745–2749.

GALLAGER, D. W., MALLORGA, P., and TALLMAN, J. F., 1980, Interaction of diphenylhydantoin and benzodiazepines in the CNS, *Brain Res.* **189**:209–220.

GALLAGER, D. W., MALLORGA, P., OERTEL, W., HENNEBERRY, R., and TALLMAN, J., 1981*a*, [^{3}H]Diazepam binding in mammalian CNS: a pharmacological characterization, *J. Neurosci.* **1**:218–225.

GALLAGER, D. W., MALLORGA, P., SWAIMAN, K. F., NEALE, E. A., and NELSON, P. G., 1981*b*, Effects of Phenytoin on [^{3}H]diazepam binding in dissociated primary cortical cell cultures, *Brain Res.* **218**:319–330.

GALLAGHER, J. P., HIGASHI, H., and NISHI, S., 1978, Characterization and ionic basis of GABA-induced depolarization recorded *in vitro* from cat primary afferent neurons, *J. Physiol.* **275**:263–282.

GALLAGHER, J. P., INOKUCHI, H., NAKAMURA, J., and SHINNICK-GALLAGHER, P., 1981, Effects of anticonvulsants on excitability and GABA sensitivity of cat dorsal root ganglion cells, *Neuropharmacology* **20**:427–433.

GALLO, V., LEVI, G., RAITERI, M., and COLETTI, A., 1981, Enhancement by GABA of glutamate-depolarization-induced release of glutamate from cerebellar nerve endings, *Brain Res.* **205**:431–435.

GAVISH, M., and SNYDER, S. H., 1980*a*, Soluble benzodiazepine receptors: GABAergic regulation, *Life Sci.* **26**:579–582.

GAVISH, M., and SNYDER, S. H., 1980*b*, Benzodiazepine recognition sites on GABA receptors, *Nature (London)* **287**:651–652.

GELLER, H. M., TAYLOR, D. A., and HOFFER, B. J., 1978, Benzodiazepines and central inhibitory mechanisms, *Naunyn-Schmiedeberg's Arch. Pharmacol.* **304**:81–88.

GERSCHENFELD, H. M., 1973, Chemical Transmission in invertebrate central nervous systems and neuromuscular junctions, *Physiol. Rev.* **53**:1–119.

GESSA, G. L., VARGIU, L., and CRABAI, F., 1966, Selective increase in brain dopamine induced by gamma-hydroxybutyrate, *Life Sci.* **5**:1921–1930.

GODFRAIND, J. M., KRNJEVIC, K., and PUMAIN, R., 1970, Doubtful value of bicuculline as a specific antagonist of GABA, *Nature (London)* **228**:675.

GODIN, Y., HEINER, L., MARK, J., and MANDEL, P., 1969, Effects of di-*n*-propylacetate, an anticonvulsive compound, on GABA metabolism, *J. Neurochem.* **16**:869–873.

GOGOLAK, G., HUCK, S., PORGES, P., and STUMPF, C., 1974, Action of strychnine and central depressants on the cerebello-rubral system, *Arch. Int. Pharmacodyn. Ther.* **207**:322–332.

GOLDBERG, O., LUINI, A., and TEICHBERG, V. I., 1981, Lactones derived from kainic acid: novel selective antagonists of amino acid-induced Na^+ fluxes in rat striatum slices, *Neuroscience Lett.* **23**:187–191.

GOLDINGER, A., and MULLER, W. E., 1980, Stereospecific interaction of bicuculline with specific [^{3}H]strychnine binding to rat spinal cord synaptosomal membranes, *Neurosci. Lett.* **16**:91–95.

Grace, A. A., and Bunney, B. S., 1979, Paradoxical GABA excitation of nigral dopaminergic cells: indirect mediation through reticulata inhibitory neurons, *Eur. J. Pharmacol.* **59:**211–218.
Grandison, L., and Guidotti, A., 1979, γ-Aminobutyric acid receptor function in rat anterior pituitary: evidence for control of prolactin release, *Endocrinology* **105:**754–759.
Gration, K. A. F., Lambert, J. J., Ramsey, R., and Usherwood, P. N. R., 1981, Non-random openings and concentration dependent lifetimes of glutamate-gated channels in muscle membrane, *Nature (London)* **291:**423–426.
Greenlee, D. V., Van Ness, P. C., and Olsen, R. W., 1978, Gamma-aminobutyric acid binding in mammalian brain: receptor-like specificity of sodium-independent sites, *J. Neurochem.* **31:**933–938.
Grossman, A., Delitalia, G., Yeo, T., and Besser, G. M., 1981, GABA and muscimol inhibit the release of prolactin from dispersed rat anterior pituitary cells, *Neuroendocrinology* **32:**145–149.
Grundfest, H., Reuben, J. P., and Rickles, W. H., 1959, The electrophysiology and pharmacology of lobster neuromuscular synapses, *J. Gen. Physiol.* **42:**1301–1323.
Gruol, D. L., Barker, J. L., and Smith, T. G., 1980, Naloxone antagonism of GABA-evoked membrane depolarizations in cultured mouse spinal cord neurons, *Brain Res.* **198:**323–332.
Guidotti, A., 1978, Synaptic mechanisms in the action of benzodiazepines, in: *Psychopharmacology: A Genberation of Progress* (M. A. Lipton, A. DiMascio, and K. F. Killam, eds.), pp. 1349–1357, Raven Press, New York.
Guidotti, A., Naik, S. R., and Kurosawa, A., 1977, Possible role of gamma aminobutyric acid (GABA) in the regulation of cAMP system in rat anterior pituitary, *Psychoneuroendocrinology* **2:**227–235.
Guidotti, A., Toffano, G., and Costa, E., 1978, An endogenous protein modulates the affinity of GABA and benzodiazepine receptors in rat brain, *Nature (London)* **275:**553–555.
Gumulka, S. W., Dinnendahl, V., and Schonhofer, P., 1981, Baclofen and cerebellar cyclic GMP levels in mice, *Pharmacology* **19:**75–81.
Haefely, W. E., 1980, GABA and the anticonvulsant action of benzodiazepines and barbiturates, *Brain Res. Bull. Suppl. 2* **5:**873–878.
Hall, J. G., Hicks, T. P., McLennan, H., Richardson, T. L., and Wheal, H. V., 1979, The excitation of mammalian central neurons by amino acids, *J. Physiol.* **286:**29–39.
Harvey, P. K. P., Bradford, H. F., and Davison, A. N., 1975, The inhibitory effect of sodium-*n*-dipropylacetate on the degradative enzymes of the GABA shunt, *FEBS Lett.* **52:**251–254.
Hayashi, T., 1952, A physiological study of epileptic seizures following cortical stimulation in animals and its application to human clinics, *Jpn. J. Physiol.* **3:**46–64.
Heidmann, T., and Changeux, J. P., 1978, Structural and functional properties of the acetylcholine receptor protein in its purified and membrane-bound states, *Annu. Rev. Biochem.* **47:**317–357.
Hicks, T. P., Hall, J. G., and McLennon, H., 1978, Ranking of excitatory amino acids by the antagonists glutamic acid diethyl ester and D-γ-amino adipic acid, *Can. J. Physiol. Pharmacol.* **56:**9012–907.
Hill, D. R., and Bowery, N. G., 1981, [^{3}]Baclofen and [^{3}H]-GABA binding to bicuculline-insensitive GABA sites in rat brain, *Nature (London)* **290:**149–152.
Hill, M. W., and Bangham, A. D., 1975, General depressant drug dependency: a biophysical hypothesis, *Adv. Exp. Biol. Med.* **59:**1–9.
Hochner, B., Spira, M. E., and Werman, R., 1976, Penicillin decreases chloride conductance in crustacean muscle: a model for the epileptic neuron, *Brain Res.* **107:**85–103.
Homma, S., and Rovainen, C. M., 1978, Conductance increases produced by glycine and γ-aminobutyric acid in lamprey interneurones, *J. Physiol.* **279:**231–252.
Horng, J. S., and Wong, D. T., 1979, γ-aminobutyric acid receptors in cerebellar membranes of rat brain after a treatment with Triton X-100, *J. Neurochem.* **32:**1379–1386.

Hosli, L., and Tebecis, A. K., 1973, Action of amino acids and convulsants on bulbar reticular neurones, *Exp. Brain Res.* **11:**111–127.

Hosli, L., Andres, P. F., and Hosli, E., 1978, Neuron–glia interactions: indirect effect of GABA on cultured glial cells, *Exp. Brain Res.* **33:**425–434.

Hosli, L., Hosli, A., Landolt, H., and Zehnter, C., 1981, Efflux of potassium from neurons excited by glutamate and aspartate causes a depolarization of cultured glial cells, *Neurosci. Lett.* **21:**83–86.

Hue, B., Pelhate, M., and Chanelet, J., 1979, Pre- and postsynaptic effects of taurine and GABA in the cockroach central nervous system, *Can. J. Neurol. Sci.* **6:**243–250.

Hunkeler, W., Mohler, H., Pieri, L., Polc, P., Bonetti, E. P., Cumin, R., Schaffner, R., and Haefely, W., 1981, Selective antagonists of benzodiazepines, *Nature (London)* **209:**514–516.

Hunt, P., and Clements-Jewery, S., 1981, A steroid derivative, R 5135, antagonizes the GABA/benzodiazepine receptor interaction, *Neuropharmacology* **20:**357–361.

Hwang, E. C., and Van, Woert, M. H., 1979, Antimyoclonic action of clonazepam: the role of serotonin, *Eur. J. Pharmacol.* **60:**31–40.

Iversen, L. L., 1972, The uptake, storage, release, and metabolism of GABA in inhibitory nerves in: *Perspectives in Neuropharmacology* (S. H. Snyder, ed.), pp. 75–93, Oxford University Press, New York.

Jakobs, C., and Loscher, W., 1978, Identification of metabolites of valproic acid in serum of humans, dogs, rat, and mouse, *Epilepsia* **19:**591–602.

James, T. A., and Starr, M. S., 1978, The role of GABA in the substantia nigra, *Nature (London)* **275:**229–230.

James, V. A., Sharma, P. R., Walker, R. J., and Wheal, H. V., 1980, Actions of glutamate kainate, dihydrokainate, and analogues on leech neurone acidic amino acid receptors, *Eur. J. Pharmacol.* **62:**35–39.

Jessen, K. R., 1981, GABA and the enteric nervous system, *Mol. Cell. Biochem.* **38:**69–76.

Johnston, G. A. R., 1979, Central nervous system receptors for glutamic acid in: *Glutamic Acid: Advances in Biochemistry and Physiology* (L. J. Filer, Jr., S. Garratini, M. R. Kare, M. A. Reynolds, and R. J. Wurtman, eds.), pp. 177–185, Raven Press, New York.

Johnston, G. A. R., and Mitchell, J. F., 1971, The effect of bicuculline, metrazol, picrotoxin, and strychnine on the release of [^{3}H]-GABA from rat brain slices, *J. Neurochem.* **18:**2441–2446.

Johnston, G. A. R., and Willow, M., 1981, Barbiturate and GABA receptors, in: *GABA and Benzodiazepine Receptors* (E. Costa, G. DiChiara, and G. L. Gessa, eds., pp. 191–198, Raven Press, New York.

Johnston, G. A. R., Kennedy, S. M. E., and Lodge, D., 1978, Muscimol uptake, release, and binding in rat brain slices, *J. Neurochem.* **31:**1519–1523.

Johnston, G. A. R., Kennedy, S. M. E., and Twitchin, B., 1979, Action of the neurotoxin kainic acid on high affinity uptake of glutamic acid in rat brain slices, *J. Neurochem.* **32:**121–127.

Kaakola, S., 1980, Circling behavior induced by intranigral injection of baclofen in rats, *Arch. Int. Pharmacodyn.* **244:**113–122.

Karobath, M., and Sperk, G., 1979, Stimulation of benzodiazepine receptor binding by γ-aminobutyric acid, *Proc. Natl. Acad. Sci. USA* **76:**1004–1006.

Karobath, M., Placheta, M., Lippitsch, M., and Krogsgaard-Larsen, P. 1979, Is stimulation of benzodiazepine binding mediated by a novel GABA receptor? *Nature (London)* **278:**748–749.

Kato, E., and Kuba, K., 1980, Inhibition of transmitter release in bullfrog sympathetic ganglia induced by γ-aminobutyric acid, *J. Physiol.* **298:**271–283.

Katz, B., and Thesleff, S., 1957, A study of the desensitization produced by acetylcholine at the motor end plate, *J. Physiol.* **138:**63–80.

Kelly, J. S., 1975, Microiontophoretic application of drugs onto single neurons, in: *Handbook of Psychopharmacology*, Vol. 2 (L. L. Iversen, S. D. Iversen, and S. H. Snyder, eds.), pp. 29–67, Plenum Press, New York.

Kelly, J. S., and Beart, P. M., 1975, Amino acid receptors in CNS. II. GABA in supraspinal regions, in: *Handbook of Psychopharmacology*, Vol. 4 (L. L. Iversen, S. D. Iversen, and S. H. Snyder, eds.), pp. 129–198, New York, Plenum.

Kendall, D. A., Fox, D. A., and Enna, S. J., 1981, Effect of γ-vinyl-GABA on bicuculline-induced seizures, *Neuropharmacology* **20:**351–355.

Kerwin, R. W., and Taberner, P. V., 1981, The mechanism of action of sodium valproate, *Gen. Pharmacol.* **12:**71–75.

Klepner, C. A., Lippa, A. S., Benson, D. I., Sano, M. C., Beer, B., 1979, Resolution of two biochemically and pharmacologically distinct benzodiazepine receptors, *Pharmacol. Biochem. Behav.* **11:**457–462.

Kohler, C., and Schwarcz, R., 1981, Monosodium glutamate: increased neurotoxicity after removal of neuronal reuptake sites, *Brain Res.* **211:**485–491.

Kohler, C., Schwarcz, R., and Fuxe, K., 1979, Hippocampal lesions indicate differences between the excitotoxic properties of acidic amino acids, *Brain Res.* **176:**366–371.

Kozhechkin, S. N., 1981, The influence of sodium gamma-hydroxybutyrate on GABA receptors and catecholamine effect potentiation, *Arch. Int. Pharmacodyn.* **250:**242–253.

Krause, D. N., Wong, E., Degener, P., and Roberts, E., 1980, GABA receptors in bovine cerebral blood vessels: binding studies with [^{3}H]muscimol, *Brain Res.* **185:**51–57.

Krnjevic, K., 1976, Inhibitory action of GABA and GABA mimetics on vertebrate neurons, in: *GABA and Nervous System Function* (E. Roberts, T. N. Chase, and D. B. Tower, eds.), Raven Press, New York.

Krnjevic, K., 1980, The role of GABA receptors in the genesis of seizures, in: *Neurotransmitters and their Receptors* (U. Z. Littauer, Y. Dudai, I. Silman, V. I. Teichberg, and Z. Voigel, eds.), pp. 405–416, Wiley, New York.

Krnjevic, K., Puil, E., and Werman, R., 1977, GABA and glycine actions on spinal motoneurons, *Can. J. Physiol. Pharmacol.* **55:**658–669.

Krogsgaard-Larsen, P., and Arnt, J., 1980, Pharmacological studies of interactions between benzodiazepines and GABA receptors, *Brain Res. Bull. Suppl. 2* **5:**867–872.

Krogsgaard-Larsen, P., and Falch, E., 1981, GABA agonists, development and interactions with the GABA receptor complex, *Mol. Cell. Biochem.* **38:**129–146.

Krogsgaard-Larsen, P., Hjeds, H., Curtis, D. R., Lodge, D., and Johnston, G. A. R., 1979, Dihydromuscimol, thiomuscimol, and related heterocyclic compounds as GABA analogues, *J. Neurochem.* **32:**1717–1724.

Krogsgaard-Larsen, P., Falch, E., Schousboe, A., Curtis, D. R., and Lodge, D., 1980, Piperidine-4-sulphonic acid, a new specific GABA agonist, *J. Neurochem.* **34:**756–759.

Krogsgaard-Larsen, P., Schultz, B., Mikkelson, H., Aaes-Jorgensen, T., and Bogeso, K. P., 1981, THIP, isoguvacine, isoguvacine oxide, and related GABA agonists, in: *Amino Acid Neurotransmitters* (F. V. DeFeudis and P. Mandel, eds.), pp. 69–76, Raven Press, New York.

Kudo, Y., 1978, The pharmacology of the amphibian spinal cord, *Prog. Neurobiol.* **11:**1–76.

Kudo, Y., Oka, J. I., and Yamada, K., 1981*a*, Anisatin, a potent GABA antagonist, isolated from *Illicium Anisatum*, *Neurosci. Lett.* **25:**83–88.

Kudo, Y., Kurachi, M., and Fukuda, H., 1981*b*, An apparent excitatory action of baclofen on the isolated perfused spinal cord of the frog, *Gen. Pharmac.* **12:**193–197.

Kuhn, W., Neuser, D., and Przuntek, H., 1981, [^{3}H]Diazepam displacing activity in human cerebrospinal fluid, *J. Neurochem.* **37:**1045–1047.

Kupersmith, M. J., and Lieberman, A. N., 1980, The effect of specific brain lesions on the high-affinity binding of GABA in the substantia nigra, *Brain Res.* **194:**536–539.

Kurioka, S., Kimura, Y., and Matsuda, M., 1981, Effects of sodium and bicarbonate ions on γ-aminobutyric acid receptor binding in synaptic membranes of rat brain, *J. Neurochem.* **37**:418–421.

LAMBERT, J. D. C., FLATMAN, J. A., and ENGBERG, I., 1981, Actions of excitatory amino acids on membrane conductance and potential in motoneurones, in: *Glutamate as a Neurotransmitter* (G. DiChiara and G. L. Gessa, eds.), pp. 205–216, Raven Press, New York.

LANGLEY, J. N., 1905, On the reaction of cells and nerve endings to certain poisons chiefly as regards the reaction of striated muscle to nicotine and to curare, *J. Physiol.* **33**:374–413.

LANGMOEN, I. A., and HABLITZ, J. J., 1981, Reversal potential for glutamate responses in hippocampal pyramidal cells, *Neurosci. Lett.* **23**:61–65.

LEACH, M. J., 1979, Effect of taurine on release of [^{3}H]-GABA by depolarizing stimuli from superfused slices of rat brain cerebral cortex *in vitro*, *J. Pharm. Pharmacol.* **31**:533–535.

LEEB-LUNDBERG, F., SNOWMAN, A., and OLSEN, R. W., 1980, Barbiturate receptor sites are coupled to benzodiazepine receptors, *Proc. Natl. Acad. Sci. USA* **77**:7468–7472.

LESTER, B. R., MILLER, A. L., and PECK, E. J., 1981, Differential solubilization of γ-aminobutyric acid receptive sites from membranes of mammalian brian, *J. Neurochem.* **36**:154–164.

LIEBMAN, J. M., PASTOR, G., BERVARD, P. S., and SAELENS, J. K., 1980, Antagonism of intrastriatal and intravenous kainic acid by *l*-nuciferine: comparison with various anticonvulsants and GABA mimetics, *Life Sci.* **27**:1991–1998.

LINGLE, C., EISEN, J. S., and MARDER, E., 1981, Block of glutamatergic excitatory synaptic channels by chlorisondamine, *Mol. Pharmacol.* **19**:349–353.

LIPPA, A. S., KLEPNER, C. A., BENSON, D. I., CRITCHETT, D. J., SANO, M. C., and BEER, B., 1980, The role of GABA in mediating the anticonvulsant properties of benzodiazepines, *Brain Res. Bull. Suppl. 2* **5**:861–865.

LLOYD, K. G., and BEAUMONT, K., 1980, Possible role of phospholipids in GABA receptor function in human and rat brain, *Brain Res. Bull. Suppl. 2* **5**:285–290.

LLOYD, K. G., and DAVIDSON, L., 1979, [^{3}H]-GABA binding in brains from Huntington's chorea patients: altered regulation by phospholipids? *Science* **205**:1147–1149.

LLOYD, K. G., and DREKSLER, S., 1978, [^{3}H]-GABA binding to membranes prepared from postmortem human brain: pharmacological and pathological investigations, in: *Amino Acids as Chemical Transmitters* (F. Fonnum, ed.), pp. 457–466, Plenum Press, New York.

LLOYD, K. G., and DREKSLER, S., 1979, An analysis of [^{3}H]Gamma-aminobutyric acid (GABA) binding in the human brain, *Brain Res.* **163**:77–87.

LLOYD, K. G., and WORMS, P., 1981, Neuropharmacological actions of GABA agonists: Predictability for their clinical usefulness, in: *Amino Acid Neurotransmitters* (F. V. DeFeudis and P. Mandel, eds.), pp. 59–67, Raven Press, New York.

LLOYD, K. G., SHEMENN, L., and HORNYKIEWICZ, O., 1977, Distribution of high-affinity sodium-independent [^{3}H]gamma-aminobutyric acid [^{3}H]-GABA binding in the human brain: alterations in Parkinson's disease, *Brain Res.* **127**:269–278.

LODGE, D., 1981, Uptake inhibitors, amino acids, and spinal neurones, in: *Amino Acid Neurotransmitters* (F. V. DeFeudis and P. Mandel, eds.), pp. 327–332, Raven Press, New York.

LODGE, D., CURTIS, D. R., and BRAND, S. J., 1977, A pharmacological study of the inhibition of ventral group I α-excited spinal interneurones, *Exp. Brain Res.* **29**:97–105.

LODGE, D., CURTIS, D. R., JOHNSTON, G. A. R., and BORNSTEIN, J. C., 1980, *In vivo* inactivation of quisqualate: studies in the spinal cord, *Brain Res.* **182**:491–495.

LONDON, E. D., and COYLE, J. T., 1979, Specific binding of [^{3}H]kainic acid to receptor sites in rat brain, *Mol. Pharmacol.* **15**:492–505.

LONDON, E. D., KLEMM, N., and COYLE, J. T., 1980, Phylogenetic distribution of [^{3}H]kainic acid receptor binding sites in neuronal tissue, *Brain Res.* **192**:463–476.

LONDON, E. D., YAMAMURA, H. I., BIRD, E. D., and COYLE, J. T., 1981, Decreased receptor binding sites for kainic acid in brains of patients with Huntington's Disease, *Biol. Psychiatry.* **16**:155–162.

LOSCHER, W., 1979, GABA in plasma and cerebrospinal fluid of different species: effects of γ-acetylenic GABA, α-vinyl GABA, and sodium valproate, *J. Neurochem.* **32:**1587–1591.
LOSCHER, W., 1981, Relationship between drug-induced changes in seizure thresholds and the GABA concentration of brain and brain nerve endings, *Naunyn-Schmiedeberg's Arch. Pharmacol.* **317:**131–134.
LUINI, A., GOLDBERG, O. and TEICHBERG, V. I., 1981, Distinct pharmacological properties of excitatory amino acid receptors in the rat striatum: study by Na^+ efflux assay *Proc. Natl. Acad. Sci. USA* **78:**3250–3254.
LYNCH, G. S., GRIBKOFF, V. K., and DEADWYLER, S. A., 1976, Long-term potentiation by a reduction in dendritic responsiveness to glutamic acid, *Nature (London)* **263:**151–153.
MAAYANI, S., and WEINSTEIN, H., 1980, Specific binding of [^{3}H]phencyclidine: artefacts of the rapid filtration method, *Life Sci.* **26:**2011–2017.
MACDONALD, J. F., and NISTRI, A., 1978, A comparison of the action of glutamate, ibotenate, and other related amino acids on feline spinal interneurones, *J. Physiol.* **275:**449–465.
MACDONALD, R. L., and BARKER, J. L., 1978, Different actions of anticonvulsant and anesthetic barbiturates revealed by use of cultured mammalian neurons, *Science* **200:**775–777.
MACDONALD, R. L., and BARKER, J. L., 1978*b*, Benzodiazepines specifically modulate GABA-mediated postsynaptic inhibition in cultured mammalian neurones, *Nature* (*London*) **271:**563–564.
MACDONALD, R. F., and BERGEY, G. K., 1979, Valproic acid augments GABA-medicated post-synaptic inhibition in cultured mammalian neurons, *Brain Res.* **170:**558–562.
MCGEER, E. G., MCGEER, P. L., and MCLENNAN, H., 1961, The inhibitory action of 3-hydroxytyramine, γ-aminobutyric-acid (GABA), and some other compounds toward the crayfish stretch receptor neurone, *J. Neurochem.* **8:**36–49.
MCGEER, E. G., JAKUBOVIC, A., and SINGH, E. A., 1980, Ethanol, baclofen, and kainic acid neurotoxicity, *Exp. Neurol.* **69:**359–364.
MACKERER, C. R., KOCHMAN, R. L., SHEN, T. F., and HERSHENSON, F. M., 1977, The binding of strychnine and strychnine analogs to synaptic membranes of rat brainstem and spinal cord, *J. Pharmacol. Exp. Ther.* **201:**326–331.
MCLENNAN, H., 1974, Actions of excitatory amino acids and their antagonism, *Neuropharmacology* **13:**449–454.
MCLENNAN, H., 1975, Excitatory amino acid receptors in the central nervous system, in: *Handbook of Psychopharmacology*, Vol. 4 (L. L. Iversen, S. D. Iversen, and S. H. Snyder, eds.), pp. 211–228, Plenum Press, New York.
MCLENNAN, H., HICKS, T. P., and HALL, J. G., 1981, Receptors for the excitatory amino acids, in: *Amino Acid Neurotransmitters* (F. V. DeFeudis and P. Mandel, eds.), pp. 213–221, Raven Press, New York.
MAGGI, A. and ENNA, S. J., 1979, Characteristics of muscimol accumulation in mouse brain after systemic administration, *Neuropharmacology* **18:**361–364.
MAILMAN, R. B., MUELLER, R. A., and BREESE, G. R., 1978, The effect of drugs which alter GABAergic function on cerebellar guanosine-3,5-monophosphate content, *Life Sci.* **23:**623–627.
MANTHEY, A. A., 1966, The effect of calcium on the desensitization of membrane receptors at the neuromuscular junction, *J. Gen. Physiol.* **49:**963–966.
MARCO, E., MAO, C. C., CHENEY, D. L., REVUELTA, A., and COSTA, E., 1976, The effects of antipsychotics on the turnover of GABA and acetylcholine in rat brain nuclei, *Nature (London)* **264:**363–365.
MARDER, E., and PAUPARDIN-TRITSCH, D. 1978, The pharmacological properties of some crustacean neuronal acetylcholine, γ-aminobutyric acid, and L-glutamate responses, *J. Physiol.* **280:**213–236.
MASSOTTI, M., and GUIDOTTI, A., 1980, Endogenous regulators of benzodiazepine recognition sites, *Life Sci.* **27:**847–854.

MASSOTTI, M., GUIDOTTI, A., and COSTA, E., 1981, Characterization of benzodiazepine and γ-aminobutyric recognition sites and their endogenous modulators, *J. Neurosci.* **1**:409–418.

MATTHEWS, W. D., INTOCCIA, A. P., OSBORNE, V. L., and MCCAFFERTY, G. P., 1981, Correlation of [^{14}C]muscimol concentration in rat brain with anticonvulsant activity, *Eur. J. Pharmacol.* **69**:249–254.

MATTSON, H., 1980, Bicyclic phosphates increase the cyclic GMP level in rat cerebellum, presumably due to reduced GABA inhibition, *Brain Res.* **181**:175–184.

MAURER, R., 1979, The GABA agonist THIP, a muscimol analogue, does not interfere with the benzodiazepine binding site on rat cortical membranes, *Neurosci. Lett.* **12**:65–68.

MAYNERT, E. W., MARCZYNSKI, T. J., and BROWNING, R. A., 1975, The role of the neurotransmitters in the epilepsies, *Adv. Neurol.* **13**:81–148.

MELDRUM, B., and HORTON, R., 1980, Effects of the bycyclic GABA agonist THIP on myoclonic and seizure responses in mice and baboons with reflex epilepsy, *Eur. J. Pharmacol.* **61**:231–237.

MELDROM, R., PEDLEY, T., HORTON, R., ANZELARK, G., and FRANKS, A., 1980, Epileptogenic and anticonvulsant effects of GABA agonists and GABA uptake inhibitors, *Brain Res. Bull. Suppl. 2* **5**:685–690.

MENDEL, C. M., and ALMON, R. R., 1979, Associations of [^{3}H]dihydroalprenolol with biological membranes, *Gen. Pharmacol.* **10**:31–40.

MENON, M. K., and VIVONIA, C. A., 1981, Muscimol-induced myoclonic jerks in mice, *Neuropharmacology* **20**:441–444.

MEYER, H., and PRINCE, D., 1973, Convulsant actions of penicillin: effects on inhibitory mechanisms, *Brain Res.* **53**:447–482.

MILLER, R. F., FRUMKES, T. E., SLAUGHTER, M., and DACHEUX, R. F., 1981, Physiological and pharmacological basis of GABA and glycine action on neurons of the mudpuppy retina: I. Receptors, horizontal cells, bipolars, and G cells, *J. Neurophysiol.* **45**:743–763.

MITCHELL, R., 1980, A novel GABA receptor modulates stimulus-induced glutamate release from cortico-striatal terminals, *Eur. J. Pharmacol.* **67**:119–122.

MITCHELL, P. R., and MARTIN, I. L., 1978, Is GABA release modulated by presynaptic receptors? *Nature (London)* **274**:904–905.

MOHLER, H., and OKADA, T., 1977*a*, GABA receptor binding with [^{3}H]-(+)-bicuculline methiodide in rat CNS, *Nature (London)* **267**:65–67.

MOHLER, H., and OKADA, T., 1977*b*, Benzodiazepine receptor: demonstration in the central nervous system, *Science* **198**:849–851.

MORIN, A. M., and WASTERLAIN, C. G., 1980, The binding of [^{3}H]-isoguvacine to mouse brain synaptic membranes, *Life Sci.* **26**:1239–1245.

MULLER, W. E., and SNYDER, S. H., 1978, Glycine high-affinity uptake and strychnine binding associated with glycine receptors in the frog central nervous system, *Brain Res.* **143**:487–498.

NADLER, J. V., EVENSON, D. A., and SMITH, E. M., 1981, Evidence from lesion studies for epileptogenic and nonepileptogenic neurotoxic interactions between kainic acid and excitatory innervation, *Brain Res.* **205**:405–410.

NAPIAS, C., BERGMAN, M. O., VAN NESS, P. C., GREENLEE, D. V., and OLSEN, R. W., 1980, GABA binding in mammalian brain: inhibition by endogenous GABA, *Life Sci.* **21**:1001–1011.

NASTUK, W. L., and PARSONS, R. L., 1970, Factos in the inactivation of postjunctional membrane receptors of frog skeletal muscle, *J. Gen. Physiol.* **56**:218–249.

NESTOROS, J. N., 1980, Ethanol specifically potentiates GABA-mediated neurotransmission in feline cortex, *Science* **209**:708–710.

NICHOLSON, S. H., SUCKLING, C. T., and IVERSEN, L. L., 1979, GABA analogues: conformational analysis of effects on [^{3}H]-GABA binding to postsynaptic receptors in human cerebellum, *J. Neurochem.* **32**:249–252.

NICKOLS, G. A., and BROOKER, G., 1979, Induction of refractoriness to isoproterenol by prior treatment of C6-2B rat astrocytoma cells with cholera toxin, *J. Cycl. Nucleotide Res.* **5**:435–447.

NICOLL, R. A., 1975, Presynaptic action of barbiturates in the frog spinal cord, *Proc. Natl. Acad. Sci. USA* **72**:1460–1463.

NICOLL, R. A., 1978, Pentobarbital: differential postsynaptic actions on sympathetic ganglion cells, *Science* **199**:451–452.

NICOLL, R. A., and WOJTOWICZ, J. M., 1980, The effects of pentobarbital and related compounds on frog motoneurons, *Brain Res.* **191**:225–237.

NIELSEN, M., BRAERSTRUP, C., and SQUIRES, R. F., 1978, Evidence for a late evolutionary appearance of brain specific benzodiazepine receptors: an investigation of 18 vertebrate and 5 invertebrate species, *Brain Res.* **141**:342–346.

NISHI, S., MINOTA, S., and KARCZMAR, A. G., 1974, Primary afferent neurones: the ionic mechanism of GABA-mediated depolarization, *Neuropharmacology* **13**:215–219.

NISTRI, A., 1981*a*, Excitatory and inhibitory actions of ibotenic acid on frog spinal motoneurones *in vitro*, *Brain Res.* **208**:397–408.

NISTRI, A., 1981*b*, New insights into the mechanism of action of inhibitory amino acids on frog spinal neurones, in: *Amino Acid Neurotransmitters*, (F. V. DeFeudis and P. Mandel, eds.). pp. 263–269, Raven Press, New York.

NISTRI, A., and CONSTANTI, A., 1979, Pharmacological characterization of different types of GABA and glutamate receptors in vertebrates and invertebrates, *Prog. Neurobiol.* **13**:117–235.

OLNEY, J. W., 1978, Neurotoxicity of excitatory amino acids, in: *Kainic Acid as a Tool in Neurobiology* (E. G. McGeer, J. W. Olney, and P. L. McGeer, eds.), pp. 95–121, Raven Press, New York.

OLNEY, J. W., DEGUBAREFF, T., and LUBRUYERE, J., 1979, α-aminoadipate blocks the neurotoxic actions of *N*-methylaspartate, *Life Sci.* **25**:537–540.

OLNEY, J. W., FULLER, T. A., and DEGUBAREFF, T., 1981, Kainate-like neurotoxicity of folates, *Nature (London)* **292**:165–167.

OLPE, H. R., and KOELLA, W. P., 1979, Inhibition of nigral and neocortical cells by α-hydroxybutyrate: a microiontophoretic investigation, *Eur. J. Pharmacol.* **53**:359–364.

OLPE, H. R., DEMIEVILLE, H., BALTZER, V., BENCE, W. L., KOELLA, W. P., WOLF, P., and HAAS, H. L., 1979, The biological activity of *d*- and *l*-baclofen, *Eur. J. Pharmacol.* **52**:133–136.

OLSEN, R. W., 1981, GABA–benzodiazepine–barbiturate receptor interactions, *J. Neurochem.* **37**:1–13.

OLSEN, R. W., and LEEB-LUNDBERG, F., 1980, Endogenous inhibitors of picrotoxinin-convulsant binding sites in rat brain, *Eur. J. Pharmacol.* **65**:101–104.

OLSEN, R. W., and LEEB-LUNDBERG, F., 1981, Convulsant and anticonvulsant drug binding sites related to GABA-regulated chloride ion channels, in: *GABA and Benzodiazepine Receptors*, (E. Costa, G. DiChiara, and G. L. Gessa, eds.), pp. 93–102, Raven Press, New York.

OLSEN, R. W., BAN, M., and MILLER, T., 1976, Studies on the neuropharmacological activity of bicuculline and related compounds, *Brain Res.* **102**:283–299.

OLSEN, R. W., LEEB-LUNDBERG, F., and NOPIAS, C., 1980, Picrotoxin and convulsant binding sites in mammalian brain, *Brain Res. Bull. Suppl. 2* **5**:217–221.

OLSEN, R. W., BERGMAN, M. O., VAN NESS, P. C., LUMMIS, S. C., WATKINS, A. E., NAPIAS, C., and GREENLEE, D. V., 1981, α-aminobutyric acid receptor binding in mammalian brain; heterogeneity of binding sites, *Mol. Pharmacol.* **19**:217–227.

PADJEN, A. L., and SMITH, P. A., 1981, Possible role of divalent cations in amino acid responses of frog spinal cord, in: *Amino Acid Neurotransmitters* (F. V. DeFeudis and P. Mandel, eds.), pp. 271–280, Raven Press, New York.

PALACIOS, J. M., YOUNG, W. S., and KUHAR, M. J., 1980, Autoradiographic localization of α-aminobutyric acid (GABA) receptors in the rat cerebellum, *Proc. Natl. Acad. Sci. USA* **77**:670–674.

PATEL, J. B., NELSON, L. R., and MALICK, J. B., 1980, Effects of selected psychoactive agents on bicuculline-induced convulsions in mice, *Brain Res. Bull. Suppl. 2* **5**:639–642.

PAUL, S. M., and SKOLNICK, P., 1978, Rapid change in brain benzodiazepine receptors after experimental seizures, *Science* **202**:892–894.

PECK, E. G., SCHAEFFER, J. M., and CLARK, J. H., 1976, In pursuit of the GABA receptor in: *GABA in Nervous System Function* (E. Roberts, T. N. Chase, and D. B. Tower, eds.), pp. 319–336, Raven Press, New York.

PELLMAR, T. C., 1981, Transmitter control of voltage-dependent currents, *Life Sci.* **28**:2199–2205.

PORTER, L. A., 1967, Picrotoxinin and related substances, *Chem. Rev.* **67**:441–464.

POTASHNER, S. J., 1979, Baclofen: effects on amino acid release and metabolism in slices of guinea pig cerebral cortex, *J. Neurochem.* **32**:103–109.

POTASHNER, S. J., and LAKE, N., 1981, Action of baclofen and pentobarbital on amino acid release, in: *Glutamate as a Neurotransmitter* (G. DiChiara and G. L. Gessa, eds.), pp. 139–145, Raven Press, New York.

REGAN, J. W., ROESKE, W. R., and YAMAMURA, H. I., 1980, The effect of GABA on the binding of [^{3}H]flunitrazepam in mouse brains during development, *Brain Res. Bull. Suppl. 2* **5**:857–860.

RENNIE, P., and GLIEMANN, J., 1981, Rapid down regulation of insulin receptors in adipocytes: artefact of the incubation buffer, *Biochem. Biophys. Res. Commun.* **102**:824–831.

ROBERTS, C. J., KROGSGAARD-LARSEN, P., and WALKER, R. J., 1981, Studies on the action of GABA, muscimol, and related compounds on *Periplaneta* and *Limulus* central neurons, *Comp. Biochem. Physiol. C* **69**:7–11.

ROBERTS, E., 1974, γ-aminobutyric acid and nervous system function—a perspective, *Biochem. Pharmacol.* **23**:2637–2649.

ROBERTS, E., and FRANKEL, S., 1950, γ-Aminobutyric acid in brain: its formation from glutamic acid, *J. Biol. Chem.* **187**:55–63.

ROBERTS, E., KRAUSE, D. N., WONG, E., and MORI, A., 1981, Different efficacies of *d*- and *l*-α-amino-β-hydroxybutyric acids in GABA receptor transport systems, *J. Neurosci.* **1**:132–140.

ROBERTS, P. J., 1974*a*, Inhibition of high-affinity glial uptake of [^{14}C]glutamate by folate, *Nature (London)* **250**:429–430.

ROBERTS, P. J., 1974*b*, Glutamate receptors in the rat central nervous system, *Nature (London)* **252**:399–401.

ROBERTS, P. J., 1981, Receptors for excitatory amino acids: binding studies and stimulation of cyclic CMP levels, in: *Amino Acid Neurotransmitters* (F. V. DeFeudis and P. Mandel, eds.), pp. 379–386, Raven Press, New York.

ROBERTS, P. J., GUPTA, H. K., and SHARGILL, N. S., 1978, The interaction of baclofen (β-4-chlorophenyl)GABA with GABA systems in rat brain: evidence for a releasing action, *Brain Res.* **155**:209–212.

ROBERTS, P. J., FOSTER, G. A., and THOMAS, E. M., 1981, Neurotoxic action of methyltetrahydrofolate in rat cerebellum unrelated to direct activation of kainate receptors, *Nature (London)* **293**:654–656.

ROBBINS, J., 1959, The excitation and inhibition of crustacean muscle by amino acids, *J. Physiol.* **148**:39–50.

RODBELL, M., 1980, The role of hormone receptors and GTP-regulatory proteins in membrane transduction, *Nature (London)* **284**:17–21.

ROTH, R. H., and NOWYCKY, M. C., 1977, Dopaminergic neurons: effects elicited by α-hydroxybutyrate are reversed by picrotoxin, *Biochem. Pharmacol.* **26**:2079–2082.

ROTH, R. H., DOHERTY, J. D., and WALTERS, J. R., 1980, Gamma-hydroxybutyrate: a role in the regulation of central dopaminergic neurons? *Brain Res.* **189**:556–560.

RUCK, A., KRAMER, S., METZ, J., and BRENNAN, J. W., 1980, Methyltetrahydrofolate is a potent and selective agonist for kainic acid receptors, *Nature (London)* **287**:852–85.

SARNE, Y., 1976, Desensitization to α-aminobutyric acid in crustacean muscle fibres, *J. Physiol.* **257**:779–790.

Schenberg, L. C., and Graeff, F. G., 1978, Role of the periaqueductal grey substance in the anti-anxiety action of benzodiazepines, *Pharmacol. Biochem. Behav.* **9:**287–295.

Schmidt, M. J., Thornberry, J. F., and Molloy, B. B., 1977, Effects of kainate and other glutamate analogues on cyclic nucleotide accumulation in slices of rat cerebellum, *Brain Res.* **121:**182–189.

Schmutz, M., Olpe, H. R., and Koella, W. P., 1979, Central actions of valproate sodium, *J. Pharm. Pharmacol.* **31:**413–414.

Schoemaker, H., Bliss, M., and Yamamura, H. I., 1981, Specific high affinity saturable binding of [^{3}H]-Ro5-4864 to benzodiazepine binding sites in the rat cerebral cortex, *Eur. J. Pharmacol.* **71:**173–175.

Schulz, D. W., and MacDonald, R. L., 1981, Barbiturate enhancement of GABA-mediated inhibition and activation of chloride ion conductance: correlation with anticonvulsant and anesthetic actions, *Brain Res.* **209:**177–188.

Schwarcz, R., and Fuxe, K., 1979, [^{3}H]Kainic acid binding: relevance for evaluating the neurotoxicity of kainic acid, *Life Sci.* **24:**1471–1480.

Schwarcz, R., Scholz, D., and Coyle, J. T., 1978, Structure–activity relationships for the neurotoxicity of kainic acid derivatives and glutamate analogues, *Neuropharmacology* **17:**145–151.

Segal, M., 1980, The action of serotonin in the rat hippocampal slice preparation, *J. Physiol.* **303:**423–439.

Sepinwall, J., and Cook, L., 1980, Relationship of γ-aminobutyric acid (GABA) to anti-anxiety effects of benzodiazepines, *Brain Res. Bull. Suppl. 2* **5:**839–848.

Sharif, N. A., and Roberts, P. J., 1981, L-aspartate binding sites in rat cerebellum: a comparison of the binding of L-[^{3}H]aspartate and L-[^{3}H]glutamate to synaptic membranes, *Brain Res.* **211:**293–303.

Shinozaki, H., 1980, The pharmacology of the excitatory neuromuscular junction in the crayfish, *Prog. Neurobiol.* **14:**121–155.

Shoulson, I., Goldblatt, D., Charlton, M., and Joynt, R. J., 1978, Huntington's disease: treatment with muscimol, a GABA-mimetic drug, *Ann. Neurol.* **4:**279–284.

Sieghart, W., and Karobath, M., 1980, Molecular heterogeneity of benzodiazepine receptors, *Nature (London)* **286:**285–287.

Simmonds, M. A., 1981, Distinction between the effects of barbituates, benzodiazepines, and phenytoin on responses to gamma-aminobutyric acid receptor activation and antagonism by bicuculline and picrotoxin, *Br. J. Pharmacol.* **73:**739–747.

Simmonds, M. A., 1980, A site for the potentiation of GABA-mediated responses by benzodiazepines, *Nature (London)* **284:**558–560.

Simon, J. R., Contreras, J. F., and Kuhar, M. J., 1976, Binding of [^{3}H]kainic acid, an analogue of L-glutamate, to brain membranes, *J. Neurochem.* **26:**141–147.

Skerritt, J. H., and Johnston, G. A. R., 1981, Uptake and release of *N*-methyl-D-aspartate by rat brain slices, *J. Neurochem.* **36:**881–885.

Skolnick, P., Paul, S. M., and Barker, J. L., 1980, Pentobarbital potentiates GABA-enhanced [^{3}H]diazepam binding to benzodiazepine receptors, *Eur. J. Pharmacol.* **65:**125–127.

Slevin, J. T., and Coyle, J. T., 1981, Ontogeny of receptor binding sites for [^{3}H]glutamic acid and [^{3}H]kainic acid in the rat cerebellum, *J. Neurochem.* **37:**531–533.

Smart, T. G., and Constanti, A., 1981, A re-examination of the GABA-inhibitory action of bicuculline in lobster muscle, *Eur. J. Pharmacol.* **70:**25–33.

Snead, O. C., and Bearden, L. J., 1980, Naloxone overcomes the dopaminergic, EEG, and behavioral effects of γ-hydroxybutyrate, *Neurology* **30:**832–838.

Snodgrass, S. R., 1978, The use of [^{3}H]muscimol for GABA receptor studies, *Nature (London)* **273:**392–394.

Snyder, S. H., and Bennett, J. P., 1976, Neurotransmitter receptors in the brain: biochemical identification, *Annu. Rev. Physiol.* **38:**153–175.

SONNHOF, U., and BUHRLE,, C. P., 1980, On the post synaptic action of glutamate on frog spinal motoneurons, *Pflugers Arch.* **388:**101–109.

SONNHOF, U., GRAFE, P., KRUMNIKL, J., LINDER, M., and SCHINDLER, L., 1975, Inhibitory postsynaptic actions of taurine, GABA, and other amino acids on motoneurons of the isolated frog spinal cord, *Brain Res.* **100:**327–341.

SPENCER, H. J., GRIBKOFF, V. K., COTMAN, C. W., and LYNCH, G. S., 1976, GDEE antagonism of iontophoretic amino acid excitations in the intact hippocampus and in the hippocampal slice preparation, *Brain Res.* **105:**471–481.

SQUIRES, R. F., 1981, GABA receptors regulate the affinities of anions required for brain specific benzodiazepine binding in: *GABA and Benzodiazepine Receptors* (E. Costa, G. DiChiara, and G. L. Gessa, eds.), pp. 129–138, Raven Press, New York.

SQUIRES, R. F., and BRAESTRUP, C., Benzodiazepine receptors in rat brain, *Nature (London)* **266:**732–734.

SQUIRES, R., NAQUET, R., RICHE, D., and BRAESTRUP, C., 1979, Increased thermolability of benzodiazepine receptors in cerebral cortex of a baboon with spontaneous seizures, case report, *Epilepsia* **20:**215–221.

STARKE, K., 1981, Presynaptic receptors, *Annu. Rev. Pharmacol. Toxicol.* **21:**7–30.

STEIN, L., WISE, C. D., and BELLUZZI, J. D., 1975, Effect of benzodiazepines on central serotonergic mechanisms, in: *Mechanisms of Action of the Benzodiazepines* (E. Costa, and P. Greengard, eds.), pp. 29–44, Raven Press, New York.

STOKES, B. T., and BIGNALL, K. E., 1977, The emergence of inhibition in the chick embryo spinal cord, *Brain Res.* **77:** 231–242.

STONE, T. W., 1973, Cortical pyramidal tract interneurones and their sensitivity to L-glutamic acid, *J. Physiol.* **233:**211–225.

STONE, W. E., 1977, Effects of alterations in the metabolism of γ-aminobutyrate on convulsant potencies, *Epilepsia* **18:**507–515.

STREIT, P., STELLA, M., and CUENOD, M., 1980, Kainate-induced lesion in the optic tectum: dependency upon optic nerve afferents or glutamate, *Brain Res.* **187:**47–57.

STRICKLAND, S., and LOEB, J. N., 1981, Obligatory separation of hormone binding and biological response curves in systems dependent upon secondary mediators of hormone action, *Proc. Natl. Acad. Sci. USA* **78:**1366–1370.

SU, Y. F., HARDEN, K., and PERKINS, J. P., 1980, Catecholamine-specific desensitization of adenylate cyclase, evidence for a multistep process, *J. Biol Chem.* **255:**7410–7419.

SUBRAMANIAN, N., WHITMORE, W. L., SEIDLER, F. J., and SLOTKIN, T. A., 1980, Histamine stimulates brain phospholipid turnover through a direct H-1 receptor-mediated mechanism, *Life Sci.* **27:**1315–1319.

SUPAVILAI, P., and KAROBATH, M., 1980, The effect of temperature and chloride ion on the stimulation of [^{3}H]flunitrazepam binding by THIP and PSA, *Neurosci. Lett.* **19:**337–341.

SUPAVILAI, P., and KAROBATH, M., 1981*a*, Action of pyrazolopyridines as modulators of [^{3}H]-flunitrazepam binding to the GABA/benzodiazepine receptor complex of the cerebellum, *Eur. J. Pharmacol.* **70:**183–193.

SUPAVILAI, P., and KAROBATH, M., 1981*b*, *In vitro* modulation by Avermectin B 1a of the GABA benzodiazepine complex of the cerebellum, *J. Neurochem.* **36:**798–803.

SWANN, J. W., and CARPENTER, D. O., 1975, Organization of receptors for neurotransmitters on Aplysia neurones, *Nature (London)* **258:**751–754.

SYAPIN, P., and SKOLNICK, P., 1978, Characterization of benzodiazepine receptors in transformed cell lines, *J. Neurochem.* **32:**1047–1051.

TAKEMOTO, T., 1978, Isolation and structural identification of naturally occurring excitatory amino acids, in: *Kainic Acid as a Tool in Neurobiology*, (E. G., McGeer, J. W. Olney, and P. L. McGeer, eds.), pp. 1–16, Raven Press, New York.

TAKEUCHI, A., and ONEDERA, K., 1975, Effects of kainic acid on the glutamate receptors of crayfish muscle, *Neuropharmacology* **14:**619–625.

TAKEUCHI, A., and TAKEUCHI, N., 1966, On the permeability of the presynaptic terminal of the crayfish neuromuscular junction during synaptic inhibition and the action of γ-aminobutyric acid, *J. Physiol.* **183:**433–449.

Tallman, J. F., Thomas. J. W., and Gallager, D. W., 1978, GABAergic modulation of benzodiazepine binding site sensitivity, *Nature (London)* **274:**383–385.

Tallman, J., Paul, S. M., Skolnick, P., and Gallager, D. W., 1980, Receptors for the age of anxiety: pharmacology of the benzodiazepines, *Science* **207:**274–281.

Tamminga, C. A., Crayton, J. W., and Chase, T. N., 1979, Improvement in tardive dyskinesia after muscimol therapy, *Arch. Gen. Psychiatry* **36:**595–598.

Tapia, R., 1975, Biochemical pharmacology of GABA in the CNS, in: *Handbook of Psychopharmacology*, Vol. 4 (L. L. Iversen, S. D. Iversen, and S. H. Snyder, eds.), pp. 1–58, Plenum Press, New York.

Tebecis, A. K., and Phillis, J. W., 1969, The use of convulsants in studying possible functions of amino acids in the toad spinal cord, *Comp. Biochem. Physiol.* **28:**1303–1315.

Thiebot, M. H., Jobert, A., and Soubrie, P., 1980, Chlordiazepoxide and GABA injections into raphe dorsalis release the conditioned behavioral suppression induced in rats by a conflict procedure without nociceptive component, *Neuropharmacology* **19:**633–641.

Ticku, M. K., 1981, Interaction of stereoisomers of barbiturates with [^{3}H]-α-dihydropicrotoxinin binding sites, *Brain Res.* **211:**127–133.

Ticku, M. K., and Burch, T., 1980, Purine inhibition of [^{3}H]-γ-aminobutyric acid receptor binding to rat brain membranes, *Biochem. Pharmacol.* **29:**1217–1220.

Ticku, M. K., and Olsen, R. W., 1978, Interaction of barbiturates with dihydropicrotoxinin binding sites related to the GABA receptor–ionophore complex, *Life Sci.* **22:**1643–1652.

Ticku, M. K., and Olsen, R. W., 1979, Cage convulsants inhibit picrotoxinin binding, *Neuropharmacology* **18:**315–318.

Ticku, M. K., Ban, M., and Olsen, R. W., 1978*a*, Binding of [^{3}H]-α-dihyrdopicrotoxinin, a -gg-aminobutyric acid synaptic antagonist, to rat brain membranes, *Mol. Pharmacol.* **14:**391–402.

Ticku, M. K., Van Ness, P. C., Haycock, J. W., Levy, W. B., and Olsen, R. W., 1978*b*, Dihydropicrotoxinin binding sites in rat brain: comparison to GABA receptors, *Brain Res.* **150:**642–647.

Toffano, G., Guidotti, A., and Coista, E., 1978, Purification of an endogenous protein inhibitor of the high-affinity binding of γ-aminobutyric acid to synaptic membranes of rat brain, *Proc. Natl. Acad. Sci. USA* **75:**4024–4028.

Towe, A. L., and Mann, M. D., 1973, Effect of strychnine on the primary evoked response and the corticofugal reflex discharge, *Exp. Neurol.* **39:**395–413

Tunnicliff, G., and Smith, J. A., 1981, competitive inhibition of γ-aminobutyric acid receptor binding by *N*-2-hydroxyethylpiperazine-*N*-2-ethanesulfonic acid and related buffers, *J. Neurochem.* **36:**1122–1126.

Tye, N. C., Iverson, S. D., and Green, A. R., 1979, The effects of benzodiazepines and serotonergic manipulations on punished responding, *Neuropharmacology* **18:**689–695.

Van Harreveld, A., 1959, Compounds in brain extracts causing spreading depression of cerebral cortical activity and contraction of crustacean muscle, *J. Neurochem.* **3:**300–315.

Van, Woert, M. H., and Hwang, E. C., 1981, Treatment of myoclonus, in: *Disorders of Movement* (A. Barbeau, ed.), pp. 59–80, Lippincott, Philadelphia.

Vincent, J. P., Vignon, J., Kartalovski, B., and Lazdunski, M., 1980, Binding of phencyclidine to rat brain membranes: technical aspect, *Eur. J. Pharmacol.* **68:**73–77.

Waddington,, J. L., and Cross, A. J., 1978, Denervation supersensitivity in the striatonigral GABA pathway, *Nature (London)* **276:**618–620.

Waddington, J. L., and Cross, A. J., 1979, Baclofen and muscimol: Behavioral and neurochemical sequelae of unilateral intranigral administration and effects on [^{3}H]-GABA receptor binding, *Naunyn-Schmiedeberg's Arch. Pharmacol.* **306:**1275–280.

Waldmeier, P. C., and Fehr, B., 1978, Effects of baclofen and γ-hydroxybutyrate on rat striatal and mesolimbic 5-HT metabolism, *Eur. J. Pharmacol.* **49:**177–184.

WALKER, R. J., 1976, The action of kainic acid and quisqualic acid on the glutamate receptors of the three identifiable neurones from the brain of the snail, *Comp. Biochem. Physiol. C* **55**:61–67.

WALKER, R. J., AZANZA, M. J., KERKUT, G. A., and WOODRUFF, G. N., 1975, The action of γ-aminobutyric acid (GABA) and related compounds on two identifiable neurones in the brain of the snail, *Helix Aspersa, Comp. Biochem. Physiol. C* **50**:147–154.

WALZ, M. A., and DAVIS, W. M., 1979, Experimental diazepam intoxication in rodents: physostigmine and naloxone as potential antagonists, *Drug Chem. Toxicol.* **2**:257–267.

WANG, Y. J., SALVATERRA, P., and ROBERTS, E., 1979, Characterization of [^{3}K]muscimol binding to mouse brain membranes, *Biochem. Pharmacol.* **28**:1123–1128.

WARBRITTEN., J. D., STEWART, R. M., and BALDESSARINI, R. J., 1978, Decreased locomotor activity and attenuation of amphetamine hyperactivity with intraventricular infusion of serotonin in the rat, *Brain Res.* **143**:373–382.

WASZCZAK, B. L., and WALTERS, J. R., 1980, Intravenous GABA agonist administration stimulates firing of A10 dopaminergic neurons, *Eur. J. Pharmacol.* **66**:141–144.

WASZCZAK, B. L., HUME, C., and WALTERS, J. R., 1981, Supersensitivity of substantia nigra pars reticulata neurons to GABAergic drugs after striatal lesions, *Life Sci.* **28**:2411–2420.

WATKINS, J. C., DAVIES, J., EVANS, R. H., FRANCIS, A. A., and JONES, A. W., 1981. Pharmacology of receptors for excitatory amino acids, in: *Glutamate as a Neurotransmitter* (G. DiChiara, and G. L. Gessa, eds.), pp. 263–273, Raven Press, New York.

WERMAN, R., 1969, Interaction of γ-aminobutyric acid with postsynaptic inhibitory receptor of insect muscle, *Fed. Proc.* **28**:831.

WERMAN, R., DAVIDOFF, R. A., and APRISON, M. H., 1967, Inhibition of motoneurones by iontophoresis of glycine, *Nature (London)* **214**:681–683.

WERMAN, R. DAVIDOFF, R. A., and APRISON, M. H., 1968, Inhibitory action of glycine on spinal neurons in the cat, *J. Neurophysiol.* **31**:81–95.,

WHEAL, H. V., and KERKUT, G. A., 1976, The action of muscimol on the inhibitory postsynaptic membrane of the crustacean neuromuscular junction, *Brain Res.* **109**:179–183.

WHITE, W. F., DICHTER, M. A., and SNODGRASS, S. R., 1982, *Brain Res.*, in press.

WHITTLE, S. R., and TURNER, A. J., 1978, Effects of the anitconvulsant sodium valproate on γ-aminobutyrate and aldehyde metabolism in ox brain, *J. Neurochem.* **31**:1453–1459.

WHITTLE, S. R., and TURNER, A. J., 1981, Anticonvulsants and brain aldehyde reductase; inhibitory characteristics of ox brain aldehyde reductase, *Biochem. Pharmacol.* **30**:1191–1196.

WILLIAMS, M., and RISLEY, E. A., 1980, Biochemical characterization of putative central purinergic receptors by using 2-chloro-[^{3}H]adenosine, a stable analog of adenosine, *Proc. Natl. Acad. Sci. USA.* **77**:6892–6896.

WILLOW, M., and JOHNSTON, G. A. R., 1980, Enhancement of GABA binding by pentobarbitone, *Neurosci. Lett.* **18**:323–327.

WILLOW, M., and JOHNSTON, G. A. R., 1981*a*, Enhancement by anesthetic and convulsant barbiturates of GABA binding to rat brain synaptosomal membranes, *J. Neurosci.* **1**:364–367.

WILLOW, M., and JOHNSTON, G. A. R., 1981*b*, Dual action of pentobarbitone on GABA binding: role of binding site integrity, *J. Neurochem.* **37**:1291–1294.

WONG, D. T., and HORNG, J. S., 1977, Na^+-independent binding of GABA to the triton X-100 treated synaptic membranes from the cerebellum of rat brain, *Life Sci.* **20**:445–452.

WORMS, P., and LLOYD, P., 1980, Biphasic effects of direct, but not indirect GABA-mimetics and antagonists on haloperidol-induced catalepsy, *Naunyn-Schmiedeberg's Arch. Pharmacol.* **311**:179–184.

WORMS, P., DEPOORTERE, H., and LLOYD, K. G., 1979, Neuropharmacological spectrum of muscimol, *Life Sci.* **25**:607–614.

YAMAMOTO, C., YAMASHITA, H., and CHUJO, T., 1976, Inhibitory action of glutamic acid on cerebellar interneurones, *Nature (London)* **262**:786–787.

YOSHIKAMI, D., 1981, Transmitter sensitivity of neurons assayed by autoradiography, *Science* **212**:929–931.

YOUNG, A. B., and SNYDER, S. H., 1973, Strychnine binding associated with glycine receptors of the central nervous system. *Proc. Natl. Acad. Sci. USA* **70**:2832–2836.

YOUNG, A. B., and SNYDER, S. H., 1974, Strychnine binding in rat spinal cord membranes associated with the synaptic glycine receptor. Cooperativity of glycine interactions, *Mol. Pharmacol.* **10**:790–809.

YOUNG, A. B., OSTER-GRANITE, M. L., HERNDON, R. M., and SNYDER, S. H., 1974, Glutamic acid: selective depletion by viral-induced granule cell loss in hamster cerebellum, *Brain Res.* **73**:1–13.

YOUNG, W. S., and KUHAR, M. J., 1979, A new method for receptor autoradiography: [^{3}H]-opioid receptors in rat brain, *Brain Res.* **179**:255–270.

ZARBIN, M. A., WAMSLEY, J. K., and KUHAR, M. J., 1981, Glycine receptor: light microscopic autoradiographic localization with [^{3}H]strychnine, *J. Neurosci.* **1**:532–547.

ZIEGLANSBERGER, W., and CHAMPAGNAT, J., 1979, Cat spinal motoneurones exhibit topographic sensitivity to glutamate and glycine, *Brain. Res.* **160**:95–104.

ZUKIN, S. R., and ZUKIN, P. R., 1979, Specific [^{3}H]phencyclidine binding in rat central nervous system, *Proc. Natl. Acad. Sci. USA* **76**:5372–5376.

ZUKIN, S. R., YOUNG, A. B., and SNYDER, S. H., 1974, Gamma-aminobutyric acid binding to receptor sites in the rat central nervous system, *Proc. Natl. Acad. Sci. USA* **71**:4802–4807.

ZUKIN, S. R., YOUNG, A. B., and SNYDER, S. H., 1975, Development of the synaptic glycine receptor in the chick embryo spinal cord, *Brain Res.* **83**:525–530.

6

THE NATURE OF MUSCARINIC RECEPTOR BINDING

Frederick J. Ehlert, William R. Roeske, and Henry I. Yamamura

1. INTRODUCTION

During the last decade, the availability of muscarinic cholinergic drugs of high specific activity has enabled a direct characterization of the way in which acetylcholine and other drugs interact with the muscarinic receptor. As might be anticipated, the advent of a new experimental method of scientific inquiry is often followed by a surge of information, and such is the relationship between the development of receptor binding techniques and muscarinic receptor pharmacology. In general, the recent knowledge derived from studies of the binding of muscarinic agonists and antagonists is consistent with classical theories of agonist–receptor interaction and competitive inhibition which were developed primarily from studies of the interaction of drugs with isolated smooth muscle preparations such as the guinea pig ileum. However, additional complicated ligand–receptor interactions have been detected by receptor binding methods which could not have been realized from the results of classical pharmacological assays on whole tissue preparations.

In this paper, the nature of agonist and antagonist binding to the muscarinic receptor will be reviewed, and, where appropriate, the rela-

Frederick J. Ehlert, William R. Roeske, and Henry I. Yamamura • Departments of Pharmacology, Psychiatry, Biochemistry, and Internal Medicine, University of Arizona Health Sciences Center, Tucson, Arizona 85724.

tionship between ligand binding and pharmacological activity will be discussed. Considerable attention will be focused on the modulation of muscarinic receptor binding properties by noncholinergic agents, since this relatively recent field of research promises new insight into mechanisms of receptor function.

2. THE BINDING OF ANTAGONISTS

2.1. Pharmacological Uses of Competitive Antagonists

Historically, receptor classification has been greatly aided by the availability of competitive antagonists. The utility of specific competitive antagonists as quantitative probes for identifying receptors stems from the relatively simple nature of competitive antagonism which contrasts with the more complex phenomenon of agonist–receptor interaction. In a given tissue the relationship between agonist–receptor occupation and a distal physiological response is complicated by an intermediate sequence of molecular events that in many cases are nonlinearly coupled. Consequently, unless the measured response is directly coupled to the receptor, its magnitude might not be proportional to receptor occupancy, particularly if the agonist is highly efficacious. Nevertheless, these complications are avoided when pharmacological antagonism is measured. Presumably, antagonists prevent agonist–receptor stimulation by forming an inert complex with the receptor, and, consequently, the extent of inhibition should be a predictable function of antagonist receptor occupancy. Based on the assumption that drug–receptor interactions are consistent with the law of mass action, a quantitative theory for competitive inhibition (Arunlakshana and Schild, 1959) was developed quite some time ago which enabled the affinity constants of antagonists to be determined by measurement of pharmacological antagonism. The success of the theory is exemplified by the remarkable agreement between estimates for the affinity constant (pA_2) of a competitive inhibitor derived from antagonism of responses to different agonists in different tissues (Arunlakshana and Schild, 1959). Thus, the values of antagonist affinity constants determined by pharmacological methods represent standards of comparison for those estimated by measurement of antagonist binding.

2.2. Characteristics of [^{3}H]Antagonist Binding

The first demonstration of [^{3}H]antagonist binding to muscarinic receptors was made by Paton and Rang (1965), who measured the binding of [^{3}H]atropine to the longitudinal muscle layer of the guinea pig ileum.

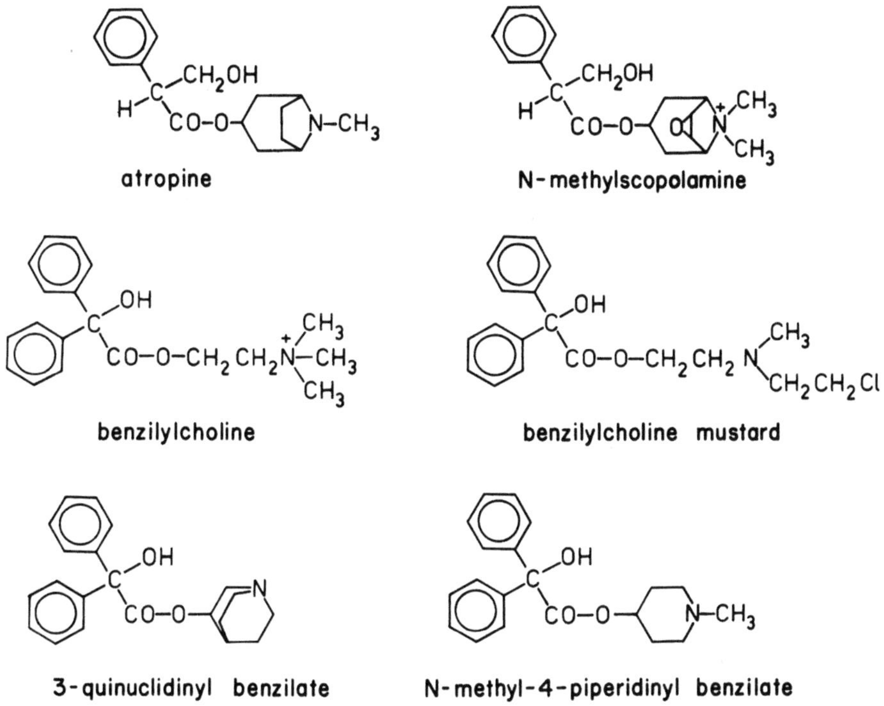

FIG. 1. Structure of some muscarinic antagonists that have been used in binding assays.

A high-affinity component of binding was detected which had a dissociation constant (1.1 nM) equivalent to that determined by measurements of pharmacological antagonism. Moreover, the binding of [^{3}H]atropine displayed a pharmacological specificity consistent with a muscarinic receptor interaction since the binding was inhibited by other muscarinic antagonists and unaffected by drugs lacking antimuscarinic activity.

Since the work of Paton and Rang (1965), the binding of several [^{3}H]-antagonists to muscarinic receptors in neuronal and muscular tissue has been demonstrated. The [^{3}H]antagonists that have been used include [^{3}H]-atropine (Farrow and O'Brien, 1973; Beld and Ariens, 1974; Schleifer and Eldefrawi, 1974), [^{3}H]-*N*-methylscopolamine (Hulme *et al.*, 1978), [^{3}H]-QNB* (Yamamura and Snyder, 1974*a*), [^{3}H]benzetimide (Soudijn *et al.*, 1973; Laduron *et al.*, 1979), [^{3}H]benzilylcholine (Birdsall *et al.*, 1976), [^{3}H]benzilylcholine mustard (Fewtrell and Rang, 1973), and [^{3}H]-*N*-methyl-4-piperidinyl benzilate (Kloog and Sokolovsky, 1977). The structure of some of these antagonists is shown in Fig. 1. The most commonly used techniques for measuring binding are the filtration and contrifugation

* [^{3}H]-QNB, [^{3}H]-3-quinuclidinyl benzilate.

methods, during which tissue homgenates or fractions thereof are incubated with labeled drug, and membrane-bound drug is subsequently trapped by filtration using glass fiber filters or by centrifugation of the membranes into a pellet. From the results of several studies of the binding of [^{3}H]antagonists, a reasonably consistent picture of the nature of antagonist interaction with the muscarinic receptor has emerged: the binding of antagonists is saturable with increasing ligand concentration and displays a remarkable consistency with the consequences of the law of mass action for a single class of receptors (Birdsall and Hulme, 1976). Moreover, the binding is stereospecific and sensitive to inhibition by drugs that are thought to interact with muscarinic receptors but not by drugs which lack direct muscarinic cholinergic effects (Paton and Rang, 1965; Beld and Ariens, 1974; Yamamura and Snyder, 1974*a*; Yamamura and Snyder, 1974*b*). The regional distribution of [^{3}H]antagonist binding is consistent with concepts of cholinergic innervation, and within the brain it generally parallels other elements of the cholinergic nervous system, including acetylcholinesterase, choline acetyltransferase, and high-affinity choline uptake (Yamamura *et al.*, 1974; Snyder *et al.*, 1975; Hiley and Burgen, 1974). Perhaps the most compelling evidence demonstrating that antagonist binding represents a specific interaction with the muscarinic receptor is the excellent agreement between the values of antagonist affinity constants determined by binding measurements and by pharmacological antagonism of smooth muscle contraction (Birdsall *et al.*, 1977; Yamamura and Snyder, 1974*b*).

As described above, the binding of antagonists is consistent with the law of mass action for a single class of receptors. This conclusion is based on the results of numerous equilibrium binding experiments in which antagonist binding has been investigated directly using [^{3}H]antagonists or by measurement of the competitive inhibition of [^{3}H]antagonist binding by nonlabeled antagonists. Transformation of the equilibrium binding data by the appropriate methods for Scatchard plots and Hill plots usually results in linearity with Hill coefficients not significantly different from one. However, there is a growing body of evidence which suggests that the kinetics of antagonist binding are inconsistent with a simple bimolecular process. Kinetic analysis of the binding of [^{3}H]-*N*-methyl-4-piperidinyl benzilate (Kloog and Sokolovsky, 1978; Kloog *et al.*, 1978; Kloog *et al.*, 1979*a*; Kloog *et al.*, 1979*b*; Jarv *et al.*, 1980) and [^{3}H]-QNB (Galper and Smith, 1978; Galper *et al.*, 1977; Schimerlik and Searles, 1980) to homogenates of the murine brain and smooth muscle and porcine and avian myocardium is consistent with a fast bimolecular binding process followed by a slow isomerization of the receptor into a high-affinity state. This phenomenon was readily demonstrable by the dissociation kinetics of [^{3}H]-QNB which was shown to be biphasic, consisting of fast- and slow-reaction components (Galper and Smith, 1978; Galper *et al.*, 1977). When the dissociation was examined after a brief period of incubation of receptor with [^{3}H]ligand, both components were obvious; however, after longer incubations of sufficient duration to allow the slow isomerization of the [^{3}H]ligand–receptor complex, only the slow reaction component was demonstrable. This behavior is consistent with the model that antagonists isomerize the receptor into a high-affinity state. It should be emphasized that although this model is incompatible with simple bimolecular reaction kinetics, it is entirely consistent with the consequences of the law of mass action at equilibrium. Although the functional significance of this process is not apparent, it may account for the discrepancies in the rates of atropine binding and pharmacological antagonism which have been noted in the guinea pig ileum (Rang, 1967).

Additional complexities in the nature of antagonist binding have been noted when the binding of the nonclassical muscarinic antagonist, pirenzepine, was examined (Hammer *et al.*, 1980). Unlike classical muscarinic antagonists which typically have similar affinities for muscarinic receptors in different tissues and species, pirenzepine displayed a 30-fold variation in its affinity for receptors in the brain and smooth muscle of rats. This variation in the potency of pirenzepine for muscarinic receptors correlates generally with the potency of pirenzepine for producing pharmacological antagonism in various tissues. Analysis of pirenzepine/[^{3}H]-*N*-methylscopolamine competition curves revealed deviations from simple mass-action behavior, and the data were reported to be consistent with the presence of multiple antagonist binding sites having different affinities for pirenzepine and nearly equal affinity for classical muscarinic antagonists (Hammer *et al.*, 1980). Presumably, it is the relative proportions of these multiple antagonist states of the muscarinic receptor which determine the overall affinity of pirenzepine in a given tissue. Antagonist receptor heterogeneity has also been detected with classical muscarinic antagonists, but the conditions required to manifest it depend primarily on the ionic composition of the incubation medium (Hulme *et al.*, 1980; Ehlert *et al.*, 1981*c*). Thus, the interaction of antagonists with the muscarinic receptor is more complicated than originally thought, although it is primarily a bimolecular mass-action-determined process when binding is measured at equilibrium in most buffer systems that are commonly used in binding assays. The regulation of the different antagonist states of the muscarinic receptor will be discussed in another section.

2.3. Cellular Localization of [^{3}H]Antagonist Binding

In the few cases in which the question has been examined, the cellular localization of [^{3}H]antagonist binding in the brain appears to be generally postsynaptic. Within the brain, the hippocampus is particularly suitable for investigating presynaptic receptors since it is thought to have an extrinsic cholinergic innervation via the septal-hippocampal pathway (Lewis and Shute, 1967; Kuhar *et al.*, 1973). When this pathway was lesioned in rats, no decreases in [^{3}H]-QNB binding were detected in various membrane fractions of the hippocampus up to two weeks after lesioning, suggesting that the majority of [^{3}H]-QNB binding sites in the hippocampus are postsynaptic (Yamamura and Snyder, 1974*c*). Nevertheless, there is what might be considered as compelling pharmacological evidence for the existence of presynaptic muscarinic receptors in the brain. Studies of the release of acetylcholine from neocortical (Bourdois *et al.*, 1974; Szerb and Somogyi, 1973) and hippocampal (Hadhazy and Szerb, 1977; Szerb, 1977) slices have shown that muscarinic agonists and cholinesterase inhibitors inhibit the release of acetylcholine whereas muscarinic antagonists potentiate the release. Although feedback inhibition of acetylcholine release in brain slices could be mediated by postsynaptic receptors via a multisynaptic pathway to cholinergic terminals, the inhibition persists in the presence of tetrodotoxin (Molenar and Polak, 1970), which would block axonal conduction through such a pathway. Moreover, inhibition of acetylcholine release by muscarinic agonists has been demonstrated in synaptosomal preparations (Weiler and Jenden, 1977) in which neuronal pathways are obviously precluded. Thus, it would appear that something on the presynaptic cholinergic nerve terminal monitors the ambient levels of acetylcholine.

It has been suggested that the affinities of classical muscarinic antagonists like QNB and atropine for presynaptic muscarinic receptors are much less than their corresponding affinities for postsynaptic receptors (Szerb, 1977). Thus, the absence of measurable changes in hippocampal [^{3}H]-QNB binding following septal lesioning has been attributed to the possible inability of [^{3}H]-QNB to interact with presynaptic muscarinic receptors at the low concentrations usually employed in binding assays. Accordingly, experiments have been performed in which the binding of a muscarinic agonist, [^{3}H]-CD,* together with that of [^{3}H]-QNB at high concentrations, was measured in the hippocampus of septal lesioned rats (Overstreet *et al.*, 1980).The results of these experiments once again showed no measurable changes in muscarinic receptor binding in the hippocampus. It might be that a change in the binding of [^{3}H]-QNB to presynaptic muscarinic receptors in the hippocampus is difficult to detect because of a large background of postsynaptic sites.

This interpretation appears more plausible in view of reports demonstrating a decrease in [^{3}H]-QNB binding in peripheral tissues following denervation. In a study in which [^{3}H]-QNB binding was measured in parasympathetically denervated rat parotid glands, 28–45% decreases in

* [^{3}H]-CD, [^{3}H]-*cis*-methyldioxiolane.

binding were noted (Talamo *et al.*, 1979). Moreover, a 50% reduction in [^{3}H]-QNB binding in crude microsomal fractions of the rat myocardium was reported following chemical sympathectomy with 6-hydroxydopamine, suggesting the existence of muscarinic receptors on adrenergic nerve terminals (Sharma and Banerjee, 1978). When the experiments were repeated, however, on whole tissue homogenates of the heart, no decreases in [^{3}H]-QNB binding were detected following chemical sympathectomy despite a 60–80% reduction in the levels of norepinephrine (Yamada *et al.*, 1980*a*). These latter data suggest that the majority of the muscarinic receptors in the heart are not located on adrenergic nerve terminals. Perhaps the most convincing evidence for muscarinic receptors on adrenergic nerve terminals comes from the results of experiments on the splenic (Laduron, 1980; Roeske *et al.*, 1981) and sciatic (Wamsley *et al.*, 1981) nerves in which axonal transport and presynaptic localization of [^{3}H]-QNB binding sites have been demonstrated by standard binding methods and audioradiographic techniques.

3. THE BINDING OF AGONISTS

3.1. Characteristics of Agonist Binding

In contrast to the simple nature of antagonist binding, agonist binding is complex and deviates significantly from mass-action behavior, especially when full agonists are considered. This phenomenon is manifested by agonist/[^{3}H]antagonist competition curves which typically extend over 3–4 log concentration units of the agonist and consequently have Hill coefficients that are less than one (Burgen and Hiley, 1974; Birdsall *et al.*, 1978; Birdsall *et al.*, 1979*a*; Fields *et al.*, 1978). An example of this phenomenon is shown in Fig. 2, which illustrates the competitive inhibition of [^{3}H]-QNB binding by the antagonist atropine and the agonist carbachol in the longitudinal muscle of the ileum. It can be seen that the carbachol competition curve is much flatter than a mass-action curve as indicated by the large systematic deviations between the observed binding values and the mass action curve (dotted line) through the IC_{50} point of the binding isotherm. In constrast, the atropine competition curve is consistent with that predicted by appliction of the law of mass action. Apparently, the shallow slopes of agonist competition curves are not the result of negatively cooperative interactions since occlusion of the majority of the receptors with an irreversible antagonist does not alter the slope of the agonist competition curve when binding is normalized on a percent occupancy scale (Birdsall *et al.*, 1978). Presumably, irreversible blockade of most of the receptor sites would greatly reduce the probability of more than one

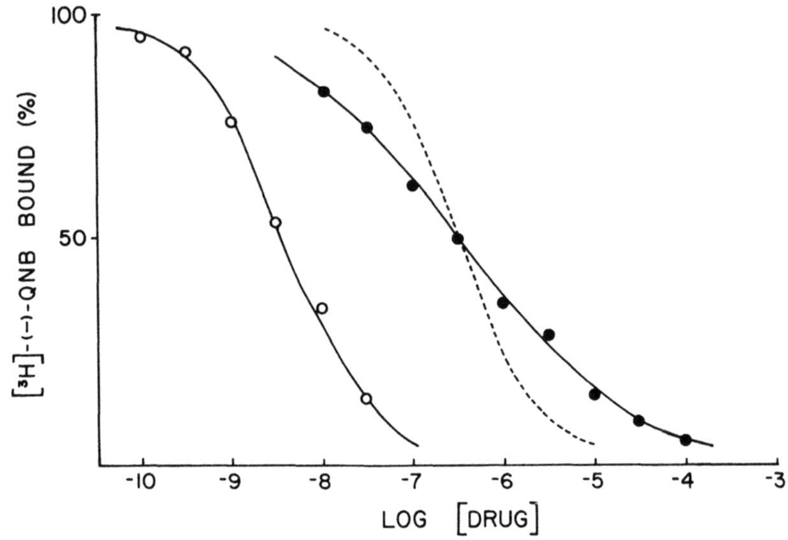

FIG. 2. Competitive inhibition of [^{3}H]-(−)-QNB binding to the longitudinal muscle of the rat ileum by atropine (○) and carbachol (●). The dashed line represents a mass-action curve intersecting the carbachol competition curve at the IC_{50} point. The radiolabeled ligand concentration was 0.4 nM.

agonist molecule interacting with the same oligomeric receptor complex. Birdsall *et al.* (1978) have demonstrated that the complexities of agonist/[^{3}H]antagonist competition curves are the result of receptor heterogeneity, with the data being sufficiently described by the contribution of three populations of muscarinic receptors, superhigh, high, and low affinity (SH, H, and L), each having different affinities for agonists and equal affinity for antagonists. Although the proportions of these independent states of the receptor remain constant in a given tissue for a variety of agonists, the distribution of these sites varies in different regions of the brain (Birdsall *et al.*, 1980*b*) and in smooth muscle (Ehlert *et al.*, 1980*a*).

The complexities of agonist binding observed in agonist/[^{3}H]antagonist competition experiments have been verified by direct measurements of the binding of radiolabeled agonists.* This was elegantly shown in a study in which the binding of ACh and oxotremorine-M was measured by competitive inhibition of [^{3}H]antagonist binding and compared with direct measurements of the binding of [^{3}H]-ACh and [^{3}H]oxotremorine-M (Birdsall *et al.*, 1978). When the [^{3}H]agonist binding isotherms were scaled to the capacity of the [^{3}H]antagonist binding sites, agonist receptor

* The structures of some muscarinic agonists are shown in Fig. 3.

occupancy was essentially the same as that determined independently by competition with an [^{3}H]antagonist. These results not only show that muscarinic [^{3}H]agonists and [^{3}H]antagonists bind competitively at the same receptor locus, but also illustrate that the complicated nature of agonist binding is an intrinsic property of agonists and not the consequence of complex agonist–antagonist interactions which have been invoked occasionally to explain non-mass-action behavior in agonist/[^{3}H]antagonist competition experiments. It should be noted, however, that complete [^{3}H]-agonist binding isotherms have not been achieved presently, the main reason being that even with the most suitable [^{3}H]agonists, ligand concentrations in excess of 10^{-7} M are required to achieve half-maximal receptor occupation in most tissues, and, consequently, high nonspecific binding prevents accurate measurements of binding at moderate to high levels of receptor occupancy.

Studies of the competitive inhibition of [^{3}H]agonist binding by nonlabeled agonists have provided further evidence for agonist receptor heterogeneity. When measured by competitive inhibition of [^{3}H]agonist binding, the potency of nonlabeled agonists is generally much greater than that determined by measurement of [^{3}H]antagonist displacement (Birdsall *et al.*, 1978; Ehlert *et al.*, 1980*c*). This behavior is to be expected since [^{3}H]-agonists selectively label the SH and H receptors at the low ligand concentrations usually employed in binding assays. In contrast, when agonist inhibition of [^{3}H]antagonist binding is measured, the proportion of [^{3}H]ligand bound to the individual agonist receptor populations is equivalent to the relative abundances of these receptors in the tissue since [^{3}H]antagonists have equal affinity for these sites. This uniformity of antagonist binding is particularly manifest in experiments of the competitive inhibition of [^{3}H]agonist and [^{3}H]antagonist binding by nonlabeled antagonists. Typically, antagonists have equal potency when determined

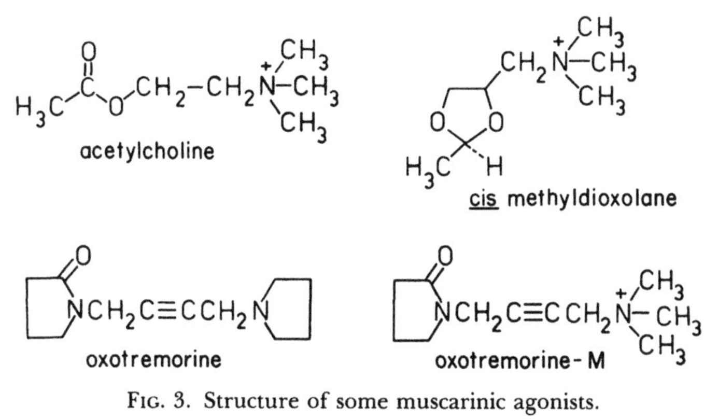

FIG. 3. Structure of some muscarinic agonists.

by the two independent methods (Birdsall and Hulme, 1976; Ehlert *et al.* 1980*c*).

Although measurements of [^{3}H]agonist binding are fraught with unfavorably low specific/nonspecific binding ratios near half maximal receptor occupation, [^{3}H]agonist ligands are invaluable probes for measurement of the SH receptor. By using a very low concentration of a suitable agonist like [^{3}H]oxotremorine-M, it is possible to determine the dissociation constant and receptor density of the SH receptors by competitively inhibiting [^{3}H]oxotremorine-M binding with nonlabeled oxotremorine-M. Technically speaking, another nonlabeled agonist could be used; however, the analysis is simpler and more accurate for self-inhibition experiments since additional independent estimates of receptor affinity are not required. Analysis of the data by an appropriate Gauss–Newton curve-fitting procedure or by Scatchard analysis results in reliable parameter estimates of the SH receptor which have been studied by implementation of this technique (Birdsall *et al.*, 1980*b*). Using the specific muscarinic agonist [^{3}H]-CD, we have determined the density of SH sites in the rat forebrain by direct measurements of [^{3}H]-CD binding, and comparison of the density of the SH receptor with that of the total density of [^{3}H]-QNB binding sites illustrates that the SH receptor accounts for only 5% of the total sites (Ehlert *et al.*, 1980*c*). Similarly, in the hippocampus and cerebral cortex, the SH site represents only 3–6% of the total receptor population (Birdsall *et al.*, 1980*b*). The low density of this site in the cerebral cortex makes it exceedingly difficult to detect by analysis of standard agonist/[^{3}H]antagonist competition curves, and the resulting data are usually consistent with a two-site model of H and L agonist receptors. Reliable binding parameters for all three agonist sites (SH, H, and L) have been estimated by analysis of agonist/[^{3}H]antagonist competition data from experiments on the rat cerebellum. Although this brain region has a sparce abundance of muscarinic receptors, the proportion of SH receptors is relatively high (31%) (Birdsall *et al.*, 1980*b*).

3.2. Explanations for the Complexities of Agonist Binding

The existence of multiple agonist binding sites raises the question of how agonists discriminate between these sites whereas potent muscarinic antagonists are unable to do so. A cursory review of the structure–activity relationships of muscarinic agonists and antagonists helps to resolve this anomally. Maximum cholinolytic potency is usually associated with the presence of alkyl or aromatic rings on the acyl moiety of anticholinergic esters (Ariens and Simonis, 1967; Ariens and Beld, 1977; Ariens and Rodrigues de Miranda, 1979). Presumably, these lipophilic groups interact with complementary hydrophobic receptor loci adjacent to the ACh binding region of the receptor. Moreover, these lipophilic groups probably bestow antagonists with high-potency since strong hydrophobic interactions are favored in an aqueous physiological environment. A hydroxyl group on the acyl moiety of antimuscarinic esters is also required for high potency (Lands *et al.*, 1949; Long *et al.*, 1956) although there are some exceptions to this general rule; *N,N*-dimethyl-4-piperidinyldiphenylacetate being one such example (Hulme *et al.*, 1978). In constrast to the structure of antagonists, classical muscarinic agonists lack sizable alkyl or aromatic substituents on the acyl part of the molecule. In general, as the number of carbon atoms in the acyl group of classical muscarinic agonists increases, there is a gradual loss of intrinsic activity (Ariens and Simonis, 1967). This conversion is associated with an initial decrease in potency followed by a gradual increase in potency as antagonistic properties prevail. The importance of the structure of the acyl moiety of muscarinic antagonists is reflected by the large difference in the potency of enantiomers having an asymmetric carbon atom in the acyl moiety (Ariens and Simonis, 1967). In contrast, the enantiomers of the benzilate ester of β-methylcholine are equally potent antagonists, whereas the enantiomers of the structurally related agonist analog, acetyl-β-methylcholine, have a large difference in pharmacological activity (Ellenbrock *et al.*, 1965). It should be noted that stereospecificity is apparent with some antagonists that have an asymmetric carbon in the amino alcohol residue. One such example is the benzilate ester of 3-quinuclidinol (QNB)(Meyerhoffer, 1972; Baumgold *et al.*, 1977; Rehavi *et al.*, 1977). This stereospecificity is retained in the structurally related agonist analog, aceclidine (3-quinuclidinyl acetate) (Cho *et al.*, 1972; Weinstein *et al.*, 1975). Interestingly, conversion of both QNB and aceclidine to their corresponding *N*-methyl quaternary salts results in a loss of stereospecificity (Cho *et al.*, 1972; Weinstein *et al.*, 1975; Hulme *et al.*, 1978). A consideration of the structure–activity relationships described above suggests that antagonists derive considerable affinity through hydrophobic interactions with receptor loci distinct from the agonist binding region of the receptor and that a high degree of complementarity between antagonists and the agonist binding region of the receptor is not essential for maximum cholinolytic potency. These structure–activity relationships are illustrated graphically in Fig. 4. Thus, the inability of classical muscarinic antagonists to distinguish between the multiple classes of agonist binding sites might be rationalized on the basis of a lack of complementarity between antagonists and the agonist binding region of the receptor.

Although the structure–activity relationships described above may provide an explanation for the inability of potent antagonists to discriminate between the different agonist states of the muscarinic receptor, an explanation for the functional significance of these independent binding sites is still lacking. If the molecular structures of the SH, H, and L receptor are mutually distinct, then the existence of selective high-affinity

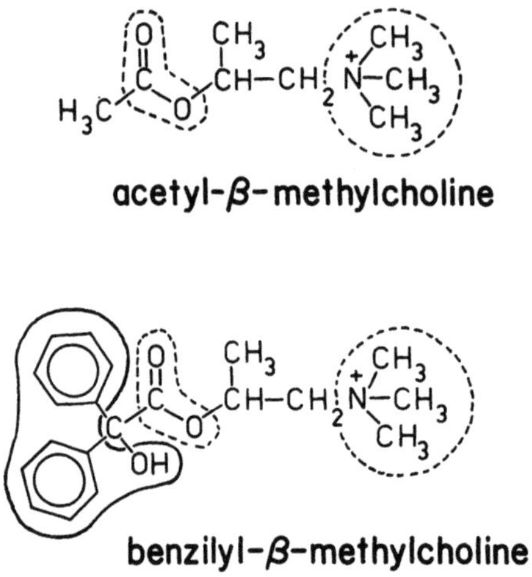

FIG. 4. Structure–activity relationships of classical muscarinic agonists and antagonists. The area enclosed by dashed lines represents hypothetical receptor loci that are thought to be important for agonist–receptor interactions. The area circumscribed by the solid line represents a hypothetic receptor surface that is thought to be important for the interaction of classical muscarinic antagonists with the receptor.

ligands for each receptor should be possible. However, the results of numerous studies suggest that this possibility is unlikely. Within a given tissue, the rank order for the affinities of the SH, H, and L receptors for various drugs is always SH $\geq$ H $\geq$ L. Moreover, the ability of muscarinic agonists to discriminate between the H and L receptors is highly correlated with the efficacy of the drug (Birdsall *et al.*, 1977, 1978). These correlations suggest that the implications of agonist receptor heterogeneity ought to be realized from a viewpoint which accounts for the fundamental ability of agonists to initiate receptor–effector transduction. This classical assumption provides the basis for a theory proposed by Birdsall *et al.* (1977) which contends that the different classes of agonist binding sites represent independent states of the same receptor molecule, and that the differences in affinity arise as the result of conformational constraints imposed by the environment or coupling state of the receptor. Briefly, this model assumes that agonist–receptor interactions are characterized by an isomerization process which converts the receptor into a high-affinity state. For those receptors that are highly constrained or coupled, the isomerization process is hindered and agonist affinity is reduced, whereas the converse is true for unconstrained or uncoupled receptors. Thus, this model assumes that the propensity of the SH, H, and L receptors to undergo agonist-induced isomerization differs but that the bimolecular interaction of a ligand with the ground state of the various receptors is the same. Consequently, this model predicts that antagonists should have equal affinity for the different classes of agonist binding sites, that pure and partial agonists should have a constant ratio (K_{SH}/K_H) of affinity constants for the SH and H sites, and that K_H/K_L should be proportional to efficacy. These predictions have been fulfilled by several agonists (Birdsall *et al.*, 1977, 1978, 1980*b*).

Since the model of Birdsall *et al.* (1977) contends that agonist receptor heterogeneity is the result of varying degrees of receptor–effector constraint, a theoretical consequence of this model is that isolated muscarinic receptors should display homogeneous properties. In this regard, when labeled muscarinic receptors from the rat brain were solubilized in sodium dodecyl sulfate and examined by polyacrylamide gel electrophoresis, only one radiolabeled peak was observed, suggesting the presence of a single receptor protein with a molecular weight of 80,000 (Birdsall *et al.*, 1979*b*). Similar characteristics have been observed for solubilized muscarinic receptors from the longitudinal muscle of the guinea pig ileum when methods are employed to avoid proteolysis (Birdsall *et al.*, 1979*b*); otherwise multiple-molecular-weight species are apparent (Fewtrell and Rang, 1973). The results of binding studies on solubilized muscarinic receptors from bovine brain were consistent with the presence of a homogeneous population of solubilized receptors in that the binding of agonists and antagonists was adequately described by simple mass-action behavior (Carson *et al.*, 1977; Haga, 1980). Thus, the results of studies on solubilized muscarinic receptors are consistent with the hypothesis that agonist receptor heterogeneity is caused by the association of the muscarinic receptor with different membrane constituents.

3.3. Comparison of Agonist Binding with Pharmacological Responses

The existence of multiple agonist binding sites might be construed as evidence for the existence of multiple effector systems with which the various agonist receptors (SH, H, and L) are coupled. Comparison of agonist dissociation constants (K_{SH}, K_H, and K_L) with ED_{50} values for pharmacological responses is complicated since the heterogeneity of agonist binding is matched by an even greater diversity of pharmacological responses that are thought to be mediated by muscarinic receptors. Ideally, estimates of agonist dissociation constants determined by binding experiments should agree with ED_{50} values for pharmacological responses. However, quantitative agreement between the two measurements is often precluded by an intervening sequence of biochemical events between the receptor and the measured response. If the intermediate steps are nonlinearly coupled, then receptor occupancy and response will not be proportional and a large receptor reserve may exist. For these reasons, measurements of biochemical events that are closely coupled to the receptor should provide the most useful information relating to the pharmacological significance of receptors.

Several experimental findings suggest that the L receptor is functionally coupled to relevant pharmacological events. Stimulation of smooth muscle contraction in the guinea pig ileum is correlated with agonist occupancy of the L receptor after elimination of spare receptors (Birdsall and Hulme, 1976; Birdsall *et al.*, 1977), and stimulation of cyclic GMP formation in striatal brain slices (Hanley and Iversen, 1978) and neuroblastomal cells (Strange *et al.*, 1977) is also correlated with the L receptor. Moreover, the maximum cyclic GMP response in neuroblastomal clones appears to be correlated with the efficacy of the agonist (Strange *et al.*, 1977). Carbachol has been shown to inhibit both GTP-stimulated adenylate cyclase activity in myocardial homogenates (Murad *et al.*, 1962; Watanabe *et al.*, 1978; Jakobs *et al.*, 1979; Birdsall *et al.*, 1980*a*) and PGE_1-stimulated adenylate cyclase activity in neuroblastomal cells (Lichtshtein *et al.*, 1979), and the K_i for inhibition of the enzyme agrees quantitatively with the affinity of carbachol for the L receptor (Birdsall *et al.*, 1980*a*; Lichtshtein *et al.*, 1979). Within the rat brain, the regional distribution of the L receptor is consistent with that predicted for functional muscarinic receptors. It has been noted that the density of this receptor is 50 times greater in the striatum as compared to the cerebellum, and 100-fold variations in the abundance of the L receptor have been observed in various regions of the cerebral cortex (Birdsall *et al.*, 1979*a*). In contrast, the maximum variations in the densities of the SH and H sites were reported to be only 3- to 6-fold, respectively (Birdsall *et al.*, 1979*a*). Thus, the distribution of the L receptor in the brain correlates generally with other cholinergic markers such as cholinesterase, choline acetyltransferase, and high-affinity choline uptake (Yamamura *et al.*, 1974; Snyder *et al.*, 1975).

Another biochemical response that is thought to be mediated by muscarinic receptor stimulation is the breakdown of phosphatidylinositol (Michell, 1975). It has been suggested that this event is closely coupled to the receptor and may be responsible for opening Ca^{2+} channels and initiating the cascade of events that are characteristic of muscarinic receptor stimulation (Michell, 1975). The notion that phosphatidylinositol breakdown is an early step in the sequence is consistent with the observation that this response is insensitive to Ca^{2+} (Trifaro, 1969; Hokin, 1966), whereas a variety of other responses to muscarinic stimulation are Ca^{2+} dependent. Although there is little doubt that muscarinic receptor stimulation enhances phosphatidylinositol turnover, it is unclear how the resulting transient changes in the concentrations of phosphatidylinositol, 1,2-diacylglycerol, and phosphatidic acid initiate the events characteristic of muscarinic stimulation. Alterations in membrane fluidity and membrane surface charge have been implicated as possible consequences of PI turnover (Triggle, 1979).

In the longitudinal muscle of the guinea pig ileum, muscarinic agonists enhance PI turnover, and the agonist dose–response curve correlates with agonist occupancy of both the H and L receptors (Jafferji and Michell, 1976). Thus, these data might suggest that both the H and L receptors are functionally important in the guinea pig ileum.

4. REGULATION OF MUSCARINIC RECEPTORS BY GUANINE NUCLEOTIDES

4.1. Introduction

Guanine nucleotides have been shown to modify the binding properties of several hormonal and neurotransmitter receptors, including opiate (Blume, 1978; Childers and Snyder, 1978), dopaminergic (Creese and Snyder, 1978; Creese *et al.*, 1978; Zahniser and Molinoff, 1978), β-adrenergic (Lefkowitz *et al.*, 1976; Maguire *et al.*, 1976; Howlett *et al.*, 1978), α-adrenergic (U'Prichard and Snyder, 1978; Tsai and Lefkowitz, 1979), glucagon (Rodbell *et al.*, 1971*a*), and muscarinic receptors (Berrie *et al.*, 1979; Rosenberger *et al.*, 1979; Wei and Sulakhe, 1979; Ehlert *et al.*, 1980*a*; Sokolovsky *et al.*, 1980; Galper and Smith, 1980). In general, this modulation by guanine nucleotides is characterized by a selective reduction in agonist potency, with the binding of antagonists being relatively unaffected. In studies of several adenylate cyclase systems, GTP has been shown to be a requirement for neurotransmitter and hormonal activation (Rodbell *et al.*, 1971*b*; Krishna *et al.*, 1972; Lefkowitz *et al.*, 1977; Ross *et al.*, 1977; Northup and Mansour, 1978; Pike and Lefkowitz, 1978; Abramowitz and Birnbaumer, 1979). Since adenylate cyclase activation requires GTP, the selective effects of guanine nucleotides on the binding of agonists may be a manifestation of receptor coupling to the cyclase itself or to another membrane component or nucleotide binding site which transfers free energy from the agonist–receptor complex to the effector. Thus, the results of studies of the regulation of muscarinic receptor binding by guanine nucleotides may have important implications regarding the relationship between agonist receptor heterogeneity and receptor–effector coupling.

4.2. Regulation of Agonist Binding by Guanine Nucleotides

Initial studies of the influence of guanine nucleotides on the binding properties of muscarinic receptors were carried out on homogenates of the rat myocardium. When measured by competitive inhibition of [^{3}H]-antagonist binding, the IC_{50} values of the agonists oxotremorine and carbachol increased by a factor of 10–20 in the presence of GTP or its nonhydrolyzable analog, Gpp(NH)p* (Berrie *et al.*, 1979; Rosenberger *et al.*, 1979; Rosenberger *et al.*, 1980*a*). In contrast, only small or insignficant effects of guanine nucleotides on the binding properties of antagonists were noted. The effects of various concentrations of Gpp(NH)p on the competitive inhibition of cardiac [^{3}H]-QNB binding by oxotremorine are

* Guanyl-5′-ylimidodiphosphate.

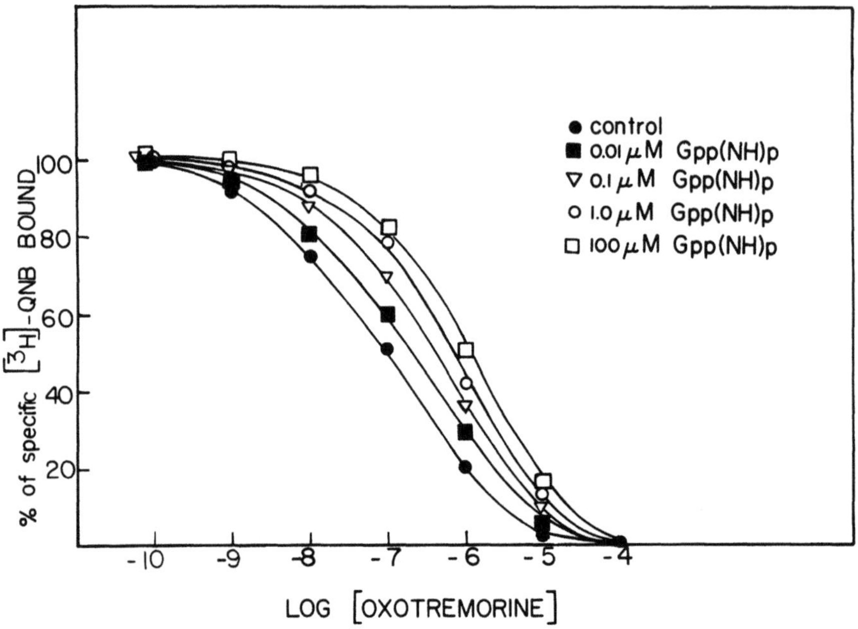

FIG. 5. Effect of Gpp(NH)p on the competitive inhibition of [^{3}H]-(−)-QNB binding to myocardial homogenates by oxotremorine. Assays were done in the absence (●) and presence of various concentrations of Gpp(NH)p: 0.01 μM (■), 0.1 μM (▽), 1.0 μM (○), and 100 μM (□). The radiolabeled ligand concentration was 0.12 nM. The data are from Rosenberger *et al.* (1980*a*).

shown in Fig. 5. It can be seen that Gpp(NH)p caused a dose-dependent reduction in the potency of oxotremorine and that the ED_{50} of Gpp(NH)p for this effect was in the 10^{-8} to 10^{-7} M range. When incubations are carried out at 30–38°C, the effects of GTP (10^{-4} M) on agonist binding are relatively short lived, and after 30–40 min of incubation the effects are absent (Berrie *et al.*, 1979). The transient nature the GTP effect is probably due to hydrolysis of the nucleotide since the effect of the nonhydrolyzable GTP analog [Gpp(NH)p] is very persistent. The short duration of the GTP effect also suggests that the interaction of guanine nucleotides with their presumed binding sites is a reversible process.

The modulation of agonist binding by guanine nucleotides is specific for compounds structurally related to GTP. When relatively standard muscarinic receptor binding assays are carried out at short incubation times (10–15 min) Gpp(NH)p is about 10 times more potent than GTP (Berrie *et al.*, 1979; Rosenberger *et al.*, 1979). GDP also modulates muscarinic receptor binding, and it is about equiactive with GTP (Berrie *et al.*, 1979; Rosenberger *et al.*, 1979). In contrast, a variety of other nucleotides

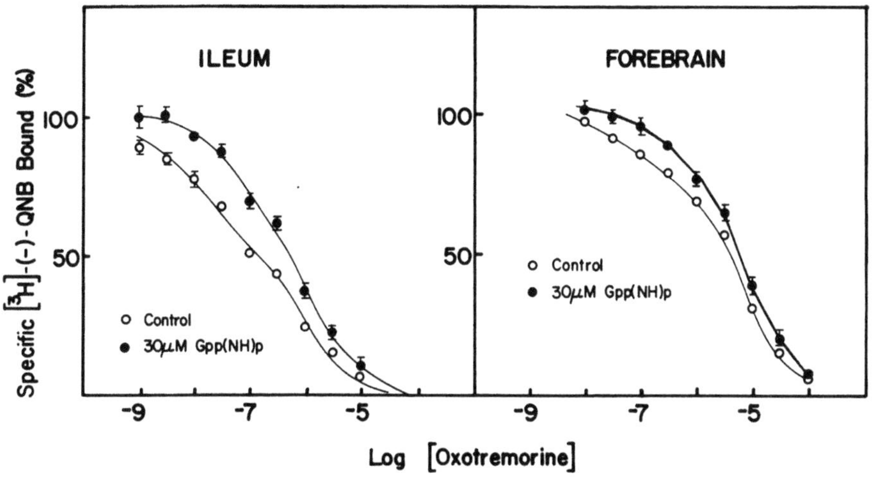

FIG. 6. Effect of Gpp(NH)p on the competitive inhibition of $[^3H]$-(−)-QNB binding by oxotremorine in the longitudinal muscle of the ileum and forebrain of the rat. Assays were done in the presence (●) and absence (○) of 30 μM Gpp(NH)p. The concentration of radiolabeled ligand was 0.4 nM. The data are taken from Ehlert *et al.*, (1980*a*).

are inactive, including ATP, ADP, AMP, App(NH)p, cyclic AMP, and cyclic GMP (Berrie *et al.*, 1979; Sokolovsky *et al.*, 1980; Rosenberger *et al.*, 1980*a*). It is possible that the true activity of GTP and GDP might actually approach or perhaps exceed that of Gpp(NH)p since the former nucleotides are susceptible to hydrolysis during standard binding assays.

Although the effects of guanine nucleotides on agonist binding in myocardial homogenates are readily demonstrable in the absence of added Mg^{2+} ions, the effect requires Mg^{2+}. This Mg^{2+} requirement was demonstrated in a study in which microsomal preparations of the rat atria were treated with EDTA and subsequently washed by centrifugation (Wei and Sulakhe, 1980*a*). When these Mg^{2+}-free homogenates were used in carbachol/$[^3H]$-QNB competitive binding assays, GTP (10^{-3} M) only caused a twofold increase in the IC_{50} of carbachol, whereas an 11-fold increase in the IC_{50} of carbachol was observed when similar experiments were carried out in the presence of $MgCl_2$ (10^{-3} M). In a subsequent study, it was demonstrated that treatment of cardiac membranes with the sulfhydryl reagent, NEM,* abolished the effects of guanine nucleotides on agonist binding (Wei and Sulakhe, 1980*b*).

Guanine nucleotides have also been shown to modulate the binding of agonists to muscarinic receptors in smooth muscle and various brain regions (Ehlert *et al.*, 1980*a*; Sokolovsky *et al.*, 1980), although the mag-

* NEM, *N*-ethylmaleimide.

nitude of the effect in these tissues is not as great as that observed in the heart. Figure 6 shows the effects of 30 μM Gpp(NH)p on the competitive inhibition of [^{3}H]-(−)-QNB binding by oxotremorine in homogenates of the rat forebrain and longitudinal muscle of the ileum. In the absence of Gpp(NH)p, the oxotremorine competition curves are relatively flat, indicating deviations from the simple mass-action relationship. Nonlinear regression analysis of the data reveals the presence of two major populations of H and L receptors, with the relative abundances of the high-affinity receptors being 21 and 44% in the forebrain and ileum, respectively. In the presence of Gpp(NH)p (30 μM) there is a modest increase in the IC_{50} values and a steepening of the competition curves. Analysis of these data is complicated by the presence of a small population of SH sites, and the results are consistent with the hypothesis that guanine nucleotides selectively reduce the affinity of the high-affinity sites or cause a conversion of high- to low-affinity sites. The data in Fig. 6 also show that Gpp(NH)p had a greater effect on the binding of oxotremorine to muscarinic receptors in the ileum as compared to the forebrain. These results are partially a consequence of the greater proportion of high-affinity sites in the ileum and also to the greater sensitivity of the ileal muscarinic receptors to the effects of guanine nucleotides.

The magnitude of the effect of guanine nucleotides on the binding of agonists to the muscarinic receptor depends upon the efficacy of the drug. In a study (Ehlert *et al.*, 1980*d*) of the influence of guanine nucleotides on the competitive inhibition of ileal [^{3}H]-(−)-QNB binding by cholinergic agonists, it was noted that Gpp(NH)p (10^{-4} M) caused a relatively large reduction in the potency of efficacious agonists like carbachol and oxotremorine, whereas Gpp(NH)p only caused a small reduction in the potency of partial agonists like pilocarpine and pentyltrimethylammonium. These results are consistent with the idea that guanine nucleotides cause a selective conversion of H to L sites. For highly efficacious agonists, which have a large geometric difference between the values of K_H and K_L ($K_L/K_H \gg 1$), a guanine nucleotide-induced conversion of H to L would cause a relatively large increase in the IC_{50} value. In contrast, the IC_{50} values of partial agonists, which have relatively smaller K_L/K_H ratios, would not be expected to change much in the presence of guanine nucleotides. The relationship between the guanine nucleotide effect and efficacy is illustrated graphically in Fig. 7.

The selective inhibitory effects of Gpp(NH)p on the binding of agonists, which have been studied extensively by measurement of agonist/[^{3}H]antagonist competition, have also been detected by direct measurement of [^{3}H]agonist binding in the presence of Gpp(NH)p (Ehlert *et al.*, 1980*b*, 1980*d*, 1981*a*). The use of [^{3}H]agonist ligands is a sensitive and convenient method for detecting guanine nucleotide effects in several tissues. A guanine nucleotide-induced shift in an agonist/[^{3}H]antagonist

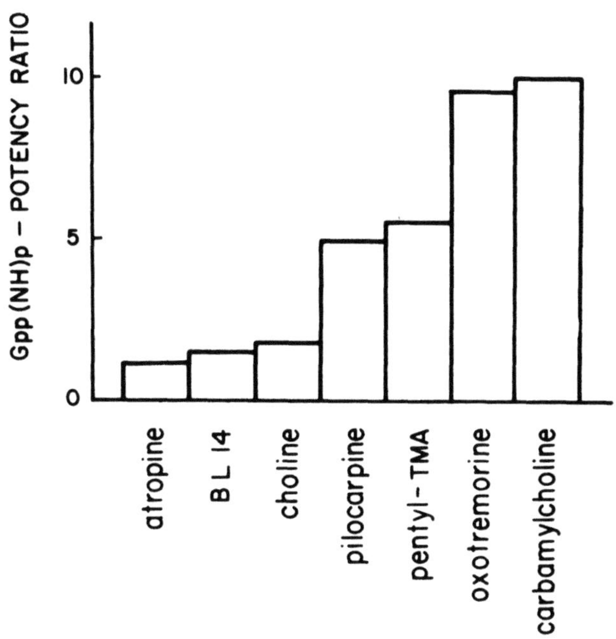

FIG. 7. Effect of Gpp(NH)p on the binding of a series of cholinergic analogs to the longitudinal muscle of the rat ileum. The competitive inhibition of [^{3}H]-(–)-QNB binding by a series of cholinergic analogs was determined in the rat ileum in the presence and absence of 100 μM Gpp(NH)p. IC_{50}'s for each drug were determined from 5–9-point competitive curves, and the ratio of IC_{50} values of a given drug, determined in the presence and absence of Gpp(NH)p, was calculated. The ordinate [Gpp(NH)p–potency ratio] refers to the magnitude of these ratios. The various cholinergic analogs are listed generally in the order of their efficacy starting on the left with the antagonist atropine and ending on the right with the highly efficacious agonist carbachol. Abbreviations used: BL14, *N*-(5-pyrrolidino-3-pentynyl)succinimide; pentyl-TMA, *N*-pentyl-*N,N,N*-trimethylammonium. The data are from Ehlert *et al.* (1980*d*).

competition curve may result in a small percentage difference between binding values in the competition curve. In contrast, the same effect may be manifested by a relatively larger percentage difference between direct measurements of [^{3}H]agonist binding to H and SH receptors. Using the agonist ligand [^{3}H]-CD, we have screened several brain regions and tissues of the rat to determine the regional sensitivity of muscarinic receptors to the effects of Gpp(NH)p (100 μM) (Ehlert *et al.*, 1981*a*). The largest effects of guanine nucleotides were measured in the heart, ileum, and cerebellum, in which [^{3}H]-CD binding was reduced by 77, 63, and 74% in the presence of 100 μM Gpp(NH)p. The next largest effects of guanine nucleotides were noted in the medulla-pons and midbrain regions where 47 and 42% reductions in [^{3}H]-CD binding were observed in the presence of Gpp(NH)p (100 μM). The smallest effects within the rat brain were detected in various diencephalic and telencephalic structures in which the inhibition of [^{3}H]-CD binding by Gpp(NH)p varied from a high value of 35% in the corpus striatum and thalamus to a low of 11% in the hypothalamus.

The data described above illustrate that the ability of guanine nucleotides to perturb agonist binding varies in different tissues. However, these data primarily reflect the variation in the maximum degree of inhibition of agonist binding by guanine nucleotides since the concentration of Gpp(NH)p used in the experiments was relatively high (100 μM). To determine the sensitivity of [^{3}H]agonist binding to the effects of guanine nucleotides, we measured [^{3}H]-CD binding in the presence of various concentrations of Gpp(NH)p. The results of these experiments are shown in Fig. 8 which illustrates the inhibition of [^{3}H]-CD binding by Gpp(NH)p in the heart, longitudinal muscle of the ileum, corpus striatum, and cerebral cortex of the rat. It can be seen that the potency of Gpp(NH)p for inhibiting agonist binding varies in the different tissues and that the heart is the most sensitive tissue whereas the forebrain regions are the

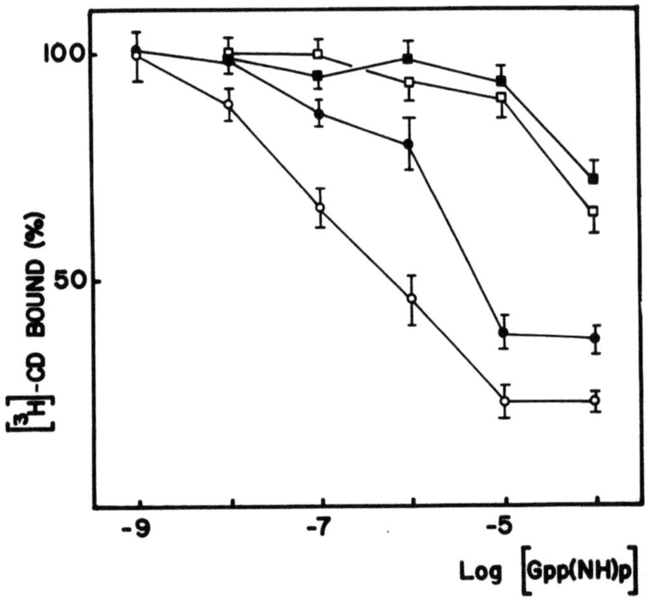

FIG. 8. Effect of Gpp(NH)p on [^{3}H]agonist binding in various tissues of the rat. [^{3}H]-CD binding was measured at a concentration of 5.0 nM in homogenates of the heart (○), longitudinal muscle of the ileum (●), corpus striatum (□), and cerebral cortex (■) in the presence of various concentrations of Gpp(NH)p. Binding assays were run in 50 mM Tris citrate buffer, pH 7.4, at 0°C, containing 10 mM $MgCl_2$. Incubations were carried out for 15 min at 37°C followed by 1 hr at 0°C. Mean binding values ± S.E.M. of at least four experiments are shown. Experimental details are similar to those described by Ehlert *et al.* (1980*c*).

least sensitive. If the ED_{50} of Gpp(NH)p is defined as the concentration that produced half of the inhibition observed at 10^{-4} M, then the ED_{50} values of Gpp(NH)p are 0.18, 1.8, 20, and 22 μM in the heart, ileum, corpus striatum, and cerebral cortex, respectively. It should be noted that the true ED_{50}'s of Gpp(NH)p in the corpus striatum and cerebral cortex are probably greater than those reported above since the data in Fig. 8 suggest that the maximum inhibition of [^{3}H]-CD binding was not achieved at the highest concentration of Gpp(NH)p (10^{-4} M).

The known requirement of GTP and its regulatory protein for activation of adenylate cyclase has led to the view that perturbations in receptor binding by guanine nucleotides are evidence for an adenylate cyclase-linked receptor. Thus, GTP has been shown to be a requirement for activation of adenylate cyclase by glucagon (Rodbell *et al.*, 1971*a*), prostaglandins (Krishna *et al.*, 1972; Lefkowitz *et al.*, 1977), β-adrenergic agonists (Pike and Lefkowitz, 1978), and dopamine (Clement-Cormier *et al.*, 1975), and the binding of each of these ligands has been shown to be inhibited by guanine nucleotides (Rodbell *et al.*, 1971*a*; Lefkowitz *et al.*, 1976; Creese *et al.*, 1978). Moreover, α-adrenergic agonists (Jakobs *et al.*, 1976; Tsai and Lefkowitz, 1978) and opiates (Sharma *et al.*, 1975; Wilkening *et al.*, 1980) have been shown to inhibit adenylate cyclase activity, and the binding of these ligands has been shown to be regulated by guanine nucleotides in a inhibitory manner (U'Prichard and Snyder, 1978; Childers and Snyder, 1978). Thus, neurotransmitters and hormones which either stimulate or inhibit adenylate cyclase activity have been shown to bind to their respective receptors in a manner which is regulated by guanine nucleotides.

A role for muscarinic receptor-mediated inhibition of adenylate cyclase activity has been established in several tissue preparations. Muscarinic agonists inhibit PGE_1 and adenosine-stimulated adenylate cyclase activity in neuroblastoma glial cells (Lichtshtein *et al.*, 1979; Blume *et al.*, 1977), and GTP-stimulated cyclase activity in murine (Jakobs *et al.*, 1979; Birdsall *et al.*, 1980*a*) and canine (Watanabe *et al.*, 1978) myocardial homogenates. In all cases in which the question has been investigated, GTP has been shown to be a requirement for muscarinic cholinergic inhibition of adenylate cyclase. Moreover, the inhibition of adenylate cyclase by α-adrenergic agonists (Jakobs *et al.*, 1978) and opiates (Wilkening *et al.*, 1980) also requires GTP. Thus, GTP-induced perturbations in the binding of agonists to muscarinic receptors may represent an activation of the linkage or a "functional" coupling of the muscarinic receptor with adenylate cyclase.

This interpretation appears plausible particularly with regard to muscarinic receptors in the heart. In cardiac homogenates of the rat, GTP (10^{-3} M) causes a 100-fold shift in the carbachol/[^{3}H]antagonist competition curve by converting a population of predominantly high-affinity sites into one of predominantly low affinity (Birdsall *et al.*, 1980*a*). Moreover, the K_i of carbachol for inhibition of GTP-stimulated adenylate cyclase activity in the heart agrees quantitatively with the dissociation constant of the low-affinity receptor (Birdsall *et al.*, 1980*a*). In cardiac homogenates, muscarinic agonists have been shown to reverse the GTP-induced reduction in the potency of isoproterenol for inhibition of [^{3}H]-dihydroalprenolol binding to β-adrenergic receptors, and this effect was correlated with a muscarinic receptor-mediated inhibition of isoproterenol-stimulated cyclase activity (Watanabe *et al.*, 1978). Both of these muscarinic effects were blocked by atropine. Interestingly, a similar adrenergic–cholinergic receptor interaction has been described with regard to α_1-adrenergic receptor binding in the heart (Yamada *et al.*, 1980*b*). The selective effects of muscarinic agonists on the binding of β-adrenergic agonists suggests that a complementary interaction may function for muscarinic receptors in the heart. Such an interaction was detected in this laboratory during a study of the effects of isoproterenol on oxotremorine/[^{3}H]-(−)-QNB competition curves in the heart (Rosenberger *et al.*, 1980*b*; Roeske and Yamamura, 1980). It was noted that isoproterenol (10^{-6} M) caused a reduction in the proportion and affinity of high-affinity sites in the rat heart. The interactions among adenylate cyclase, GTP-regulatory proteins, β-adrenergic receptors, and muscarinic receptors illustrate a "functional" coupling between these macromolecules. As discussed below, this concept of functional coupling does not necessarily imply a direct linkage of all the elements in the system to one another.

Receptor systems that inhibit adenylate cyclase activity have some characteristics that are different from those of receptor-stimulated adenylate cyclase systems. For example, sodium ions have been shown to be a requirement for inhibition of adenylate cyclase activity by muscarinic agonists in the rabbit myocardium and mouse neuroblastoma × glioma hybrid cells (Jakobs *et al.*, 1979; Lichtshtein *et al.*, 1979), by α-adrenergic agonists and prostaglandins in hamster fat cells (Aktories *et al.*, 1979*a*; Aktories *et al.*, 1979*b*), and by opiates in neuroblastoma × glioma cells (Blume and Boone, 1979; Lichtshtein *et al.*, 1979). Also, the concentrations of GTP required for agonist activation of adenylate cyclase are generally 5–10 times smaller than those required for receptor-mediated inhibition of the enzyme (Jakobs *et al.*, 1979; Lichtshtein *et al.*, 1979; Jakobs, 1979). These findings suggest that functionally distinct GTP-regulatory proteins mediate agonist-induced stimulation and inhibition of adenylate cyclase activity. Another difference between excitatory and inhibitory cyclase systems is manifested by the relationship between the ED_{50}'s of agonists for activation (K_{act}) or inhibition (K_i) of the cyclase and the dissociation constant (K_D) of agonists for binding to the receptor. Direct-binding studies of receptor systems that stimulate adenylate cyclase activity, like the glucagon-sensitive adenylate cyclase of the liver (Rodbell *et al.*, 1971*b*;

Rodbell *et al.*, 1974; Lin *et al.*, 1977) and the β-adrenergic-sensitive adenylate cyclase of turkey erythrocytes (Pike and Lefkowitz, 1980; Stadel *et al.*, 1980), revealed that the concentrations of GTP required for activation of adenylate cyclase converted a heterogeneous population of high- ($K_D = K_H$) and low- ($K_D = K_L$) affinity receptors into a state of predominantly low affinity with a dissociation constant (K_L) much greater than K_{act}. Curiously, the K_{act} of glucagon and β-adrenergic agonists is approximately equal to K_H. In contrast, the muscarinic receptor inhibitory cyclase system of the heart behaved in a different fashion. As described above, GTP at concentrations required for muscarinic agonist-mediated inhibition of adenylate cyclase converts a population of predominantly H sites into a state of low affinity with a dissociation constant (K_L) equivalent to K_i (Birdsall *et al.*, 1980*a*).

A cursory review of models for adenylate cyclase stimulation by hormones and neurotransmitters will illustrate differences between the nature of activation and inhibition of adenylate cyclase activity and may provide insight for the development of a model for muscarinic receptor-mediated adenylate cyclase inhibition. Several models have been proposed for the mechanism by which GTP and receptor agonists stimulate adenylate cyclase (DeHaen, 1976; Abramowitz *et al.*, 1979; Rodbell, 1980; DeLean *et al.*, 1980; Swillens and Dumont, 1980), and a plausible explanation is given below in the sequence of reactions which represents a simplified version of the ternary model of Lefkowitz (1980):

$$\mathrm{H + R \rightleftharpoons HR} \qquad \text{low affinity} \tag{1}$$

$$\mathrm{HR + N \rightleftharpoons HRN} \qquad \text{high affinity} \tag{2}$$

$$\mathrm{HRN + GTP \rightleftharpoons HR + N{-}GTP} \tag{3}$$

$$\mathrm{C + N{-}GTP \rightleftharpoons CN{-}GTP} \tag{4}$$

Here the binding of the hormone (H) to its receptor (R) leads to the formation of a low-affinity hormone–receptor complex (HR). Occupation of R by H facilitates the association of HR with a guanine nucleotide regulatory protein (N) to form the thermodynamically stable ternary complex (HRN) having high affinity for H. Unlike N, the HRN complex preferentially interacts with GTP. The binding of GTP to HRN results in an unstable HRN–GTP complex which decays readily to yield the low-affinity HR complex and a N–GTP complex which preferentially interacts with adenylate cyclase (C) to form an activated enzyme complex (CN–GTP). In this model, R is not coupled directly to C, and cyclase activity in the presence of a given concentration of GTP is dependent upon the formation of the high-affinity HRN complex. Hence, this model accounts for the agreement between K_H and K_{act}. Also, this model can explain the known temporal differences in the effects of guanine nucleotides on receptor binding and cyclase activity without the need for the existence of distinct molecular forms of N (Welton *et al.*, 1977; Ross *et al.*, 1977). These differences can be rationalized by the association of a single type of N with other components of the system to form a heterogeneous population of N complexes (RN, NC).

In contrast to the model outlined above, muscarinic receptor-mediated inhibition of adenylate cyclase in the heart is proportional to receptor occupancy of the low-affinity agonist state of the receptor ($K_L \simeq K_i$) (Birdsall *et al.*, 1980*a*). This relationship suggests that the GTP-induced low-affinity agonist state of the muscarinic receptor is directly coupled to C. A possible reaction scheme is given below:

$$\text{high affinity} \qquad \text{HRN} + \text{GTP} \rightleftharpoons \text{HRN–GTP} \qquad \text{low affinity} \tag{5}$$

$$\underset{\substack{|\\ \text{GTP}}}{\text{HRN}} + \text{C} \rightleftharpoons \underset{\substack{|\\ \text{GTP}}}{\text{HRNC}} \tag{6}$$

Here the GTP-bound HRNC complex represents a low-affinity agonist state of the muscarinic receptor and an inactive form of the cyclase. Although this model is incomplete since it does not describe the equilibrium between the various macromolecular components of the system (R, N, and C), it does predict that the dissociation of GTP from the HRNC complex should result in a simultaneous reversal of the effects of GTP on agonist binding and adenylate cyclase inhibition. In contrast, the effects of guanine nucleotides on the binding of agonists to receptors that stimulate adenylate cyclase are readily reversible as compared to the duration of cyclase activation by guanine nucleotides (Welton *et al.*, 1977; Ross *et al.*, 1977).

Although it seems convincing that the effects of guanine nucleotides on cardiac muscarinic receptor binding indicate a functional coupling of these receptors to an adenylate cyclase system, there is little evidence for muscarinic effects on cyclase systems in various brain regions despite the fact that [^{3}H]-CD binding is modulated by Gpp(NH)p in all major regions of the rat brain. Moreover, guanine nucleotides have been shown to regulate opiate receptors in brain homogenates (Childers and Snyder, 1978; Blume *et al.*, 1979), angiotension receptors in the adrenal cortex (Glossman *et al.*, 1974), and α-adrenergic receptors in the rat brain and liver (Blackmore *et al.*, 1978); yet no effects of these receptors on a cyclase system have been detected in the respective tissues listed above. It may be that technical and methodological limitations preclude the detection of receptor-mediated changes in cyclase activity in these tissues. Alternatively, the demonstration of guanine nucleotide effects on the binding of hormones and neurotransmitters to their receptors might not constitute sufficient evidence to conclude that a given receptor is coupled to adenylate cyclase. Indeed, it has been suggested that GTP-regulatory proteins may be involved generally in the process of membrane signal transduction between surface receptors and their effectors (Rodbell, 1980).

4.3. Regulation of [^{3}H]Antagonist Binding by Guanine Nucleotides

An interesting property of the muscarinic receptor is that the binding of both agonists and antagonists is regulated by guanine nucleotides. Unlike the nature of the modulation of agonist binding, the binding of classical antagonists is enhanced by guanine nucleotides (Hulme *et al.*, 1980; Ehlert *et al.*, 1981*a*). The magnitude of this guanine nucleotide effect is much less than that of the inhibitory effect observed with pure agonists, and it appears to be more demonstrable in buffers of low ionic strength which lack inorganic ions (Hulme *et al.*, 1980). Under these conditions, the binding of the classical muscarinic antagonist [^{3}H]-*N*-methylscopolamine to cardiac muscarinic receptors is inconsistent with the simple Langmuir isotherm, but the data can be explained in terms of receptor heterogeneity with 20% of the sites having high affinity and the remaining sites having an unusually low affinity (Hulme *et al.*, 1980). Hulme *et al.* (1981) have demonstrated that guanine nucleotides enhance antagonist binding in the heart by an apparent conversion of low- to high-affinity antagonist binding sites.

As is the case in the heart, muscarinic receptors in various other tissues and brain regions are also affected by guanine nucleotides in a manner which enhances antagonist binding. Among the tissues that have been examined, the largest effects were noted in the heart and ileum in which the binding of [^{3}H]-(−)-QNB at a concentration of 0.05 nM was stimulated by 73 and 50%, respectively, in the presence of 10 μM Gpp(NH)p (Ehlert *et al.*, 1981*a*). Application of the same methodology in a study (Ehlert *et al.*, 1981*a*) of the effects of guanine nucleotides on rat brain muscarinic receptors revealed that Gpp(NH)p (10 μM) caused the greated enhancement of [^{3}H]-(−)-QNB binding in caudal regions of the brain such as the cerebellum and medulla-pons, which displayed 30 and 16% increases in binding. Somewhat smaller increases in [^{3}H]-(−)-QNB binding (8–11%) were measured in the mesencephalon and diencephalon, while virtually no significant effects of Gpp(NH)p on antagonist binding were detected in the cerebral cortex, hippocampus, and corpus striatum. It is conceivable that guanine nucleotides enhance [^{3}H]antagonist binding in the forebrain under assay conditions different from those used in the study described above (50 mM Tris HCl, 1 mM $MgCl_2$). Unlike the behavior of cardiac receptors, antagonist binding in the cerebral cortex is enhanced by low-ionic-strength buffers (Birdsall *et al.*, 1979*c*). Consequently, additional enhancement of binding by guanine nucleotides might not be achievable in a moderately low-ionic-strength buffer like Tris HCl (50 mM).

In the longitudinal muscle of the ileum, guanine nucleotides enhance [^{3}H]-QNB binding by increasing the affinity of muscarinic receptors without a significant change in their number. The effect of 10 μM Gpp(NH)p on the equilibrium binding isotherm of [^{3}H]-(−)-QNB in the ileum is illustrated by Scatchard analysis in Fig. 9. It can be seen that Gpp(NH)p caused a reduction in the dissociation constant of [^{3}H]-(−)-QNB from a control value of 0.4 to 0.14 nM in the presence of Gpp(NH)p. Presumably, guanine nucleotides enhance antagonist binding in other tissues and brain regions in a similar fashion.

In general, the potency of guanine nucleotides for enhancement of antagonist binding in the heart and ileum is greater than that for inhibition of agonist binding (Hulme *et al.*, 1980; Ehlert *et al.*, 1980*d*). This relation-

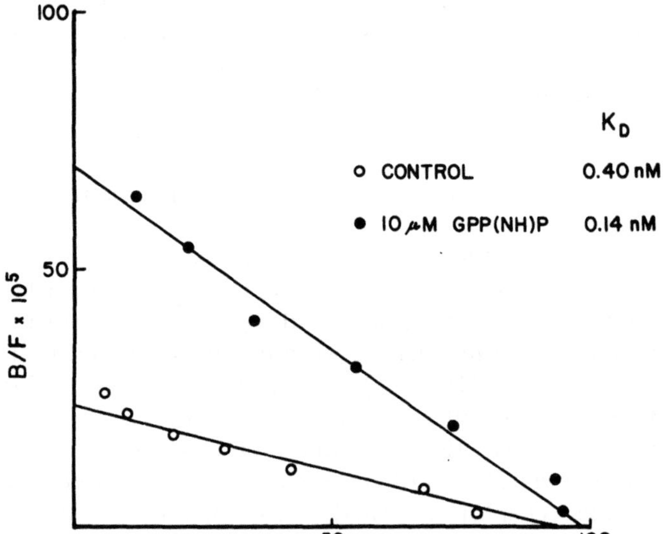

FIG. 9. Effect of Gpp(NH)p on [^{3}H]-(−)-QNB binding in the longitudinal muscle of the rat ileum. [^{3}H]-(−)-QNB binding was measured in the presence (●) and absence (○) of 10 μM Gpp(NH)p. The plot represents Scatchard analysis of mean binding values from four individual experiments, each done in triplicate. Incubations lasted 2 hr at 25°C in 50 mM Tris HCl, pH 7.4, containing 1 mM $MgCl_2$. The concentrations of [^{3}H]-(−)-QNB ranged from 0.024–3.1 nM. The assay conditions are the same as those described by Ehlert *et al.* (1980*d*).

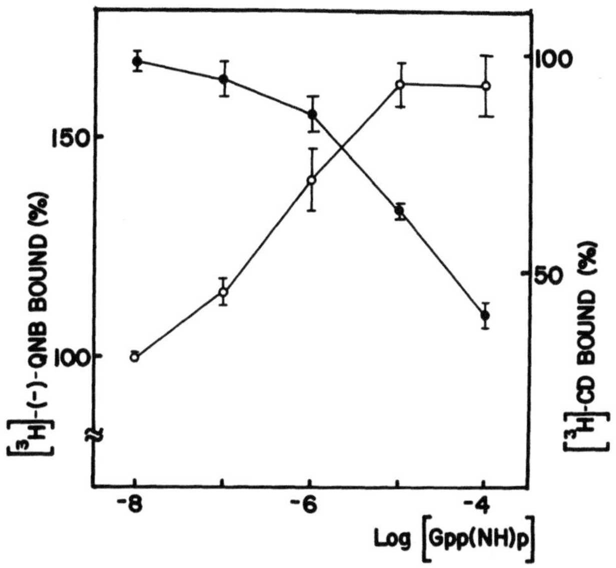

FIG. 10. Effects of Gpp(NH)p on the binding of [^{3}H]-(−)-QNB and [^{3}H]-CD to homogenates of the longitudinal muscle of the rat ileum. The binding of [^{3}H]-(−)-QNB (○) and [^{3}H]-CD (●) was measured under identical conditions in the presence of various concentrations of Gpp(NH)p. Incubations lasted 2 hr at 25°C in Tris HCl buffer containing 1 mM $MgCl_2$. The details of the assay are given by Ehlert *et al.* (1980*d*).

ship is illustrated by the data in Fig. 10 which show the effects of various concentrations of Gpp(NH)p on the binding of [^{3}H]-(−)-QNB and [^{3}H]-CD to the rat ileum under identical assay conditions. Gpp(NH)p caused a maximal 65% stimulation of [^{3}H]-(−)-QNB binding and a maximal inhibition of [^{3}H]-CD binding of at least 60%, with the ED_{50}'s for these effects being approximately 0.4 and 5.6 μM, respectively. Thus, the potency of Gpp(NH)p for perturbation of agonist and antagonist binding in the ileum differs by a factor of 14, suggesting that different GTP-regulatory proteins may be involved in these processes. Alternatively, it is conceivable that the same GTP binding site mediates an allosteric enhancement of antagonist binding and inhibition of agonist binding. For example, if the agonist-induced conformational change in the receptor is associated with a concerted conformational change in a GTP binding site from a state of high to low affinity, then the binding of agonists and GTP will appear to be competitive. There is evidence that antagonists induce a slow isomerization of the receptor into a high-affinity antagonist state, a process which is surely distinct from agonist activation of the receptor. If the antagonist-induced conformation is associated with a high-affinity GTP binding site, then the potency of guanine nucleotides for enhancement of antagonist binding will be greater than that for inhibition of agonist binding. Moreover, if the agonist and antagonist isomerization steps are subject to independent environmental constraints, then the existence of antagonist receptor heterogeneity that is independent of agonist receptor heterogeneity is possible. This condition has been noted for cardiac muscarinic receptors (Hulme *et al.*, 1980). Although the functional significance of the guanine nucleotide regulation of antagonist binding is not known, it demonstrates that the ground state of the muscarinic receptor can exist in different conformations which might not have been predicted on the basis of classical theories of drug action.

5. THE INFLUENCE OF SULFHYDRYL REAGENTS ON MUSCARINIC RECEPTOR BINDING

Several studies have shown that the binding properties of muscarinic receptors are affected by sulfhydryl reagents. Aronstam *et al.* (1978) found that treatment of homogenates of the rat forebrain with NEM (1.0 mM) for 20 min at 35°C caused an increase in the potency of carbachol for inhibition of [^{3}H]-QNB binding when binding assays were carried out for 40 min at 20°C following removal of the NEM. This enhancement of agonist affinity took place without measurable changes in the binding of the antagonist [^{3}H]-QNB (Aronstam *et al.*, 1977, 1978). In a subsequent study (Aronstam and Eldefrawi, 1979) it was observed that treatment of neural membranes with the oxidizing agents, potassium ferricyanide and 5,5′-dithiobis-(2-nitrobenzoic acid), as well as the sulfhydryl-alkylating agent, NEM, enhanced agonist potency, whereas the sulfhydryl-reducing agents, dithiothreitol and 2-mercaptoethanol, caused a reduction in agonist potency. Scatchard analysis of the carbachol/[^{3}H]-QNB competitive curves was consistent with the interpretation that the changes in agonist potency caused by the sulfhydryl reagents were due to reversible conversions of the high- and low-affinity agonist states of the muscarinic receptor. Thus, oxidation and reduction of sulfhydryl–disulfide groups on the muscarinic receptor probably increases and decreases, respectively, the proportion of high-affinity agonist binding sites.

The effects of NEM on the binding of agonists to the muscarinic receptor have also been observed when direct measurements of the binding of an [^{3}H]agonist were made. In a study of the effect of NEM treatment on the binding of [^{3}H]-CD to rat forebrain homogenates, it was noted that NEM treatment enhanced [^{3}H]-CD binding in a manner that was consistent with an increase in the density of superhigh-affinity agonist binding sites (Ehlert *et al.*, 1980*d*). Thus, these latter data indicate that the superhigh-affinity agonist binding sites are subject to change by sulfhydryl reagents.

The nature of the perturbation in muscarinic receptor binding properties resulting from exposure to sulfhydryl reagents is highly dependent upon the incubation conditions of both the reaction with sulfhydryl reagents and the binding assay itself. In a study in which assays were carried out in Krebs–Henseleit, NEM treatment caused a decrease in the binding of both agonists and antagonists (Hedlund and Bartfai, 1979). Moreover, when binding assays are carried out at 22–25°C in Na-K/phosphate buffer, the result of NEM treatment is an increase in agonist affinity, whereas little or no effects are detectable when binding assays are run at 37°C (Carson, 1980). It is clear from the foregoing that careful consideration should be given to assay conditions when the effects of sulfhydryl reagents on the muscarinic receptor are studied.

6. IONIC PERTURBATION OF MUSCARINIC RECEPTOR BINDING

In general, common physiological monovalent and divalent cations inhibit both agonist and antagonist binding in a manner that is correlated with ionic strength (Birdsall *et al.*, 1979*c*). The binding of agonists appears to be more susceptible to the effects of ionic strength than the binding of classical antagonists. For example, in the longitudinal muscle of the rat ileum, NaCl (140 mM) causes a 4- to 5-fold reduction in the potency of a series of cholinergic analogs for inhibition of $[^3H]$-(−)-QNB binding in a manner that is independent of the efficacy of the cholinergic agonist (Ehlert *et al.*, 1981*b*). In contrast, the binding of classical antagonists like QNB and atropine is only slightly affected by 140 mm NaCl (Ehlert *et al.*, 1981*b*). It has been suggested that hydrophobic interactions between the muscarinic receptor and classical muscarinic antagonists are less susceptible to the effects of ionic strength (Ehlert *et al.*, 1981*b*). In regions of the forebrain, intermediate to high concentrations (0.2–2.0 M) of monovalent cations cause a nonselective reduction in the affinity or proportion of high-affinity agonist binding sites such that agonist receptor heterogeneity is reduced under these conditions (Birdsall *et al.*, 1979*c*; Ehlert *et al.*, 1980*a*). In the cerebral cortex, antagonist affinity is highest in buffered media lacking inorganic ions, whereas cardiac muscarinic receptors relax into a state of lower affinity in low-ionic-strength buffers as compared to the affinity that is measured in standard buffers of greater ionic strength (Hulme *et al.*, 1980). Thus, antagonist binding to cardiac muscarinic receptors is first stimulated by low concentrations of monovalent ions followed by a nonselective depression of binding at higher concentrations of ionic strength.

In addition to the nonselective ionic pertubations described above, some selective effects have been noted in studies of the ionic regulation of muscarinic receptor binding. Perhaps the most profound selective ionic effects that have been reported are those of Tl^+ in the cerebral cortex (Birdsall *et al.*, 1979*c*). At ion concentrations of 0.2 M, Tl^+ causes a 22-fold greater reduction in the affinity of the low-affinity agonist receptor as compared to that caused by Na^+, and the effects of Tl^+ on the affinity of the high-affinity agonist receptor are even greater (170-fold). It has been suggested that this might implicate a K^+ conductance mechanism for the muscarinic receptor since Tl^+ is often regarded as similar to K^+ (Birdsall *et al.*, 1979*c*). Indeed, small selective ionic effects of K^+ have been noted (Birdsall *et al.*, 1979*c*).

Selective effects of divalent cations on muscarinic receptor binding have also been reported. In the medulla-pons, Mg^{2+} has been reported to cause a selective enhancement of agonist binding (Birdsall *et al.*, 1980*c*). It has also been reported that Mg^{2+} (10 mM) reduces the IC_{50} (increases affinity) of carbachol for competitive inhibition of [^{3}H]-(−)-QNB binding to cardiac homogenates (Wei and Sulakhe, 1980*b*); however, interpretation of these results is complicated since a high saturating concentration of [^{3}H]-QNB was used in the competition experiments, and it is not clear whether the reduction in agonist IC_{50} was due to an increase in agonist potency or a decrease in the potency of [^{3}H]-QNB. The latter possibility might be a contributing factor since we have found that in the presence of Mg^{2+} (10 mM), 43% of the muscarinic receptors in the corpus striatum are converted into a state of anomalously low affinity for [^{3}H]-(−)-QNB (K_D = 0.64 nM), with the rest of the sites maintaining the typical high-affinity state (K_D = 0.03 nM) that is usually noted in our [^{3}H]-QNB binding experiments on rat brain muscarinic receptors (Ehlert *et al.*, 1981*c*). Moreover, we have observed that Mg^{2+} inhibits the binding of a fixed concentration of [^{3}H]-(−)-QNB with a similar potency in the corpus striatum, heart, and longitudinal muscle of the ileum (unpublished observations). At the present time, it is unclear whether other divalent cations can satisfy the Mg^{2+} requirement for the manifestation of antagonist receptor heterogeneity.

7. REGULATION OF ANTAGONIST BINDING BY DOPAMINERGIC AGONISTS

Under conditions in which antagonist receptor heterogeneity is readily demonstrable, dopaminergic agonists enhance the binding of [^{3}H]-(−)-QNB by an apparent conversion of low- to high-affinity antagonist sites (Ehlert *et al.*, 1981*c*). This phenomenon is illustrated by the data in Fig.

11 which shows the effects of 1 μM apomorphine on [^{3}H]-(−)-QNB binding in 50 mM Na HEPES buffer containing 10 mM Mg^{2+} and 10 μM Gpp(NH)p. The control binding isotherm deviates from a simple Langmuir isotherm and has a Hill coefficient of approximately 0.69. Weighted nonlinear regression analysis showed that the data were consistent with the presence of two binding sites with 57% of the sites having a high-affinity dissociation constant of 0.03 nM and the remaining sites having a low-affinity dissociation constant of 0.64 nM. Under the assay conditions described above, this manifestation of antagonist receptor heterogeneity required Mg^{2+}. Inclusion of 1 μM apomorphine into the incubation media caused an enhancement of [^{3}H]-(−)-QNB binding by an apparent conversion of low- to high-affinity sites. The resulting data were entirely consistent with that predicted for a single population of high-affinity sites having a dissociation constant (K_D = 0.025 nM) that is not significantly different from the value for the subpopulation of high-affinity sites shown in the control curve.

The results of experiments investigating the ability of a series of agonists to enhance specific [^{3}H]-(−)-QNB binding revealed a pattern that was generally consistent with a dopaminergic effect, although evidence

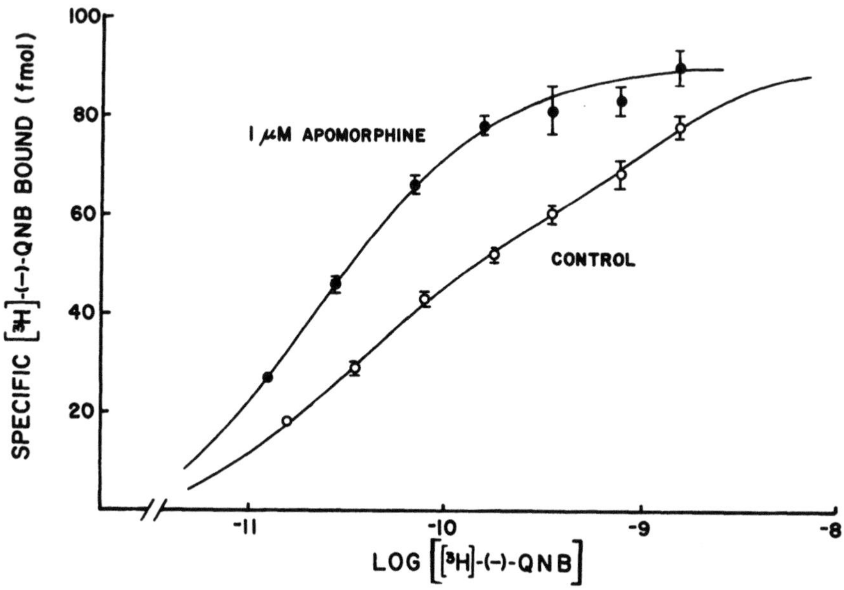

FIG. 11. Effect of apomorphine on [^{3}H]-(−)-QNB binding to homogenates of the rat stiatum. The binding of [^{3}H]-(−)-QNB was measured in the presence (●) and absence (○) of 1 μM apomorphine. Incubations were carried out for 2 hr at 25°C in 50 mM Na HEPES, pH 7.4, containing 10 mM $MgCl_2$ and 10 μM Gpp(NH)p. Mean binding values ± S.E.M. of four individual experiments are shown. The data are from Ehlert *et al.* (1981*c*).

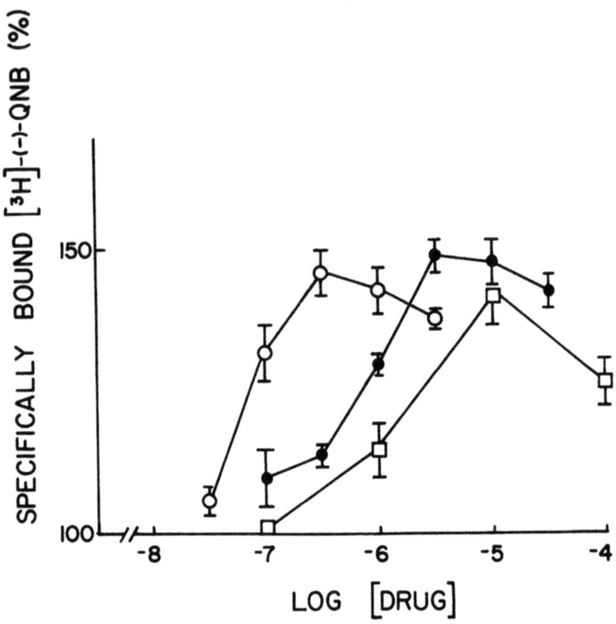

FIG. 12. Effects of apomorphine, dopamine, and isoproterenol on [^{3}H]-(−)-QNB binding in homogenates of the rat striatum. The binding of [^{3}H]-(−)-QNB was measured at a concentration of 0.05 nM in the presence of various concentrations of apomorphine (○), dopamine (●), and isoproterenol (□). Assay conditions were the same as those in the legend of Fig. 11. Mean binding values ± S.E.M. of three individual experiments are shown.

for a β-adrenergic effect was also demonstrable (Ehlert *et al.*, 1981*d*). Figure 12 shows the effects of various concentrations of apomorphine, dopamine, and isoproterenol on the binding of [^{3}H]-(−)-QNB measured at a concentration of 0.05 nM. It can be seen that all three drugs caused a similar maximal enhancement of [^{3}H]-(−)-QNB binding (50%) and that the most potent drug was apomorphine which had an ED_{50} of approximately 0.063 μM. The ED_{50}'s of dopamine and isoproterenol were 0.71 and 2.4 μM, respectively. Other drugs that also stimulated [^{3}H]-(−)-QNB binding to a similar maximum as that shown in Fig. 12 included *N*-propylnorapomorphine (ED_{50} = 0.14 μM) and the catecholamines epinephrine and norepinephrine which were approximately one-third as potent as dopamine. A variety of other drugs, including γ-aminobutyric acid, clonazepam, histamine, morphine, phenylephrine, serotonin, clonidine, methoxamine, metaproterenol, and buspirone caused no significant enhancement of binding at concentrations as high as 100 μM.

Evidence which suggests a different mechanism of action of isoproterenol compared to that of apomorphine comes from the results of experiments carried out on striatal membrane preparations that were prepared in 0.32 M sucrose and washed once by centrifugation at 48,000

$\times$ *g* for 10 min followed by resuspension in fresh 0.2 M sucrose. In these homogenates, apomorphine only caused a maximal 25% enhancement of binding, with the ED_{50} for this effect being 0.1 μM, whereas isoproterenol caused no enhancement of binding at concentrations between 0.1–100 μM (Ehlert *et al.*, 1981*c*).

A variety of dopaminergic antagonists were screened for their ability to block the enhancement of [^{3}H]-(−)-QNB binding caused by 1.0 μM apomorphine (Ehlert *et al.*, 1981*c*). Haloperidol, sulperide, and spiperone had no specific effects at concentrations as high as 10–100 μM. In contrast, α-flupenthixol and fluphenazine specifically blocked the enhancement of [^{3}H]-(−)-QNB binding with IC_{50} values of approximately 10^{-6} M, which corresponds to a calculated K_i of 10^{-7} M assuming competitive inhibition between apomorphine and the antagonists. The pharmacologically inactive geometric isomer, β-flupenthixol, and the adrenergic antagonists phentolamine, prazosin, and (−)-propranolol caused no significant blockade of the apomorphine enhancement of binding at concentrations as high as 10–100 μM.

The implications of the dopaminergic regulation of muscarinic receptor binding are not readily apparent. The effect appears to be primarily specific for dopaminergic agonists since, of all the drugs tested, apomorphine, *N*-propylnorapomorphine, and dopamine were the most potent. Within the brain, dopamine receptors have been classified into at least two subtypes: those that are thought to be coupled to adenylate cyclase (D1 receptors) and those that are independent of the cyclase (D2 receptors) (Iversen, 1975). The fact that the potent inhibitors of the dopamine-stimulated adenylate cyclase, α-flupenthixol and fluphenazine (Iversen, 1975), blocked the apomorphine enhancement of [^{3}H]-(−)-QNB binding, whereas haloperidol, sulperide, and spiperone did not, suggests that D1 receptors are responsible for the dopaminergic effect. However, the potency of flupenthixol and fluphenazine for inhibition of dopaminer-stimulated adenylate cyclase activity in the striatum has been reported to be in the 10 nM range. Thus, the low potency of these drugs (K_i = 100 nM) for blockade of apomorphine-stimulated [^{3}H]-(−)-QNB binding does not point conclusively to a D1 receptor-mediated mechanism. The regional distribution of the apomorphine enhancement of [^{3}H]-(−)-QNB binding in the brain (Fig. 13) correlates generally with the distribution of the dopamine-sensitive adenylate cyclase in that the effect is greatest in the striatum, intermediate in the cerebral cortex, and negligible in caudal regions of the brain. However, the data in Fig. 13 also illustrate a large effect in the hypothalamus, which is a brain region having little dopamine-stimulated adenylate cyclase activity (Iversen *et al.*, 1980).

In contrast to the effects of apomorphine on the binding of the antagonist [^{3}H]-(−)-QNB, no effects were detected when muscarinic receptor binding was measured using the agonist, [^{3}H]-CD. Apparently,

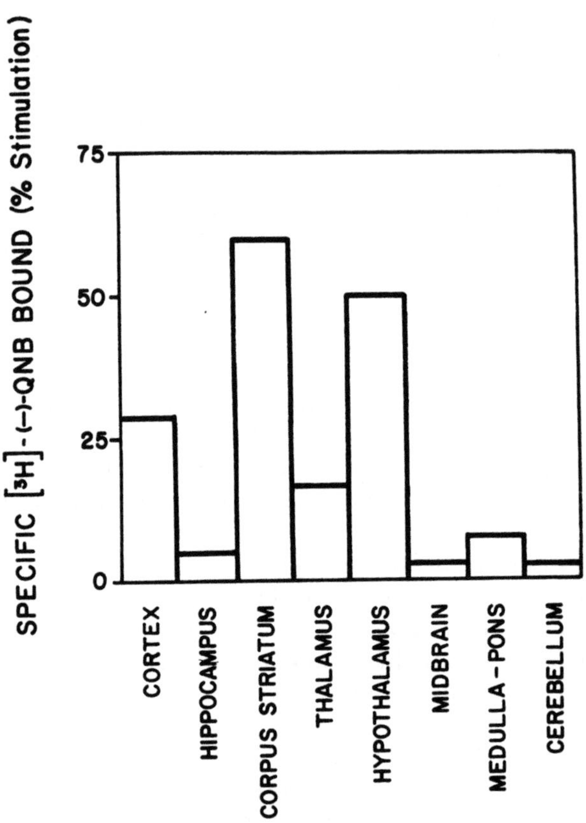

FIG. 13. Effects of apomorphine on [^{3}H]-(−)-QNB binding in various regions of the rat brain. The binding of [^{3}H]-(−)-QNB was measured at a concentration of 0.05 nM in homogenates of various regions of the rat brain in the presence and absence of 1 μM apomorphine. [^{3}H]-(−)-QNB binding in the presence of apomorphine is expressed as percent stimulation over control binding. Assay conditions were the same as those indicated in the legend of Fig. 11. Mean binding values of two experiments are shown.

dopaminergic analog modulate antagonist receptor heterogeneity without causing a change in the binding properties of the various agonist states of the receptor. Preliminary evidence indicates that guanine nucleotides potentiate the effects of dopaminergic agonists on [^{3}H]-(−)-QNB binding. Collectively, these data indicate complex interactions between the ground states of the muscarinic receptor and Mg^{2+}, guanine nucleotides, and dopaminergic agonists. Future research in this area may elucidate the significance of these interactions and may reveal mechanisms whereby different neurotransmitter systems interact with one another at the receptor level.

8. CONCLUSION

The application of receptor binding techniques to the study of muscarinic receports has expanded the field of muscarinic receptor pharmacology and has enabled detection of complex ligand–receptor interactions that were not apparent in studies of whole tissue assay systems like the guinea pig ileum. These complex ligand–receptor interactions are manifest by an apparent heterogeneous population of binding sites. The relationship between the efficacy of agonists and the binding parameters of the various agonist states of the receptor suggests that agonist receptor heterogeneity is caused by factors influencing the microenvironment of the receptor, such as varying degrees of receptor–effector coupling, and not by the existence of different molecule species of the muscarinic receptor. The ability of guanine nucleotides, ions, and sulfhydryl reagents to cause a conversion of one agonist state of the receptor into another provides support for the concept of molecular homogeneity of the muscarinic receptor and suggests that these agents may be useful for studying mechanisms of receptor–effector coupling. Perhaps the most unexpected results of recent muscarinic receptor binding studies are the complexities of antagonist binding, including the apparent heterogeneity of binding sites that are independent of agonist states of the receptor and the antagonist-induced isomerization of the receptor. The fact that guanine nucleotides, ions, and dopaminergic agonists convert a heterogeneous population of antagonist binding sites into a homogeneous one suggests that the differences among the various antagonist states of the receptor are due to environmental differences and not molecular heterogeneity. An explanation for the existence of multiple antagonist states (ground states) of the muscarinic receptor cannot be found in classical theories of drug action, and it is possible that future investigations of the complexities of antagonists binding may yield new insights into mechanisms of receptor function.

Acknowledgments

We thank Isabelle Preiss for secretarial assistance. Much of the work described in this chapter was supported by United States Public Health Service grants MH-2757, MH-30626, and Program Project Grant HL-20984. William R. Roeske is a recipient of a United States Public Health Service Research Scientist Development Award (HL-00776) from the National Heart, Lung, and Blood Institute, and Henry I. Yamamura is a recipient of a United States Public Health Service Research Scientist

Development Award, Type II (MH-00095), from the National Institute of Mental Health.

9. REFERENCES

Abramowitz, J., and Birnbaumer, L., 1979, Prostacyclin activation of adenylyl cyclase in rabbit corpus luteum membranes: comparison with 6-keto prostaglandin $F_1\alpha$ and prostaglandin E_1, *Biol. Reprod.* **21**:509–616.

Abramowitz, J., Iyengar, R., and Birnbaumer, L., 1979, Guanyl nucleotide regulation of hormonally responsive adenylyl cyclases, *Mol. Cell. Endocrinol.* **16**:129–146.

Aktories, K., Jacobs, K. H., and Schultz, G., 1979*a*, Influence of sodium chloride on the inhibition of hamster fat cell adenylate cyclase by GTP and on the inhibitory effects of alpha-adrenergic agonists and prostaglandin E_1, *Naunyn-Schmiedeberg's Arch. Pharmacol.* **308**:Suppl. R15.

Aktories, K., Schultz, G., and Jakobs, K. H., 1979*b*, Inhibition of hamster fat cell adenylate cyclase by prostaglandin E_1 and epinephrine requirement for GTP and sodium ions, *FEBS Lett.* **107**:100–104.

Ariens, E. J., and Beld, A. J., 1977, The receptor concept in evolution, *Biochem. Pharmacol.* **26**:913–918.

Ariens, E. J., and Rodrigues de Miranda, J. F., 1979, The receptor concept: recent experimental and theoretical developments, in: *Recent Advances in Receptor Chemistry* (F. Gualtieri, M. Giannella, and C. Melchiorre, eds.), pp. 1–6, Elsevier/North Holland Biomedical Press, Amsterdam, New York, and Oxford.

Ariens, E. J., and Simmons, A. M., 1967, Cholinergic and anticholinergic drugs: do they act on common receptors? *Ann. N.Y. Acad. Sci.* **144**:842–867.

Aronstam, R. S., and Eldefrawi, E., 1979, Reversible conversion between affinity states for agonists of the muscarinic acetylcholine receptor from rat brain, *Biochem. Pharmacol.* **28**:701–703.

Aronstam, R. S., Hoss, W., and Abood, L. G., 1977, Conversion between configurational states of the muscarinic receptor in rat brain. *Eur. J. Pharmacol.* **46**:279–282.

Aronstam, R. S., Abood, L. G., and Hoss, W., 1978, Influence of sulfhydryl reagents and heavy metals on the funtional state of the muscarinic acetylcholine receptor in rat brain, *Mol. Pharmacol.* **14**:575–586.

Arunlakshana, O., and Schild, H. O., 1959, Some quantitative uses of drug antagonists, *Br. J. Pharmacol.* **14**:48–58.

Baumgold, J., Abood, L. G., and Aronstam, R. S. 1977, Studies on the relationship of binding affinity to psychoactive and anticholinergic potency of a group at psychotomimetic glycolates, *Brain Res.* **124**:331–340.

Beld, A. J., and Ariens, E. J., 1974, Stereospecific binding as a tool in attempts to localize and isolate muscarinic receptors. II. Binding of (+)-benzetimide, (−)-benzetimide, and atropine to a fraction from bovine tracheal smooth muscle and to bovine caudate nucleus, *Eur. J. Pharmacol.* **25**:203–209.

Berrie, C. P., Birdsall, N. J. M., Burgen, A. S. V., and Hulme, E. C., 1979, Guanine nucleotides modulate muscarinic receptor binding in the heart, *Biochem. Biophys. Res. Commun.* **87**:1000–1005.

Birdsall, N. J. M., and Hulme, E. C., 1976, Biochemical studies on muscarinic acetylcholine receptors, *J. Neurochem.* **27**:7–16.

Birdsall, N. J. M., Burgen, A. S. V., Hiley, C. R., and Hulme, E. C., 1976, Binding of agonists and antagonists to muscarinic receptors, *J. Supramol. Struct.* **4**:367(327)–371(331).

Birdsall, N. J. M., Burgen, A. S. V., and Hulme, E. C., 1977, Correlation between the binding properties and pharmacological responses of muscarinic receptors, in: *Cholinergic Mechanisms and Psychopharmacology* (D. J. Jenden, ed.), pp. 25–33, Plenum Press, New York.

Birdsall, N. J. M., Burgen, A. S. V., and Hulme, E. C., 1978, The binding of agonists to brain muscarinic receptors, *Mol. Pharmacol.* **14:**723–736.

Birdsall, N. J. M., Burgen, A. S. V., and Hulme, E. C., 1979*a*, Muscarinic receptors: biochemical and binding studies, in: *Recent Advances in Receptor Chemistry* (F. Gualtieri, M. Giannella, and C. Melchiorre, eds.,) pp. 71–96, Elsevier/North-Holland Biomedical Press, Amsterdam, New York, and Oxford.

Birdsall, N. J. M., Burgen, A. S. V., and Hulme, E. C., 1979*b*, A study of the muscarinic receptor by gel electrophoresis, *Br. J. Pharmacol.* **66:**337–342.

Birdsall, N. J. M., Burgen, A. S. V., Hulme, E. C., and Wells, J. W., 1979*c*, The effects of ions on the binding of agonists and antagonists to muscarinic receptors, *Br. J. Pharmacol.* **67:**371–377.

Birdsall, N. J. M., Berrie, C. P., Burgen, A. S. V., and Hulme, E. C., 1980*a*, Modulation of the binding properties of muscarinic receptors: evidence for receptor–effector coupling, in: *Receptors for Neurotransmitters and Peptide Hormones* (G. Pepeu, M. J. Kuhar, and S. J. Enna, eds.), pp. 107–116, Raven Press, New York.

Birdsall, N. J. M., Hulme, E. C., and Burgen, A. S. V., 1980*b*, The character of muscarinic receptors in different regions of the rat brain, *Proc. R. Soc. London* **207:**1–12.

Birdsall, N. J. M., Hulme, E. C., Hammer, R., and Stockton, J. S., 1980*c*, Subclasses of muscarinic receptors, in: *Psychopharmacology and Biochemistry of Neurotransmitter Receptors* (H. I. Yamamura, R. W. Olson, and E. Usdin, eds.), pp. 97–100, Elsevier/North Holland Biomedical Press, New York.

Blackmore, P. F., Brumley, F. T., Marks, J. L., and Exton, J. H., 1978, Studies on α-adrenergic activation of hepatic glucose output, *J. Biol. Chem.* **253:**4851–4858.

Blume, A. J., 1978, Interactions of ligands with opiate receptors of brain membrane: regulation by ions and nucleotides, *Proc. Natl. Acad. Sci. USA* **73:**1713–1717.

Blume, A. J., and Boone, G., 1979, Effective coupling of opiate receptors to adenylate cyclase: requirement for nucleoside triphosphates and monovalent cations, *Fed. Proc.* **38:**628.

Blume, A. J., Chen, C., and Foster, C. J., 1977, Muscarinic regulation of cAMP in mouse neuroblastoma, *J. Neurochem.* **29:**625–632.

Blume, A. J., Lichtshtein, D., and Boone, G., 1979, Coupling of opiate receptors to adenylate cyclase: requirement for Na^+ and GTP, *Proc. Natl. Acad. Sci. USA* **76:**5626–5630.

Bourdois, P. S., Mitchell, J. F., Somogyi, G. T., and Szerb, J. C., 1974, The output per stimulus of acetylcholine from cerebral cortical slices in the presence or absence of cholinesterase inhibition, *Br. J. Pharmacol.* **52:**509–517.

Burgen, A. S. V., and Hiley, C. R., 1974, Two populations of acetylcholine receptors in guinea pig ileum, *Proc. Br. Pharmacol. Soc.* 127p.

Carson, S., 1980, Differential effect of *N*-ethylmaleimide on muscarinic agonist binding in rat and bovine brain membranes, *FEBS Lett.* **109:**81–84.

Carson, S., Godwin, S., Massoulie, J., and Kato, G., 1977, Solubilization of atropine-binding material from brain, *Nature (London)* **266:**176–178.

Childers, S. R., and Snyder, S. H., 1978, Guanine nucleotides differentiate agonist and antagonist interactions with opiate receptors, *Life Sci.* **23:**759–762.

Cho, A. K., Jenden, D. J., and Lamb, S. I., 1972, Rates of alkaline hydrolysis and muscarinic activity of some aminoacetates and their quaternary ammonium analogs, *J. Med. Chem.* **15:**391–394.

Clement-Cormier, Y. C., Parrish, R. G., Petzold, G. L., Kebabian, J. W., and Greengard, P., 1975, Characterization of a dopamine-sensitive adenylate cyclase in the rat caudate nucleus, *J. Neurochem.* **15:**143–149.

CREESE, I., and SNYDER, S. H., 1978, Dopamine receptor binding of ADTN regulated by guanine nucleotides, *Eur. J. Pharmacol.* **50**:459–461.

CREESE, I., PROSSER, T., and SNYDER, S. H., 1978, Dopamine receptor binding: specificity, localization, and regulation by ions and guanyl nucleotides, *Life Sci.* **23**:495–500.

DEHAEN, C., 1976, The nonstoichiometric floating receptor model for hormone-sensitive adenylyl cyclase, *J. Theor. Biol.* **58**:383–400.

DELEAN, A., STADEL, J. M., and LEFKOWITZ, R. J., 1980, A ternary complex model explains the agonist-specific binding properties of adenylate cyclase-coupled β-adrenergic receptor, *J. Biol. Chem.* **255**:7108–7117.

EHLERT, F. J., ROESKE, W. R., ROSENBERGER, L. B., and YAMAMURA, H. I., 1980*a*, The influence of guanyl-5′-ylimidodiphosphate and sodium on muscarinic receptor binding in the rat brain and longitudinal muscle of the rat ileum, *Life Sci.* **26**:245–252.

EHLERT, F. J., YAMAMURA, H. I., TRIGGLE, D. J., and ROESKE, W. R., 1980*b*, The influence of guanyl-5′-ylimidodiphosphate and sodium chloride on the binding of the muscarinic agonist, [^{3}H]-*cis*-methyldioxolane, *Eur. J. Pharmacol.* **61**:317–318.

EHLERT, F. J., DUMONT, Y., ROESKE, W. R., and YAMAMURA, H. I., 1980*c*, Muscarinic receptor binding in rat brain using the agonist, [^{3}H]-*cis*-methyldioxolane, *Life Sci.* **26**:961–967.

EHLERT, F. J., ROESKE, W. R., and YAMAMURA, H. I., 1980*d*, Regulation of muscarinic receptor binding by guanine nucleotides and *N*-ethylmaleimide, *J. Supramol. Struct.* **14**:149–162.

EHLERT, F. J., ROESKE, W. R., and YAMAMURA, H. I., 1981*a*, Muscarinic receptor: regulation by guanine nucleotides, ions, and *N*-ethylmaleimide, *Fed. Proc.* **40**:153–159.

EHLERT, F. J., ROESKE, W. R., and YAMAMURA, H. I., 1981*b*, The regulation of the muscarinic receptor by guanine nucleotides, ions, and dopaminergic agonists, in: *Pharmacological and Biochemical Aspects of Neurotransmitter Receptors* (H. Yoshida and H. I. Yamamura, eds.), Wiley, New York, in press.

EHLERT, F. J., ROESKE, W. R., and YAMAMURA, H. I., 1981*c*, Striatal muscarinic receptors: regulation by dopaminergic agonists, *Life Sci.* **28**:2441–2448.

EHLERT, F. J., ROESKE, W. R., and YAMAMURA, H. I., 1981*d*, Dopaminergic regulation of muscarinic receptor binding in the corpus striatum, *Proc. West. Pharmacol. Soc.* **24**:93–95.

ELLENBROCK, B. W. J., NIVARD, R. J. F., VAN ROSSUM, J. M., and ARIENS, E. J., 1965, Absolute configuration and parasympathetic action: pharacodynamics of enantiomorphic and diastereoisomeric esters of β-methylcholine, *J. Pharm. Pharmacol.* **17**:393–404.

FARROW, J. T., and O'BRIEN, R. D., 1973, Binding of atropine and muscarone to rat brain fractions and its relation to the acetylcholine receptor, *Mol. Pharmacol.* **9**:33–40.

FEWTRELL, C. M. S., and RANG, H. P., 1973, The labelling of cholinergic receptors in smooth muscle, in: *Drug Receptors* (H. P. Rang, ed.), pp. 211–224, Macmillan, New York.

FIELDS, J. Z., ROESKE, W. R., MORKIN, E., and YAMAMURA, H. I., 1978, Cardiac muscarinic cholinergic receptors, *J. Biol. Chem.* **253**:3251–3258.

GALPER, J. B., and SMITH, T. W., 1978, Properties of muscarinic acetylcholine receptors in heart cell cultures, *Proc. Natl. Acad. Sci. USA* **75**:5831–5835.

GALPER, J. B., and SMITH, T. W., 1980, Agonist and guanine nucleotide modulation of muscarinic cholinergic receptors in cultured heart cells, *J. Biol. Chem.* **255**:9571–9579.

GALPER, J. B., and KEIN, W., and CATTERALL, W. A., 1977, Muscarinic Acetylcholine receptors in developing chick heart, *J. Biol. Chem.* **252**:8692–8699.

GLOSSMAN, H., BAUKAL, A., and CATT, K. J., 1974, Angiotensin II receptors in bovine adrenal cortex, *J. Biol. Chem.* **249**:664–666.

HADHAZY, P., and SZERB, J. C., 1977, The effect of cholinergic drugs on [^{3}H]acetylcholine release from slices of rat hippocampus, striatum, and cortex, *Brain Res.* **123**:311–322.

HAGA, T., 1980, Solubilization of muscarinic acetylcholine receptors by L-α-lysophosphatidylcholine, *Biomed. Res.* **1**:265–268.

HAMMER, R., BERRIE, C. P., BIRDSALL, N. J. M., BURGEN, A. S. V., and HULME, E. C., 1980, Pirenzepine distinguishes between different subclasses of muscarinic receptors, *Nature (London)* **283**:90–92.

Hanley, M. R., and Iversen, L. L., 1978, Muscarinic cholinergic receptors in rat corpus striatum and regulation of guanosine cyclic 3',5'-monophosphate, *Mol. Pharmacol.* **14**:246–255.

Hedlund, B., and Bartfai, T., 1979, The importance of thiol- and disulfide groups in agonist and antagonist binding to the muscarinic receptor, *Mol. Pharmacol.* **15**:531–544.

Hiley, C. R., and Burgen, A. S. V., 1974, The distribution of muscarinic receptor sites in the nervous system of the dog, *J. Neurochem.* **22**:159–162.

Hokin, L. E., 1966, Effects of calcium omission on acetylcholine-stimulated amylase secretion and phospholipid synthesis in pigeon pancreas slices, *Biochim. Biophys. Acta.* **115**:219–221.

Howlett, A. C., Van Arsdale, P. M., and Gilman, A. G., 1978, Efficiency of coupling between the beta-adrenergic receptor and adenylate cyclase, *Mol. Pharmacol.* **14**:531–539.

Hulme, E. C., Birdsall, N. J. M., Burgen, A. S. V., and Mehta, P., 1978, The binding of antagonists to brain muscarinic receptors, *Mol. Pharmacol.* **14**:737–750.

Hulme, E. C., Berrie, C. P., Birdsall, N. J. M., and Burgen, A. S. V., 1980, Interactions of muscarinic receptors with guanine nucleotides and adenylate cyclase, in: *Drug Receptors and Their Effectors* (N. J. M. Birdsall, ed.), pp. 23–34, Macmillan, London.

Hulme, E. C., Berrie, C. P., Birdsall, N. J. M., Burgen, A. S. V., 1981, Two populations of binding sites for muscarinic antagonists in the rat heart, *Eur. J. Pharmacol.* **73**:137–142.

Iversen, L. L., 1975, Dopamine receptors in the brain, *Science* **188**:1084–1089.

Iversen, L. L., Quik, M., Emson, P. C., Dowling, J. E., and Watling, K. J., 1980, Further evidence for the existence of multiple receptors for dopamine in the central nervous system, in: *Receptors for Neurotransmitters and Peptide Hormones* (G. Pepeu, M. J. Kuhar, and S. J. Enna, eds.), pp. 193–202, Raven Press, New York.

Jafferji, S. S., and Michell, R. H., 1976, Muscarinic cholinergic stimulation of phosphatidylinositol turnover in the longitudinal smooth muscle of guinea pig ileum, *Biochem. J.* **154**:653–657.

Jakobs, K. H., 1979, Inhibition of adenylate cyclase by hormones and neurotransmitters, *Mol. Cell. Endocrinol.* **16**:147–156.

Jakobs, K. H., Saur, W., and Schultz, G., 1976, Reduction of adenylate cyclase activity in lysates of human platelets by the alpha-adrenergic component of epinephrine, *J. Cyclic Nucleotide Res.* **2**:381–392.

Jakobs, K. H., Saur, W., and Schultz, G., 1978, Inhibition of platelet adenyl cyclase by epinephrine requires GTP, *FEBS Lett.* **85**:167–170.

Jakobs, K. H., Aktories, K., and Schultz, G., 1979, GTP-dependent inhibition of cardiac adenylate cyclase by muscarinic cholinergic agonists, *Naunyn-Schmiedeberg's Arch. Pharmacol.* **310**:113–119.

Jarv, J., Hedlund, B., and Bartfai, T., 1980, Kinetic studies on muscarinic antagonist–agonist competition, *J. Biol. Chem.* **255**:2649–2651.

Kloog, Y., and Sokolovsky, M., 1977, Muscarinic acetylcholine receptor interactions: competition binding studies with agonists and antagonist, *Brain Res.* **134**:167–172.

Kloog, Y., and Sokolovsky, M., 1978, Studies on muscarinic acetylcholine receptors from mouse brain: characterization of the interaction with antagonists, *Brain Res.* **144**:31–48.

Kloog, Y., Egozi, Y., and Sokolovsky, M., 1978, Characterization of muscarinic acetylcholine receptors from mouse brain: evidence for regional heterogeneity and isomerization, *Mol. Pharmacol.* **15**:545–558.

Kloog, Y., Egozi, Y., and Sokolovsky, M., 1979*a*, Regional heterogeneity of muscarinic receptors of mouse brain, *FEBS Lett.* **97**:265–268.

Kloog, Y., Heron, D. S., Korczyn, A. D., Sachs, D. I., and Sokolovsky, M., 1979*b*, Muscarinic acetycholine receptors in albino rabbit iris-ciliary body, *Mol. Pharmacol.* **15**:581–587.

Krishna, G., Harwood, J. P., Barber, A. J., and Jamieson, G. A., 1972, Requirement for guanosine triphosphate in the prostaglandin activation of adenyl cyclase platelet membranes, *J. Biol. Chem.* **247**:2253–2254.

Kuhar, M. J., Sethy, V. H., Roth, R. H., and Aghajanian, G., 1973, Choline: selective accumulation by central cholinergic neurons, *J. Neurochem.* **20:**581–593.

Laduron, P., 1980, Dopamine and muscarinic receptors: *in vivo* identification, axonal transport, and solubilization, in: *Psychopharmacology and Biochemistry of Neurotransmitter Receptors* (H. I. Yamamura, R. W. Olsen, and E. Usdin, eds.), pp. 115–132, Elsevier/North Holland Biomedical Press, New York.

Laduron, P. M., Verwimp, M., and Leysen, J. E., 1979, Stereospecific *in vitro* binding of [^{3}H]-dexetimide to brain muscarinic receptors, *J. Neurochem.* **32:**421–427.

Llands, A. M., Anenenko, E., Jones, G., Hoppe, J. O., and Becker, T. J., 1949, The antispasmodic action of basic nitriles, *J. Pharmacol. Exp. Ther.* **96:**1–10.

Lefkowitz, R. J., 1980, Modification of Adenylate cyclase activity by alpha- and beta-adrenergic receptors, in: *Psychopharmacology and Biochemistry of Neurotransmitter Receptors* (H. I. Yamamura, R. W. Olsen, and E. Usdin, eds.), pp. 155–170, Elsevier/North Holland, New York.

Lefkowitz, R. J., Mullikin, D., and Caron, M. G., 1976, Regulation of beta-adrenergic receptors by guanyl-5′-ylimidodiphosphate and other purine nucleotides, *J. Biol. Chem.* **251:**4686–4692.

Lefkowitz, R. J., Mullikin, D., Wood, C. L., Gore, T. B., and Mukherjee, C., 1977, Regulation of prostaglandin receptors by prostaglandins and guanine nucleotides in frog erythrocytes, *J. Biol. Chem.* **252:**5295–5303.

Lewis, P. R., and Shute, C. C. D., 1967, The cholinergic limbic system, *Brain* **90:**521–540.

Lichtshtein, D., Boone, G., and Blume, A., 1979, Muscarinic receptor regulation of NG108-15 adenylate cyclase: requirement for Na^+ and GTP, *J. Cyclic Nucleotide Res.* **367**–375.

Lin, M. C., Nicosia, S., Lad, P. M., and Rodbell, M., 1977, Effects of GTP on binding of [^{3}H]glucagon to receptors in rat hepatic plasma membranes, *J. Biol. Chem.* **252:**2790–2792.

Long, J. P., Luduena, F. P., Tullar, B. F., and Lands, A. M., 1956, Stereochemical factors involved in cholinolytic activity, *J. Pharmacol.* **117:**29–38.

Maguire, M. D., Van Arsdale, P. M., and Gilman, A. G., 1976, An agonist-specific effect of guanine nucleotides on binding to the beta-adrenergic receptor, *Mol. Pharmacol.* **12:**335–339.

Meyerhoffer, A., 1972, Absolute configuration of 3-quinuclidinyl benzilate and the behavioral effect in the dog of the optical isomers, *J. Med. Chem.* **15:**994–995.

Michell, R. H., 1975, Inositol phospholipids and cell surface receptor function, *Biochim. Biophys. Acta* **415:**81–147.

Molenaar, P. C., and Polak, R. L., 1970, Stimulation by atropine of acetylcholine release and synthesis in cortical slices from rat brain, *Br. J. Pharmacol.* **40:**406–417.

Murad, F., Chi, Y. M., Rall, T. W., and Sutherland, E. W., 1962, The effect of catecholamines and choline esters on the formation of adenosine 3′,5′-phosphate by preparations from cardiac muscle and liver, *J. Biol. Chem.* **237:**1233–1238.

Northup, J. K., and Mansour, T. E., 1978, Adenylate cyclase from *Fascicola hepatica, Mol. Pharmacol.* **14:**820–833.

Overstreet, D. H., Speth, R. C., Hruska, R. E., Ehlert, F., Dumont, Y., and Yamamura, H. I., 1980, Failure of septal lesions to alter muscarinic cholinergic or benzodiazepine binding sites in hippocampus of rat brain, *Brain Res.* **195:**203–207.

Paton, W. D., and Rang, H. P., 1965, The uptake of atropine and related drugs by intestinal smooth muscle of the guinea pig in relation to acetylcholine receptors, *Proc. R. Soc. London Ser. B* **163:**1–44.

Pike, L. J., and Lefkowitz, R. J., 1978, Agonist-specific alterations in receptor binding affinity associated with solubilization of turkey erythrocyte membrane beta-adrenergic receptors, *Mol. Pharmacol.* **14:**370–375.

Pike, L. R., and Lefkowitz, R. J., 1980, Activation and desensitization of β-adrenergic receptor-coupled GTPase and adenylate cyclase of frog and turkey erythrocyte membranes, *J. Biol. Chem.* **255:**6860–6867.

RANG, H. P., 1967, The uptake of atropine and related compounds by smooth muscle, *Ann. N.Y. Acad. Sci.* **144:**756–765.

REHAVI, M., MAAYANI, S., SOKOLOVSKY, M., 1977, Enzymatic resolution and cholinergic properties of (±)3-quinuclidinol derivatives, *Life Sci.* **21:**1293–1302.

RODBELL, M., 1980, The role of hormone receptors and GTP-regulatory proteins in membrane transduction, *Nature (London)* **284:**17–22.

RODBELL, M., KRANS, H. M. S., POHL, S. L., and BIRNBAUMER, L., 1971*a*, The glucagon-sensitive adenylate cyclase system in plasma membranes of rat liver, *J. Biol. Chem.* **246:**1861–1871.

RODBELL, M., KRANS, H. M. J., POHL, S. L., and BIRNBAUMER, L., 1971*b*, The glucagon-sensitive adenylate cyclase system in plasma membranes of rat liver, *J. Biol. Chem.* **246:**1872–1876.

RODBELL, M., LIN, M. C., and SALOMON, Y., 1974, Evidence for interdependent action of glucagon and nucleotides on hepatic adenylate cyclase system, *J. Biol. Chem.* **249:**59–65.

ROESKE, W. R., and YAMAMURA, H. I., 1980, Muscarinic cholinergic receptor regulation, in: *Psychopharmacology and Biochemistry of Neurotransmitter Receptors* (H. I. Yamamura, R. W. Olson, and E. Usdin, eds.), pp. 101–114, Elsevier/North-Holland Biomedical Press, New York.

ROESKE, W. R., YAMAMURA, H. I., and YAMADA, S., 1981, Postsynaptic β-adrenergic receptors increase and presynaptic muscarinic receptors decrease in splenic tissue after 6-hydroxydopamine treatment, *Fed. Proc.* **40:**260.

ROSENBERGER, L. B., ROESKE, W. R., and YAMAMURA, H. I., 1979, The regulation of muscarinic cholinergic receptors by gaunine nucleotides in cardiac tissue, *Eur. J. Pharmacol.* **56:**179–180.

ROSENBERGER, L. B., YAMAMURA, H. I., and ROESKE, W. R., 1980*a*, Cardiac muscarinic cholinergic receptor binding is regulated by Na^+ and guanyl nucleotides, *J. Biol. Chem.* **255:**820–823.

ROSENBERGER, L. B., YAMAMURA, H. I., and ROESKE, W. R., 1980*b*, The regulation of cardiac muscarinic cholinergic receptors by isoproterenol, *Eur. J. Pharmacol.* **65:**129–130.

ROSS, E. M., MAGUIRE, M. E., STURGILL, T. W., BILTONEN, R. L., and GILMAN, A. G., 1977, Relationship between the β-adrenergic receptor and adenylate cylase, *J. Biol. Chem.* **252:**5761–5775.

SCHIMERLIK, M. I., and SEARLES, R., 1980, Ligand interactions with membrane-bound porcine atrial muscarinic receptors, *Biochemistry* **19:**3413–3407.

SCHLEIFER, L. S., and ELDEFRAWI, M. E., 1974, Identification of the nicotinic and muscarinic acetylcholine receptors of mouse brain, *Neuropharmacology* **13:**53–63.

SHARMA, V. K., and BANERJEE, S. P., 1978, Presynaptic muscarinic cholinergic receptors, *Nature (London)* **272:**276–278.

SHARMA, S. K., NIRENBERG, M., and KLEE, W. A., 1975, Morphine receptors as regulators of adenylate cyclase activity, *Proc. Natl. Acad. Sci. USA* **72:**590–594.

SNYDER, S. H., CHANG, K. J., KUHAR, M. J., and YAMAMURA, H. I., 1975, Biochemical identification of the mammalian muscarinic cholinergic receptor, *Fed. Proc.* **34:**1915–1921.

SOKOLOVSKY, M., GURWITZ, D., and GALRON, R., 1980, Muscarinic receptor binding in mouse brain: regulation by guanine nucleotides, *Biochem. Biophys. Res. Commun.* **94:**487–492.

SOUDIJN, W., VANWIJNGAARDEN, I., and ARIENS, E. J., 1973, Dexetimide, a useful tool in acetycholine-receptor localization, *Eur. J. Pharmacol.* **24:**43–48.

STADEL, J. M., DELEAN, A., and LEFKOWITZ, R. J., 1980, A high-affinity agonist β-adrenergic receptor complex is an intermediate for catecholamine stimulation of adenylate cyclase in turkey and frog erythrocyte membranes, *J. Biol. Chem.* **25:**1436–1441.

STRANGE, P. G., BIRDSALL, N. J. M., and BURGEN, A. S. V., 1977, Occupancy of muscarinic acetylcholine receptors stimulates a guanylate cyclase in neuroblastoma cells, *Biochem. Soc. Trans.* **5:**189–191.

SWILLENS, S., and DUMONT, J. E., 1980, A unifying model of current concepts and data on adenylate cyclase activation by β-adrenergic agonists, *Life Sci.* **27:**1013–1028.

SZERB, J. C., 1977, Characterization of presynaptic muscarinic receptors in central cholinergic neurons, in: *Cholinergic Mechanisms and Psychopharmacology* (D. J. Jenden, eds.), pp. 49–60, Plenum Press, New York.

SZERB, J. C., and SOMOGYI, G. T., 1973, Depression of acetylcholine release from cerebral cortical slices by cholinesterase inhibition and by oxotremorine, *Nature New Biol.* **241:**121–122.

TALAMO, B. R., ADLER, S. C., and BURT, D. R., 1979, Parasympathetic denervation decreases muscarinic receptor binding in rat parotid, *Life Sci.* **24:**1573–1580.

TRIFARO, J. M., 1969, The effect of Ca^{2+} omission on the secretion of catecholamines and the incorporation of orthophosphate-^{32}P into nucleotides and phospholipids of bovine adrenal medulla during acetylcholine stimulation, *Mol. Pharmacol.* **5:**424–427.

TRIGGLE, D. J., 1979, The mucarinic receptor: structural, ionic, and biochemical implications, in: *Recent Advances in Receptor Chemistry* (F. Gualtieri, M. Giannella, and C. Melchiorre, eds.), pp. 127–146, Elsevier/North-Holland Biomedical Press, Amsterdam, New York, and Oxford.

TSAI, B. S., and LEFKOWITZ, R. J., 1978, Agonist-specific effects of monovalent and divalent cations on adenylate cyclase-coupled alpha adrenergic receptors in rabbit platelets, *Mol. Pharmacol.* **14:**540–548.

TSAI, B. S., and LEFKOWITZ, R. J., 1979, Agonist-specific effects of guanine nucleotides on alpha-adrenergic receptors in human platelets, *Mol. Pharmacol.* **16:**61–68.

U'PRICHARD, D. C., and SNYDER, S. H., 1978, Guanyl nucleotide influences of [^{3}H]ligand binding to α-noradrenergic receptors in calf brain membranes, *J. Biol. Chem.* **253:**3444–3452.

WAMSLEY, J. K., ZARBIN, M. A., and KUHAR, M. J., 1981, Muscarinic cholinergic receptor flow in the sciatic nerve, *Brain Res.* **217:**155–161.

WATANABE, A. M., MCCONNAUGHEY, M. M., STRAWBRIDGE, R. A., FLEMING, J. W., JONES, L. R., and BESCH, H. R., 1978, Muscarinic cholinergic receptor modulation of β-adrenergic receptor affinity for catecholamines, *J. Biol. Chem.* **253:**4833–4836.

WEI, J. W., and SULAKHE, P. V., 1979, Agonist–antagonist interactions with rat atrial muscarinic cholinergic receptor sites: differential regulation by guanine nucleotides, *Eur. J. Pharmacol.* **58:**91–92.

WEI, J. W., and SULAKHE, P. V., 1980*a*, Cardiac muscarinic cholinergic receptor sites: opposing regulation by divalent cations and guanine nucleotide of receptor–agonist interaction, *Eur. J. Pharmacol.* **62:**345–347.

WEI, J. W., and SULAKHE, P. V., 1980*b*, Requirement for sulfhydryl groups in the different effects of magnesium ion and GTP on agonist binding of muscarinic cholinergic receptor sites in rat atriaal membrane fraction, *Naunyn-Schmiedeberg's Arch. Pharmacol.* **314:**51–59.

WEILER, M. H., and JENDEN, D. J., 1977, Modulation of acetylcholine (ACh) release from synaptosomes by oxotremorine (OT) and oxotremorine methiodide (OTMI), *Proc. Soc. Neurosci.* **3:**417.

WEINSTEIN, H., MAAYANI, S., SREBRENIK, S., COHEN, S., and SOKOLOVSKY, M., 1975, A theoretical and experimental study of the semirigid cholinergic agonist 3-acetoxyquinuclidine, *Mol. Pharmacol.* **11:**671–689.

WELTON, A. F., LAD, P. M., NEWBY, A. C., YAMAMURA, H., NICOSIA, S., and RODBELL, M., 1977, Solubilization and separation of the glucagon receptor and adenylate cyclase in guanine nucleotide-sensitive states, *J. Biol. Chem.* **253:**5947–5950.

WILKENING, D., SABOL, S. L., and NIRENBERG, M., 1980, Control of opiate receptor–adenylate cyclase interactions by calcium ions and guanosine-5′-triphosphate, *Brain Res.* **189:**459–466.

YAMADA, S., YAMAMURA, H. I., and ROESKE, W. R., 1980*a*, Alterations in cardiac autonomic receptors following 6-hydroxydopamine treatment in rats, *Mol. Pharmacol.* **18:**185–192.

YAMADA, S., YAMAMURA, H. I., and ROESKE, W. R., 1980*b*, The regulation of cardiac α_1-adrenergic receptors by guanine nucleotides and muscarinic cholinergic agonists, *Eur. J. Pharmacol.* **63:**239–241.

YAMAMURA, H. I., and SNYDER, S. H., 1974*a*, Muscarinic cholinergic binding in rat brain, *Proc. Natl. Acad. Sci. USA* **71:**1725–1729.

YAMAMURA, H. I., and SNYDER, S. H., 1974*b*, Muscarinic cholinergic receptor binding in the longitudinal muscle of the guinea pig ileum with [^{3}H]quinuclidinyl benzilate, *Mol. Pharmacol.* **10:**861–867.

YAMAMURA, H. I., and SNYDER, S. H., 1974*c*, Postsynaptic localization of muscarinic cholinergic receptor binding in rat hippocampus, *Brain Res.* **78:**320–326.

YAMAMURA, H. I., KUHAR, M. J., GREENBERG, D., and SNYDER, S. H., 1974, Muscarinic cholinergic receptor binding: regional distribution in monkey brain, *Brain Res.* **66:**541–546.

ZAHNISER, N. R., and MOLINOFF, P. B., 1978, Effect of guanine nucleotides on striatal dopamine receptors, *Nature (London)* **275:**453–455.

7

BENZODIAZEPINE RECEPTORS

Claus Braestrup and Mogens Nielsen

1. INTRODUCTION

1,4-Benzodiazepines are among the most widely used drugs. In 1973, approximately 15% of the adult population in the West admitted to having used benzodiazepines (BZ) at least once within the previous year (Balter *et al.*, 1974) to achieve anticonvulsant, anxiolytic, or hypnotic effects. The total amount of benzodiazepine sold in a recent 6-month period in Denmark, calculated as number of *d*efined *d*aily *d*oses (DDD), corresponded to 8% of the whole population using benzodiazepines daily (Schou, 1980).

Some benzodiazepines are very potent drugs. Clinical effects are achieved after as little as 3.0 μg/kg of triazolam and, 5 mg of diazepam to nontolerant adults often gives manifest responses.

Clues to the way benzodiazepines work were obtained in 1975 when it was reported that they enhanced the effects of the inhibitory neurotransmitter GABA (Haefely *et al.*, 1975; Costa *et al.*, 1975). GABA is the major inhibitory neurotransmitter of the mammalian CNS, and several effects of benzodiazepines can be explained by enhanced GABAergic function.

In 1977, it was demonstrated that specific receptors exist for benzodiazepines on neurons in the CNS of all higher vertebrates (Squires and Braestrup, 1977; Moehler and Okada, 1977*a*; Mackerer *et al.*, 1978), and

Claus Braestrup • Research Laboratories, A/S Ferrosan, DK-2860 Soeborg, Denmark; and Skt. Hans Mental Hospital, DK-4000 Roskilde, Denmark. *Mogens Nielsen* • Psychopharmacological Research Laboratories, Department E, Skt. Hans Mental Hospital, DK-4000 Roskilde, Denmark.

in 1978 it became clear that benzodiazepine receptors and GABA receptors at the molecular level are coupled in a supramolecular GABA receptor/benzodiazepine receptor complex (Tallman *et al.*, 1978), providing the means for a functional interaction between benzodiazepines and GABA. The link to GABA receptors may further explain the distinct features of some convulsant β-carboline BZ receptor ligands and of BZ receptor antagonists (Braestrup *et al.*, 1982*a*).

Furthermore, recent biochemical studies have provided evidence for the idea that another major class of minor tranquilizers, the barbiturates, may also exert their effects at the GABA/benzodiazepine receptor complex. Barbiturates bind to picrotoxinin sites on GABA receptor-related chloride channels, and in turn enhance benzodiazepine receptor affinity (Olsen and Leeb-Lundberg, 1981).

This review of benzodiazepine receptor mechanisms also deals with the numerous attempts that have been made to isolate and characterize presumed endogenous ligands for benzodiazepine receptors. Several agents have been isolated and identified by means of their affinity for benzodiazepine receptors, but conclusive evidence for the exclusive role of any of these as "the endogenous ligand" is not available.

2. BIOCHEMICAL CHARACTERISTICS OF BENZODIAZEPINE RECEPTOR BINDING

2.1. [^{3}H]Benzodiazepine Radioligands

2.1.1. *Affinity, Saturability, and Subcellular Distribution*

Three different benzodiazepine radioligands are available for labeling benzodiazepine receptors, [^{3}H]flunitrazepam ([^{3}H]-FNM), [^{3}H]diazepam, and [^{3}H]clonazepam (Fig. 1). The characteristics of [^{3}H]diazepam and [^{3}H]-FNM binding have been investigated in detail, whereas [^{3}H]clonazepam binding has been less thoroughly characterized (Baraldi *et al.*, 1979). [^{3}H]-FNM is technically superior to [^{3}H]diazepam as a radioligand, because its binding affinity is higher and its blank values (nonspecific binding) are lower. Furthermore, the "off-rates" for [^{3}H]-FNM are lower than those for [^{3}H]diazepam, thereby reducing the possible occurrence of dissociation artifacts when free radioligand is separated from bound radioligand. Recently the water-soluble benzodiazepine [^{3}H]flurazepam and the benzodiazepine antagonists [^{3}H]-Ro 15–1788 and [^{3}H]-CGS 8216 have also become available (Moehler and Richards, 1981; Czernik *et al.*, 1982).

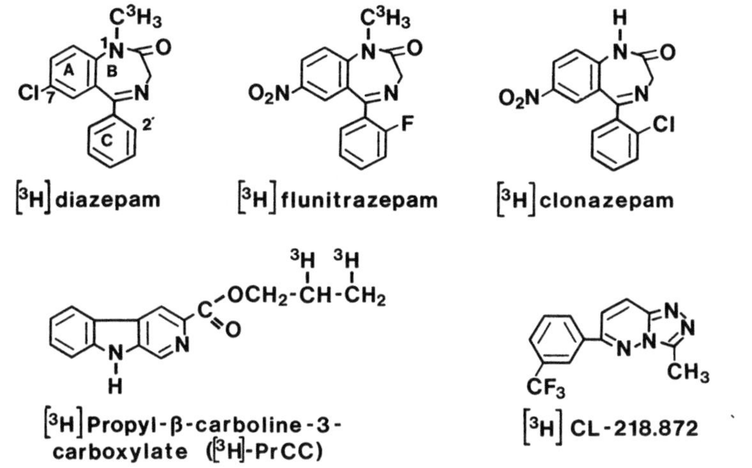

FIG. 1. Molecular structures of five radioligands for BZ receptor binding.

$[^3H]$-FNM and $[^3H]$diazepam bind to the same classes of binding sites in mammalian brain tissue (Braestrup and Squires, 1978*a*), they have a similar regional distribution, and their binding to benzodiazepine receptors is equally affected by other benzodiazepines (see, for example, Braestrup and Squires, 1978*a*) and by GABA analogs (Karobath *et al.*, 1979; Braestrup *et al.*, 1979*d*). $[^3H]$-FNM, $[^3H]$diazepam, and $[^3H]$clonazepam can probably be used interchangeably for studies in mammalian CNS. However, in certain other tissues it is not clear whether or not these three ligands bind to the same classes of binding sites. These sites occur in membranes of primary astroglial cells, glioma cells, cell nuclei, kidney, and liver, on fish brain membranes, and on serum albumin (Table 4). $[^3H]$Clonazepam will probably not bind to these "peripheral binding sites."

$[^3H]$-FNM and $[^3H]$diazepam are commercially available at high specific activities, ca. 80 Ci/mmol, close to the theoretical upper limit for three tritium atoms per molecule (29 Ci/mAtom tritium). Tritium atoms are introduced in the 1-methyl group. Loss of tritium by demethylation does not occur under normal conditions *in vitro*.

$[^3H]$-FNM and $[^3H]$diazepam bind to rat brain membranes with high affinity. The dissociation equilibrium constant for $[^3H]$-FNM at 0°C is ca. 1 nM (dissociation equilibrium constant K_D = affinity constant = binding constant) (Table 1). This value is in the same range as values obtained for other high-affinity radioligands for several brain neurotransmitter receptors. The association and dissociation rate constants at 0°C for $[^3H]$-FNM and $[^3H]$diazepam are shown in Table 1. The dissociation equilibrium constant determined from the respective initial rate constants, $K_D^k = k_{-1}/k_{+1}$ = 1.0 nM for $[^3H]$-FNM, agrees closely with values determined by saturation experiments, K_D = 1.1 nM. The observation of polyphasic dissociation curves has been reported (Squires *et al.*, 1979*a*). When data from saturation experiments are transformed and plotted according to the principle described by Scatchard (1949) or according to the Hill equation, linear plots are obtained which conform closely to simple Langmuir binding isotherms. Thus, each benzodiazepine apparently interacts with equal affinity with all class of benzodiazepine receptors in a noncooperative way.

The binding of [^{3}H]-FNM and [^{3}H]diazepam to brain membranes is saturable (Fig. 2), indicating the presence of a limited number of binding sites. There is a marked regional variation in the number of binding sites (see Section 4). The highest levels, approximately 100 pmol/g tissue, are found in cerebral cortex; this level represents ca. 0.005% of the total membrane-bound protein content of the brain.

The benzodiazepine binding site is proteinaceous; the binding is susceptible to proteolytic enzymes such as trypsin, chymotrypsin, and carboxypeptidase (Braestrup and Squires, 1977; Moehler and Okada, 1977*a*). The binding protein is membrane bound, the subcellular distribution being characterized by the presence of high levels in the so-called synaptosomal "P_2 fraction" (Braestrup *et al.*, 1978*b*; Bosmann *et al.*, 1978). However, the "microsomal" fraction (P_3) contains, in addition, large amounts of binding sites for benzodiazepines (Braestrup *et al.*, 1978*b*). Several neurotransmitter receptors occur in high concentrations in the microsomal fraction (Blas and Mahler, 1978; Laduron *et al.*, 1978). Sub-

TABLE 1

Characteristics of [^{3}H]Flunitrazepam, [^{3}H]Diazepam, and [^{3}H]Propyl-β-carboline-3-carboxylate ([^{3}H]-PrCC) Binding to Rat Whole Brain Membranes[a]

Parameter	[^{3}H]Flunitrazepam	[^{3}H]Diazepam	[^{3}H]-PrCC
K_D, M	1.1×10^{-9}	3.2×10^{-9}	0.9×10^{-9}
k_{+1}, M^{-1} min^{-1}	4.4×10^7	6.8×10^7	2.3×10^8
k_{-1}, min^{-1}	0.044[b]	0.16[b]	0.22
k_{-1}/k_{+1}	1.0×10^{-9}	2.4×10^{-9}	1.0×10^{-9}
B_{max}, pmol/g	120	34[c]	81[d]

[a] Values are adapted from Speth *et al.* (1979*a*), Mackerer *et al.* (1978), and Nielsen *et al.* (1981) for [^{3}H]flunitrazepam, [^{3}H]diazepam, and [^{3}H]-PrCC, respectively. These values are representative of data presented elsewhere (Squires and Braestrup, 1977; Braestrup and Squires, 1978*a*; Braestrup *et al.*, 1980*a*; Braestrup and Nielsen, 1981*b*; Damm *et al.*, 1978; Dudai and Sherman-Gold, 1980; Moehler and Okada, 1977*a*, 1978; Regan *et al.*, 1980*a*; Tallman *et al.*, 1978; Speth *et al.*, 1978, 1979*a*). Crude whole brain membranes ([^{3}H]flunitrazepam, [^{3}H]-PrCC) or P_2 fractions ([^{3}H]diazepam) were investigated. Note that these membrane preparations ([^{3}H]flunitrazepam, [^{3}H]diazepam, [^{3}H]-PrCC) contain appreciable amounts of endogenous GABA.

[b] Dissociation curves are slightly polyphasic; the presented values reflect initial rate constants.

[c] The number of binding sites for [^{3}H]diazepam and [^{3}H]flunitrazepam is equal when assayed in identical tissue preparations (Fig. 2).

[d] Unpublished.

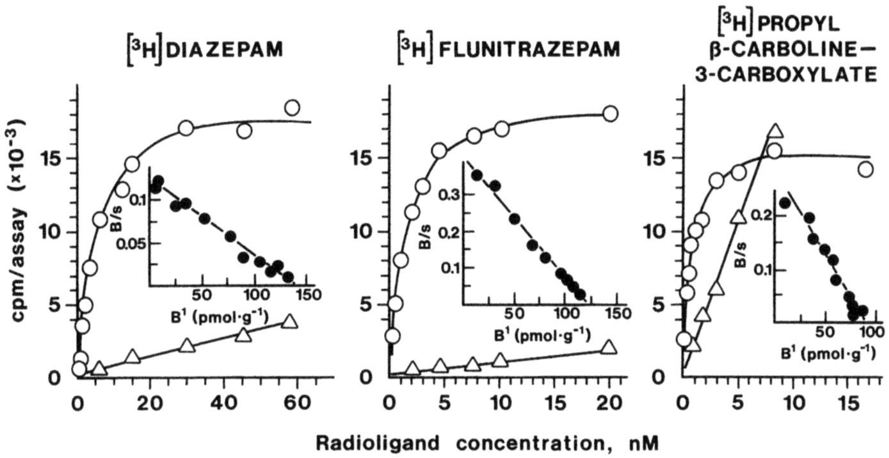

FIG. 2. Saturation experiments. Binding of [^{3}H]diazepam, [^{3}H]flunitrazepam ([^{3}H]-FNM), and [^{3}H]propyl-β-carboline-3-carboxylate ([^{3}H]-PrCC) to fresh rat cortex membranes. Brain tissue was homogenized in 2 × 10 ml ice-cold KH_2PO_4, 25 mM, pH 7.1, by an Ultra-Turrax homogenizer. The homogenate was centrifuged at 48,000 × *g* for 10 min and the pellet was resuspended in another portion of ice-cold buffer (250 ml/g of original tissue for [^{3}H]-diazepam or [^{3}H]FNM; 500 ml/g for [^{3}H]-PrCC). Aliquots of the crude membrane preparation (0.5 ml for [^{3}H]diazepam or [^{3}H]-FNM, 2.5 ml for [^{3}H]-PrCC) were incubated with [^{3}H]-diazepam (76 Ci/mmol NEN, 0.15–58 nM final concentration), [^{3}H]-FNM (86.4 Ci/mmol NEN, 0.25–20 nM), and [^{3}H]-PrCC (44.6 Ci/mmol, Ferrosan A/S, 0.14–16 nM; all final concentrations) for 60 min in an ice bath. The samples were filtered through Whatman GF/C glass-fiber filters and washed immediately with ice-cold buffer. Radioactivity on the filters was determined by conventional liquid scintillation counting at 42% efficiency. Specific binding (○) was obtained by subtracting nonspecific binding (△), which is binding in the presence of diazepam (3 × 10^{-6} M), from total binding. Inserts show the data transformed to Scatchard plots (●) (B^1, specific binding in picomoles per gram original tissue; B, specific binding in cpm/assay; s, free ligand concentration in cpm/assay at equilibrium). K_D = 4.4 nM for [^{3}H]diazepam, K_D = 1.2 nM for [^{3}H]-FNM, and K_D = 0.52 nM for [^{3}H]-PrCC. (C. Braestrup and M. Nielsen, unpublished results.)

fractionation of synaptosomal P_2 fractions after osmotic shock at pH 8, using discontinuous sucrose gradients, indicated that the binding sites are located on plasma (cell) membranes (Braestrup *et al.*, 1978*b*), and not, for example, on vesicular or mitochondrial membranes.

2.1.2. *Selecting Assay Conditions*

There are basically three ways of preparing brain tissue for benzodiazepine receptor binding.

Early investigations of benzodiazepine receptors were done on brain "P_2" fractions obtained by differential centrifugation of homogenates prepared in isotonic (0.32 M) sucrose. One advantage of the P_2 fraction is that of improved homogeneity, which is, however, by no means absolute.

Postsynaptic elements are attached to the presynaptic particles; numerous nonsynaptosomal elements are also present in the P_2 fraction. P_2 fractions are useful for qualitative studies, such as, for example, investigations of the chemical selectivity of BZ receptors (Braestrup and Squires, 1978*b*). For quantitative studies, however, the P_2 fraction is less appropriate, because small changes in synaptosomal buoyancy or stability may affect membrane recovery in unpredictable ways. Furthermore, the percentage recovery of BZ receptors in the P_2 fraction varies from one brain region to another, the recovery in rat cerebellum, for example, being less than in rat hippocampus (Nielsen *et al.*, 1981).

Using a "total membrane preparation" reduces the possibility of receptor losses during membrane fractionation. Crude membrane preparations contain several irrelevant elements, such as mitochondria, cell nuclei, etc., but if the selectivity of the radioligand is sufficiently high and the agent used to define nonspecific binding is properly selected, these elements will not interfere with specific binding in mammalian brain tissue. Crude membrane preparations are usually obtained by homogenization of brain tissue in a hypotonic buffer (Tris HCl, KH_2PO_4, or Tris citrate, pH 7.1–7.4) or occasionally in physiological buffers, followed by a single centrifugation ($48{,}000 \times g \times 10$ min) which causes sedimentation of virtually all membraneous elements. A crude unwashed membrane preparation is made by resuspension of the pellet so obtained. This preparation can be used directly for binding assays (see Fig. 2).

It is important to realize that the crude unwashed preparation contains endogenous brain GABA which may interfere with benzodiazepine binding (see Section 8.2.1). The problem of unknown GABA contamination can be overcome by adding supramaximal concentrations of exogenous GABA or a GABA agonist such as muscimol to ensure that the receptors are maximally stimulated (Karobath *et al.*, 1980).

In the third method of tissue preparation, endogenous GABA and other interfering solutes are removed by extensive washing of the membranes. Washing by homogenization and recentrifugation at least five times with fresh buffer appears necessary to remove endogenous GABA (see the legend to Fig. 9). Washed membranes are used by most investigators. The nature of the buffer used, the assay temperature, and the number of freezing/thawings seems important parameters.

Receptor binding assays are conducted by incubating tissue in the presence of radioligand until equilibrium is reached, usually after 20–60 min at 0°C, but faster at higher temperatures. Low incubation temperature gives higher binding and an improved ratio of specific/nonspecific binding. However, high incubation temperatures (30–37°C) may unmask effects that are not evident at 0°C (Supavilai and Karobath, 1980*c*; Karobath *et al.*, 1981*a*).

Separation of free and bound radioactivity can be achieved by filtration through glass-fiber filters. Brain membranes are retained on these filters in a saturable way, probably by adsorption. The capacity of, for example, Whatman GF/C glass-fiber filters is ca. 7.5 mg of crude tissue (ca. 750 μg protein per filter). A few authors prefer centrifugation assays to filtration assays. The radioligands for BZ receptors are lipophilic and may adsorb onto plastic centrifugation tubes, giving rise to excessive nonspecific binding. Published values for affinity constants tend to be higher for centrifugation assays as compared to values obtained by filtration assays (Damm *et al.*, 1978).

BZ receptors are quite stable proteins; tissue can be stored for months without loss of receptors. Stability of receptor preparation is dependent on the medium; for example, in 50 mM Tris HCl at pH 7.1 receptors are lost within a few minutes at 60°C, whereas in 25 mM KH_2PO_4 stability increases 5- to 10-fold (Squires *et al.*, 1979*a*; Braestrup *et al.*, 1980*b*).

2.1.3. Irreversible Binding

a. Photoaffinity Labeling. [^{3}H]-FNM binding to BZ receptors is fully and readily reversible under normal binding conditions. However, when flunitrazepam is illuminated with ultraviolet light (Multiband, 250–360 nm) in the presence of brain tissue, [^{3}H]-FNM will react with tissue components and form covalent bonds (Battersby *et al.*, 1979; Moehler *et al.*, 1980). The mechanism by which flunitrazepam is photoactivated has not been elucidated, although the nitro group in the 7 position of FNM is probably involved; two additional 7-nitro benzodiazepines (nitrazepam and clonazepam) also cause an irreversible block of BZ receptors upon UV exposure. This is not true of three other benzodiazepines without nitro groups (diazepam, lorazepam, and Ro 5-3027) (Johnson and Yamamura, 1979). The covalent incorporation of tritium into brain proteins upon UV irradiation of receptor-bound [^{3}H]-FNM is inhibited by pharmacologically active benzodiazepines with potencies corresponding to their affinity for BZ receptors. The K_D value for photoaffinity labeling with [^{3}H]-FNM is 1.4 nM, which is in close agreement with the affinity constant for reversible binding of [^{3}H]-FNM. The occurrence of specific photoaffinity labeling in kidney membranes has not been described.

The stoichiometry of [^{3}H]-FNM photoaffinity labeling is anomalous. It has not been possible, even by prolonged UV exposure, to reduce the number of remaining binding sites by more than 90%, indicating that not all receptors are available for photolabeling. Furthermore, when [^{3}H]-FNM is used as the photoaffinity label, the amount of specifically incorporated radioactivity is less than the corresponding loss of binding sites. UV exposure in the absence of FNM does not change BZ receptor binding.

These findings indicate either that one molecule of irreversibly attached flunitrazepam can "cover" more than one receptor site, which would require the presence of receptor sites close to each other, or that irreversible binding of FNM to one site transforms adjacent sites into a particular conformation which has very low affinity for benzodiazepines (Karobath and Supavilai, 1982). Solubilization in sodium dodecyl sulfate (SDS) under reducing conditions followed by polyacrylamide electrophoresis reveal a molecular weight for the [^{3}H]-FNM binding proteins of 50,000–55,000 daltons (Sieghart and Karobath, 1980; Moehler *et al.*, 1980; Tallman *et al.*, 1981; Braestrup *et al.*, 1981).

b. Irazepine and Kenazepine. Irazepine and kenazepine (see Table 9) are benzodiazepine derivatives which, under certain conditions, show noncompetitive inhibition of BZ receptors (Rice *et al.*, 1979; Williams *et al.*, 1980). Irazepine is an isothicyanobenzodiazepine which can react with nucleophilic reagents such as diethylamine and methylmercaptoacetate. Apparent irreversibility of binding was described only for a limited time span (20 min).

2.2. Nonbenzodiazepine Radioligands

Recently, radioligands have been described which bind to BZ receptors without being derivatives of benzodiazepines. Some of these ligands, [^{3}H]-PrCC (see Section 2.2.1) and [^{3}H]-CL 218.872 (Yamamura *et al.*, 1981), may interact selectively with BZ receptor subclasses.

2.2.1. *[^{3}H]Propyl-β-carboline-3-carboxylate*

β-Carboline-3-carboxylic acid esters which act as inhibitors of BZ receptors were discovered during a search for endogenous ligands (see Section 11.1). One member of this class of compounds, propyl-β-carboline-3-carboxylate, which possesses high affinity for BZ receptors, has been tritiated to high specific activity, ca. 105 Ci/mmol, and applied as a radioligand for BZ receptors. [^{3}H]Propyl-β-carboline-3-carboxylate ([^{3}H]-PrCC) binds to the BZ_1 subclass of benzodiazepine receptors (Section 7) with high affinity; the affinity constant determined by saturation analysis is ca. 0.9 nM; the number of [^{3}H]-PrCC binding sites in the rat cerebellum, but not in other brain regions (see Fig. 2), is equal to the number of [^{3}H]-FNM binding sites; the binding of each of these radioligands is exclusive in the sense that binding of one ligand excludes the subsequent binding of the other (Nielsen *et al.*, 1981).

Specific binding of [^{3}H]-PrCC to rat forebrain membranes is inhibited only by agents that also inhibit the binding of [^{3}H]-FNM and [^{3}H]diazepam

(Nielsen *et al.*, 1981). [^{3}H]-PrCC apparently binds to the benzodiazepine recognition site of BZ_1 receptors.

Technically [^{3}H]-PrCC is inferior to [^{3}H]-FNM as a radioligand because nonspecific binding is higher (13–40% of total binding) for [^{3}H]-PrCC than for [^{3}H]-FNM (3–9%); [^{3}H]-PrCC is less stable in aqueous solutions, and the "on-rates" (k_{+1}) and the "off-rates" (k_{-1}) are very high (Table 1). [^{3}H]Methyl-β-carboline-3-carboxylate (Braestrup and Nielsen, 1981*c*) and [^{3}H]ethyl-β-carboline-3-carboxylate (Marangos and Patel, 1981) likewise bind to brain BZ receptors; their subclass selectivity and GABA sensitivity are not identical to that of [^{3}H]-PrCC.

2.2.2. [^{3}H]-CL 218.872

[^{3}H]-CL 218.872 may exhibit binding to benzodiazepine receptor subclasses. In addition, binding is not chloride dependent. Nonspecific binding represents the major part of total binding (Yamamura *et al.*, 1982; M. Kuhar, unpublished results).

2.2.3. [^{3}H]-CGS 8216

The potent benzodiazepine receptor antagonist CGS 8216 binds to benzodiazepine receptors. [^{3}H]-CGS 8216 has high affinity for BZ receptors ($K_D \cong 0.1$ nM), and nonspecific binding is low. The rate constants are remarkably low as compared to other ligands. Under some conditions interaction between [^{3}H]-CGS 8216 and benzodiazepines is not fully competitive (Czernik *et al.*, 1982).

2.3. Binding Thermodynamics

Thermodymanic differences between the binding of agonists and antagonists to receptor sites may provide new insights into the molecular consequences of receptor–ligand interaction (Weiland *et al.*, 1979). Binding of [^{3}H]diazepam and [^{3}H]-FNM is highly temperature dependent, decreasing at higher temperatures. Binding of [^{3}H]diazepam is more temperature dependent than that of [^{3}H]-FNM binding (Braestrup and Squires, 1978*a*). Due to technical difficulties, accurate thermodynamic studies of [^{3}H]-FNM binding are not easily conducted by conventional binding assay techniques. Although the reported thermal dependence of the K_D values for FNM binding varies slightly in published reports, it is generally agreed that binding of FNM to BZ receptors at temperatures 0–20°C is driven by a combination of enthalpy ($\Delta H \simeq -5$ to -10 kcal/mol) and entropy (ΔS = 5–20 cal/mol/deg), whereas binding at higher temperatures (20–35°C) is mainly by enthalpy.

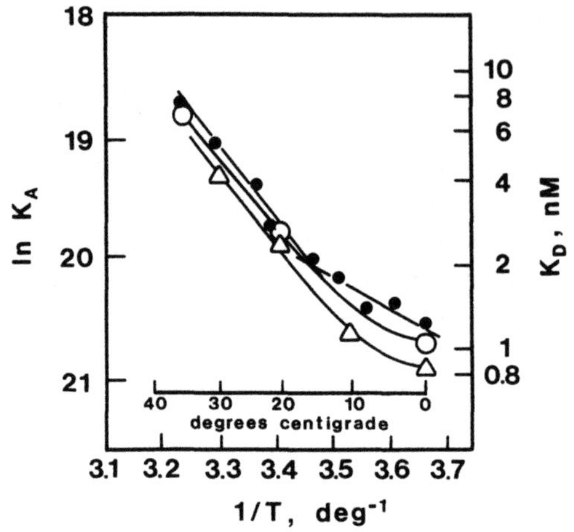

FIG. 3. Van't Hoff plot of K_D for [^{3}H]-FNM (●, from Speth *et al.*, 1979*a*; ○, from Braestrup and Squires, 1978*a*) and [^{3}H]-PrCC (△, unpublished results, method according to Nielsen *et al.*, 1981). Binding of both radioligands to rat brain exhibits a similar temperature dependence, the affinity being higher at low temperatures.

The binding of [^{3}H]-PrCC is likewise highly temperature dependent. The contributions of ΔH and ΔS to [^{3}H]-PrCC binding do not differ from the values for [^{3}H]-FNM (Fig. 3). This was unexpected, because some β-carboline-3-carboxylates and benzodiazepines can be viewed as receptor antagonists and agonists respectively (see Section 8.6). It has been reported that β-receptor agonists and antagonists differ markedly in their thermodynamic binding parameters (Weiland *et al.*, 1979).

The ΔH component of [^{3}H]-FNM and [^{3}H]-PrCC binding may indicate their firm attachment to the recognition site. Several factors may contribute to the small positive entropy of binding. Removal of an orderly arranged layer of water molecules around the ligand (and the receptor) tends to increase ΔS to higher positive values. Conformational changes in the receptor as a consequence of binding, however, may produce a more ordered situation accompanied by a negative entropy change. As the temperature increases, ΔS approaches zero probably because of a combination of these contingencies.

3. BZ RECEPTOR SOLUBILIZATION

To clarify the interaction among BZ receptors, GABA receptors, and ionophores, and to characterize further the benzodiazepine binding proteins, solubilization and purification of BZ receptors have been undertaken. These studies, which are still preliminary, have shown that BZ receptors can be solubilized from brain membranes by several ionic and nonionic detergents (see Table 2).

TABLE 2
Characteristics of Solubilized Rat or Calf Brain Benzodiazepine Receptors

Detergent	Assay	Nonspecific binding (% ot total)	[^{3}H]FNM binding, K_D (nM)	Solubilization (%)	Apparent MW (daltons)	Enhancement by GABA[f] (%)	Appropriate pharmacology	References[i]
0.2% DOC[h] + 1 M KCl	Amberlite XAD-2	15	1.8 ± 0.3	30–35	ca. 200,000[a]	23	Yes	1
0.7% Triton X-100[c]	—	—	[g]	60	115,000[b]	[g]	[g]	2
0.2–2% DOC[h]	—	—	1.7 ± 0.06	—	—	0<	Yes	3
0.7% Triton X-100[d]	Whatman DE 81 filter	10	12 ± 0.4	60–70	ca. 250,000[a]	—	—	5
1% digitonin	$(NH_4)_2SO_4$ precipitate	12	1.5	50	—	75	Yes	4
2.3% SDS[c]	—	—	[g]	—	49,500	[g]	[g]	6
0.3 DOC[h] + 0.5 M KCl	$(NH_4)_2SO_4$ precipitate	4	1.5 ± 0.2	30–40	ca. 200,000[a]	30	Yes	7
0.5% Lubrol-Px	PEG-6000[h] precipitate	16	[^{3}H]diazepam	60	220,000[e]	<10	Yes	8

[a] Ultracentrifugation in a sucrose gradient, sedimentation coefficients from 10–12.9S.
[b] Gelfiltration on Ultrogel AcA 34.
[c] [^{3}H]-FNM photoaffinity-labeled prior to solubilization.
[d] EDTA, leupeptin, pepstatin, bacitracin, benzamidine, and PMSF was added to inhibit proteases.
[e] Gelfiltration on Sephadex G-200.
[f] GABA agonists were added in a supramaximal concentration to assays and the percentage increase of [^{3}H]diazapam or [^{3}H]flunitrazepam binding was determined.
[g] Cannot be evaluated.
[h] DOC is deoxycholate; PEG is polyethylene glycol.
[i] References: (1) Asano and Ogasawara, 1980; (2) Braestrup *et al.*, 1981; (3) Bymaster and Wong, 1980; (4) Gavish *et al.*, 1979; Gavish and Snyder, 1980*a*; (5) Lang *et al.*, 1979; (6) Moehler *et al.*, 1980; (7) Sherman-Gold and Dudai, 1980; (8) Yousufi *et al.*, 1979.

There is no general agreement concerning choice of detergent; differences in assay principles may be an important factor (see Table 2). Nonionic detergents yield high blank values when ammonium sulfate precipitation is used for the separation of free and receptor bound [^{3}H]-FNM (Sherman-Gold and Dudai, 1980). Sodium dodecylsulfate, although a very efficient solubilization agent, denatures proteins and can only be used when receptors are prelabeled with [^{3}H]-FNM (or when receptor antibodies become available). Other means of separating free from bound radioligand include first, Amberlite XAD-2, a polystyrene resin, which adsorbs unbound [^{3}H]-FNM while receptor-bound [^{3}H]-FNM freely passes through; second, Whatman DE 81 filters (DEAE cellulose) which retain solubilized acidic BZ receptors but not free [^{3}H]-FNM; and third, polyethylene glycol 6000 (PEG)/γ-globulin which denatures solubilized BZ receptors, so that they are retained on GF/B glass-fiber filters. The latter assay technique is very useful.

Solubilized receptors were retained on a concanavalin 4-B column, indicating the presence of sugar moieties (glucosyl or mannosyl) (Fong and Goldstein, 1980). The isoelectric point is quite low, pI = 5.5, indicating that BZ receptors or attached entities carry acidic groupings (Tallman *et al.*, 1981). Different values for the apparent molecular weight of solubilized BZ receptors have been reported (Table 2). It remains unknown whether this variability represents the presence of dimer or tetramer complexes of the 50,000-dalton receptor subunits, whether nonreceptor proteins are cosolubilized in aggregates with BZ receptors, or whether other receptors, for example GABA receptors and chloride channels, are solubilized in supramolecular complexes with BZ receptors. Nonionic detergents such as Triton X-100 produce only slight dissociation of protein/protein complexes existing in the membrane, and such complexes are, therefore, retained as single units in solution (Bjerrum, 1977). However, a molecular weight for GABA receptors of 800,000 daltons has been reported (Greenlee and Olsen, 1979). Target size analyses using irradiation inactivation of membrane receptors yields molecular weights of 215,000, 90,000, and 57,000 daltons, respectively, depending on the assay conditions (Chang *et al.*, 1981; Doble and Iversen, 1982; Paul *et al.*, 1981*b*). Saturation experiments using [^{3}H]-FNM or [^{3}H]diazepam do not indicate the presence of more than a single class of binding sites after solubilization. The affinity constants for solubilized receptors are similar but slightly higher than for membrane-bound receptors. The appropriate pharmacology of the receptors is retained when solubilization is carried out. These findings indicate that the BZ receptor apparently retains its fundamental properties when water-soluble detergents are substituted for the lipids of the cell membrane. Martini *et al.* (1982) purified the BZ receptor 5200-fold by affinity chromatography on a Sepharose 4B column coupled with the potent BZ ligand delorazepam.

The ability of GABA agonists to stimulate BZ receptors is retained to a varying degree upon solubilization. Full enhancement, comparable to that observed for membrane-bound receptors, has been obtained under some experimental conditions (Gavish and Snyder, 1980*a*); partial enhancement, comparable to that observed in heat-inactivated membranes (Braestrup *et al.*, 1980*b*), or no enhancement has been reported under other experimental conditions (for references, see Table 2). Apparently freezing/thawing of membranes before the solubilization procedure improve GABA enhancement of [^{3}H]diazepam binding to solubilized receptors (unpublished results).

4. OCCURRENCE OF BZ RECEPTORS

4.1. Brain versus Periphery

Benzodiazepine receptors are found in the central nervous system, including the spinal cord and retina. Initial investigations into the binding of [^{3}H]diazepam indicated that specific (displaceable) binding could be obtained not only in brain tissue, but also in membranes derived from liver, kidney, and other peripheral tissues (Braestrup and Squires, 1977; Gallager *et al.*, 1981). The binding sites on kidney membranes, although possessing a high affinity for [^{3}H]diazepam ($K_D \sim 40$ nM, which is only 10 times less than for brain membranes), showed fundamentally different

TABLE 3
Differential Inhibition of [^{3}H]Diazepam Binding by Clonazepam and Ro 5-4864 in Brain and Periphery[a]

	Inhibition of specific [^{3}H]diazepam binding, IC_{50} (nM)	
Tissue	Clonazepam	Ro 5-4864
Brain	5	163,000
Kidney	2900	4.7
Liver	5100	4.1
Lung	7900	4.9

[a] Tissues were homogenized with an Ultra Turrax homogenizer in 50 mM Tris-HCl, pH 7.4. The P_2 pellet was obtained, resuspended in 25 volumes of fresh buffer, and incubated in the presence of 1.6 nM [^{3}H]diazepam at 0°C. Total membrane-bound [^{3}H]diazepam was determined by filtration through Whatman GF/C glass-fiber filters. Diazepam, 3×10^{-6}, defined nonspecific binding. (From Braestrup and Squires, 1977.)

TABLE 4

Occurrence of Brain Type and Peripheral Type of Specific [^{3}H]Diazepam or [^{3}H]-FNM Binding

Tissue	Classification[a]	[^{3}H]Diazepam, K_D (nM)	Inhibition by[b] Clonazepam	Ro 5-4864	References[g]
Rat brain	B	2.6	++	—	4
Hen brain	B	3.6	++	—	14
Turtle brain	B	2.7	++	—	14
Frog brain	B	1.7	++	—	14
Codfish brain	B?	6[c]	+	—	14
Retina	B	4	++	—	2, 11, 15, 16
Neuronal cells, primary cultures	B	9.9	++	—	7, 12, 13
Nuclear membranes	?	28	+	—	3, 18
Glial cell fractions[d]	?	4	++	[f]	8, 9
Kidney membranes	P	40	—	++	4, 17, 18, 19
Astrocytes, primary cultures	P	30	—	++	5, 6, 7, 10
NB_2A Neuroblastoma[e]	P?	9	—	+	1, 18
$NB_{41}A_3$ Neuroblastoma[e]	P	20	—	++	17
C_6 Glioma[e]	P	5–12	—	++	1, 17, 18
Peritoneal mast cells	P	90	—	++	19
Serum albumin	?	2000	—	[f]	20

[a] B = Brain type; P = Peripheral type (see text).
[b] Inhibition of specific [^{3}H]diazepam or [^{3}H]-FNM binding is designated the following way: —, absent; +, present; ++, present and potent.
[c] Curvilinear Scatchard plot.
[d] Cells from fresh rat brain tissue were fractionated by differential centrifugation and other procedures.
[e] Cell lines.
[f] Not determined.
[g] References: (1) Baraldi *et al.*, 1979; (2) Borbe *et al.*, 1980; (3) Bosmann *et al.*, 1980; (4) Braestrup and Squires, 1977; (5) Braestrup *et al.*, 1978*c*; (6) Chang *et al.*, 1980; Dudai *et al.*, 1979; (8) Henn and Henke, 1978; (9) Henn *et al.*, 1980; (10) Hertz and Mukerji, 1980; (11) Howells *et al.*, 1979; (12) Huang *et al.*, 1980; (13) Mallorga *et al.*, 1980*b*; (14) Nielsen *et al.*, 1978; (15) Paul *et al.*, 1980; (16) Regan *et al.*, 1980*b*; (17) Syapin and Skolnick, 1979; (18) Tallman *et al.*, 1981; (19) Taniguchi *et al.*, 1980; (20) Wong and Sellers, 1979.

pharmacological specificity from the brain. The differences are most marked for two benzodiazepine derivatives, clonazepam and Ro 5-4864 (see Table 9). Clonazepam is highly pharmacologically active; it inhibits brain [^{3}H]diazepam binding at low concentrations but does not inhibit kidney binding even at high concentrations. On the other hand, Ro 5-4864 is a pharmacologically inactive agent; it does not inhibit [^{3}H]diazepam binding to brain receptors at low concentrations, but it inhibits kidney [^{3}H]diazepam binding at nanomolar concentrations (Table 3). The weak activity of Ro 5-4864 on brain BZ receptors is particularly striking, because its structure is quite similar to diazepam, differing from diazepam only by possessing a *p*-chloro substituent in ring C.

The selectivity of clonazepam and Ro 5-4864 for CNS and peripheral receptors, respectively, makes them very useful agents for distinguishing brain receptors from peripheral binding sites. Table 4 illustrates the classification of benzodiazepine binding sites using these two agents. Note that the brain receptor can be subdivided into at least two subclasses, BZ_1 and BZ_2 (see Section 7). The presence of the peripheral type of [^{3}H]-diazepam binding sites has also been demonstrated in several tissue cultures of neuronal origin (NB_2A and $NB_{41}A_3$ neuroblastoma cell lines), whereas the brain type was present in primary neuron cultures. Well differentiated primary astrocyte cultures and primary peritoneal mast cell cultures exhibit specific [^{3}H]diazepam binding of the peripheral type. Specific binding of [^{3}H]diazepam to membranes derived from nuclei of brain cells has not been fully classified (Bosmann *et al.*, 1980). The affinity constant for the nuclear sites is above the value obtained for brain receptors (Table 4). On the other hand, when differentiation was attempted using clonazepam/Ro 5-4864, the nuclear binding site showed properties reminiscent of brain receptors (Tallman *et al.*, 1981).

The peripheral type of benzodiazepine binding is probably not involved in known pharmacological and clinical effects of benzodiazepines. Interestingly, however, Strittmatter *et al.* (1979) reported that occupancy of kidney-type benzodiazepine binding sites by Ro 5-4864, and to a lesser extent by diazepam, increased membrane phospholipid methylation. Phospholipid methylation changes the fluidity of membrane lipids, but the physiological consequence is not clear.

[^{3}H]-Ro 5-4864 is a useful radioligand for detecting the "peripheral kind" of [^{3}H]diazepam binding sites (note the "peripheral sites" are present in the brain, for example, on glial cells) (Schoemaker *et al.*, 1981).

4.2. Neuronal Localization

It is now generally agreed that the vast majority of brain BZ receptors are localized on neurons.

4.2.1. Lesion Studies

Autoradiographic histological examinations (Section 4.3) did not achieve the desired resolution for demonstrating a neuronal localization, and even electron-microscopic examination of autoradiograms of covalently bound [^{3}H]-FNM is inherently uncertain, although a neuronal localization was indicated. Antibodies for BZ receptors are not available for immunohistochemical studies.

One of the most straightforward biochemical approaches to demonstrate a neuronal localization is to investigate animals having selective degenerations of CNS neurons. Studies with neurotoxic agents clearly showed a neuronal BZ receptor localization when it became clear that a certain degeneration period or a certain administration-paradigm was necessary before membrane-bound BZ receptors disappeared from brains of kainic acid-lesioned animals (Sperk and Schloegl, 1979) (Table 5). Furthermore, it became clear that restricted neurotoxic procedures affecting only selected neuron types or pathways never produced a substantial reduction in the number of receptors. BZ receptors apparently reside on many different neurons, but not necessarily on all neurons. For example, cerebellar granule cells may be devoid of BZ receptors (see Section 4.2.2).

Kainic acid is generally believed to destroy glutamate-innervated neuron cell bodies selectively (McGeer *et al.*, 1978). Cerebellar granule cells are themselves glutamatergic, having no known glutamate input. The selectivity of kainic acid as a neurotoxic agent is low, however, after administration of massive doses, such as those indicated in Table 5, and it is not possible to make conclusive statements regarding the detailed localization of BZ receptors in the cerebellum based upon kainic acid treatments alone. The results obtained, however, agree closely with experiments done on primary astrocyte cultures (Table 4) showing that BZ receptors [and high-affinity GABA receptors (Ossola *et al.*, 1980)] are not located to any major extent on glial cells (see, however, Henn and Henke, 1978, and Henn *et al.*, 1980). Investigations on brain microvessels also failed to indicate a major vascular localization of BZ receptors (Peroutka *et al.*, 1980).

The loss of BZ receptors in rat retina reported after treatment with monosodium glutamate or kainic acid (Skolnick *et al.*, 1980*c*) suggests that BZ receptors are localized on the neuronal elements of the inner plexiform layer and/or ganglion cells in the retina. Mimaki *et al.* (1980) used daily administration of diphenylhydantoin at high doses (200 mg/kg) as a neurotoxic agent to evoke Purkinje cell degeneration in rats and found a 25–48% reduction in the number of cerebellar but not hippocampal BZ receptors after 14–28 days of treatment. Fuxe *et al.* (1981) used the glutamate-related neurotoxic agent, ibotenic acid, injected directly into the

TABLE 5
Evidence for a Neuronal Localization of BZ receptors[a]

Neuronal lesion, origin of[b]	Brain region investigated	Days after lesion	BZ-receptors (% of control)	References[g]
Kainic acid, 2–2.5 μg	Corpus striatum	35–45	40–55	5, 12
Kainic acid, 0.75–1 μg	Substantia nigra	25–28	40–59	1, 6
Kainic acid, 2–4 μg	Cerebellum	9–100	35–76[c]	3, 5, 15
Ibotenic acid, 10 μg	Dorsal hippocampus	5	70	7
Diphenylhydantoin[d]	Cerebellum	—	51[c]	14
Diphenylhydantoin[d]	Cerebral cortex	—	66[c]	14
Huntington's chorea	Putamen		45	8, 10
Mouse mutants				
Weaver (wv/wv)[f]	Cerebellum	20–70[e]	125–143	3, 5, 13
Nervous (nr/nr)	Cerebellum	53–70[e]	60–80[c]	3, 5, 11, 13, 16
Purkinje-cell deficient (pcd/pcd)	Cerebellum	30–45[e]	84	4
Staggerer (sg/sg)	Cerebellum	20[e]	51–54	13, 14

[a] Specific binding of [^{3}H]-FNM or [^{3}H]diazepam was determined in lesioned tissue and expressed in percentage of values obtained in appropriate control tissue. The percentages presented are different from control at $p < 0.05$–0.001 in all cases.
[b] No significant changes of benzodiazepine receptors were reported after the following procedures, known to cause selective cell losses in rats: 3-Acetylpyridine (60–75 mg/kg i.p.), evaluated in cerebellum on day 50 (refs. 2, 5, 14); X-ray (200 r) on days 0, 1, 3, and 5 after birth, evaluated in hippocampus on day 14 (ref. 2); hemisection of medial forebrain bundle, evaluated in corpus striatum on days 7–8 (refs. 2, 5); 6-OH-dopamine in corpus striatum, evaluated in corpus striatum on day 8 (refs. 5, 9).
[c] Reduced B_{max} values.
[d] Diphenylhydantoin (200 mg/kg i.p.) was administered once daily for 28 days. Animals were sacrificed 6 hr following the last dose.
[e] Age in days after birth.
[f] Included for comparison.
[g] References: (1) Braestrup and Squires, 1978*a*; (2) Braestrup *et al.*, 1978*a*; (3) Braestrup *et al.*, 1979*c*; (4) C. Braestrup, unpublished results; (5) Chang *et al.*, 1980; (6) Costa *et al.*, 1979*a*; (7) Fuxe *et al.*, 1981; (8) Moehler *et al.*, 1978*a*; (9) Reisine *et al.*, 1979*a*; (10) Reisine *et al.*, 1979*b*; (11) Skolnick *et al.*, 1979*a*; (12) Sperk and Schloegl, 1979; (13) Speth and Yamamura, 1979; (14) Speth *et al.*, 1981; (15) Biggio *et al.*, 1980; (16) Lippa *et al.*, 1978*b*.

rat hippocampus, to show a 30% decrease in BZ receptors concomitantly with degeneration of intrinsic neurons.

4.2.2. *Mouse Mutants*

In an attempt to determine the cellular distribution of BZ receptors, [^{3}H]diazepam and [^{3}H]flunitrazepam binding were studied in the cerebellum of mutant mice with various neurological disorders (Table 5).

The number of BZ receptors is slightly lower in the cerebellum of the "nervous" mouse. This mutant undergoes a spontaneous degeneration of cerebellar Purkinje cells which is totally developed 50–60 days after birth (Landis, 1973). The "nervous" phenotype is only expressed in homozygote (genotype nr/nr) animals; phenotypic normal animals (genotype nr/+) serve as controls. The Purkinje cell-deficient mouse (pcd/pcd) undergoes

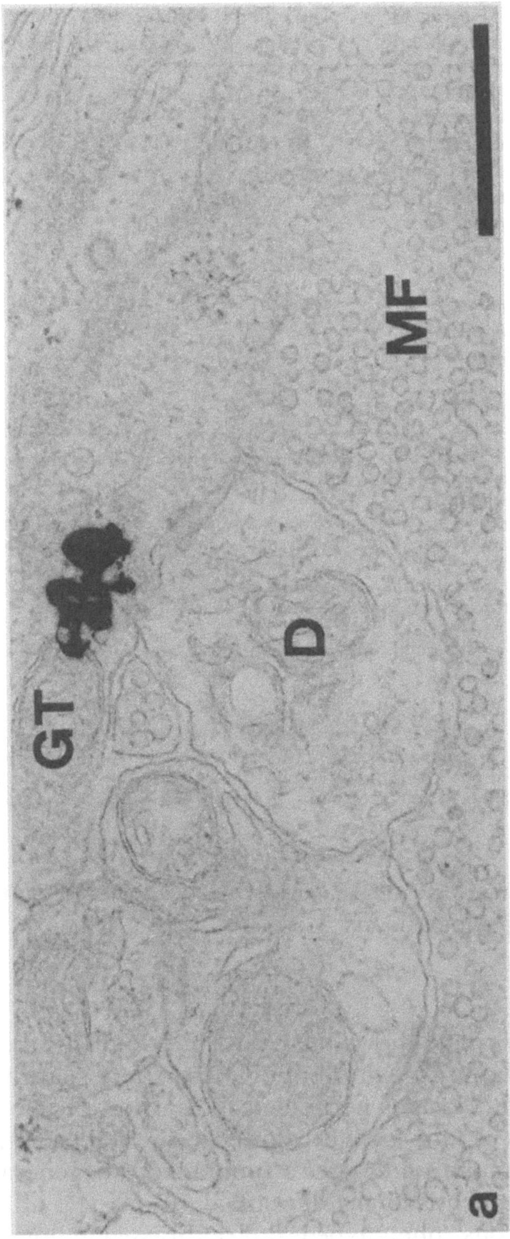

FIG. 4. Electron-microscopic autoradiographs of rat cerebellar cortex slices were obtained after intravenous injection of 170 μCi/kg [^{3}H]-FNM followed 2 min later by perfusion with glutaraldehyde/formaldehyde fixative and photolabeling (a). The silver grains are probably located in a Golgi nerve terminal (GT) because it is adjacent to a dendrite (D) which is contacted by a mossy fiber (MF). (b) The photolabeled nerve ending shows an immunocytochemical reaction with glutamic acid decarboxylase (GAD) antiserum using the peroxidase–antiperoxidase method. Magnification × 56,000, bar 0.5 μm. (From Moehler *et al.*, 1981.)

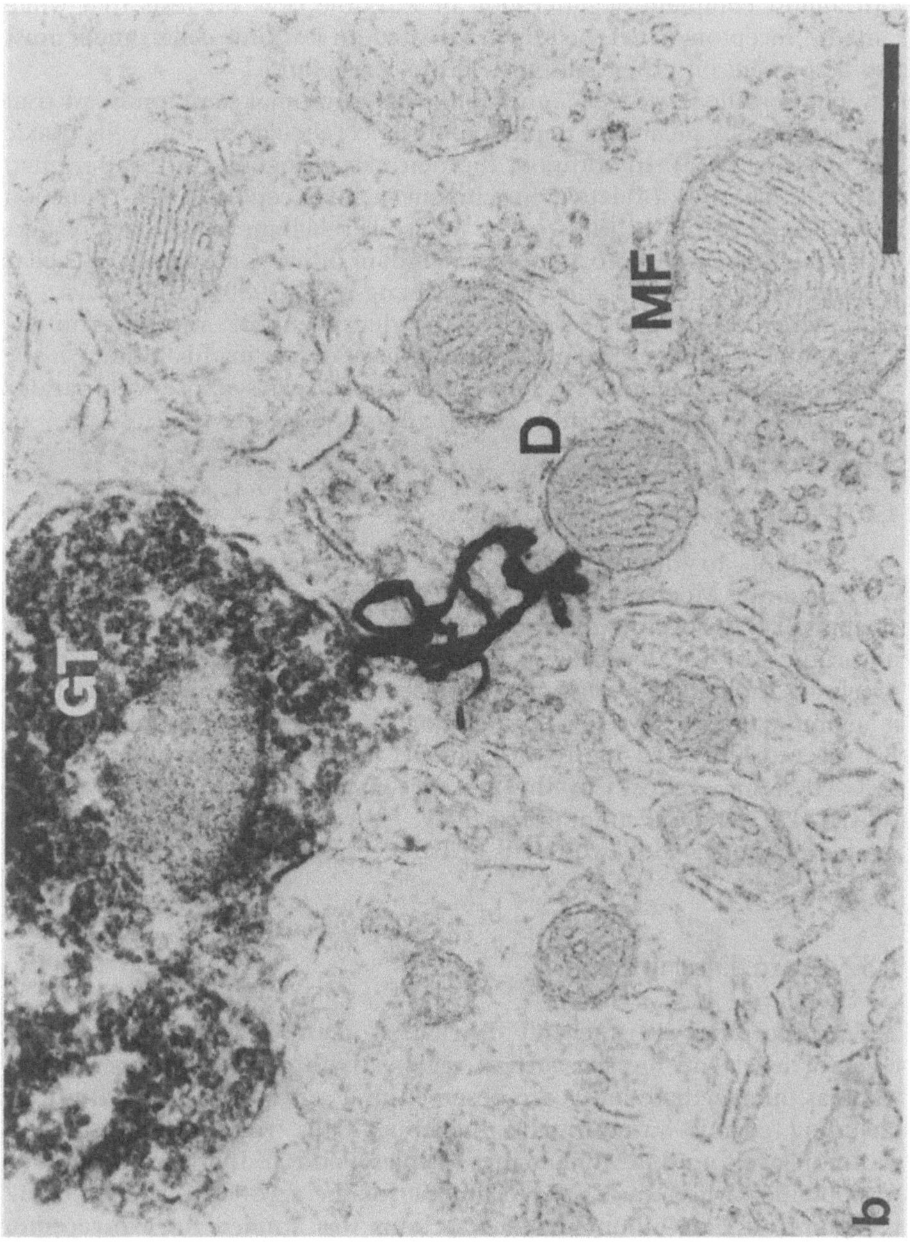
GT
D
MF
b

progressive cerebellar Purkinje cell degeneration 15–30 days after birth (Mullen *et al.*, 1976). The partial reduction in BZ receptors in animals with almost complete degeneration of Purkinje cells suggests that while some BZ receptors in cerebellum are located on Purkinje cells, others must also be present on other cell types in the cerebellum.

The so-called "weaver" mutation is an autosomal semidominant trait characterized by near total depletion of the cerebellar granule cells (Rakic and Sidman, 1973). In addition, there are abnormalities in the Bergman glial cells. There is a relative enrichment in BZ receptors but a reduction in GABA receptors in the weaver mouse cerebellum at 20 days of age, indicating that BZ receptors are less abundant or even absent on cerebellar granule cells (Braestrup *et al.*, 1979*c*; Chang *et al.*, 1980).

There is a sizable decrease in the density of BZ receptors in the cerebellum of "staggerer" mice. The primary defect in this mutant is the absence of dendritic spines, where the Purkinje cells synapse with parallel fibers arising from the granule cells (Landis and Reese, 1977).

4.2.3. EM Autoradiography

Attempts to localize BZ receptors at the electron-microscopic level were made by Moehler *et al.* (1980, 1981) using [^{3}H]-FNM photoaffinity labeling of BZ receptors in brain slices (see Fig. 4). Although the inherent uncertainty of the localization of silver grains on EM autoradiograms precludes a detailed description, it is clear from these studies that there is an abundance of silver grains in regions of synaptic contact. Some of the BZ receptor-enriched neuronal elements have been positively identified as GABAergic synapses by the presence of glutamic acid decarboxylase (GAD) (EC 4.1.1.15) immunoreactivity. However, GAD was not demonstrated in contact with two-thirds of BZ receptor-bound [^{3}H]-FNM.

4.3. Brain Distribution

BZ receptors are unevenly distributed in the brain. In general, the receptor levels are high in cortical regions, both cerebral and cerebellar, whereas intermediate levels are observed in a series of subcortical nuclei, including corpus striatum and thalamus. Only very low densities of receptors are found in white matter (Tables 6 and 7). Increased resolution was obtained when an *in vitro* technique for the reversible labeling of BZ receptors on slide-mounted tissue sections was applied for BZ receptor autoradiography at the light-microscope level (Fig. 5). This technique revealed small but significant variations in the regional distribution of BZ sites from one species to another. For example, in both rat and man there is a high density of BZ receptors in the molecular layer of the cerebellum;

TABLE 6
Benzodiazepine Receptor Binding in Regions of Human Brain[a]

Brain region	Specific [^{3}H]diazepam binding: B_{max} (pmol/g protein)	Specific [^{3}H]diazepam binding: K_D (nM)
Frontal lobe cortex	1030	5.3
Occipital lobe cortex	1020	6.2
Temporal lobe cortex	890	6.0
Cerebellar vermis	740	6.9
Cerebellar hemisphere cortex	655	5.0
Hippocampus	640	5.4
Amygdala	615	7.9
Hypothalamus	485	7.8
Nucleus caudatus	410	6.3
Thalamus	370	6.7
Putamen + globus Pallidus	320	5.7
Substantia nigra	290	11
Tegmentum	180	10
Medulla oblongata	175	17
Nucleus dentatus	160	7.9
Olive	160	14
Pons	160	9.6
Corpus callosum	80	13

[a] Values represent the means of data from Braestrup *et al.* (1977) and Moehler and Okada (1978). [^{3}H]Diazepam saturation analyses were done on crude membrane preparations from tissue, which was dissected and frozen 17–72 hr post mortem. Similar studies in man (Speth *et al.*, 1978) and rat (Braestrup and Squires, 1977; Moehler and Okada, 1977*b*; Mackerer *et al.*, 1978) have revealed similar profiles throughout the brain. The regional distribution of BZ_1 versus BZ_2 receptors is not available for human brain.

but compared with that of the human, the granule cell layer of the rat cerebellum has a relatively reduced concentration of BZ receptors (Young and Kuhar, 1979). Furthermore, it appeared that in certain regions, such as lamina II and X of the rat spinal cord, the retina, Island of Calleja, and parts of the limbic system, high BZ receptor concentrations may be present in restricted areas.

An important achievement in studies of the distribution of BZ receptors in the living brain was recently achieved by positron emission tomography (PET) analysis of [^{11}C]-FNM binding in the brain of the living baboon (Comar *et al.*, 1979; Mazière *et al.*, 1981). This highly sophisticated approach still possesses only limited spatial resolution (see Fig. 6) but shows promise for future clinical investigations of brain receptors in disease and in various environmental conditions.

One must be very cautious in inferring a causal connection between

TABLE 7
Prevalence of BZ_1 Receptors in Rat Brain[a]

Brain region	BZ_1 (% of total BZ receptor population)	Specific [^{3}H]-FNM binding[b] (pmol/g)
Cerebellum	100 (definition)	25 ± 2
Medial cortex	84 ± 2(4)[c]	47 ± 7
Pons	81 ± 3(5)	21 ± 2
Occipital cortex	81 ± 1(4)	48 ± 7
Thalamus	78, 81 (2)	21, 24
Frontal cortex	78 ± 1(6)	51 ± 5
Bulbus olfactorius	78(1)	66
Corpus striatum	62 ± 3(4)	25 ± 3
Hippocampus	59 ± 2(5)	41 ± 4
Nucleus accumbens	57 ± 2(4)	28 ± 4
Hypothalamus	56, 59(2)	23
Forebrain, whole	69 ± 3(3)	45 ± 4

[a] Rat brains were dissected according to Glowinski and Iversen (1966) and homogenized in 10 ml 25 mM KH_2PO_4, pH 7.1. The homogenate was centrifuged at 48,000 × *g* for 10 min and resuspended in 500 volumes of fresh buffer. Specific binding of [^{3}H]-PrCC (0.28 nM) and [^{3}H]-FNM (1.0 nM) was determined according to Braestrup and Nielsen (1981*b*); the percentage of BZ_1 receptors was calculated as the proportion of specific binding of [^{3}H]-PrCC to [^{3}H]-FNM multiplied by the cerebellar proportion of [^{3}H]-FNM to [^{3}H]-PrCC binding.
[b] Specific binding of [^{3}H]-FNM at 1.0 nM. This value is proportional to B_{max} for the total BZ receptor population; provided that K_D is constant throughout the brain.
[c] Mean ± S.E.M. of (*N*) samples investigated in duplicate assays.

the distribution of drug receptors in brain and clinical effects. However, it is tempting to speculate that BZ receptors in the limbic system and part of the cerebral cortex may be involved in mediating the anxiolytic actions of benzodiazepines (Young and Kuhar, 1980). The complex limbic system, which involves hippocampus, amygdala, and parts of the frontal cortex and associated areas (Papez, 1937), has been implicated as the anatomical substrate for emotion, and benzodiazepines do suppress activity in this system. Benzodiazepines efficiently prevent the spread of electric activity in the cortex associated with seizures, and it is possible that the anticonvulsant effect is mediated via cortical BZ receptors. Ataxia and motor incoordination may be related to cerebellar BZ receptors, and part of the muscle relaxant effect may be mediated via BZ receptors in the spinal cord dorsal horn.

Differences in BZ receptors exist among certain strains of rats and mice (Robertson, 1980*a*; Robertson *et al.*, 1978); for example, "emotional" mice had slightly but significantly lower BZ receptor binding in their brains when compared to three other "nonemotional" strains (Robertson, 1979).

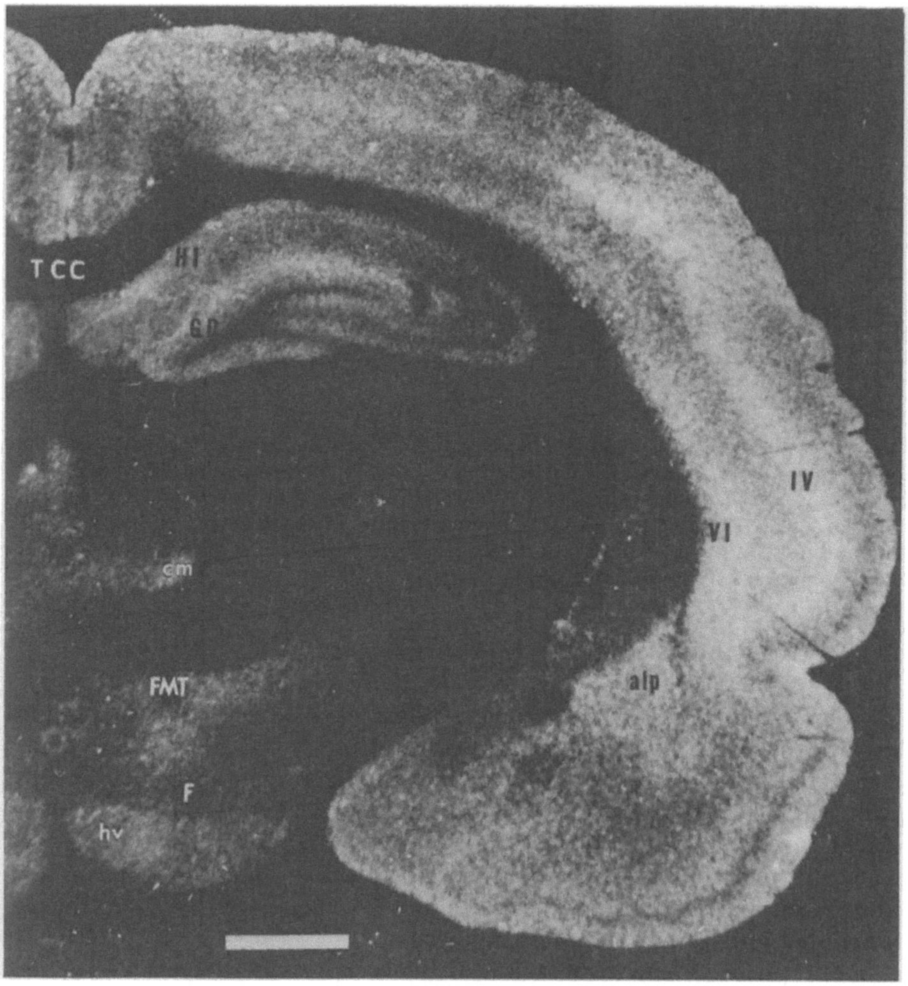

FIG. 5. Dark-field photomicrograph of BZ receptors in rat brain at A 3990 μm (König and Klippel, 1963). Note the particularly high amounts of BZ receptors in Lamina IV and VI of the cortex, as well as in the nucleus amygdaloideus lateralis (alp) and parts of the hippocampal formation (GD, gyrus dentatus; HI, hippocampus). cm, nucleus center medium; F, columna fornicis; FMT, fasciculus mamillothalamicus; hv, nucleus ventromedialis; Tcc, truncus corporis callosi. Tissue slices were incubated with 1 nM [^{3}H]-FNM at 0°C *in vitro* for 2 min and processed for autoradiography. Nonspecific binding was only ca. 2%. Bar, 1000 μm. (From Young and Kuhar, 1980.)

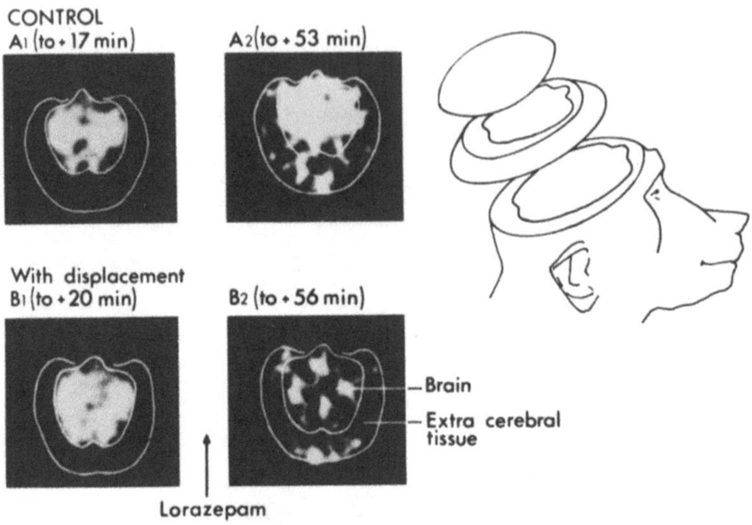

FIG. 6. Visualization of BZ receptors in the brain of the living baboon by positron emission tomography (PET) of [^{11}C]-FNM binding. A_1 and A_2 were recorded 17 and 53 min respectively, after i.v. administration of [^{11}C]-FNM. Bright regions have high ^{11}C concentrations. B_1 and B_2, 5 mg lorazepam was administered to a baboon 29 min after [^{11}C]-FNM administration; at t = 56 min (B_2) nonspecific binding is visible; the difference between A_2 and B_2 represents specific binding. (From Mazière *et al.*, 1981.)

4.4. Ontogenesis and Phylogenesis

Ontogenetic and phylogenetic studies of BZ receptor distributions have been carried out in several laboratories. It was anticipated that a detailed knowledge of the ontogeny and phylogeny might form a basis for evaluating the physiological function of these receptors. For example, it would be interesting to determine whether BZ receptors and GABA receptors always occur in a fixed ratio, as this would indicate a firm connection between these two receptors as suggested by their mutual functional interactions.

The phylogenetic distribution of BZ receptors is remarkable because their occurrence throughout the animal kingdom is restricted to the brain of tetrapods and higher bony fishes (osteichthyes). BZ receptors have not been demonstrated in invertebrates (Nielsen *et al.*, 1978; Fernholm *et al.*, 1979) (Table 8). Specific [^{3}H]diazepam binding of the brain type was not detected in cyclostomes (for example, the hagfish) or in chondrichthyes (for example, the shark). This means that BZ receptors have a recent phylogenetic appearance, which is fundamentally different from the distribution of GABA receptors. GABA receptors, as defined by high-affinity binding of [^{3}H]-GABA and [^{3}H]muscimol, are present in the brains of cyclostomes and chondrichthyes (Mann and Enna, 1980). Electrophysiological responses to GABA in, for example, crustaceae clearly show the presence of GABAergic transmission in invertebrates (Takeuchi and Takeuchi, 1975). GABAergic transmission is not potentiated by chlordiazepoxide in invertebrates (Matthews and Wickelgren, 1979).

The ontogeny of BZ receptors is not fully agreed upon. Some investigators find that BZ receptors are demonstrable at 14 days of gestation in rats and that BZ receptors, in contrast to GABA receptors, are rapidly developed at the just after birth (Braestrup and Nielsen, 1978; Candy and Martin, 1979; Palacios *et al.*, 1979). Other investigators, however, find a gradual increase in BZ receptor density during the first 3–4 weeks of life both in rats (Mallorga *et al.*, 1980*a*; Massotti *et al.*, 1980) and in mice (Regan *et al.*, 1980*a*). The gradual increase was not different from the ontogeny of GABA receptors. Benzodiazepine receptors can be affected by GABA already from birth (Regan *et al.*, 1980*a*; Palacios *et al.*, 1979; Mallorga *et al.*, 1980*a*). Benzodiazepines are also potent inhibitors of pentylenetetrazole-evoked seizures in the newborn (Garattini *et al.*, 1973).

5. STRUCTURAL SELECTIVITY OF BZ RECEPTORS

There are several reasons for believing that the biochemically identified recognition sites for benzodiazepines on brain membranes (Sections 2 and 3) represent the pharmacologically relevant benzodiazepine receptor in the brain. The binding sites show a high degree of chemical specificity, with appropriate rank order of affinity of benzodiazepines in relation to their *in vivo* potencies in animal tests. In addition, BZ receptors show the expected functional coupling with the GABA neurotransmission system, localized at the receptor/effector level (see Section 8).

5.1. Benzodiazepines

5.1.1. Structure/Activity

Except for some recently discovered agents (Table 10), the only group of chemicals interacting with BZ receptors at reasonably low concentrations (<100 μM) are benzodiazepines. The ability of numerous benzodiazepine derivatives to inhibit [^{3}H]diazepam or [^{3}H]flunitrazepam binding to BZ receptors has been reported. Table 9 shows IC_{50} values for 58 benzodiazepines; there is a 10^4- to 10^5-fold range between Ro 5-3448, triazolam, and KC-2846, which are the most potent agents, and Ro 5-4864, which is the least potent agent. IC_{50} values for clinically used benzodiazepines range from 2–50 nM with a few exceptions.

TABLE 8

Specific [^{3}H]Diazepam and [^{3}H]-GABA Binding in Cephalic Ganglia of Invertebrates and in Brains of Vertebrates

Species	Specific binding of [^{3}H]-GABA (0.5 nM), mean ± S.E.M.[a] pmol/g protein	Specific [^{3}H]diazepam binding[b]: B_{max} (pmol/g tissue[c])	Specific [^{3}H]diazepam binding[b]: Affinity constant K_D (nM)	Specific [^{3}H]diazepam binding[b]: IC_{50} for clonazepam (nM)
Invertebrates				
Annelida				
Oligochaeta				
Earthworm	n.d.[d]	<3	—	—
Mollusca				
Cephalopoda				
Squid	n.s.[e]	<3	—	—
Arthropoda				
Chrustacea				
Woodlouse	n.d.	<3	—	—
Lobster	n.d.	<3	—	—
Blue crab	n.s.	n.d.	—	—
Insecta				
Locust	n.d.	<3	—	—
Cockroach	n.s.	n.d.	—	—
Vertebrates				
Agnatha				
Cyclostomata				
Hagfish	60 ± 5	<3	—	—
Lamprey	n.d.	<3	—	—
Chondrichthyes				
Shark	n.d.	<3	—	—
Rabbit-fish	n.d.	<3	—	—
Spiny dogfish	202 ± 33	n.d.	—	—

Pisces				
Osteichthyes				
Reedfish	n.d.	14.5	2.4	n.d.
Eel	n.d.	9	5.8	1.1
Plaice	n.d.	22	5.9	2.3
Codfish	n.d.	13	6.3	1.7
Goldfish	235 ± 48	n.d.	n.d.	n.d.
Tetrapoda				
Amphibia				
Toad	n.d.	16	1.7	n.d.
Frog	141 ± 27	14	3.4	1.9
Reptilia				
Turtle	101 ± 13	20	2.7	n.d.
Lizard	n.d.	18	1.9	1.1
Aves				
Pigeon	n.d.	34	4.1	2.7
Hen	n.d.	28	3.6	2.3
Gull	n.d.	21	3.4	1.9
Chicken	438 ± 25	n.d.	n.d.	n.d.
Turkey	387 ± 42	n.d.	n.d.	n.d.
Mammalia				
Mouse	422 ± 10	35	2.7	2.7
Hamster	n.d.	41	3.6	2.6
Rat	386 ± 28	27	2.8	n.d.
Cat	n.d.	38	3.7	n.d.
Dog	n.d.	14	2.8	1.8
Pig	n.d.	37	3.1	2.9
Cow	n.d.	28	3.6	n.d.

[a] From Mann and Enna (1980).
[b] From Fernholm *et al.* (1979) and Nielsen *et al.* (1978).
[c] Crude homogenates contain ca. 10% protein of original brain weight.
[d] n.d., not done.
[e] n.s., not significant.

TABLE 9
Inhibition of [^{3}H]Diazepam Binding by Benzodiazepines[a]

Compound	R_1	R_2	R_3	[^{3}H]Diazepam binding, IC_{50} (nM)	Pentylenetetrazole,[b] ED_{50} (mg/kg)
Ro 5-3448	–Cl	*o*-Cl	$-CH_3$	1.3 (2)	0.4
Clonazepam	$-NO_2$	*o*-Cl	–H	4 ± 2 (6)	0.3
Ro 5-3027	–Cl	*o*-Cl	–H	5 ± 4 (4)	0.9
Flunitrazepam	$-NO_2$	*o*-F	$-CH_3$	5 ± 2 (3)	0.1
Ro 7-1986/1	–Cl	*o*-F	$-CH_2CH_2NH_2 \cdot 2HCl$	8 (1)	3
Ro 5-3590	$-NO_2$	*o*-CF_3	–H	9 ± 5 (3)	0.5
Ro 5-2180	–Cl	—	–H	11 (2)	1.0
Diazepam	–Cl	—	$-CH_3$	16 ± 11(10)	2.0
Nitrazepam	$-NO_2$	—	–H	17 ± 12 (6)	0.7
Kenazepine	–Cl	*o*-F	$-CH_2CH_2NHCOCH_2Br$	19 (1)	—
N-desmethyl FNM	$-NO_2$	*o*-F	–H	20[d]	—
Flurazepam	–Cl	*o*-F	$-CH_2CH_2N(C_2H_5)_2$	25 ± 16 (5)	2.0
Bromazepam	–Br	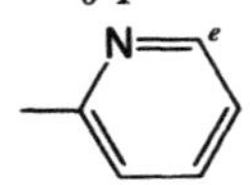 [e]	–H	34 ± 17 (4)	0.7
Ro 5-2904	$-CF_3$	—	–H	37 ± 34 (3)	0.9
Ro 5-6227	–Cl	*o*-F	$-(CH_2)_3N(CH_3)_2 \cdot 2HCl$	47 (1)	6.0
Irazepine	–Cl	*o*-F	$-CH_2CH_2NCS$	100 (1)	—

Prazepam	–Cl	—	$-CH_2-\triangleleft$	128 (1)	2.0
Halazepam	–Cl	—	$-CH_2CF_3$	250 (1)	8.0
Pinazepam	–Cl	—	$-CH_2C{\equiv}CH$	270 (1)	ca. 2
Ro 5-4528	–CN	—	$-CH_3$	400 (2)	1.0
Ro 5-5807	–Cl	—	$-CH_2CONHCH_3$	3500 (3)	1.0
Ro 5-4864	–Cl	*p*-Cl	$-CH_3$	160,500 (2)	>200

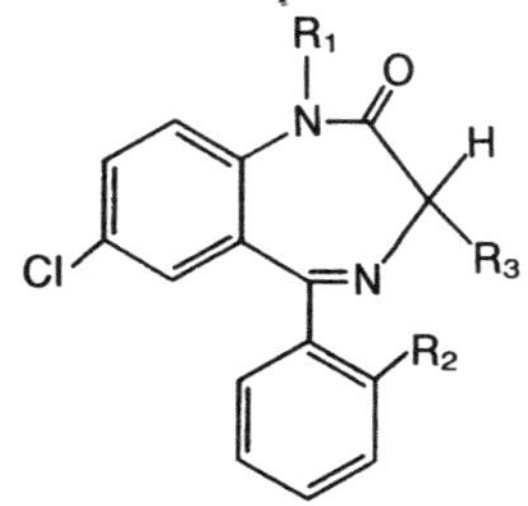

	R_1	R_2	R_3		
Lormetazepam	$-CH_3$	$-CH_3$	–OH	4 (1)	—
Lorazepam	–H	–Cl	–OH	5 ± 2 (3)	0.2
Oxazepam	–H	–H	–OH	38 ± 29 (4)	0.7
Chlorazepate	–H	–H	–COOK·KOH	52 ± 10 (4)	2.0
Oxazepam hemisuccinate	–H	–H	$-OCOCH_2CH_2COOH$	69 (1)	2[c]

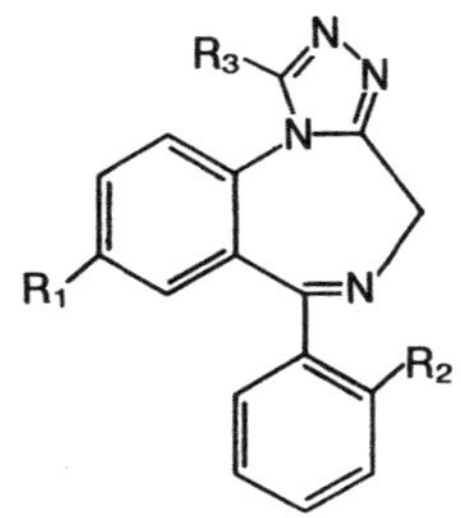

	R_1	R_2	R_3		
Triazolam	–Cl	–Cl	$-CH_3$	1.9 (1)	0.04
U-35,005	–H	–Cl	$-CH_3$	3.4 (1)	0.28
Alprazolam	–Cl	–H	$-CH_3$	4.7 (1)	0.2

(continued)

TABLE 9 (*Continued*)

Compound	Structure			[^{3}H]Diazepam binding, IC_{50} (nM)	Pentylenetetrazole,[b] ED_{50} (mg/kg)
Estazolam	–Cl	–H	–H	8–19 (2)	0.7
U-31,957	–H	–H	$-CH_3$	94 (1)	1.4
U-39,219	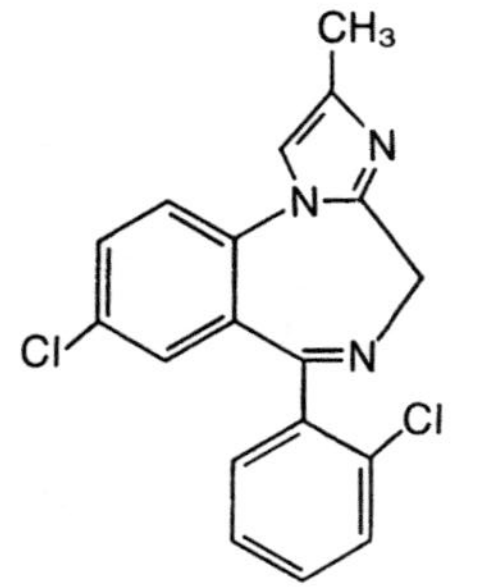			3.3 (1)	—
Ro 21-8384	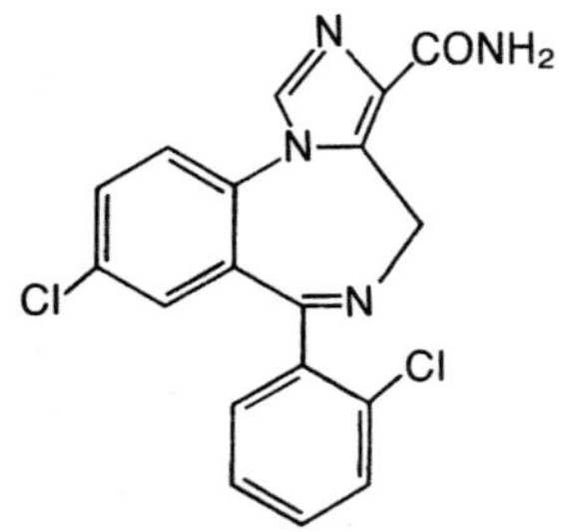			3.5 (1)	—
Midazolam	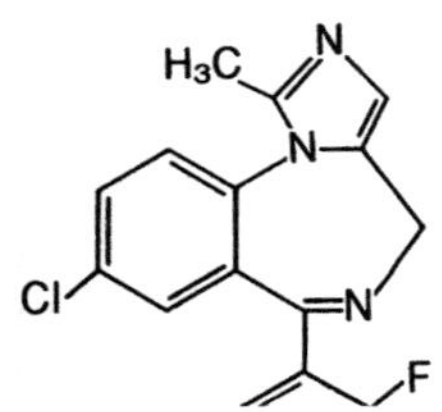			4.5 (1)	0.7[c]

KC-4-2846	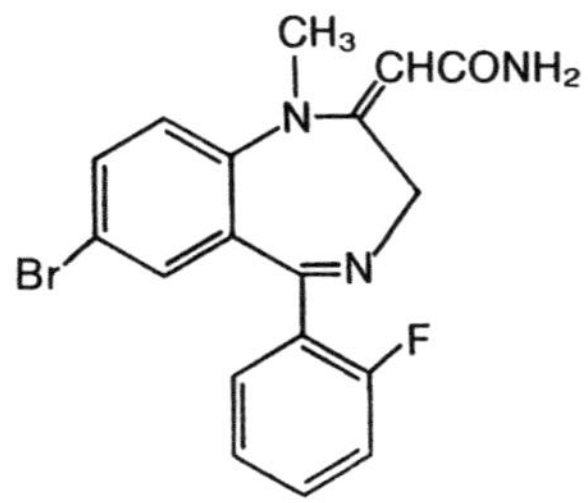	2.0 (1)	0.4
Ro 11-7800	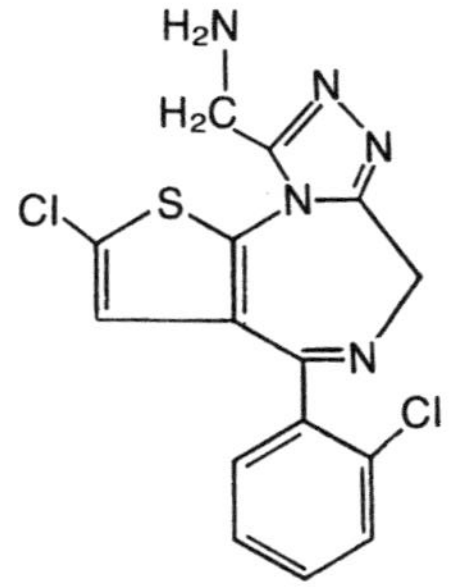	2.7 (1)	—
HR-158	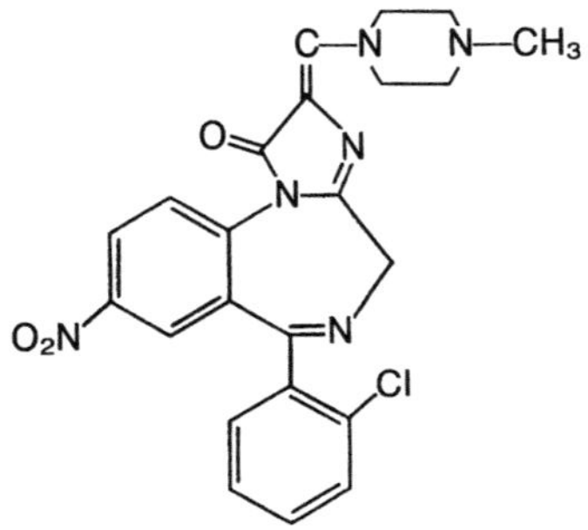	7.2 (1)	0.03

(continued)

TABLE 9 (*Continued*)

Compound	Structure		[^{3}H]Diazepam binding, IC_{50} (nM)	Pentylenetetrazole,[b] ED_{50} (mg/kg)
SC 31,312	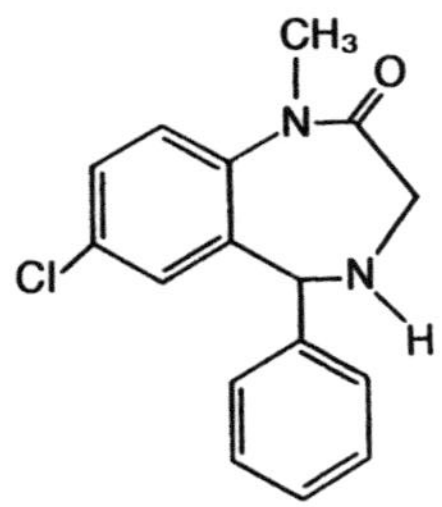		67 (1)	—
Clobazam *N*-Desmethylclobazam	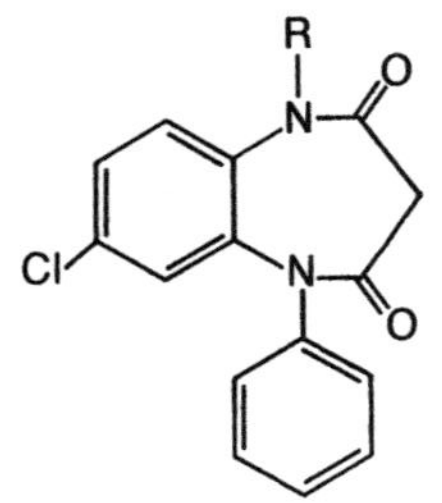	R –CH_3 R –H	170 (2) 210 (1)	7.4 13[c]
Chlordiazepoxide *N*-Desmethyl CDZ	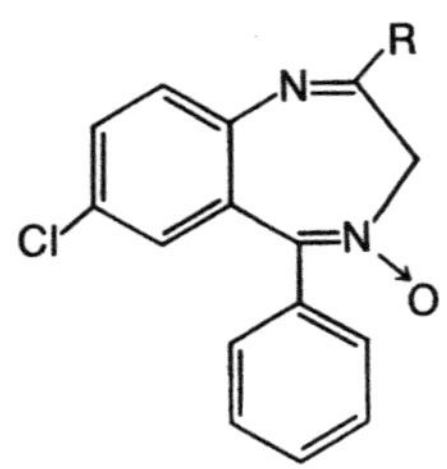	R –$NHCH_3$ R –NH_2	640 ± 460 (9) 900 (1)	8 10

Ripazepam	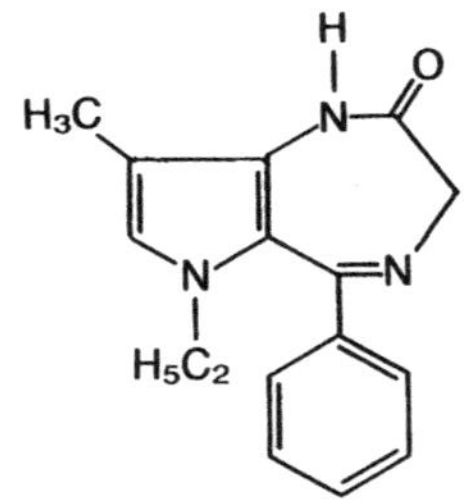	350 (2)	15
Medazepam	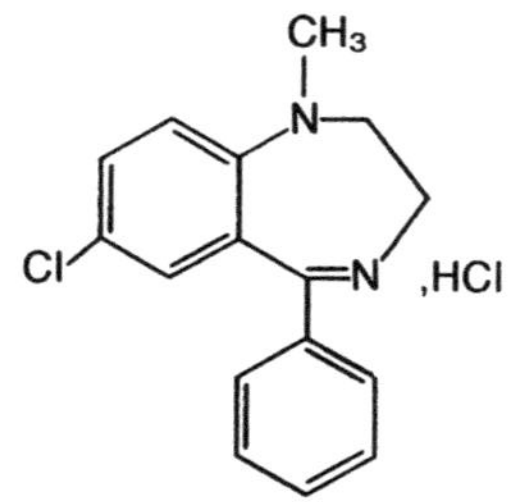	1600 ± 700 (5)	7
Ro 5-2181	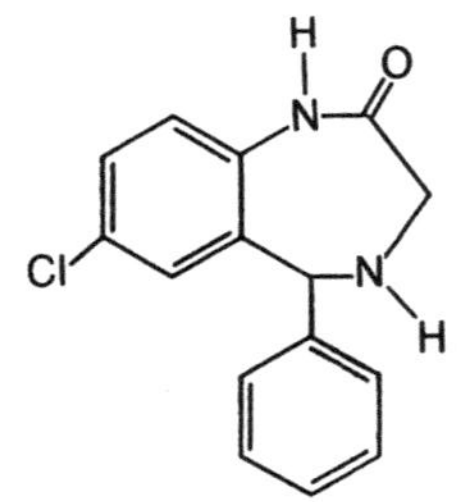	>1000 (1)	9

(continued)

TABLE 9 (*Continued*)

Compound	Structure	$[^3H]$Diazepam binding, IC_{50} (nM)	Pentylenetetrazole,[b] ED_{50} (mg/kg)
Ro 5-3785	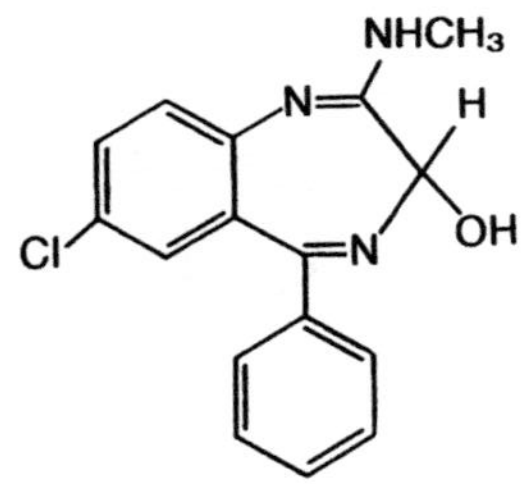	5200 (2)	300
Ro 5-3636	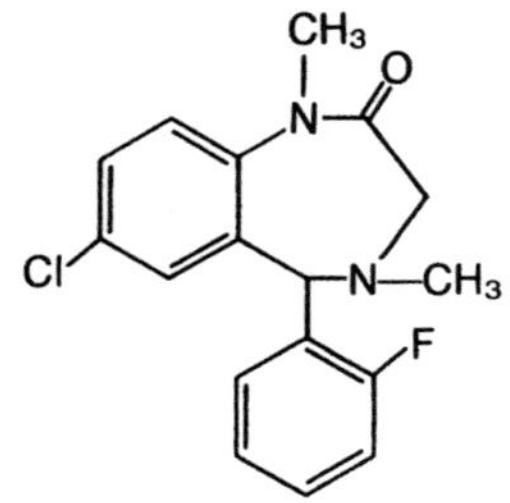	7200 (2)	10
Oxazolam Cloxazolam	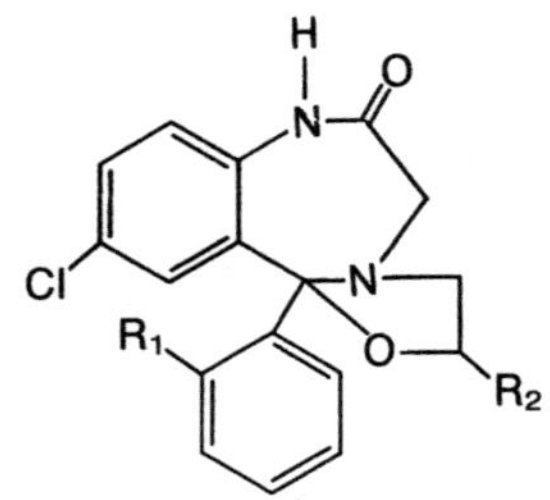 R_1 –H R_2 –CH_3 R_1–Cl R_2–H	ca. 10.000 ca. 10.000	— 0.8[d]

Ro 5-4933	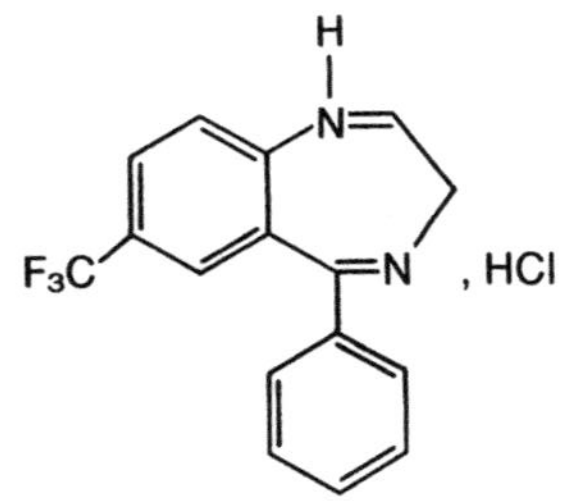	>30.000	4.0
Ro 5-3663	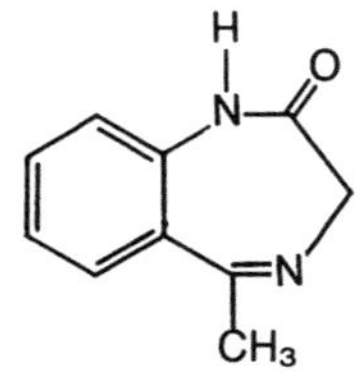	<20% inhibition at 10^{-5} M	—
Ro 11-5073; B11 (+) Ro 11-5231; B11 (−)	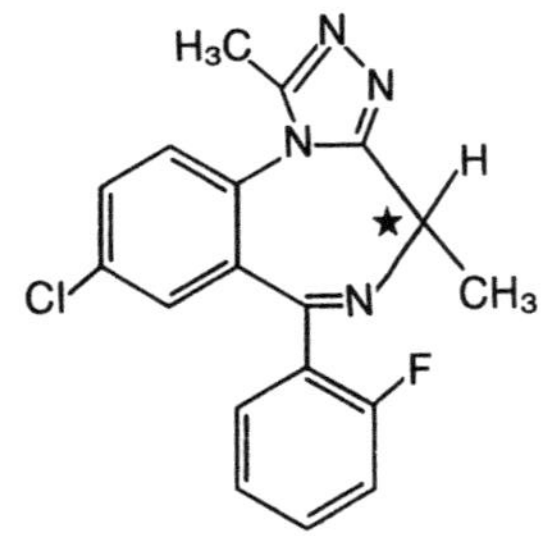	3.3 (1) 290 (1)	— —

(continued)

TABLE 9 (*Continued*)

Compound	Structure	[^{3}H]Diazepam binding, IC_{50} (nM)	Pentylenetetrazole,[b] ED_{50} (mg/kg)
Ro 11-3129; B9 (+)	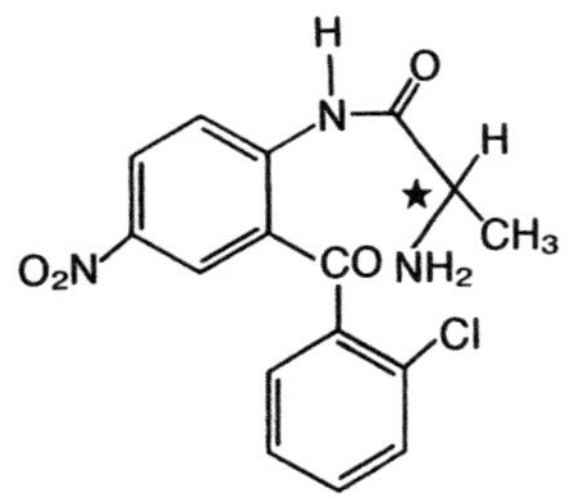	5.1 (1)	—
Ro 11-3625; B9 (−)		609 (1)	—
Ro 11-6896; B10 (+)	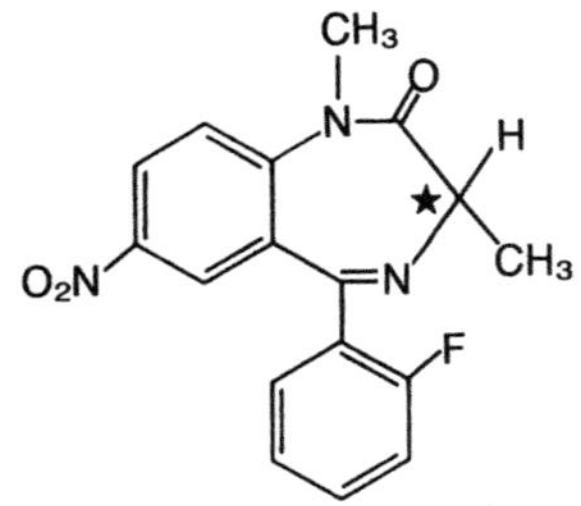	6.8 (1)	—
Ro 11-6893; B10 (−)		1470 (1)	—

[a] Compounds were added to standard benzodiazepine receptor assay, and the concentration causing 50% inhibition of specific binding was determined. The concentration of radioligand chosen in most studies was close to or below the respective K_D value, which means that K_I values are ca. 1- to 2-fold lower than the IC_{50} value. Mean ± S.E.M. of (*N*) values. (From Aaltonen *et al.*, 1979; Asano and Ogasawa, 1980; Blanchard *et al.*, 1979; Borbe *et al.*, 1980; Braestrup and Squires, 1978*a*; Braestrup and Squires, 1978*b*; Damm *et al.*, 1978; Fujimoto *et al.*, 1980*b*; Gervasi *et al.*, 1978; Hunt *et al.*, 1979; Lippa *et al.*, 1979*b*; Mackerer *et al.*, 1978; Malick and Enna, 1979; Moehler and Okada, 1977*b*; Moehler *et al.*, 1980; Mueller *et al.*, 1978; Nielsen *et al.*, 1981; O'Brien and Spirt, 1980; Rice *et al.*, 1979; Speth *et al.*, 1979*a*; Squires and Braestrup, 1977; Waddington and Owen, 1978; Williams *et al.*, 1980.)

[b] ED_{50} is the dose (s.c.) that inhibits cloniotonic seizures in 50% of mice receiving a supramaximal dose of pentylenetetrazole. Various investigators have used slightly different experimental conditions. (From Randall *et al.*, 1974; Mackerer *et al.*, 1978; Meldrum and Horton, 1979; Ager *et al.*, 1977; Caccia *et al.*, 1980; Rudzik *et al.*, 1973; Kamioka *et al.*, 1972; Angelis *et al.*, 1972; Malick and Enna, 1979; Fujimoto *et al.*, 1980*b*.)

[c] Administered i.p.

[d] Administered p.o.

[e] The whole 5-phenyl group is replaced by piperidyl in bromazepam.

Table 9 allows for some structure/activity considerations. Several types of benzodiazepines interact with BZ receptors with high affinity; these include "conventional" 1,4-benzodiazepines (such as diazepam), imidazolobenzodiazepines (such as midazolam), triazolobenzodiazepines (such as triazolam), and 1,5-benzodiazepines (such as clobazam).

Conventional benzodiazepines are composed of two aromatic rings (A and C rings) linked by a 7-membered ring (B ring) having two nitrogen atoms. Within this group of agents an electron-withdrawing substituent in position 7 of the A ring greatly increases potency in pharmacological tests in living animals (Sternbach, 1979), and almost all benzodiazepines investigated on BZ receptors are substituted with electron-withdrawing groups in the 7 position, most notably chloro and nitro groups. Additional substitution with electron-withdrawing groups (such as fluoro and chloro) in position 2′ in the C ring markedly increases the potency. On the other hand, substituents in the 4′ position of the C ring markedly reduce potency in pharmacological tests and almost completely eliminate affinity for BZ receptors (for example, Ro 5-4864). Methyl groups in position 1 in the B ring increase potency in pharmacological tests, but only slightly on BZ receptors *in vitro*; bulky substituents in the 1 position reduce potency [for example, 1-*tert*-butyl derivative (not shown) and prazepam].

5.1.2. *Stereoselectivity*

Position 3 of the B ring can be substituted without loss of activity (see, for example, lorazepam). Interestingly, the effect of substituting in the 3 position of the benzodiazepine nucleus is stereospecific. (+)-Enantiomers of 3-substituted benzodiazepines have 10–100 times higher affinity than (−)-enantiomers for BZ receptors. Thus, the receptor stereoselectively recognizes the pharmacologically active enantiomer. However, the significance of stereoselectivity should not be exaggerated. Stereoselective binding to nonreceptor materials can also occur; for example, some brands of glass-fiber filters stereospecifically bind [^{3}H]naloxone (Snyder *et al.*, 1975), and (+)-oxazepam is more avidly bound to serum albumin than (−)-oxazepam (Mueller and Wollert, 1975). Evidently, stereospecific binding, while necessary, is not a sufficient criterion for receptor binding.

5.1.3. *Pharmacological Correlates*

The IC_{50} values (or K_I values derived thereof) for benzodiazepine inhibition of BZ receptors have been correlated with the pharmacological and clinical potencies of benzodiazepines *in vivo*. There is a significant correlation between receptor affinity and the ability of benzodiazepines: (1) to antagonize seizures evoked by pentylenetetrazole (Braestrup and Squires, 1978*b*; Moehler and Okada, 1977*b*; Mackerer *et al.*, 1978; Malick and Enna, 1979), (2) to relax cat muscles, (3) to inhibit mouse motor performance on a revolving rod (rotarod), (4) to inhibit fighting in mice evoked by electric shock, (5) to elicit anxiolytic effect in a bioassay for human anxiety (Braestrup and Squires, 1978*b*), (6), to restore punishment-inhibited performance in rats (Braestrup and Squires, 1978*a*; Cook and Sepinwall, 1980) and (7) to relieve anxiety in patients as evaluated by recommended daily doses (Braestrup *et al.*, 1977; Moehler and Okada, 1978). These findings strongly indicate that the BZ receptor is the physiological target upon which benzodiazepines exert their effects.

Nevertheless, it is clear from Table 9 that although there is a good overall correlation between the anticonvulsant effects of benzodiazepines and their affinities for BZ receptor, there are individual exceptions. This is not unexpected, because benzodiazepines are subject to individual metabolism and possess different pharmacokinetic characteristics. Medazepam and cloxazolam, for example, appear to be more active clinically than predicted from their *in vitro* IC_{50} values. Medazepam is metabolized *in vivo* to desmethyldiazepam (see Randall *et al.*, 1974), which accumulates and is clinically active. Cloxazolam is rapidly metabolized *in vivo* to Ro 5-3027, which is highly active (T. Kamioka, unpublished) (see Table 9). Prazepam and pinazepam are also rather weak on BZ receptors, but their high activity in the pentylenetetrazole test in living mice is achieved by conversion to the major metabolite, desmethyldiazepam (see Bellantuono *et al.*, 1980). Chlorazepate is hydrolyzed in the intestine (Bellantuono *et al.*, 1980). Flurazepam appears to be less clinically active than predicted from its affinity for receptors; this may be due to rapid metabolism and rapid elimination *in vivo* (see Randall *et al.*, 1974). The influence of such pharmacokinetic differences in the correlation between receptor affinity and pharmacological effects can be eliminated by studying BZ receptor occupancy in the living brain (see Section 6).

5.2. Nonbenzodiazepine Inhibitors

A few agents without structural resemblance to benzodiazepines have high affinity for BZ receptors. β-Carboline-3-carboxylates were discovered during a search for endogenous ligands for BZ receptors. This group of compounds possesses a very high affinity and are among the most potent agents available for BZ receptors (Table 10). The presence of a fully aromatic C ring in β-carbolines and of substituents in the 3 position seems to be necessary to achieve high affinity; substituents in the 1 position reduce affinity. Ro 15-1788, an imidazobenzodiazepine carboxylic acid ethyl ester, and CGS 8216, a pyrazoloquinoline (structural formula in

Czernick *et al.*, 1982), both BZ receptor antagonists, have high affinity for BZ receptors (Hunkeler *et al.*, 1981; Czernik *et al.*, 1982).

Zopiclone (Blanchard *et al.*, 1979) and some triazolopyridazines (for example, CL 218.872) (Lippa *et al.*, 1979*a,b*) interact with BZ receptors at fairly low concentrations and exert some benzodiazepine-like properties. Both agents inhibit the ability of pentylenetetrazole to evoke seizures, and CL 218.872 is effective in restoring shock-suppressed drinking by thirsty rats in the water lick paradigm. This test is believed to reflect anxiolytic properties in man. Zopiclone interacts with brain BZ receptors after systemic administration (see Section 6). β-Carboline-3-carboxylates and triazolopyridazines interact preferentially with benzodiazepine BZ_1 receptors, whereas no selectivity for BZ receptor subclasses has been reported for zopiclone (Section 7).

Forminoben (Noleptan®) is a respiratory stimulant with antitussive properties (Pueschmann and Engelhorn, 1973) which interacts with BZ receptors *in vitro* (Antoniadis *et al.*, 1980*a*). The chemical structure bears some relationship to an "open-ring" benzodiazepine. Although fominoben seems to cross the blood–brain barrier quite easily (Pueschmann and Engelhorn, 1973), it is uncertain whether or not sufficient concentrations are obtained in the CNS *in vivo* for occupation of BZ receptors (see Table 12).

Several of the substances listed in Table 10 are related to proposed endogenous ligands for BZ receptors, either as analogs (purines) or as adenosine uptake inhibitors (dipyridamole, hexobendine, and papaverine), and are described again in Sections 11.2 and 11.3.

5.3. Ineffective Agents

The benzodiazepine receptor possesses a high specificity for benzodiazepines and related agents (see Sections 5.1 and 5.2). Other groups of drugs and agents do not occupy BZ receptors at relevant concentrations. The limit for relevant concentrations is set somewhat arbitrarily to ca. 100 μM in an attempt to exclude the wealth of diverse agents that inhibit binding at 10^3- to 10^6-fold higher concentrations than active benzodiazepines.

More than 300 substances, representing more than 23 different pharmacological classes, have been investigated and found to be ineffective (Table 11), including nonbenzodiazepine anxiolytics and sedatives such as barbiturates, ethanol, meprobamate, and various convulsants and anticonvulsants such as isoniazid, picrotoxin, diphenylhydantoin, and carbamazepine. Agents acting on the GABA receptors, such as GABA, muscimol, THIP, isoguvacine, and bicuculline, do not inhibit binding to the BZ

TABLE 10
Inhibition of [^{3}H]Diazepam and [^{3}H]-FNM[a]

Compound	Structure			IC$_{50}$ (μM)	
				[^{3}H]-FNM	[^{3}H]Diazepam
	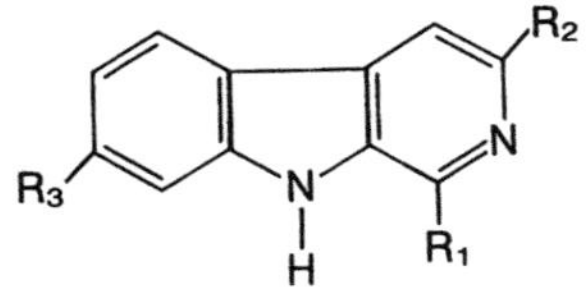				
	R_1	R_2	R_3		
FG 7109	–H	–COO-Phenyl	–H	0.001(7)[b]	—
FG 7098 (β-CCE)	–H	–$COOC_2H_5$	–H	0.007(5)	0.005(6, 16, 20)
FG 7106 (β-CCM)	–H	–$COOCH_3$	–H	0.008(7)	0.006(16)
FG 7115 (PrCC)	–H	–$COOC_3H_7$	–H	0.012(7)	—
FG 7108	–H	–$COOC_5H_{11}$	–H	0.045(7)	—
	–CH_3	–$COOC_2H_5$	–H	4.9(5)	—
Harman	–CH_3	–H	–H	7.2(18)	—
Norharman	–H	–H	–H	8.2(18)	12.5(16)
β-CC	–H	–COOH	–H	31(5)	—
Harmol	–CH_3	–H	–OH	111(18)	—
Harmine	–CH_3	–H	–OCH_3	134(18)	—
	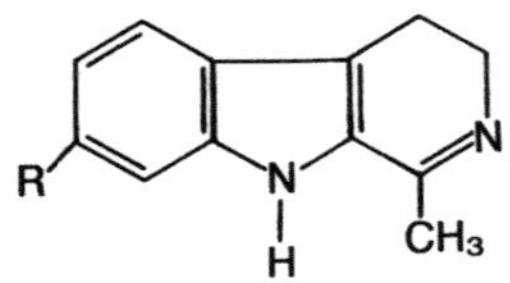				
	R				
Harmalol	–OH			305(18)	—
Harmaline	–OCH_3			500(17, 18)	2650(6)

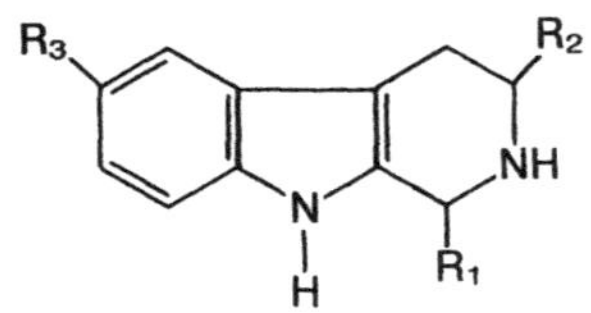

	R_1	R_2	R_3		
	–H	$-COOC_2H_5$	–H	4.9(5)	—
	–H	$-COOCH_3$	–H	17(5)	—
	–H	–COOH	–H	84(5)	—
	$-CH_3$	$-COOC_2H_5$	–H	400(5)	—
	$-CH_3$	–COOH	–H	400–6300(5, 18)	—
Tetrahydronorharman	–H	–H	–H	920(18)	—
Tetrahydroharman	$-CH_3$	–H	–H	1450(18)	—
6-Hydroxytetra-hydroharman	–H	–H	–OH	3620(18)	—

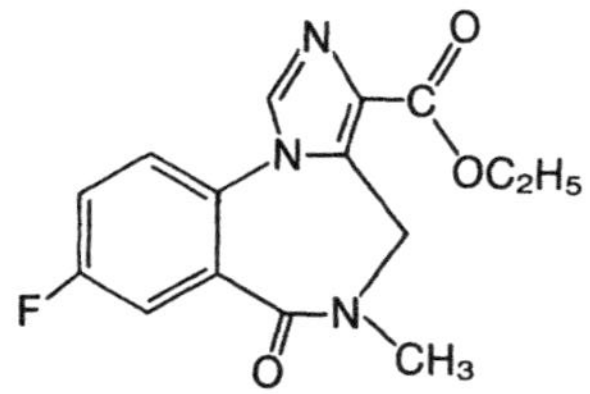

Ro 15-1788	—	0.002(9)

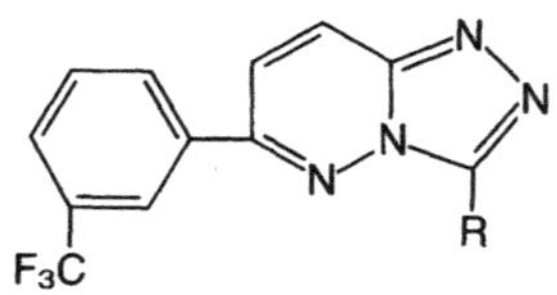

	R		
CL 218,872	$-CH_3$	0.14(11, 19)	0.11(10, 19)
CL 219,884	$-CH_2Cl$	0.50(19)	0.23(19)
CL 218,873	–H	0.81(19)	0.51(19)

(*continued*)

TABLE 10 (*Continued*)

Compound	Structure	IC50 (μM)	
		[3H]-FNM	[3H]Diazepam
6-(2-Hydroxy-5-nitro-benzyl)thioguanosine	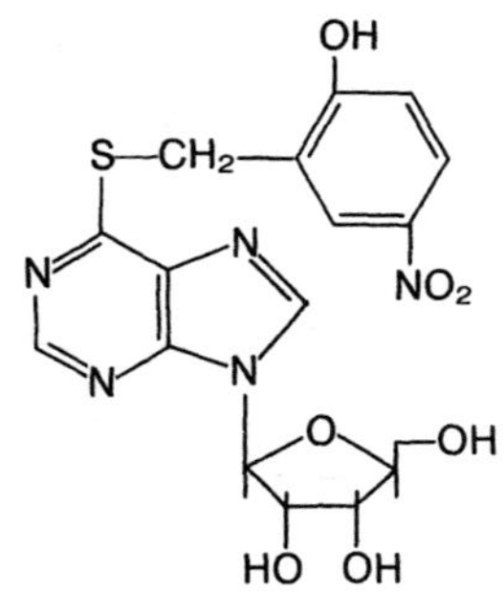	—	10(22)
1-Methylisoguanosine	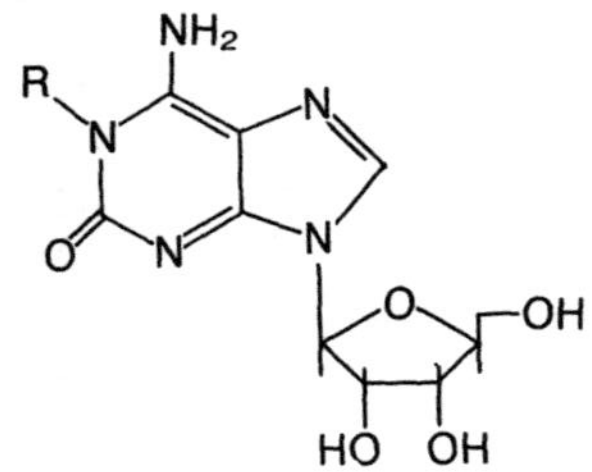 R: $-CH_3$	—	19(8)
1-Ethylisoguanosine	R: $-C_2H_5$	—	41(8)
6-dimethyl-aminopurine	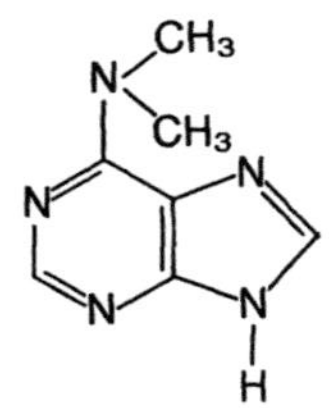	—	84(8)

1-Methylisoguanine	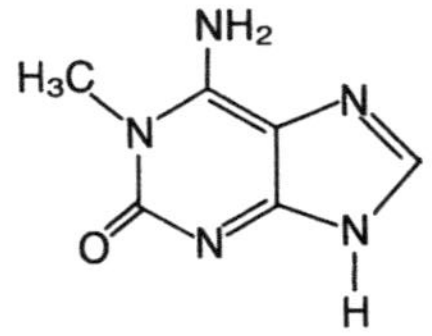	—	122(8)
2′,3′-*O*-Isopropylidene-1-methylisoguanosine	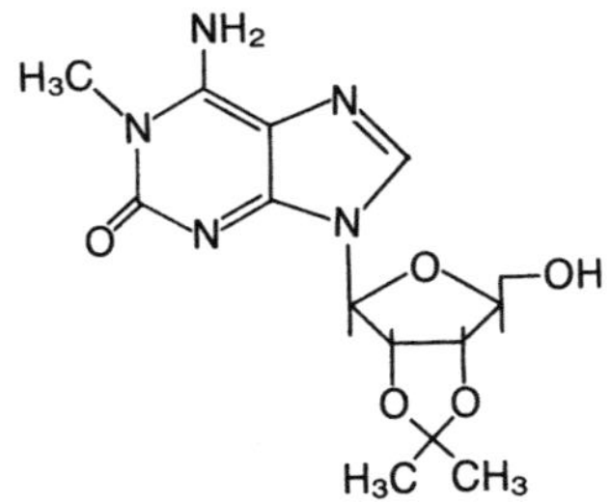	—	140(8)
Cartazolate (SQ 65,396)	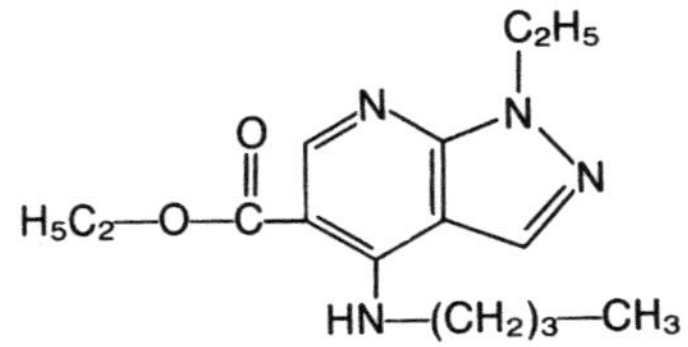	—	140(7, 12)
Hypoxanthine	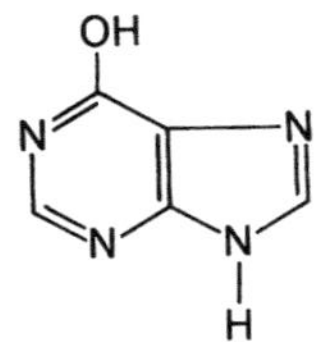	2300(18)	700–3700 (2, 6, 13, 15)

(continued)

TABLE 10 (*Continued*)

Compound	Structure	IC_{50} (μM)	
		[3H]-FNM	[3H]Diazepam
Inosine	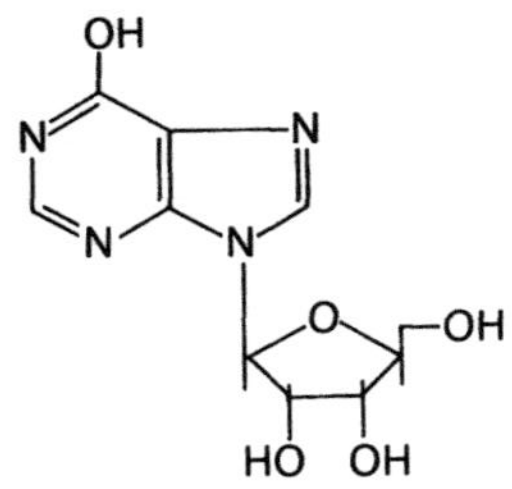	1800(4, 18)	400–1300 (2, 13, 15)
Nicotinamide	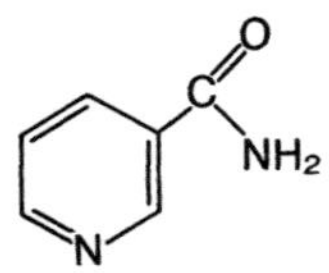	—	4300 (6, 13)
Nepenthin	MW ca. 15,000	—	0.074(21)
Zopiclone	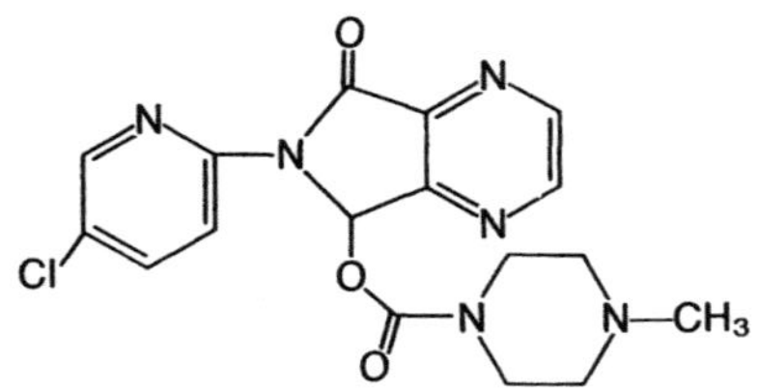	0.036(3)	0.082(3, 16)

Dipyridamole	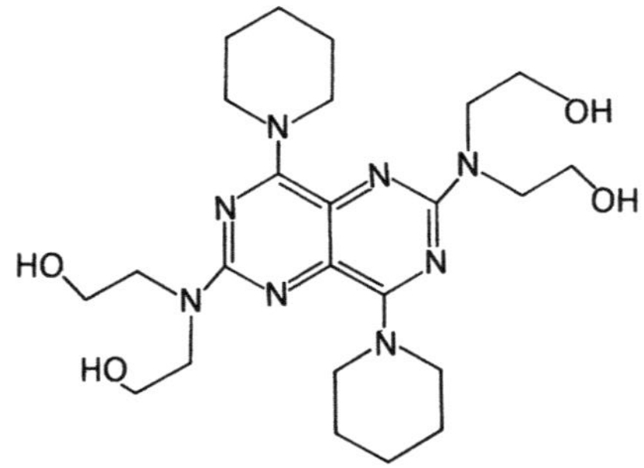	—	0.3(8, 22)
Fominoben	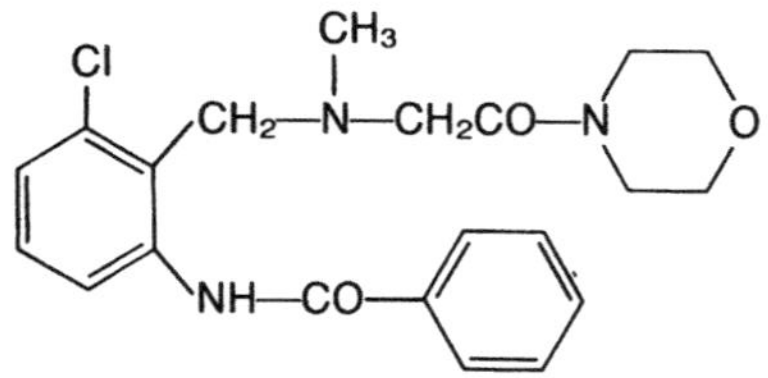	1.5(1)	—
Hexobendine	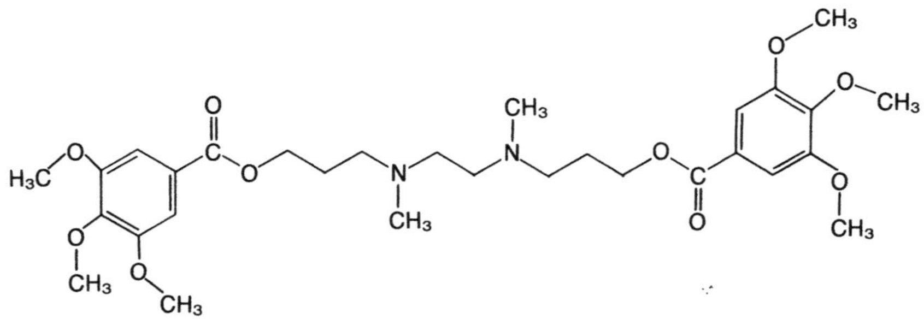	—	5(22)

(*continued*)

TABLE 10 (*Continued*)

Compound	Structure	IC_{50} (μM)	
		[^{3}H]-FNM	[^{3}H]Diazepam
PK 8165	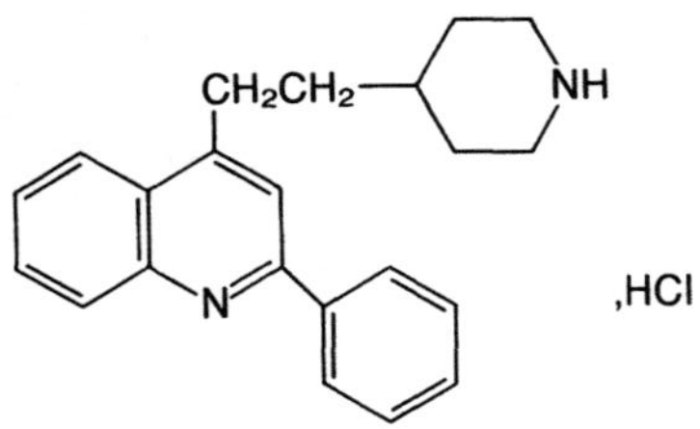	—	33(7)
Papaverine	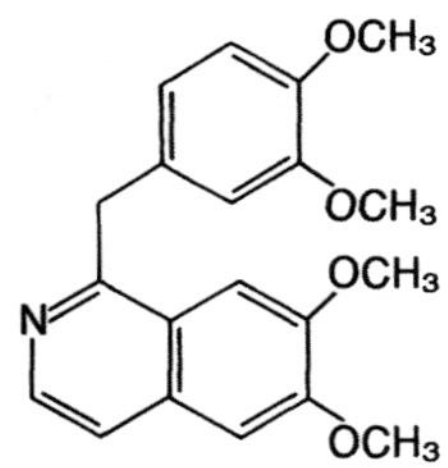	—	40(22)

Glutethimide	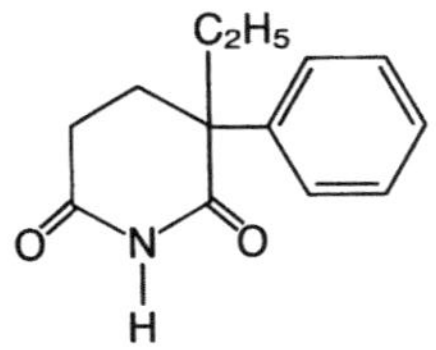	42(1)	
Methaqualone	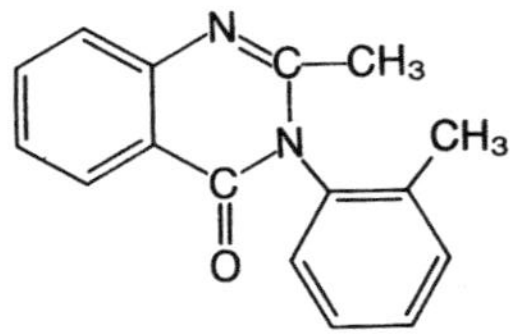	330(1, 4)	150(14)
Chlorzoxazone	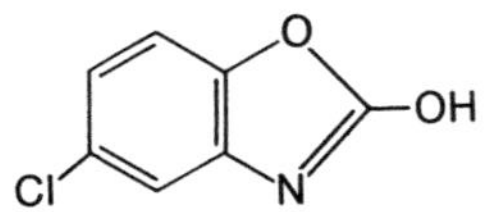		180

[a] Compounds were added to standard benzodiazepine receptor binding assays (see footnote to Table 9).
[b] Numbers in parentheses indicate references: (1) Antoniadis *et al.*, 1980*a*; (2) Asano and Spector, 1979; (3) Blanchard *et al.*, 1979; (4) Borbe *et al.*, 1980; (5) Braestrup *et al.*, 1980*a*; (6) Braestrup and Nielsen, 1980*c*; (7) C. Braestrup and M. Nielsen, unpublished results; (8) Davies *et al.*, 1980; (9) Hunkeler *et al.*, 1981; (10) Lippa *et al.*, 1979*b*; (11) Lippa *et al.*, 1980; (12) Malick and Enna, 1979; (13) Moehler *et al.*, 1979; (14) Mueller *et al.*, 1978; (15) Nielsen *et al.*, 1979; (16) Nielsen *et al.*, 1981; (17) Robertson, 1980*b*; (18) Rommelspacher *et al.*, 1980*a*; (19) Squires *et al.*, 1979*a*; (20) Tenen and Hirsch, 1980; (21) Woolf and Nixon, 1981; (22) Wu *et al.*, 1980.

TABLE 11
Ineffective Agents on BZ Receptor Recognition Sites[a]

Acetylcholine (100)[6,17,31]
Acetylsalicylate (300)[6]
Acetyl-2-thiohydantoin (31% at 10)[33]
ACTH (3)[6]
Adenine (50% at 630)[19]
Adenosine (20% at 1050)[19]
Adenosine 3′,5′-diphosphate (26% at 1000)[13]
Adenosine 5′-monophosphate (100)[31]
ADP (100)[17]
Agglutinin (wax-bean) (0.1 mg/ml)[14]
Agglutinin (wheat-germ) (0.2 mg/ml)[14]
L-α-Alanine (100)[17]
β-Alanine (100)[17]
Allantoin (100)[12]
Allopurinol (58% at 1000)[13]
Amantadine (800)[10]
γ-Aminobutyrate (GABA) (100)[6,17,31]
2-Aminobutyrate (?)[15]
ε-Aminocaprolate (100)[14]
4-Amino-2-hydroxybutyric acid (20)[15]
8-Amino-1-methylisoguanosine (68% at 1000)[13]
Aminooxyacetic acid (10)[32,34]
Aminophylline (21% at 100)[12]
3-Aminopropanesulfonic acid (300)[7]
Aminopyrine (100)[17]
δ-Aminovaleric acid (10)[15]
Amiphenazole (20% at 100)[1]
Amitriptyline (50% at 290)[24]
Amobarbital (100)[14]
AMP (5000)[3]
Apomorphine (100)[17]
L-Arginine (100)[17]
L-Aspartate (100)[6,17]
ATP (100)[12,17]
Atropine (100)[6,14,17,31]
Avermectin B_{1a} (10)[35]
Bacitracin (10 μg/ml)[6]
Baclofen (58% at 500)[24]
Barbituric acid (10)[33]
Bemegride (100)[1]
Benzoctamine (53% at 500)[24]
Benztropine (3)[6]
Benzylpenicillin (50% at 41,000)[2]
Bicuculline (100)[7,14,31]
(+)-Bicuculline-methiodide (100)[10]
Bombesin (3)[6]
Bromocriptine (3)[6]
Bufotenine (100)[17]
α-Bungarotoxin (100)[14]
Butanol (10,000)[30]
1-Butylisoguanosine (37% at 1000)[13]
Caffeine (50% at 470)[3,19]
Carbachol (200)[10]
Carbamazepine (26% at 100)[1]
Carbenicillin (5000)[2]
Carisoprodol (100)[17,31]
Cartazolate (SQ 65.396) (50% at 140)[10,18]
CF 25-397 (3)[6]
Chlormethiazole (3)[6]
Chlormezanone (3)[6]
2-Chloroadenosine (50% at 750)[3,13]
Chlorpheniramine (50% at 5000)[29]
Chlorpromazine (50% at 140)[6]
Chlorprotixene (300)[10]
Chlorzoxazone (50% at 180)[6]
Cimetidine (50% at 120)[29]
Clomethiazol (500)[24]
Clomipramine (3)[6]
Clonidine (300)[6]
Cloxacillin (50% at 5300)[2]
Clozapine (100)[31]
Cocaine (3)[6]
Coenzyme A (1000)[13]
Concanavalin-A (0.5 mg/ml)[14]
Cyclic AMP (1000)[3,13]
Cyclic GMP (19% at 1000)[13]
Cyclic IMP (21% at 1000)[13]
Cyproheptadine (3)[6]
L-Cysteine (100)[17]
Cytosine (100)[3,31]
DABA (10)[32]
Decamethonium (300)[10]
2′-Deoxyadenosine (49% at 1000)[13]
2′-Deoxyguanosine (50% at 280)[19]
2′-Deoxyinosine (50% at 395)[19]
2′-Deoxyribose (2000)[19]
Desmethylimipramine (3)[6]
Dibutyryl cyclic GMP (10)[34]
Dicloxacillin (50% at 5000)[2]
Dihydromuscimol (?)[7]
2-Dimethylamino-6-hydroxypurine (71% at 1000)[13]
1-N^6-Dimethylisoguanosine (73% at 1000)[13]
Diphenhydramine (100)[31]
Diphenylhydantoin (100)[31,b]
Diprenorphine (3)[6]
Dithiothreitol (100)[31]
L-DOPA (100)[17]

TABLE 11 (*Continued*)

Dopamine (100)[6, 17, 31]
Doxepin (100)[17]
EHNA (400)[25]
Eledoisin (3)[6]
EMD 28422 (100)[28]
Epinephrine (100)[17]
Etazolate (SQ 20.009) (100)[34]
Ethamivan (100)[1]
Ethanol (2 × 10^6)[11]
Ethosuximide (100)[1, 17]
Etomidate (100)[1]
Etorphine (3)[6]
Femoxetine (3)[6]
Fenobam (10)[18]
Fentanyl (3)[6]
α-Flupenthixol (3)[6]
Fluphenazine (300)[6]
GDP (100)[17]
L-Glutamate (100)[6, 14, 17, 31]
Glycine (100)[6, 17, 31]
Glycylglycine (100)[17]
GMP (50% at 5600)[3]
GTP (50% at 5600)[3]
3-Guanidinopropionic acid (100)[16]
Guanine (400)[25]
Guanosine (50% at 1000)[3, 13, 19]
5′-Guanylylimidodiphosphate (10)[34]
Haloperidol (100)[6, 17]
Harmaline (50% at 400)[26]
Harmalol (50% at 305)[26]
Harmine (50% at 130)[26]
Hemicholinium-3 (31% at 100)[31]
Hexamethonium (500)[10]
n-Hexobarbital (100)[1]
Histamine (100)[6, 17, 31]
L-Histidine (100)[17, 31]
Histone (3)[6]
Homocysteate (600)[10]
8-Hydrazino-l-methylisoguanosine (69% at 1000)[13]
Hydrocortisone (3)[6]
3-Hydroxy-GABA (?)[7]
6-Hydroxytetrahydronorharmane (50% at 3620)[26]
Hydroxyzine (44% at 500)[24]
Hypoxanthine (50% at 1300)[3, 19]
Imidazoleacetic acid (100)[7]
Imipramine (100)[17]
IMP (5000)[3]
Inosine (50% at 1300)[19, 25]
Iprindole (300)[10]

3-Isobutyl-l-methylxanthine (50% at 270)[3]
Isoguanosine (50% at 2300)[3]
Isoguvacine (300)[7]
L-Isoleucine (100)[17]
Isomuscimol (1000)[9]
Isoniazid (100)[17]
2′,3′-Isopropylidene adenosine (20% at 340)[19]
2′,3′-Isopropylidene guanosine (50% at 325)[19]
2′,3′-Isopropylidene inosine (50% at 360)[19]
Isoproterenol (100)[31]
Kainic acid (100)[14]
Ketamine (55% at 1000)[1]
Ketobemidone (100)[6]
Kojic amine (10)[34]
L-Leucine (100)[17]
LH-RH (3)[6]
LSD (1)[22]
L-Lysine (100)[17]
Melatonin (100)[17]
Melperone (100)[6]
Mepazine (300)[6]
Meperidine (100)[6]
Mephenesin (3)[6]
Meprobamate (50% at 700)[18]
Met-enkephalin (200)[6]
Methadone (3)[6]
Methanol (10,000)[30]
Methaqualone (50% at 150)[24]
Metergoline (300)[10]
Methiamide (3)[6]
Methocarbamol (100)[17]
2-Methoxyadenosine (28% at 1000)[13]
Methsuximide (18% at 100)[1]
1-Methyladenine (34% at 1000)[13]
1-Methyl adenosine (49% at 1000)[13, 19]
8-Methylaminoadenosine (50% at 860)[3]
6-Methylaminopurine (79% at 1000)[13]
α-Methyldopa (100)[17]
(R)-(+)-4-Methyl-GABA (400)[9]
(S)-(−)-4-Methyl-GABA (500)[9]
1-Methylguanosine (20% at 160)[19]
7-Methylguanosine (2000)[19]
1-Methylhypoxanthine (20% at 250)[19]
1-Methylinosine (2000)[19]
7-Methylinosine (2000)[19]
1-Methylisoguanosine 3′,5′-cyclic monophosphate (45% at 1000)[13]
1-Methylisoguanosine-5′-phosphate (28% at 1000)[13]

(*continued*)

TABLE 11 (*Continued*)

N-1-Methyl-*N*-3-methylaminocarbonyl-8-(β-D-ribofuranosyl)-7(9)*H*-isoguanine (25% at 1000)[13]
(R)-(+)-5′-Methylmuscimol (2300)[9]
(S)-(−)-5′-Methylmuscimol (200)[9]
Methylprylon (100)[1]
6-Methylpurine (58% at 1000)[13]
1-Methyl-8-(β-D-ribofuranosyl)-7 (9)*H*-isoguanine (19% at 1000)[13]
(R)-(+)-4-Methyl-*trans*-aminocrotonic acid (1000)[5]
(S)-(−)-4-Methyl-*trans*-aminocrotonic acid (1000)[5]
1-Methylxanthosine (1000)[13]
Methysergide (3)[6]
Metoclopramide (3)[6]
Mianserine (3)[6]
Molindone (100)[17]
Morphine (300)[6]
Muscimol (100)[31]
Nabilone (3)[6]
NAD (5000)[3]
Nalorphine (3)[6]
Naloxone (3)[6]
Naltrexone (100)[31]
Nicotine (100)[14]
Nicotinamide (50% at 3900)[23]
Nikethamide (100)[1]
Nipecotic acid (10)[34]
L-Norepinephrine (100)[6, 17, 31]
Nortriptyline (100)[31]
Oxacillin (50% at 16,000)[2]
6-Oxo-$PGF_{1\alpha}$ (3)[6]
Oxotremorine (100)[31]
Ouabain (100)[31]
Pargyline (?)[27]
Pavulon (150)[10]
Pentobarbital (100)[17]
Pentostatin (60% at 1000)[13]
Pentylenetetrazole (20% at 620)[19]
Perphenazine (500)[24]
PF 257 (3)[6]
Phenacetin (300)[6]
Phenaglycodol (10)[18]
Phencyclidine (100)[31]
Phenethicillin (50% at 26,000)[2]
Phenobarbital (100)[1, 31]
Phenoxybenzamine (3)[6]
Phensuximide (3)[6]
Phentolamine (3)[6]
L-Phenylalanine (100)[17]
Phenylethanol (1000)[30]
1-Phenylisoguanosine (1000)[13]
Phenylmethanol (1000)[30]
Phenyl-5-ureido-oxadiazole (29% at 10)[33]
Physostigmine (100)[30, 31]
Picolinic acid (3)[6]
Picrotoxin (100)[14, 31]
Piflutixol (3)[6]
Pimozide (3)[6]
Pindolol (3)[6]
Piperidine-4-sulfonic acid (100)[7]
PM 33 (50% at 700)[18]
Poly-arginine (3)[6]
Poly-lysine (3)[6]
L-Proline (100)[17]
Propanediol (100)[1]
Propoxyphene (300)[6]
Propranolol (1)[21]
Prostacyclin (3)[6]
Prostaglandin E_1 (3)[6]
Prostaglandin E_2 (100)[17]
Prostaglandin $F_{2\alpha}$ (3)[6]
Protamine sulfate (3)[6]
Purine (69% at 1000)[13]
Purine riboside (17% at 1000)[13]
Pyrilamine (50% at 120)[29]
Pyrithyldione (100)[1]
2-Pyrrolidone (100)[17]
Reserpine (100)[17]
Riboflavin (100)[17]
Ribonucleic acid (yeast; 0.5 mg/ml)[13]
Salicylate Na (1000)[30]
Scopolamine (3)[6]
L-Serine (100)[17]
Serotonin (100)[6, 17, 31]
Serum albumin (50% at 560 μg/ml)[6]
SKF-525A (100)[17]
Somatostatin (3)[6]
Spiramide (3)[6]
Spiroperidol (3)[6]
Strychnine (100)[14, 17, 31]
Substance P (100)[6]
Succinic acid (10)[34]
Succinylcholine (250)[10]
Sulpiride (3)[6]
Taurine (100)[17, 31]
Tegretol (3)[6]
Tetrahydroharmane (50% at 1500)[26]
Tetrahydronorharmane (50% at 900)[26]

TABLE 11 (*Continued*)

Tetrahydronorharmane-3-carboxylic acid (50% at 400)[8]	Trihexylphenidyl (300)[6]
Theobromine (20% at 425)[19]	Trimethadione (1000)[1]
Theophylline (50% at 660)[3, 19]	Tryptamine (50% at 1800)[26]
Thiamine (100)[17]	L-Tryptophan (50% at 1900)[26]
Thiomuscimol (?)[7]	D-Tubocurarine (100)[14, 31]
Thioproperazine (3)[6]	Tybamate (100)[31]
Thioridazine (3)[6]	L-Tyrosine (100)[17]
THIP (5000)[20]	Uracil (2000)[3, 19]
L-Threonine (100)[17]	Urea (100)[17]
Thymine (2000)[19]	Uric acid (100)[12, 17]
Ticarcillin (50% at 41,000)[2]	Uridine-5′monophosphate (100)[31]
Tolbutamide (100)[17]	L-Valine (100)[17]
trans-Aminocrotonic acid (?)[7]	Vasoactive intestinal peptide (3)[6]
TRH (10,000)[36]	Xanthine (100)[4, 12, 25, 31]
Triamterene (3)[6]	Xanthosine (1000)[13]
Tridione (3)[6]	Yohimbine (100)[6, 17]
Trifluoperazine (3)[6]	Zoxazolamine (3)[6]

[a] The table lists the compounds, concentrations in parentheses (μM or as indicated), and references in superscripts. The compounds in this table did not inhibit specific binding at a concentration of 100 μM when added to standard brain [^{3}H]diazepam or [^{3}H]flunitrazepam binding assays. Some compounds inhibit specific binding in high concentrations (>100 μM), and for these compounds the percentage inhibition of specific binding at a specified concentration (μM) is shown. For example, cartazolate (50% at 140) means that cartazolate inhibits 50% of specific benzodiazepine binding at 140 μM.; (?) means that the concentration of this compound was not specified. References: (1) Antoniadis *et al.*, 1980*a*; (2) Antoniadis *et al.*, 1980*b*; (3)Asano and Spector, 1979; (4) Bosmann *et al.*, 1980; (5) Braestrup and Nielsen, 1980*b*; (6) Braestrup and Squires, 1978*b*; (7) Braestrup *et al.*, 1979*d*; (8) Braestrup *et al.*, 1980*a*; (9) Braestrup *et al.*, 1980*b*; (10) C. Braestrup and M. Nielsen, unpublished; (11) Colello *et al.*, 1978*a*; (12) Damm *et al.*, 1979; (13) Davies *et al.*, 1980; (14) Dudai and Sherman-Gold, 1980; (15) Karobath and Sperk, 1979; (16) Karobath *et al.*, 1979; (17) Mackerer *et al.*, 1978; (18) Malick and Enna, 1979; (19) Marangos *et al.*, 1979*b*; (20) Maurer, 1979; (21) Moehler and Okada, 1977*b*; (22) Moehler *et al.*, 1978*c*; (23) Moehler *et al.*, 1979; (24) Mueller *et al.*, 1978; (25) Nielsen *et al.*, 1979; (26) Rommelspacher *et al.*, 1980*a*; (27) Skolnick *et al.*, 1979*c*; (28) Skolnick *et al.*, 1980*d*; (29) Speeg *et al.*, 1979; (30) Speeg *et al.*, 1980; (31) Speth *et al.*, 1978; (32) Tallman *et al.*, 1978; (33) Tunnicliff *et al.*, 1979; (34) Williams and Risley, 1979; (35) Williams and Yarbrough, 1979; (36) Vogel *et al.*, 1980.

[b] Using a centrifugation assay, Tunnicliff *et al.* (1979) have reported that diphenylhydantoin inhibits [^{3}H]diazepam binding with K_I = 0.9 μM at 23°C.

receptor, confirming that this receptor is not identical to the GABA receptor recognition site. Other known and putative neurotransmitters, such as norepinephrine, dopamine, serotonin, acetylcholine, glutamate, glycine, Met-enkephalin, substance P, somatostatin, and VIP are all inactive. The BZ receptor thus appears unrelated to the recognition sites of known and putative neurotransmitters.

Some of the agents that inhibit benzodiazepine binding at very high concentrations act by interfering with the receptor binding assay. Serum albumin, for example, binds the radioligand, thereby reducing the free radioligand concentration without interfering with the receptor recognition site. Triton X-100 is a detergent which inhibits receptor binding by solubilization of BZ receptors.

6. BZ RECEPTOR BINDING *IN VIVO*

6.1. Methodological Considerations

Under *in vitro* conditions, benzodiazepine receptors are deprived of their normal microenvironment composed of a specific ionic milieu, a normal intra/extra cellular membrane potential, and the presence of hypothetical labile membrane constituents with regulatory or coupling functions.

Furthermore, drug affinities for receptors *in vitro* may not always correlate with pharmacological effects *in vivo* due to variable penetration of the drugs into the relevant sites in the CNS. Determination of drug effects on receptor binding in the brains of intact animals partly overcomes these difficulties.

[^{3}H]Diazepam can be used for *in vivo* labeling of BZ receptors (Williamson *et al.*, 1978*a,b*; Tallman *et al.*, 1979), but [^{3}H]-FNM is superior as radioligand due to its slower "off-rates" (k_{-1}) (Chang and Snyder, 1978; Blanchard *et al.*, 1979; Duka *et al.*, 1979) (see the legend to Table 12). Low k_{-1} values (and k_{+1} values) ensure that new equilibria are not obtained in the tissue or in test tubes postmortem. The chemical group carrying the radioactivity for [^{3}H]diazepam and[^{3}H]flunitrazepam, the methyl group in the 1 position, is eliminated by liver enzymes; ca. 20% demethylation of [^{3}H]-FNM occurs in mice within 20 min after i.v. injection (Duka *et al.*, 1979). Demethylation of diazepam is a major metabolic pathway in mice but not in rats (Garattini *et al.*, 1973); rats, however, require high amounts of radioligand for *in vivo* binding experiments.

TABLE 12
Inhibition in Vivo of [^{3}H]Flunitrazepam Binding to Brain BZ Receptors[a]

Compound	ED_{50} (mg/kg i.p.)[b]	References[i]
Triazolam	0.12	2
Ro 11-6896, B10 (+)	0.14[c]	4
Clonazepam	0.2	2–4
Lorazepam	0.3	2
Flunitrazepam	0.6	3, 4
Nitrazepam	1.25[d]	1
Pinazepam	ca. 1	5
CGS 8216	1.1	2
Diazepam	1.1	1–4
Midazolam	1.3	2
Ro 15-1788	4	2

(*continued*)

TABLE 12 (*Continued*)

Compound	ED_{50} (mg/kg i.p.)[b]	References[i]
Oxazepam	7	2
Zopiclone	10	1, 2
CL 218.872	18	2
Chlordiazepoxide	21	1–3
β-CCM	22[e]	2
β-CCE	60[e]	2
PrCC	300[e]	2
Bicuculline	n.s. 5[g]	3
Fominoben	n.s. 60[f]	2
Harmaline	n.s. 50	2
Meprobamate	n.s. 300	2
Oxolinic acid	n.s. 100	2
Pentylenetetrazole	n.s. 100[f]	2
Strychnine	n.s. 1	3
Ro 5-5807	n.s. 100[c,h]	6
Ro 11-6893, B10 (÷)	n.s. 10[c]	4

[a] [^{3}H]-FNM (ca. 4 μCi per animal) was administered intravenously to mice 20 min before decapitation. Whole forebrains were rapidly excised and homogenized in ca. 12 ml (40 × vol.) ice-cold buffer, pH 7.1–7.7. Aliquots (1 ml) were then filtered through Whatman glass-fiber filters and washed with 2 × 5 ml ice-cold buffer. Tritium on the filters, which represents total membrane-bound [^{3}H]-FNM, ca. 50% ot total radioactivity, was determined by conventional scintillation counting. Specific binding is total membrane-bound [^{3}H]-FNM minus nonspecific binding, which was obtained after systemic administration of a supramaximal dose of clonazepam (10–25 mg/kg i.p.) 30 min before [^{3}H]-FNM administration. Nonspecific binding represents ca. 10% of total membrane-bound [^{3}H]-FNM (ref. 3). A slightly different procedure was applied in experiments using intravenous drug administration (ref. 4).

[b] The ED_{50} value is the dose of the compound that inhibits by 50% the specific binding of [^{3}H]-FNM to mouse forebrain BZ receptors. Test substances were administered by the intraperitoneal route 35 min before decapitation.

[c] Intravenously, 20 min before decapitation.

[d] Perorally, 50 min before decapitation.

[e] Subcutaneously, 35 min before decapitation.

[f] Subcutaneously, 50 min before decapitation.

[g] n.s., no significant inhibition at the indicated dose.

[h] This experiment was conducted with [^{3}H]diazepam as radioligand instead of [^{3}H]-FNM; the animals were decapitated 5 min after [^{3}H]diazepam.

[i] References: (1) Blanchard *et al.*, 1979; (2) C. Braestrup, unpublished results, method according to Chang and Snyder (1978); (3) Chang and Snyder, 1978; (4) Duka *et al.*, 1979; (5) Gervasi *et al.*, 1978; (6) Tallman *et al.*, 1979.

The apparent magnitude of the K_D value for [^{3}H]-FNM in the brain of living mice has not been determined. It is clear, however, from data presented by Chang and Snyder (1978) that the apparent K_D value is above 100 nmol/kg brain tissue, which is far above the K_D value of ca. 6 nM determined for [^{3}H]-FNM *in vitro* at 37°C (Braestrup and Squires, 1978*a*; Dudai and Sherman-Gold, 1980; Speth *et al.*, 1979*a*). Benzodiazepines, however, are highly lipophilic agents and most of the brain content of benzodiazepines is retained in lipid material, reducing the fraction available in solution at the receptor sites. In plasma, the free concentration of benzodiazepines is reduced by a 90–99% binding to plasma proteins. Therefore, it is understandable that the total concentrations of benzodiazepines in blood, CSF, or brain from patients or animals receiving adequate medication are considerably higher than the respective IC_{50} values measured *in vitro* (Braestrup *et al.*, 1978*b*).

6.2. Structural Selectivity

The ability of benzodiazepines and other agents to inhibit the binding of [^{3}H]-FNM to BZ receptors *in vivo* can be expressed as ED_{50} values. The ED_{50} value is the dose that inhibits specific binding of [^{3}H]-FNM *in vivo* by 50%. In other words, 50% of available BZ receptor sites are occupied by the test agent after administrating a dose equal to the ED_{50} value. This is approximately true because the concentration of [^{3}H]-FNM present in brain after i.v. administration of the radioligand is very low as compared to a K_D value for [^{3}H]-FNM *in vivo* (for example, 1 nmol/kg brain tissue). Table 12 shows that pharmacologically active benzodiazepines inhibit the specific binding of [^{3}H]-FNM *in vivo*. Furthermore, the effect is stereoselective (Duka *et al.*, 1979). Some β-carboline-3-carboxylates in high dosages are active *in vivo*, whereas this is not true of several other agents. Esters of β-carboline-3-carboxylic acid are less potent on BZ receptors in living animals than expected from their *in vitro* potencies, because esterases in the liver, kidney, and blood rapidly hydrolyze the ester function to the free acid, which has only a low receptor affinity (Simonsen *et al.*, 1982).

The results compiled in Fig. 7 show a remarkably strict correlation between the ability of several benzodiazepines to inhibit *in vivo* binding of [^{3}H]-FNM to BZ receptors in mouse brain, and the ability of these drugs to antagonize seizures evoked by pentylenetetrazole, also in mice. This finding is a strong argument in favor of the receptor status of specific [^{3}H]-FNM binding sites in the CNS. It is interesting to note that the correlation line shown in Fig. 7 is not superimposable on the diagonal. They are parallel, but displaced by a factor of ca. 3. Thus, when considering this particular effect of benzodiazepines, the ability to antagonize seizures evoked by 80–150 mg/kg pentylenetetrazole, half of the maximal pharmacological effect is obtained when only 20–25% of the receptors are

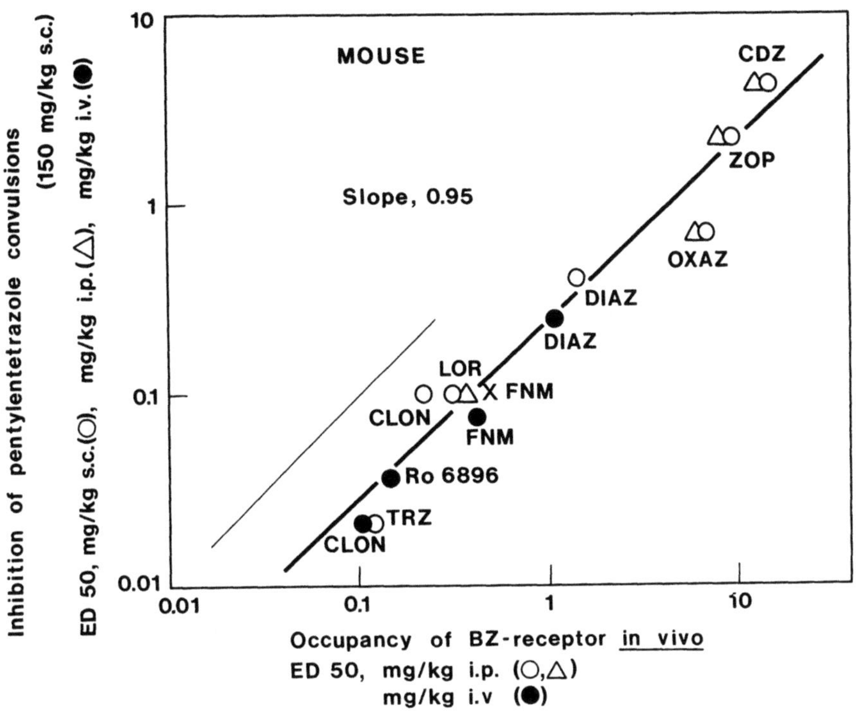

FIG. 7. Correlation between the *in vivo* occupancy of BZ receptors and antagonism of pentylenetetrazole-evoked seizures. ED_{50} for BZ receptors *in vivo* is the dose that inhibits by 50% the specific binding of [^{3}H]-FNM in the brain of living mice (see footnote to Table 12); ED_{50} for pentylenetetrazole is the dose that inhibits clonic convulsions and death in 50% of mice receiving 80–150 mg/kg s.c. of pentylenetetrazole. All tests were evaluated 20–50 min after administration of benzodiazepines (●, from Duka *et al.*, 1979; ×, from Chang and Snyder, 1978; ○, △, from Braestrup *et al.*, 1982*b*) CLON, clonazepam; CDZ, chlordiazepoxide; DIAZ, diazepam; FNM, flunitrazepam; LOR, lorazepam; OXAZ, oxazepam; TRZ, triazolam; ZOP, sopiclone.

occupied (Duka *et al.*, 1979). Other pharmacological effects, such as inhibition of minimal electroshock seizures and induction of muscle relaxant effect in mice, are elicited at 10–50 times higher doses (Randall *et al.*, 1974) and occur at higher degrees of receptor occupancy. Anticonflict effects in a water lick paradigm occur at ca. 60% receptor occupancy (Braestrup *et al.*, 1982*b*; Petersen and Buus Lassen, 1981).

7. MULTIPLE BRAIN BZ RECEPTORS

The initial characterization of benzodiazepine receptors was not in favor of the presence of more than a single class of noninteracting sites in

TABLE 13
Characteristics of BZ_1 and BZ_2 Receptors

BZ_1	BZ_2
MW of protomer ca. 51,000 daltons	MW of protomer ca. 55,000 daltons
Probably composed of di- or tetramers	Probably composed of di- or tetramers
Most abundant in cerebellum[a]	Most abundant in hippocampus and basal ganglia[a]
Least abundant in the hippocampus (50%)	Absent in the cerebellum
Coupled to GABA receptors	Coupled to GAB receptors
Coupled to chloride channels	Not evaluated
Benzodiazepines show no preference for BZ_1 or BZ_2	
Preference for β-carboline-3-carboxylates and triazolopyridazines	No preferences known
Pharmacological and clinical relation unknown for BZ_1 or BZ_2	

[a] The relative abundance of BZ_1 to total BZ receptor population is shown in Table 7.

the CNS. The presence of strictly linear Scatchard plots for all available BZ receptor radioligands ([^{3}H]diazepam, [^{3}H]clonazepam, and [^{3}H]-FNM) in all brain regions (>30) and mammalian species (>9) investigated gave no clues for the existence of more than a single class. In addition, the displacement of radioligands with unlabeled benzodiazepines yielded curves in accordance with the law of mass action with Hill coefficients close to unity. Furthermore, each of the 16 benzodiazepines investigated had similar affinity for receptors in human frontal cortex, cerebral cortex, cerebellar cortex, and putamen, thus lending no support to the idea that qualitative differences exist among benzodiazepines and among BZ receptors (Braestrup *et al.*, 1977; Reisine *et al.*, 1979*b*; Owen *et al.*, 1979). On the other hand, it is the rule rather than the exception that multiple receptors do exist (Snyder and Goodman, 1980). There are multiple acetylcholine receptors, adrenergic receptors, serotonin receptors, opiate receptors, GABA receptors, and histamine receptors. Multiple glutamate receptors may exist. Multiplicity has not yet been demonstrated for several peptide receptors, such as substance P, insulin, growth hormone, neurotensin, and VIP, or for imipramine binding sites.

Results incompatible with the presence of a single class of noninteracting BZ receptors were obtained when a new class of agents, the triazolopyridazines, was found to inhibit [^{3}H]diazepam binding with shallow dose response curves, yielding Hill coefficients below unity (Squires *et al.*, 1979*a*). These findings indicated the presence of more than one class of BZ receptors. Other anomalous findings, such as the presence of polyphasic dissociation curves for bound [^{3}H]-FNM and polyphasic thermal inactivation curves for BZ receptors, may also be related to receptor multiplicity (Squires *et al.*, 1979*a,b*).

Also incompatible with the presence of a single class of BZ receptors with noninteracting sites was the discovery of mixed-type competitive inhibition of [^{3}H]diazepam and [^{3}H]-FNM binding by ethyl-β-carboline-3-carboxylate (β-CCE) (Braestrup *et al.*, 1980*a*). Another striking feature of β-CCE and triazolopyridazines was that they exhibited regional selectivity for various brain areas. For example, these compounds displayed a very high affinity for cerebellar BZ receptors (K_I ca. 1 nM for β-CCE), whereas their potencies in the hippocampus was 4–7 times less (Nielsen and Braestrup, 1980; Lippa *et al.*, 1980). Hoffstee analyses (Fig. 8) and computerized curve-fitting procedures of the inhibition of [^{3}H]-FNM binding by β-CCE and triazolopyridazines in various regions of the rat brain indicate that with respect to these ligands the hippocampus contains 50–60% of a "high-affinity" BZ_1 receptor, and 50–40% of a "low-affinity" BZ_2 receptor (see Table 13). Benzodiazepines possess about equal affinity for both sites (Braestrup and Nielsen, 1981*b*; Supavilai and Karobath, 1980*b*; Klepner *et al.*, 1979). Cerebral cortex contains ca. 75% BZ_1 receptors (see Fig. 8), while the cerebellum contains almost exclusively BZ_1 receptors (Lippa *et al.*, 1980; Braestrup and Nielsen, 1981*b*). The forbrain of newborn rats contains predominantly BZ_2 receptors (Lippa *et al.*, 1981, and unpublished results).

[^{3}H]-PrCC may serve as a selective radioligand for brain BZ_1 receptors (Braestrup and Nielsen, 1981*b*) and has been used to investigate the

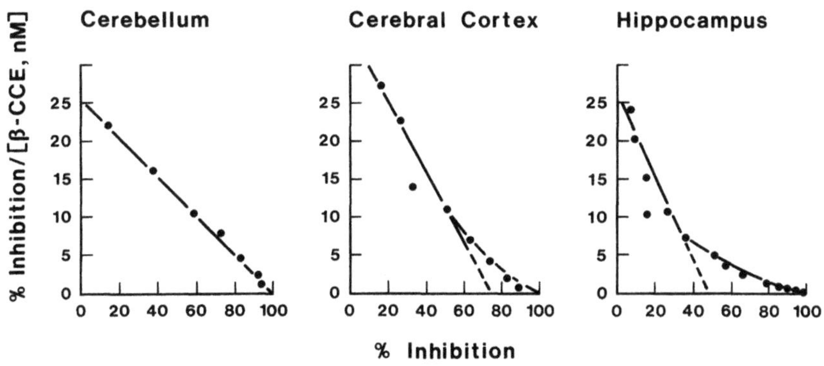

FIG. 8. Hoffstee analyses of β-CCE inhibition of [^{3}H]-FNM binding in rat brain. Increasing concentrations of β-CCE were added to standard [^{3}H]-FNM binding assays (Fig. 2) containing 1 nM [^{3}H]-FNM. Percentage of inhibition of specifc binding at each concentration of inhibitor was determined and plotted as shown. Nonlinear concave-upwards Hoffstee plots suggest multiple types of binding sites or alternatively, negative cooperativity. BZ_1 receptors comprise 100%, ca. 75%, and ca. 55% of the total BZ receptor population in cerebellum, cerebral cortex, and hippocampus, respectively. (C. Braestrup and M. Nielsen, unpublished results).

properties and occurrence of this receptor subclass. The regional brain distributions of BZ_1 and BZ_2 receptors as determined by [^{3}H]-PrCC binding is shown in Table 7. A similar regional distribution was demonstrated by an autoradiographic technique (Young *et al.*, 1981) using CL 218.872 as a selective inhibitor of [^{3}H]-FNM binding.

Sieghart and Karobath (1980) used a different approach to demonstrate the presence of at least two classes of BZ receptors. They used the unique procedure of photoaffinity labeling of BZ receptors (see Section 2.1.3) so that they could investigate the molecular weight of the receptor ligand complex in various brain areas. Rat cerebellum revealed almost exclusively one [^{3}H]-FNM binding protein with a molecular weight of 51,000 daltons (BZ_1 receptor), while the hippocampus and some other brain regions contained an additional 55,000-dalton binding protein, probably the BZ_2 receptor. Other researchers, using slightly different procedures, including washing of the photoaffinity-labeled membranes, have only characterized the 51,000-dalton receptor species (Moehler *et al.*, 1980; Braestrup *et al.*, 1981; J. F. Tallman, unpublished). Minor [^{3}H]-FNM binding proteins with molecular weights of 53,000 and 59,000 daltons and with unknown relations to BZ_1 and BZ_2 receptors were observed in some brain regions (Sieghart and Karobath, 1980). Further biochemical evidence in favor of multiple BZ receptors was provided by Fong and Goldstein (1980) who recovered solubilized brain benzodiazepine receptors in two separate fractions from hydroxyapatite chromatography. These fractions were both stimulated by GABA.

Which pharmacological and clinical features of the benzodiazepines are mediated by the two postulated receptor subclasses has not been fully evaluated. Both BZ_1 and BZ_2 receptors appear to be functionally coupled to GABA receptors (Sieghart and Karobath, 1980). The preferring BZ_1 receptor agent CL 218.872, possesses anti-conflict activity (animal test for anxiolytic activity) as well as anticonvulsant activity, whereas the sedative effect may be diminished (Lippa *et al.*, 1979*a,b*). The rather limited selectivity of β-CCE and CL 218.872 for BZ_1 versus BZ_2 receptors (ca. 10 times), however, precludes firm conclusions based on pharmacological experiments in particular because BZ_2 receptors are highly occupied when CL 218.872 elicits anticonflict activity (Braestrup *et al.*, 1982*b*). BZ_2 selective agents are not yet available.

Chiu *et al.* (1982) recently showed that the "off-rates" for [^{3}H]-FNM in rat forebrain decreased when [^{3}H]-FNM had been present at the receptors for longer time periods. The authors argued that the apparent receptor multiplicity was due to the presence of two conformations of one class of receptors. Gee and Yamamura (1982) argued that the apparent BZ receptor heterogeneity was an artifact due to nonphysiological assay temperatures (0°C) and that the multiplicity disappeared when assayed at

37°C; on the other hand, Hirsch and Lydigsen (1981) showed evidence for receptor heterogeneity in the brain of the living mouse.

8. BZ RECEPTORS AND GABA

8.1. Physiology and Pharmacology

We often try to understand the mode of action of psychotropic drugs by relating their CNS action to one or the other of the known neurotransmitter systems. The presumed neurotransmitter serotonin may be involved in some way in mediating the effects of benzodiazepines, although there is no evidence which directly relates serotonin to the way benzodiazepines act. On the other hand, there is persuasive evidence that the anticonflict potential ("anxiolytic effect") can be mimicked by agents that reduce serotonergic transmission (Stein, 1980; Cook and Sepinwall, 1975; Tye *et al.*, 1977; Iversen, 1980*b*).

However, the hypothesis that the psychopharmacological effects of the benzodiazepines can be explained by their ability to potentiate and prolong the synaptic actions of the inhibitory neurotransmitter GABA remains a most attractive idea (Iversen, 1980*a*). Application of benzodiazepines to mammalian neuronal systems increases the electrophysiological response to GABA (see Haefely *et al.*, 1979, and Tallman *et al.*, 1980, for a review).

In behavioral terms, it is well known that benzodiazepines are particularly able to protect experimental animals against seizures induced by convulsants that inhibit GABA synthesis or antagonize GABA receptors (Costa and Guidotti, 1979). Furthermore, the benzodiazepines potentiate the ability of the GABA agonists, muscimol and THIP, to enhance the behavioral effects of methylphenidate, an indirectly acting dopamine mimetic (Arnt *et al.*, 1979). Treatment of anxiety is the favorite among the potential uses of benzodiazepines. Unfortunately, it remains unclear whether the anxiolytic effect of benzodiazepines is mediated via a potentiation of GABA neurotransmission in the CNS. The conclusion of several studies is in favor of a GABAergic involvement (Stein *et al.*, 1975; Zakusov *et al.*, 1977; Haefely *et al.*, 1979; Ostrovskaya and Voronina, 1977; Billingsley and Kubena, 1978; Guidotti, 1980), whereas others have not been able to confirm a GABAergic involvement (Cook and Sepinwall, 1975; Sepinwall and Cook, 1978; Sullivan *et al.*, 1978; File and Hyde, 1977; Thiebot *et al.*, 1979). It is noteworthy that GABA is probably involved in approximately 30% of all synapses in the CNS. A complicated phenomenon, such as anxiety, is not likely to occur without the involvement of at least some GABAergic synapses.

8.2. GABAergic Influence on BZ Receptors

8.2.1. GABA Agonists Enhance [^{3}H]Diazepam Binding

A direct functional link between GABA and BZ receptors has been demonstrated in brain membranes at the molecular level. This was demonstrated as an enhanced binding of [^{3}H]diazepam when either GABA, or one of its structurally related analogs, was included in the *in vitro* binding assay for brain benzodiazepine receptors (Tallman *et al.*, 1978; Iversen, 1978; see the footnotes to Table 14 for further references). Binding to BZ receptors in retinal membranes is also enhanced by GABA (Howells and Simon, 1980; Regan *et al.*, 1980*b*). Enhancement of [^{3}H]-FNM binding in rat brain slices by GABA was observed in an autoradiographic study (Palacios *et al.*, 1981). Thorough washing of brain membranes, for example by centrifugation and rehomogenization, is imperative for obtaining good effects with GABA. Endogenous GABA is present in brain tissue in appreciable amounts, and the concentration of endogenous GABA in crude membrane preparations may be as high as 10^{-5} M (Chiu and Rosenberg, 1979*a*), which is close to the maximum effective concentration of GABA. The presence of endogenous GABA in membrane preparations may explain why the inclusion of GABA antagonists sometimes reduces [^{3}H]diazepam binding (Chiu and Rosenberg, 1979*a*; Squires *et al.*, 1980).

Manipulation of GABA levels *in vivo* by lesions or enzyme inhibitors changes BZ receptor binding accordingly (Gallager *et al.*, 1978; Biggio *et al.*, 1979; Shibuya *et al.*, 1980); see, however, Rosenberg and Chiu (1979*a*).

GABA and muscimol increase the affinity of BZ receptors for [^{3}H]-FNM and [^{3}H]diazepam by affecting k_{+1} and k_{-1}, whereas the number of binding sites remain constant (Tallman *et al.*, 1978; Regan *et al.*, 1980*a*).

One way of investigating whether the effect observed for GABA is mediated by a GABA receptor or whether the effect is "unspecific," irrelevant, and artifactual is to study the structural requirements for enhancing [^{3}H]diazepam binding. Within a certain series of GABA analogs (full agonists), there is a striking correlation between the concentration of agonist that enhances [^{3}H]diazepam binding by 50% (the EC_{50} value) and the concentration that inhibits [^{3}H]-GABA binding to presumed GABA receptors on brain membranes by 50% (the IC_{50} value) (Table 14). Furthermore, there is a fair correlation between the potencies of agonists exerting GABA-like effects on cat spinal cord motoneurons and the enhancement of [^{3}H]diazepam binding *in vitro*. Note that Table 14 includes three pairs of stereoisomeric GABA analogs, two of which present stereoselective activity. These findings suggest that GABA receptors are involved. β-Alanine may enhance [^{3}H]diazepam binding under certain conditions (Dudai, 1979).

Distinct GABA receptors with quite different structural requirements have been identified in brain slices (Bowery *et al.*, 1980) and presynaptically on brain synaptosomes (Brennan and Cantrill, 1979). These GABA receptors are activated by baclofen and δ-aminolevulinic acid, respectively, a property not shared by benzodiazepine receptor-linked GABA receptors (see footnote, Table 15).

8.2.2. *Effects of GABA Antagonists*

Another way of determining whether the GABA receptors involved in enhancing [^{3}H]diazepam binding are related to known GABA receptors is to investigate the effects of GABA receptor antagonists. Bicuculline is generally accepted as the most selective GABA antagonist available and is often used to identify GABA receptor-mediated electrophysiological events (Curtis *et al.*, 1971; Curtis and Johnston, 1974). Bicuculline is a potent inhibitor of the effects of muscimol and GABA on [^{3}H]diazepam binding. The effect is stereoselective: (+)-bicuculline methiodide is a more potent inhibitor than (−)-bicuculline methiodide (Tallman *et al.*, 1978). These findings suggest that the GABA recognition sites involved in enhancing [^{3}H]diazepam binding are indeed similar, if not identical, to "classical," electrophysiologically identified GABA receptors.

The ability of several other agents to antagonize the GABA agonist muscimol has been investigated under chloride-free conditions at 0°C (Table 15). IsoTHAZ (see structural formula in Table 14) is a bicyclic GABA analog without GABA-like properties; based on behavioral experiments, it was proposed that IsoTHAZ is a GABA receptor antagonist (Arnt and Krogsgaard-Larsen, 1979). Indeed, IsoTHAZ in high concentrations is a competitive antagonist of the GABA/benzodiazepine receptor complex (Table 15) (Karobath *et al.*, 1981*b*).

RU 5135 is a steroid derivative that inhibits both GABA and glycine receptor binding (Hunt and Clement-Jewery, 1981). RU 5135 is the most potent inhibitor of GABA and muscimol available in relation to BZ receptors (Hunt and Clement-Jewery, 1981) (Table 15).

Strychnine is believed to be a selective inhibitor of glycine receptors in the spinal cord (Curtis and Johnston, 1974). However, in the hippocampus strychnine has been shown to inhibit the electrophysiological effects of both glycine and GABA. Strychnine is a potent, competitive inhibitor of the effects of muscimol on the GABA/benzodiazepine receptor complex (Braestrup and Nielsen, 1980*b*) (Table 15).

Benzylpenicillin (Davidoff, 1972; Curtis and Johnston, 1974) and pentylenetetrazole (McDonald and Barker, 1977) can antagonize some of the electrophysiological effects of GABA. These two agents may affect GABA transmission "downstream" from the receptors (Pickles and Sim-

TABLE 14
GABA Analogs and BZ Receptors

Compound	Structure	Depression of firing of cat spinal neurons[a]	Enhancement of[b] [³H]diazepam binding: Maximum enhancement %	Enhancement of[b] [³H]diazepam binding: EC_{50} (μM)	Inhibition of [³H]-GABA binding[a] IC_{50} (μM)	$\frac{IC_{50}}{EC_{50}}$
Muscimol[d]	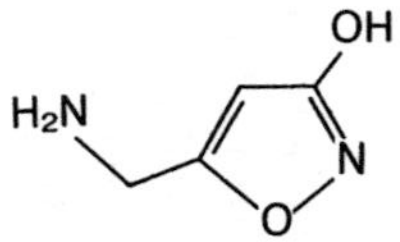	+ + + +	100	0.60	0.006	100
Dihydromuscimol	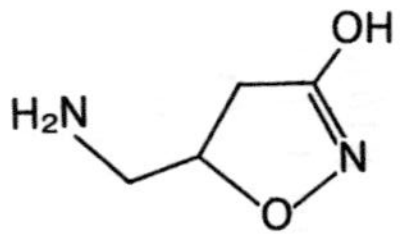	+ + + +	105	1.9	0.009	210
Thiomuscimol	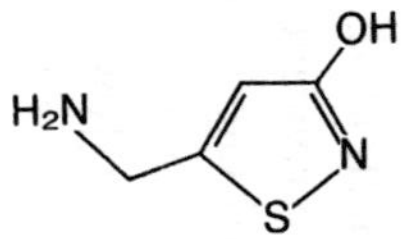	+ + + +	105	4.2	0.019	220
GABA		+ + +	100	3.8	0.033	115
trans-4-Aminocrotonic acid	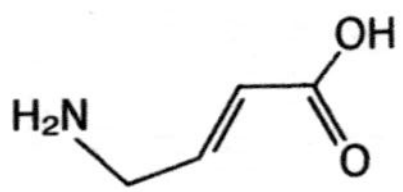	+ + + +	100	7.4	ca. 0.025	ca. 300
(RS)-4-Amino-3-hydroxy-butyric acid	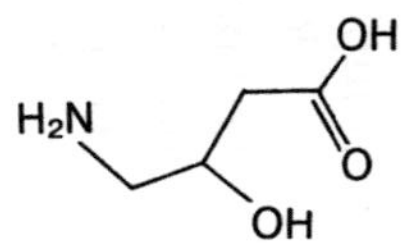	+ + (+)[c]	100	18		

(S)-(−)-5′-Methylmuscimol	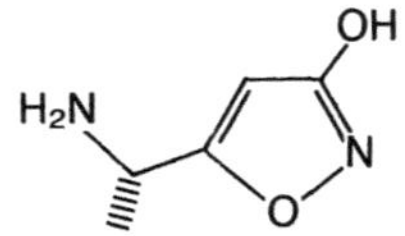	+ + (+)[c]	81	225	0.64	350
Homomuscimol	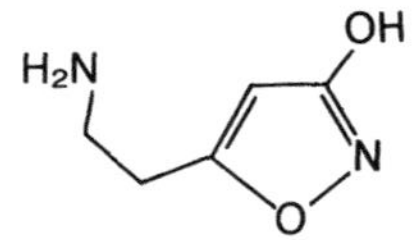	+ + +		290	ca. 3.2	90
(S)-(−)-4-Methyl-TACA	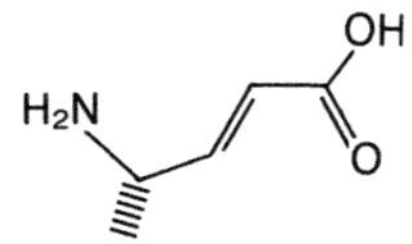	+ + (+)	>80	390	4.1	95
(R)-(+)-4-Methyl-GABA	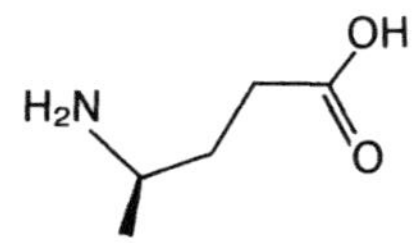		>70	410	5.0	82
(S)-(−)-4-Methyl-GABA	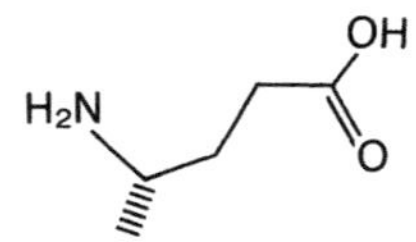		>53	550	4.7	117
(R)-(+)-5′-Methylmuscimol	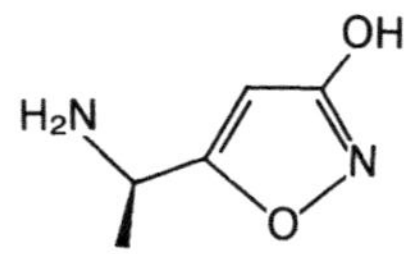	+ + (+)[c]	>30	2300	19	121
Isomuscimol	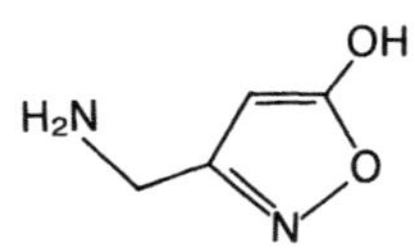	+	>20	>1000	29	>10

(*continued*)

TABLE 14 (*Continued*)

Compound	Structure	Depression of firing of cat spinal neurons[a]	Enhancement of[b] [^{3}H]diazepam binding: Maximum Enhancement %	Enhancement of[b] [^{3}H]diazepam binding: EC_{50} (μM)	Inhibition of [^{3}H]-GABA binding[a] IC_{50} (μM)	$\frac{IC_{50}}{EC_{50}}$
(R)-(+)-4-Methyl-TACA	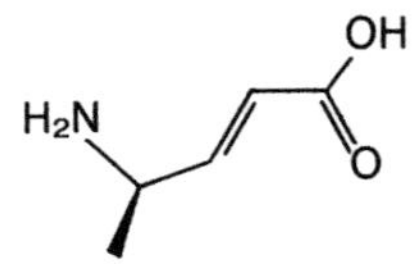			>2000	148	>13
APS	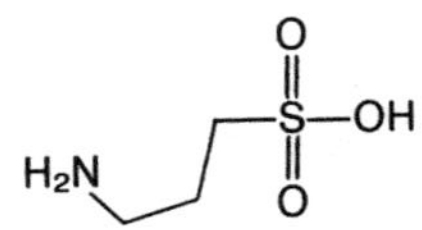	+ + +	42	>1000	ca. 0.006	>10,000
Piperidine-4-sulfonic acid	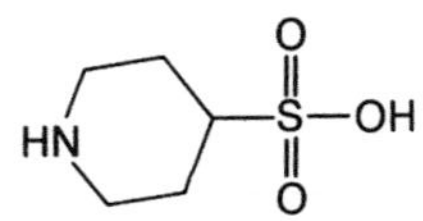	+ + + +	0	>1000	0.034	>10,000
Isoguvacine	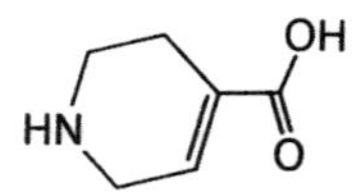	+ + + +	42	>400	0.037	>10,000
Imidazoleacetic acid	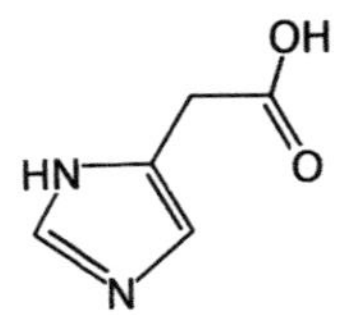		26	>1000	ca. 0.1	>10,000

THIP	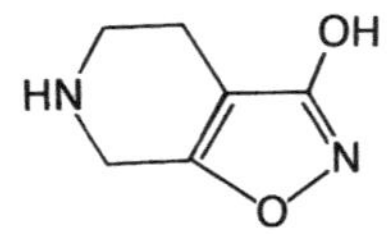	+ + +(+)	11	>1000	0.13	>7,500
β-Guanidinopropionic acid	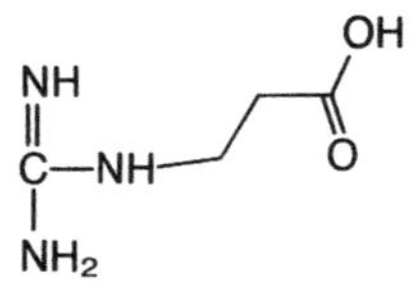		45[c]	>1000		
Kojic amine	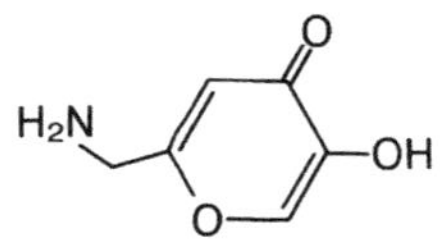		40[c]	>2500		
IsoTHAZ	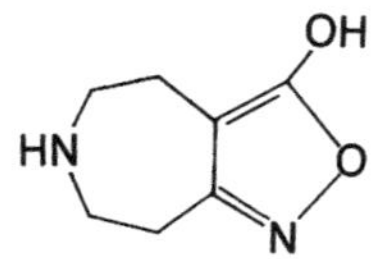	÷[e]	0	>1000	15	>60

[a] The effects of the GABA agonists on the firing of cat spinal neurons were studied using microelectrophoretic techniques. The compounds were administered near interneurons and Renshaw cells for times sufficient to obtain maximum effects of the particular rate of injection. The approximate potencies (+, + +, etc.) of the compounds relative to that of GABA were determined on the basis of electrophoretic currents required to produce equal submaximal inhibition of the firing induced by DL-homocysteate (from Krogsgaard-Larsen *et al.*, 1979; Krogsgaard-Larsen and Arnt, 1980). IsoTHAZ does not inhibit spinal neurons. IC_{50} values for [^{3}H]-GABA binding were determined in rapidly frozen, thoroughly washed rat brain membranes (from Krogsgaard-Larsen *et al.*, 1979; Krogsgaard-Larsen and Arnt, 1980; Karobath *et al.*, 1979; Beaumont *et al.*, 1978). Values preceded by ca. are obtained in other membrane preparations and normalized to muscimol.

[b] The EC_{50} value is the concentration of the agnoist that enhances [^{3}H]diazepam binding by 50% as compared to the maximal enhancement obtainable in that tissue preparation by addition of an optimal concentration of muscimol (10^{-5} M) (from Braestrup *et al.*, 1979*d*; Braestrup and Nielsen, 1980*b*).

[c] Unpublished results.

[d] Enhanced [^{3}H]diazepam binding after addition of GABA or muscimol has been reported in several studies (Briley and Langer, 1978; Chiu and Rosenberg, 1979*a*; Dudai, 1979; Gallager *et al.*, 1978; Gavish and Snyder, 1980*a*; Karobath and Sperk, 1979; Martin and Candy, 1978; Maurer, 1979; Wastek *et al.*, 1978; Williams and Risley, 1979; Massotti *et al.*, 1980; Braestrup *et al.*, 1980*b*).

[e] Tested, but no effect.

TABLE 15

Muscimol-Enhanced [^{3}H]Diazepam Binding; Inhibitors of Muscimol[a,g]

Agent	Inhibition of muscimol enhancement	
	IC_{50} (μM)	K_I (μM)
RU 5135[f]	0.03	
(±)-Bicuculline	1.8	0.5
d-Tubocurarine	2.4	[b]
Piperidine-4-sulfonic acid	12	1.2
Strychnine	6.3	1.9
Bicuculline-methiodide		4.4
Imidazoleacetic acid	19	
THIP[c]	69	7.4
IsoTHAZ[c]		100
Pseudobrucine	32	
Isostrychnine	40	
Laudanosine	75	
Ro 5-3663	1–100[d]	
Aminostrychnine	170	
Kojic amine	100–300	
Gelsemine[e]	>300	
C.O.P.[e]	>300	

[a] K_I values were determined by Schild analysis (see Fig. 9). IC_{50} values were determined the following way: [^{3}H]diazepam binding was submaximally enhanced by muscimol (3×10^{-6} M). Inhibitors were added to membranes in the presence or absence of muscimol, and the concentration of inhibitor at which muscimol only elicited half the percentage increase as compared to uninhibited membranes was determined. Values are the mean of two to five determinations. (From Braestrup and Nielsen, 1980*b*, and unpublished results.)

[b] Not competitive inhibition.

[c] Structural formulas in Table 14.

[d] Ro 5-3663 inhibited GABA-enhanced [^{3}H]diazepam binding in a non-dose-related fashion (O'Brien and Spirt, 1980).

[e] Structural formulas in Curtis and Johnston (1974).

[f] Structural formula in Hunt and Clement-Jewery (1981).

[g] The following agents neither enhanced [^{3}H]diazepam binding nor inhibited muscimol enhancement of [^{3}H]diazepam binding to rat forebrain membranes in chloride-free Tris citrate buffer (unpublished results, method according to Braestrup and Nielsen, 1980*b*) (maximal concentration tested, μM, which did not by itself inhibit [^{3}H]diazepam binding):

Acetylcholine (150)
Amantadine (800)
Apomorphine (250)
Baclofen (400)
Bemegride (100)
Benztropine (250)
Benzylpenicillin (200)
Carbachol (200)
Carisoprodol (350)
Cartazolate (SQ 65,396) (10)
Chlormezanone (350)
Chlorpromazine (300)
Chlorprotixene (300)
Clomipramine (300)
Clonidine (500)
Clozapine (300)
Cocaine (300)
Cyproheptadine (300)
Decamethonium (300)
Diprenorphine (300)
Doxepin (350)
α-Flupenthixol (300)
Glutamate (700)
Glycine (1000)
Haloperidol (250)
Harmaline (400)
Harmine (400)
Hexamethonium (500)
Homocysteate (600)
Hypoxanthine (100)
Iprindole (300)
Isoniazid (100)
IPTBO (500)
Melperone (300)
Metergoline (300)
Morphine (300)
Naloxone (300)
Nicotine (200)
Nikethamide (500)
Norharman (150)
Oxotremorine (200)
Papaverine (350)
Pavulon (150)
Pentylenetetrazole (750)
Phenoxybenzamine (500)
Phentolamine (370)
Picrotoxin (3500)
Piflutixol (300)
Pimozide (300)
Pindolol (300)
Propranolol (400)
Scopolamine (300)
Succinylcholine (250)
Sulpiride (400)
Substance P (20)
Taurine (800)
Thioridazine (200)
Yohimbine (250)

monds, 1980) since the GABA receptor site is not inhibited by benzylpenicillin and pentylenetetrazole (see footnote, Table 15). If the GABA receptors involved in [^{3}H]diazepam enhancement were related to "classical" GABA receptors, it would not be expected that agents unrelated to GABAergic transmission would inhibit the effect of muscimol and GABA. The only agents among many tested (Table 15) that inhibit the effect of muscimol without being directly related to GABA are *d*-tubocurarine and some strychnine analogs. However, antagonism of GABA by *d*-tubocurarine in an electrophysiological paradigm has been reported (Hill *et al.*, 1972; Kelly and Beart, 1975); see, however, Curtis and Johnston (1974).

Ro 5-3663 is a convulsant benzodiazepine that reduces GABA-mediated inhibition in a spinal cat preparation (Schlosser and Franco, 1979). Ro 5-3663 is a potent inhibitor of [^{3}H]picrotoxinin binding to chloride channels (Olsen and Leeb-Lundberg, 1981) and is a weak inhibitor of GABA enhancement of [^{3}H]diazepam binding (O'Brien and Spirt, 1980). It is not clear precisely at what level of the GABA/benzodiazepine complex this interaction occurs.

The pharmacology of the GABA/benzodiazepine receptor complex is dependent on assay parameters. In chloride-containing buffer at 35°C, for example, picrotoxin and IPTBO are capable of reducing GABA-enhanced [^{3}H]-FNM binding in rat cerebellum. The mechanism of action of these agents probably involves chloride channels (Karobath *et al.*, 1981*a*).

The selectivity of the GABA/benzodiazepine receptor complex for agents directed at the GABA receptor recognition site is apparently reduced when chloride is included in the assay medium.

8.2.3. *Partial Agonists*

Surprisingly, a small series of GABA analogs exert GABA-like effects on cat spinal cord motorneurons and inhibit [^{3}H]-GABA binding sites; yet these analogs enhance [^{3}H]diazepam binding only slightly or not at all. This group consists of aminopropanesulfonic acid, isoguvacine, imidazoleacetic acid, β-guanidinopropionic acid, THIP, and piperidine-4-sulfonic acid (Table 15) (Braestrup *et al.*, 1979*d*; Karobath *et al.*, 1979; Maurer, 1979). These analogs have previously been classified as potent GABA agonists because of their ability to cause bicuculline-sensitive inhibition of cat spinal cord neurons, and because of the behavioral effects of, for example, THIP. Interestingly, however, all of these agonists were capable of antagonizing the effects of GABA and muscimol on benzodiazepine binding (Braestrup *et al.*, 1979*d*; Karobath and Lippitsch, 1979) (see Fig. 9). The GABA receptor involved in the increase in [^{3}H]diazepam binding thus recognizes and binds all tested relevant chemical structures even though some agents were not capable of eliciting effects.

Partial agonists were originally defined as those which, even when

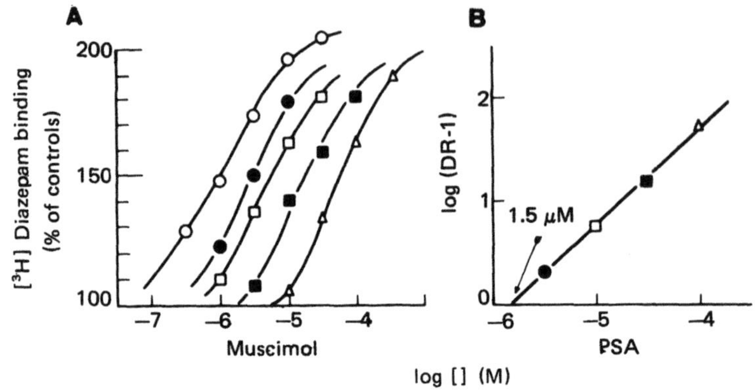

FIG. 9. Schild analysis of piperidine-4-sulfonic acid (PSA) interaction with GABA receptors. (A) Dose–response curve for the effect of muscimol on [^{3}H]diazepam binding to washed rat forebrain membranes. Various concentrations of PSA were added (○, no addition; ●, 3 × 10^{-6} M; □, 10^{-5} M; ■, 3 × 10^{-5} M; △, 10^{-4} M). (B) Schild plot constructed from the data on the left. The dose ratio (DR) is the proportion between EC_{50} values for muscimol with and without PSA added. The almost parallel displacement of dose–response curves by PSA (left) and the slope close to unity (right) indicates that muscimol and PSA compete for the same site in a first-order interaction. Crude membranes from rat forebrain were washed five times by homogenization and recentrifugation in Tris citrate buffer. (From Braestrup *et al.*, 1979*d*).

they occupied all binding sites, were incapable of producing the full response of which tissue was capable (Stephenson, 1956). The subgroup of GABA mimetics mentioned above act as partial agonists or even as competitive antagonists with respect to the conformational change in the GABA receptor that affects benzodiazepine receptors. They appear to be full agonists as far as some other conformational changes are concerned, as, for example, the change in GABA receptors that opens chloride channels in the spinal cord. β-Guanidinopropionic acid, aminopropanesulfonic acid, THIP, and isoguvacine exert partial agonistic properties (or low efficacy) in some electrophysiological paradigms (Dudel, 1965; Takeuchi and Takeuchi, 1975; James *et al.*, 1978; Pickles, 1979; Feltz, 1971).

The explanation for the apparently aberrant manner by which partial agonists interact with the GABA receptors remains unclear. Some of the partial agonists have a more rigid molecular structure than the full agonist, for example THIP, but this is not the case for aminopropanesulfonic acid and β-guanidinopropionic acid. Partial agonists evoke shorter open times for chloride channels in spinal cord neuron cultures than full agonists (Barker and Mathers, 1981), but the significance of this feature is unclear. High-affinity binding studies do not indicate, for example, that [^{3}H]-isoguvacine binds to different sites on brain membranes from those labeled by [^{3}H]-GABA (Krogsgaard-Larsen and Arnt, 1980).

The issue of partial agonists is further complicated by the fact that the effects of partial agonists are dependent on the assay conditions. The presence of chloride in the assay and the use of a physiological temperature increase the efficacy of piperidine-4-sulfonic acid from nil to almost 70% (Supavilai and Karobath, 1980*c*).

Another puzzling aspect is that GABA analogs enhance [^{3}H]diazepam binding at much higher concentrations than those that inhibit high-affinity [^{3}H]-GABA binding to presumed GABA receptors in brain membranes. There is, however, no evidence available to suggest that the high-affinity binding sites for GABA and muscimol obtained *in vitro* by freezing, thawing, and detergent treatment represent the functional state of the GABA receptor *in situ* in physiological conditions. On the contrary, the concentrations of GABA analogs needed to enhance benzodiazepine binding (Table 14 and Fig. 9) are in the micromolar range; and this is in accordance with the effective concentration range for GABA agonists and antagonists in various electrophysiological studies (for references, see Braestrup *et al.,* 1980*b*). Apparently, BZ receptors are coupled to low-affinity GABA receptors.

The GABA/benzodiazepine receptor complex, under chloride-free conditions, may be a relevant *in vitro* model for GABA receptor recognition sites.

8.3. Benzodiazepine Effects on GABA Receptors

While it is well documented that GABA enhances benzodiazepine affinity, it is less clear how benzodiazepines enhance the effects of GABA. Under certain conditions, however, such a reverse interaction can occur, and benzodiazepines enhance [^{3}H]-GABA binding to rat brain membranes (Guidotti *et al.,* 1978; Johnston and Willow, 1981). The concentration of diazepam which enhances [^{3}H]-GABA binding by 50% was estimated to be 13 nM (Johnston and Willow, 1981), which is similar to the concentration which affects BZ receptors. Barbiturates and the phosphodiesterase inhibitors cartazolate and etazolate also enhanced [^{3}H]-GABA and [^{3}H]muscimol binding (Johnston and Willow, 1981; Placheta and Karobath, 1980).

The mechanism by which benzodiazepines enhance [^{3}H]-GABA binding is not completely clear. Costa and co-workers reported that treatment of brain membranes with Triton X-100, which increases the affinity of GABA for GABA receptors (Enna and Snyder, 1977), released a heat-stable 15,000-dalton-molecular-weight protein into the medium (Costa *et al.,* 1978). They named this material "GABA-modulin" (Baraldi *et al.,* 1979) and suggested that GABA-modulin represented an endogenous modulator of GABA receptors (Toffano *et al.,* 1978). In addition to inhibiting GABA, GABA-modulin appears to inhibit benzodiazepine binding competitively by a mechanism which is dependent on the assay conditions (Guidotti *et al.*, 1978; Massotti and Guidotti, 1980). Furthermore, when GABA-modulin was added to Triton X-100 treated membranes, it inhibited high-affinity GABA binding, and this effect was stereospecifically antagonized by benzodiazepines. Thus a model was proposed wherein GABA-modulin interacts with both GABA and benzodiazepine receptors as part of a post synaptic GABA/benzodiazepine/ionophore complex to keep the GABA receptors in a low-affinity conformation (Costa, 1980). Uncertainty about the chemical nature of GABA-modulin persist, however, and endogenous GABA, in a tightly tissue-bound form, has been found to exhibit GABA-modulin-like activity (Napias *et al.*, 1980).

Massotti *et al.* (1981) reported that [^{3}H]diazepam inhibiting activity (DBI) could be separated from GABA-modulin, acting on GABA receptors.

Additional evidence for a mutual interaction between BZ receptors and GABA receptors includes the ability of benzodiazepines to protect GABA receptors against thermal denaturation, and the similar protective effects of GABA on BZ receptors (Squires, 1981; Gavish and Snyder, 1980*b*).

8.4. Barbiturates, Chloride Channels, and BZ Receptors

The binding of [^{3}H]diazepam to BZ receptors is enhanced by certain anions, including iodide, bromide, nitrite, chloride, and thiocyanate. Inactive anions include formate, lactate, acetate, propionate, oxalate, succinate, maleate, fluoride, azide, chlorate, and sulfate (Costa *et al.*, 1979*b,c*; Martin and Candy, 1978, 1980; Squires, 1981). These results suggest that BZ receptors are coupled, either directly or indirectly, to anionic recognition sites, such as the chloride channel. The selectivity of the effective anions on binding, partially agrees with their selectivity for penetrating the activated inhibitory postsynaptic membrane of cat motoneurons measured electrophysiologically (see, however, Martin and Candy, 1980). This correlation has lent additional support to the speculation that benzodiazepine receptors occur in close proximity to chloride channels.

Chloride channels in brain membranes may be labeled by [^{3}H]-α-dihydropicrotoxinin (Ticku and Olsen, 1978). A number of agents compete for these sites, including barbiturates, the convulsant benzodiazepine Ro 5-3663, and some pyrazolopyridines (etazolate and others). Interestingly, all these agents affect benzodiazepine receptor affinity in distinct patterns of interaction without directly interacting at the benzodiazepine recognition site. For example, pentobarbital (10–1000 μM) enhances [^{3}H]diazepam binding to brain BZ receptors twofold in chloride-containing buffers (Leeb-Lundberg *et al.*, 1980; Olsen and Leeb-Lundberg, 1981; Skolnick *et al.*, 1980*b*, 1981; Johnston and Willow, 1981; Ticku, 1981). The effect of pentobarbital is additive to that of muscimol; picrotoxin and bicuculline, but not THIP, partially inhibits the effect of pentobarbital (Leeb-Lundberg *et al.*, 1981). These findings indicate that pentobarbital does not act directly on GABA receptor recognition sites but that GABA receptors are indirectly involved (Olsen and Leeb-Lundberg, 1981; Ticku, 1981). The activity of a series of barbiturates to enhance [^{3}H]diazepam binding was found to correlate well with their activity as anesthetics and their ability to reverse the antagonism of GABA responses by bicuculline. Some barbiturates which show preanesthetic excitation, such as (+)-DMBB and CHEB, enhance [^{3}H]diazepam binding (for a review, see Olsen *et al.*, 1982).

The pyrazolopyridines etazolate (1–10 μM), cartazolate (1 μM), and tracazolate likewise enhance benzodiazepine binding to rat forebrain and cerebellar membranes (Beer *et al.*, 1978; Williams and Risley, 1979; Meiners and Salama, 1982), also in a chloride-dependent, picrotoxin-, and IPTBO-sensitive fashion, probably again involving chloride channels (Supavilai and Karobath, 1979, 1980*a*; Karobath *et al.*, 1981*a,b*; Olsen and Leeb-Lundberg, 1981). Ro 5-3663 inhibits the effect of GABA on [^{3}H]-diazepam binding but not in a competitive way (O'Brien and Spirt, 1980).

Coupling of BZ receptors to chloride channels has been demonstrated in biochemical experiments as described above in the cerebellum, where only the BZ_1 receptor subclass is present. BZ_2 receptors are present in other brain regions in conjunction with BZ_1 receptors. Since the effect of agents acting on chloride channels were observed but reduced in magnitude in brain regions other than the cerebellum, for example, in the hippocampus (Supavilai and Karobath, 1980*b*), it is not clear whether BZ_2 receptors too are coupled to chloride channels.

8.5. Benzodiazepine Receptor/GABA Receptor/Chloride Channel Complex

There are multiple BZ receptors (Section 7) and multiple GABA receptors (Bowery *et al.*, 1980; Brennan and Cantrill, 1979; Nistri *et al.*, 1980). However, there is no firm demonstration of any class of BZ receptors without functional coupling to GABA receptors. On the other hand, the regional distribution of BZ receptors in brain is not identical to the distribution of GABA receptors (Palacios *et al.*, 1981; Placheta and Karobath, 1979). For example, GABA receptors are present in high density in the rat cerebellar granule cell layer, whereas BZ receptors are low in this layer and enriched in the molecular layer (Young and Kuhar, 1979); BZ receptor densities are higher than those of GABA receptors in the thalamus and hypothalamus. Furthermore, in the cerebellum of Weaver mice (Chang *et al.*, 1980), and in the substantia nigra of patients with Parkinson's disease (Moehler and Riederer, 1979), there is a selective reduction in GABA receptors without any concomitant loss of BZ receptor. Triton X-100 in certain concentrations may preferentially solubilize BZ receptors, leaving GABA receptors intact (Chiu and Rosenberg, 1979*b*). Finally, the phylogenetic distribution of BZ and GABA receptors is different (see Table 8). Since there is no fixed stoichiometric molar ratio between BZ and GABA receptor numbers, it is probable that the GABA receptor recognition site does not reside on the same protein as the BZ receptor recognition site, and that at least some GABA receptors exist without accompanying BZ receptors (Unnerstall *et al.*, 1981); whether the opposite occurs is not clear. The above arguments obviously rely on all BZ and GABA receptors being detectable with the techniques applied.

Figure 10 illustrates a model for a BZ receptor/GABA receptor/chloride channel complex (Braestrup and Nielsen, 1980*a*); see also Bowery *et al.* (1980), Olsen and Leeb-Lundberg (1981), and Squires *et al.* (1980).

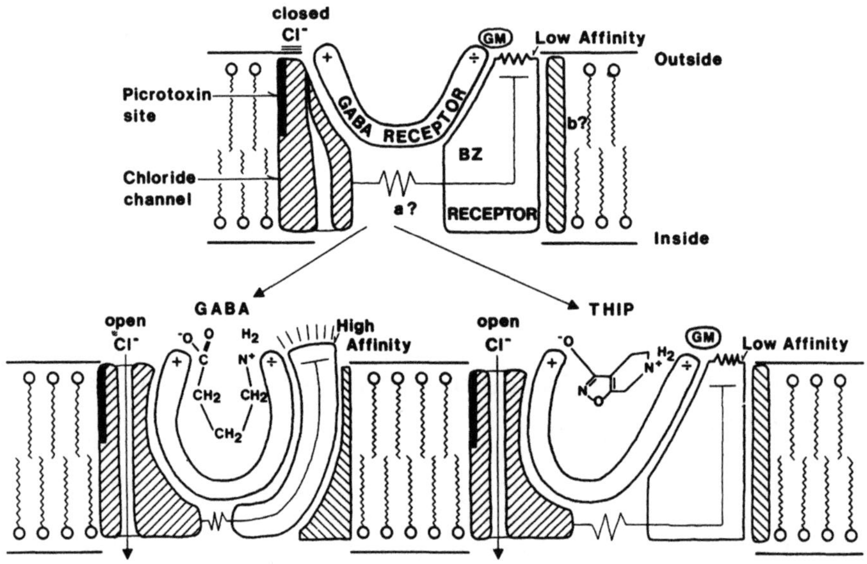

FIG. 10. A model to illustrate a possible coupling among GABA receptors, chloride channels, and benzodiazepine receptors in the double lipid layer of neuronal plasma membranes. GABA-modulin (GM) may interfere with both the GABA and the benzodiazepine receptors. There is a poorly characterized link between the chloride channels and the benzodiazepine receptor (a). Since not all the effects of benzodiazepines are clearly linked to the GABA system, it is also likely that the benzodiazepine receptor is coupled to a third, yet unidentified mechanism (b). The difference in the conformational change in GABA receptors evoked by a full GABA agonist (GABA) and a partial GABA agonist (THIP) is illustrated. Barbiturates may interfere with the picrotoxin site on chloride channels. (From Braestrup and Nielsen, 1980*a*).

The exact number of proteins as well as their precise mutual arrangement is not known. Target size analysis by electron bombardment shows that the target size of the GABA and the BZ receptor is almost identical, corresponding to 215,000 daltons (Chang *et al.*, 1981). GABA receptors and BZ receptors copurify in several fractionation systems (Gavish and Snyder, 1981).

Electrophysiological studies have shown that benzodiazepines increase the frequency by which GABA opens chloride channels (Study and Barker, 1981); single channel conductance or channel open time is not changed.

8.6. Pharmacological Efficacy of BZ Receptor Ligands; Relation to GABA

For some time it was believed that only pharmacologically active benzodiazepines, and a few benzodiazepine-like agents, interacted with high affinity with BZ receptors. Recently, however, agents such as ethyl-β-carboline-3-carboxylate (β-CCE) (Braestrup *et al.*, 1981), Ro 15-1788 (Hunkeler *et al.*, 1981), and CGS 8216 (Czernik *et al.*, 1982) have been discovered which have high affinity for BZ receptors but which lack the characteristic anticonvulsant and anticonflict effects of benzodiazepines. On the other hand, these apparently neutral agents were capable of reversing or preventing benzodiazepines from eliciting their usual sedative, anticonvulsant, and anticonflict effects in animals (Tenen and Hirsch, 1980; Hunkeler *et al.*, 1981; Cowen *et al.*, 1981; Cepeda *et al.*, 1981; Mitchell and Martin, 1980; O'Brien *et al.*, 1981; Polc *et al.*, 1981) and man (Darragh *et al.*, 1981*a,b*), indicating that their interaction with BZ receptors was indeed functional, albeit of a new type. The agents acted as antagonists at the BZ receptor.

In the course of structural modifications of β-CCE, it was observed that methyl-6,7-dimethoxy-β-carboline-3-carboxylate (DMCM, Fig. 11) was a potent convulsant in mice and rats. This effect, which is exactly the opposite of the effects of benzodiazepines, is probably produced by interaction with BZ receptors in a particular way (Braestrup *et al.*, 1982*a*).

Agents such as β-CCE, Ro 15-1788, and CGS 8216 antagonize not only benzodiazepines but also convulsive ligands such as DMCM (Braestrup *et al.*, 1982*a*). These agents thus are not selective benzodiazepine atagonists but rather selective BZ receptor antagonists in the sense that they antagonise effects elicited via BZ receptors. The BZ receptor thus represents a unique example of a receptor for which ligands exist that can affect the receptor in two apparently opposite directions. The ligands can be described as having either increasing positive efficacy (being more and more like benzodiazepines) or negative efficacy (having effects opposite to benzodiazepines) (see Ariens, 1964). True competitive receptor antag-

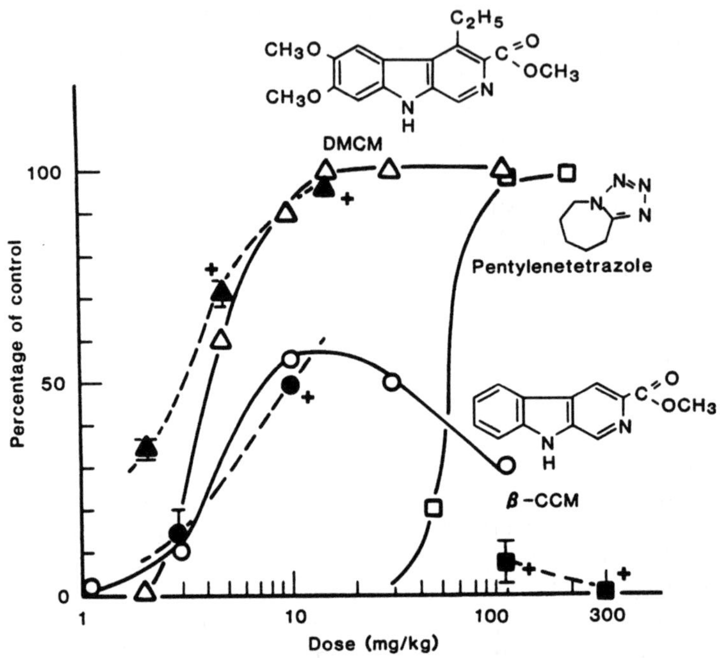

FIG. 11. Dose–response relationship for production of clonic convulsions in mice by DMCM, β-CCM, and pentylenetetrazole (open symbols). Relation to BZ receptor occupancy as measured at a time point when convulsions occur, 7–15 min after administration (closed symbols). [^{3}H]Flunitrazepam ([^{3}H]-FNM) binding *in vivo* in convulsing animals cannot be determined; therefore, we determined those values marked by "+" in animals pretreated with phenobarbital sodium (200 mg/kg p.o., 45 min before decapitation). This dose of phenobarbital sodium did not by itself affect specific [^{3}H]-FNM binding *in vivo*. [^{3}H]-FNM was administered i.v. 7 min before decapitation. (Reproduced from Braestrup *et al.*, 1982*a*.)

onists will have no efficacy. Ro 15-1788 may have a slight positive efficacy (unpublished results) while β-CCE may have a slight negative efficacy (Nutt *et al.*, 1982).

The efficacy of BZ ligands is probably related to the coupling to GABA receptors. Strikingly, it appears that GABA enhances the affinity of one group of benzodiazepine ligands, the benzodiazepine-like agents all having positive efficacy [see references in Section 8.2.1 and Fujimoto *et al.* (1982), Ehlert *et al.* (1982), Moehler and Richards (1981), and Braestrup *et al.* (1982*a*)]; GABA does not greatly affect the affinity of a second group of ligands, BZ receptor antagonists having little or no efficacy (β-CCE, Ro 15-1788, CGS 8216) (Nielsen *et al.*, 1981; Ehlert *et al.*, 1981; Mohler and Richards, 1981; Ehlert *et al.*, 1982; Fujimoto *et al.*, 1982; Skolnick *et al.*, 1982; Braestrup *et al.*, 1982*a*). and GABA reduced the affinity of a third group of ligands having actions opposite to benzodiazepines and having negative efficacy (DMCM, β-CCM) (Braestrup and Nielsen, 1981*c*; Braestrup *et al.*, 1982*a,b*).

9. MISCELLANEOUS MODULATORS

A variety of compounds increase radioligand binding to benzodiazepine receptors. Avermectin B_{1a} is a macrocyclic lactone anthelmintic compound, which increases chloride conductance at the lobster neuromuscular junction probably by a slowly reversible picrotoxin-sensitive opening of GABA receptor-associated chloride channels (Fritz *et al.*, 1979). Avermectin B_{1a} (10 μM) increases [^{3}H]diazepam binding through an effect on K_D and B_{max} (Williams and Yarborough, 1979), and [^{3}H]-β-CCE binding through an effect on B_{max} (Williams and Risley, 1982).

EMD 28422 [N^6-(2-(4-chlorophenyl)-bicyclo-2.2.2-octyl-(3))-adenosine] is a complicated synthetic derivative of adenosine. Enhancement of [^{3}H]diazepam binding to brain BZ receptors by an effect on B_{max} has been reported after *in vitro*, as well as *in vivo*, administration of EMD 28422 (Skolnick *et al.*, 1980*a*).

Various cations, Ni^{2+} and Cd^{2+} in particular, enhance the binding of [^{3}H]diazepam to rat brain membranes by an effect on the affinity constant (Mackerer and Kochman, 1978).

10. *IN VIVO* BZ RECEPTOR MODULATION

10.1. Subchronic Treatment with Benzodiazepines

Continued administration of benzodiazepines induces tolerance to the sedative and the anticonvulsant effects of these drugs. However, the anxiolytic action of the benzodiazepines is persistent (for references, see Braestrup *et al.*, 1979*b*). For these reasons, it seemed of interest to investigate whether repeated benzodiazepine administration would induce receptor changes. The published reports are difficult to interpret due to apparent discrepancies among the findings. A slight (10–20%) decrease (Chiu and Rosenberg, 1978; Rosenberg and Chiu, 1979*b*), an increase (DiStefano *et al.*, 1979), or no change (Massotti *et al.*, 1980; Moehler *et al.*, 1978*b*; Braestrup *et al.*, 1979*b*; Braestrup and Nielsen, 1981*a*) in BZ receptor binding *in vitro* have all been reported in response to prolonged benzodiazepine administration. Chiu and Rosenberg obtained their results after administration of flurazepam, while diazepam and lorazepam were used in the other studies.

However, the extent of coupling between benzodiazepine receptors and GABA receptors may be reduced by prolonged benzodiazepine exposure. The ability of muscimol to enhance [^{3}H]diazepam binding was significantly diminished in mice and newborn rats treated for long periods with diazepam or lorazepam (Massotti *et al.*, 1980; Braestrup and Nielsen, 1981*a*). These results suggest that a functional decoupling between GABA and benzodiazepine receptors may underlie the development of tolerance. Speth *et al.* (1979*b*, and unpublished results) observed that acute treatment with diazepam stabilized BZ receptors during the membrane-washing procedure.

10.2. Barbiturates

Moehler *et al.* (1978*b*) and Rosenberg and Chiu (1979*b*) did not observe changes in benzodiazepine receptor binding *in vitro* after prolonged barbiturate treatment. However, Sonawane *et al.* (1980) have reported that phenobarbital treatment (60 mg/kg, i.p., daily for 4 days) caused a significant reduction in the number of brain binding sites for [^{3}H]diazepam (B_{max}) as well as in the affinity constant in both sexes of mice.

10.3. Ethanol

Benzodiazepines are useful drugs in the management of acute alcohol withdrawal states. In particular, delirium tremens and convulsions can be successfully treated. Karobath *et al.* (1980) treated rats with ethanol in an inhalation paradigm for 19 days and observed no alterations in the total numbers of receptors or in the affinity for [^{3}H]flunitrazepam. The ability of GABA to enhance [^{3}H]flunitrazepam binding was also unchanged. Hemmingsen *et al.* (1982) treated rats with ethanol p.o. every 6 hr for 78 hr (mean blood level of ethanol 4 g/liter) and likewise found no changes in BZ receptor number or affinity and no change in the receptor stimulation by muscimol and GABA, either during treatment or during severe withdrawal reactions 14 hr after the final dose.

It cannot be ruled out, however, that ethanol intoxication for extended periods of time may induce irreversible losses of BZ receptors in the brain. When mice were fed ethanol-containing diets for 7 months and then solid food for 1 month, the density of BZ receptors in whole brain and receptor affinity were marginally lowered by 13% and 30%, respectively (Freund, 1980).

10.4. Diphenylhydantoin

The antiepileptic drug diphenylhydantoin does not interact directly with the benzodiazepine recognition site of benzodiazepine receptors. The mechanism of action of diphenylhydantoin, however, may be mediated via the GABA/benzodiazepine receptor complex. Administration of diphenylhydantoin (100 mg/kg i.p.) to adult rats increased the binding of [^{3}H]diazepam *in vivo* within 1 hr (Gallager *et al.*, 1980; Gallager and Mallorga, 1980). Subsequent *in vitro* studies indicated that the increase in [^{3}H]diazepam binding was attributable to an increase in the number of binding sites, and not to an increase in their affinity. Mimaki *et al.* (1980), however, failed to show any acute effects of diphenylhydantoin on the total number of BZ receptors.

10.5. Seizures and Other "Stress"

Benzodiazepines are effective anticonvulsants in various generalized and partial seizure disorders in man. Benzodiazepine receptors probably participate in processes involved in seizures.

Paul and Skolnick (1978) observed a rapid, 18–23%, increase in the number of available binding sites for [^{3}H]diazepam in rat cerebral cortical membranes following seizures induced by electroshock or pentylenetetrazole. McNamara *et al.* (1980) reported a slowly developing long-lasting increase in [^{3}H]diazepam binding to rat hippocampal membranes after repeated seizures induced by kindling (electric or chemical). This enhancement was associated with an increased number of binding sites (30–40%) rather than by an increase in affinity.

In a case report of a photoepileptic baboon, which died during generalized seizures, it was observed that the heat stability of the benzodiazepine receptor *in vitro* was reduced (Squires *et al.*, 1979*b*) as compared to one nonconvulsing photoepileptic baboon and two baboons without photoepilepsia.

"Stress" forms other than seizures have been investigated. Several exposures on consecutive days to electric foot shock stress, immobilization stress, cold water swim stress, and drug stress (continuous amphetamine intoxication for 5 days) evoked only marginal alterations in benzodiazepine receptors (Braestrup *et al.*, 1979*a*). Daily short periods of separation of newborn rats from their nursing mothers, however, induced a small but long-lasting decrease in the number of receptors in whole cortical membranes of the offspring (Braestrup *et al.*, 1979*a*).

Experimental anxiety in rats, produced by a conflict situation reminiscent of the experimental designs in which benzodiazepines are highly efficient anxiety relievers, decreased [^{3}H]diazepam binding transiently in the frontal cortex (Lippa *et al.*, 1978*a*).

The findings presented in this section show that benzodiazepine receptors are capable of being altered by external stimuli. The alterations have not yet been incorporated into a general theoretical framework, and it has not been determined which type of benzodiazepine receptors are involved or whether the observed effects (or lack of effects) are dependent on GABA.

11. ENDOGENOUS LIGANDS?

Certainly one of the most intriguing questions concerning the benzodiazepine receptor concerns the nature of the presumed endogenous ligand. Benzodiazepine receptors must remain classified as drug receptors until the discovery of endogenous agents (neurotransmitters or neuromodulators, which exert their effects via BZ receptors. The discovery of endogenous ligands for opiate receptors was aided by the fact that appropriate bioassays, such as the guinea pig ileum, and *in vitro* receptor assays were available for testing for opiate activity in numerous samples during fractionation procedures; in searching for endogenous benzodiazepine-like factors, the ability of tissue extracts to inhibit [^{3}H]diazepam or [^{3}H]-FNM binding was used to detect putative ligands.

All of the substances described in this section have been isolated from biological material, and characterized by virtue of their interaction with BZ receptors; they all reveal some interesting features in relation to possible endogenous ligands, but there is no compelling evidence for the exclusive role of any of them.

On the other hand, the direct electrophysiological effects of benzodiazepines, when applied directly to neurons, is remarkably weak. Only in a few studies have direct conductance changes been demonstrated (MacDonald *et al.,* 1979). The pharmacological effects of benzodiazepines appear to involve instead a modulatory action on GABAergic transmission. Thus, the existence of an additional transmitter-like substance may not be a necessary requirement to account for the presence of benzodiazepine receptors.

11.1. β-Carbolines

β-Carboline-3-carboxylic acid as the ethyl ester (β-CCE) was isolated from 1800 liters of human urine by extraction with Chromosorb® (high-surface-area polystyrene material which adsorbs many aromatic compounds), treatment with hot ethanol and purification by Sephadex LH-20 partition chromatography (Nielsen *et al.,* 1979), and preparative HPLC on silica-bonded C_{18} and silica columns (Braestrup *et al.,* 1980*a*). Identification was accomplished by high-resolution mass spectrometry and NMR spectrometry.

From an affinity point of view, β-CCE is very attractive because it possesses an even higher affinity for BZ receptors than diazepam. However, β-CCE is not present, as such, in any measurable concentration in brain or peripheral tissues *in vivo.* Furthermore, it is clear that manipulations during the isolation procedure, which were at first believed to cause hydrolysis of urinary conjugates, were in fact leading to esterifications. When extraction of β-CCE from brain tissue was attempted, the compound was only obtained when esterification with ethanol was included in the isolation procedure.

Brain tissue contains high activities of carboxymethylase enzymes, which convert certain proteins into methyl esters using *S*-adenosylmethionine as the methyl donor (Diliberto and Axelrod, 1974; Wasserman *et al.*, 1975; Kim and Li, 1979). The methyl ester of β-carboline-3-carboxylic acid is also active on BZ receptors (Table 10), but methylation of the free acid with *S*-adenosylmethionine in the presence of brain enzymes has not been demonstrated. Derivatives of β-carboline-3-carboxylic acid other than the esters, for example amides or peptides, may play a role at brain BZ receptors. The presence of such substances in the brain has not been demonstrated, although peptides with N-terminal tryptophan occur in some brain regions (Edvinsson *et al.*, 1973). β-CCE is quite specific for benzodiazepine receptors. GABA receptors, dopamine receptors, acetylcholine receptors, and norepinephrine receptors, identified by high-affinity binding techniques, are not affected by β-CCE (Braestrup *et al.*, 1980*a*).

The presence of tetrahydro-β-carbolines and β-carbolines *in vivo* in animals is difficult to demonstrate unequivocally because of the ease with which such compounds can be formed when tryptophan and tryptamine condense with aldehydes *in vitro* at physiological pH and temperature (McIsaac, 1961). On the other hand, recent studies have provided evidence for the presence of low concentrations (20 ng/g range) of tetrahydronorharman, harman, tetrahydroharman, and 6-hydroxytetrahydronorharman in blood platelets and brain tissue (Honecker and Rommelspacher, 1978; Rommelspacher *et al.*, 1980*b*; Shoemaker *et al.*, 1980; Barker *et al.*, 1979; Kari *et al.*, 1979; Bidder *et al.*, 1979). Carboxylic acid derivatives of tetrahydro-β-carboline have been isolated from cataractous human eye lenses, but not from normal lenses (Dillon *et al.*, 1976). The biosynthetic pathways that might lead to the formation of β-carbolines are uncertain. There is general agreement, however, that tryptophan and tryptamine are involved (Abramovitch and Spenser, 1964). Harman and norharman are much less active on BZ receptors than the 3-carboxy derivatives (Table 10). On the other hand, the presence of harman in the CNS had been claimed, and against this background, Rommelspacher *et al.* (1980*a*) suggested that harman might represent an endogenous ligand for BZ receptors.

11.2. Purines

11.2.1. Hypoxanthine and Inosine

Isolation of an endogenous diazepam-like material from brain tissue was accomplished initially using acidified acetone and methanol (Marangos *et al.*, 1978; Skolnick *et al.*, 1978; Moehler *et al.*, 1979; Asano and Spector, 1979). After further purification, several discrete inhibitory substances were obtained. Two of these inhibitors were identified as the purine-base derivatives hypoxanthine and inosine. Identification was accomplished by

HPLC and mass spectrometry. The possible relation of purines to endogenous ligands for BZ receptors has been reviewed by Marangos *et al.* (1979*c*) and Paul *et al.* (1981*a*). Inhibition of [^{3}H]diazepam binding to brain BZ receptors by hypoxanthine, inosine, and several purine analogs is competitive in nature (Marangos *et al.,* 1979*b*). The affinity of hypoxanthine and inosine for BZ receptors is very weak, approximately 250,000 times less than for diazepam. Hypoxanthine and inosine inhibit [^{3}H]-GABA binding sites in rat brain membranes noncompetitively (Ticku and Burch, 1980).

Although both inosine and hypoxanthine bind to benzodiazepine binding sites, it should be noted that adenosine and guanosine in high concentrations also interact with these sites. Most purines in the brain exist as nucleotides, and adenine nucleotides are the major ones. The sum of ATP, ADP, and AMP is about 2.5 μmol/g wet weight in various mammalian brains (Drewes and Gilboe, 1973; Kleihues *et al.,* 1974). Inosine and hypoxanthine are formed mainly from the adenine nucleotides, normally at low levels. The sum of inosine and hypoxanthine in the cat brain is about 0.05 mM (Kleihues *et al.,* 1974). During ischemia, there is a marked increase in brain hypoxanthine and inosine. It has not been demonstrated whether inosine and hypoxanthine are available in the brain in sufficient concentrations to affect BZ receptors.

A cardinal pharmacological feature of the benzodiazepines is their ability to inhibit completely all phases of seizures evoked by pentylenetetrazole. Intracerebroventricular injections of inosine to particular strains of mice caused a dose-dependent increase in the latency for cloniotonic seizures evoked by intraperitoneal administration of pentylenetetrazole (Skolnick *et al.,* 1979*b*; Lapin, 1980). These results seem to classify inosine and hypoxanthine as "benzodiazepine-like" although with a different profile from the benzodiazepines. On the other hand, other behavioral approaches have shown that inosine and hypoxanthine completely prevented the action of diazepam after intracerebral injection into globus pallidus (Slater and Longman, 1979). It is not clear from electrophysiological studies whether inosine should be classified as an agonist or an antagonist. Both inosine and the benzodiazepine flurazepam produce a similar, rapidly desensitizing excitatory response of rat spinal neurons in tissue culture (MacDonald *et al.,* 1979). In the same cultures, flurazepam blocked another effect of inosine, the slowly nondesensitizing inhibitory response. Thus, inosine may activate two different conductances on spinal cord neurons, whereas flurazepam can activate one of the conductances and antagonize the other.

11.2.2. Adenosine

Adenosine, which is a selective agonist for P_1 purine receptors (Burnstock, 1979), is almost as potent as inosine and hypoxanthine on BZ receptors. Diazepam and the (+)-isomer of oxazepam hemisuccinate inhibit the uptake of [^{3}H]adenosine in brain synaptosomes, slices, and glial cells (Mah and Daly, 1976; Hertz *et al.*, 1979; Phillis *et al.*, 1980; Traversa and Newman, 1979) and potentiate the depressant action of iontophoretically applied adenosine on rat cerebral cortical neurons (Phillis, 1979). The adenosine antagonist, theophylline, reduces flurazepam-induced suppression of cerebral cortical neurons (Phillis *et al.*, 1979). Two inhibitors of adenosine uptake, dipyridamole and hexobendine, are among the few nonbenzodiazepines which inhibit BZ receptors. Phillis and Wu (1980) suggested that BZ receptors are related to adenozine reuptake sites. However, glial cells possess [^{3}H]adenosine uptake sites, but probably not brain-type BZ receptors. There is no evidence available to suggest that benzodiazepines interact directly at the adenosine receptor; diazepam (100 μM) did not inhibit specific binding of [^{3}H]chloradenosine to rat-brain-presumed P_1 adenosine receptors (Williams and Risley, 1980).

11.2.3. Others

Caffeine, which is a purine with low affinity for BZ receptors, reduces the increase of segmental dorsal root potentials and dorsal root reflexes induced by diazepam at the spinal level of cats (Polc, 1980). Thus, in this model, caffeine exerted benzodiazepine antagonistic effects.

1-Methylisoguanosine is a marine natural product isolated from the sponge *Tedania digitala* (see Davies *et al.*, 1980) and from the digestive glands of *Anisodonis nobilis* (Fuhrman *et al.*, 1980). It has a weak activity *in vitro* on BZ receptors (Table 10). Like benzodiazepines, 1-methylisoguanosine has muscle-relaxant activity after oral administration to mice and rats (Baird-Lambert *et al.*, 1980).

11.3. Nicotinamide

Nicotinamide was isolated from bovine and rat brain by virtue of its affinity for BZ receptors (Moehler *et al.*, 1979). Nicotinamide has very low affinity for BZ receptors (Table 10), less than many other chemicals and drugs which are believed not to exert their effects through BZ receptors. Nicotinamide is present in brain tissue in low concentrations (0.1 μmol/g). However, it has been reported that nicotinamide enhances presynaptic inhibition in dorsal root preparations of the cat spinal cord; these effects are mimicked by intravenous injection of diazepam. Nicotinamide restored punishment-suppressed behavior (anticonflict activity) in rats (Moehler *et al.*, 1979). Some investigators, however, have been unable to show benzodiazepine-like activity of nicotinamide (Lapin, 1980; Slater and Longman, 1979; Petersen and Buus-Lassen, 1981). It does not seem likely that nicotinamide is involved in the physiological functions of BZ receptors.

11.4. Miscellaneous Agents and Extracts

A peptide isolated from the small intestine and bile ducts of the rat has been reported to inhibit competitively [^{3}H]diazepam binding to brain receptors (Woolf and Nixon, 1981). This peptide (nepenthin) has a molecular weight of 15,800 daltons and receptor activity remains after proteolytic digestion. Its affinity for brain receptors is quite high (K_I approximately 50 nM). It has not been isolated from brain tissue, but neurons throughout the brain contain nepenthin-immunoreactive material, notably in deep layers of the cerebral cortex (Woolf and Nixon, 1981).

Several unidentified substances in brain extracts having low molecular weights (Karobath *et al.*, 1978; Marangos *et al.*, 1979*a*) or high molecular weights (30,000–70,000 daltons) (Marangos *et al.*, 1979*a*; Davis and Cohen 1980*a,b*; Colello *et al.*, 1978*b*) and which inhibit brain BZ receptors have been described. Some brain extracts are reported to act more potently in inhibiting BZ receptors when GABA is present in the assay (Paul *et al.*, 1981*a*).

DBI might be an endogenous modulating ligand for BZ receptors (Massotti *et al.*, 1981).

Based on observed pharmacological interaction between diazepam and thromboxane A_2, it was suggested that thromboxane A_2 might be a candidate for an endogenous ligand for BZ receptors (Ally *et al.*, 1978).

Further purification and characterization is required before the substances described in this section can be evaluated as potential endogenous ligands for BZ receptors.

12. RADIORECEPTOR ASSAY

Simple, rapid, and sensitive radioreceptor assays have been developed for quantitative determinations of benzodiazepines in biological materials, including serum, blood, saliva, urine, CSF, and brain tissues (Chang and Snyder, 1978; Hunt *et al.*, 1979; Owen *et al.*, 1979; Skolnick *et al.*, 1979*c*; Williamson *et al.*, 1978*a*; Rosenblatt *et al.*, 1979; Fujimoto *et al.*, 1980*a*; Lund, 1981). An advantage of this technique is that it gives a measure of the total pharmacological activity in a sample toward BZ receptors, including pharmacologically active BZ metabolites. Furthermore, a single analytical method can be used for measurements of all clinically active benzodiazepine anxiolytics. The sensitivity of the method increases with the potency of the drug, thus facilitating assays of low-dose benzodiazepines, such as triazolam. A disadvantage of the radioreceptor method is its failure to discriminate between metabolites in drug metabolism studies and its failure to identify the precise nature of the measured entity. Brain

BZ receptors are availble in dry, storage-resistant pharmaceutical preparations that allow for routine determinations of 100 benzodiazepine samples a day with a minimum of equipment and without access to brain tissue (Lund, 1981).

13. REFERENCES

AALTONEN, L., LAMMINTAUSTA, R., KANGAS, L., and NIEMINEN, L., 1979, Benzodiazepine binding to receptors in rat brain synaptosomal membranes; characterization of benzodiazepines and their metabolites using [^{3}H]flunitrazepam, *Acta Physiol. Scand. Suppl.* **473**:68.

ABRAMOVITCH, R. A., and SPENSER, I. D., 1964, The carbolines, in: *Advances in Heterocyclic Chemistry* (A. R. Katritizky, A. J. Boulton, and J. M. Lagowski, eds.), Vol. 3, pp. 79–210, Academic Press, New York.

AGER, I. R., DANSWAN, G. W., HARRISON, D. R., KAY, D. P., KENNEWELL, P. D., and TAYLOR, J. B., 1977, Central nervous system activity of a novel class of annelated 1,4-benzodiazepines, aminomethylene-2,4-dihydro-1*H*-imidazo(1,2-*a*)(1,4)benzodiazepin-1-ones, *J. Med. Chem.* **20**:1035–1041.

ALLY, A. I., MANKU, M. S., HORROBIN, D. F., KARMALI, R. A., MORGAN, R. O., and KARMAZYN, M., 1978, Thromboxane A_2 as a possible natural ligand for benzodiazepine receptors, *Neurosci. Lett.* **7**:31–34.

ANGELIS, L., DE, PREDOMINATO, M., and VERTUA, R., 1972, Stereostructure–activity relationships for oxazepam hemisuccinate, *Arzneim.-Forsch.* **22**:1328–1333.

ANTONIADIS, A., MUELLER, W. E., and WOLLERT, U., 1980*a*, Central nervous system stimulating and depressing drugs as possible ligands of the benzodiazepine receptor, *Neuropharmacology* **19**:121–124.

ANTONIADIS, A., MUELLER, W. E., and WOLLERT, U., 1980*b*, Benzodiazepine receptor interactions may be involved in the neurotoxicity of various penicillin derivatives, *Ann. Neurol.* **8**:71–73.

ARIËNS, E. J., ed., 1964, *Molecular Pharmacology. The Mode of Action of Biologically Active Compounds*, Vol. 1, Academic Press, New York and London.

ARNT, J., and KROGSGAARD-LARSEN, P., 1979, GABA agonists and potential antagonists related to muscimol, *Brain Res.* **177**:395–400.

ARNT, J., CHRISTENSEN, A. V., and SCHEEL-KRUEGER, J., 1979, Benzodiazepines potentiate GABA-dopamine stereotyped dependent gnawing in mice, *J. Pharm. Pharmacol.* **31**:56–58.

ASANO, T., and OGASAWARA, N., 1980, Solubilization of the benzodiazepine receptor from rat brain, *Life Sci.* **26**:607–613.

ASANO, T., and SPECTOR, S., 1979, Identification of inosine and hypoxanthine as endogenous ligands for the brain benzodiazepine-binding sites, *Proc. Natl. Acad. Sci. USA* **76**:977–981.

BAIRD-LAMBERT, J., MARWOOD, J. F., DAVIES, L. P., and TAYLOR, K. M., 1980, 1-Methylisoguanosine: an orally active marine natural product with skeletal muscle and cardiovascular effects, *Life Sci.* **26**:1069–1077.

BALTER, M. B., LEVINE, J., and MANHEIMER, D. I., 1974, Cross-national study of the extent of anti-anxiety/sedative drug use, *N. Engl. J. Med.* **290**:769–774.

BARALDI, M., GUIDOTTI, A., SCHWARTZ, J. P., and COSTA, E., 1979, GABA receptors in clonal cell lines: a model for study of benzodiazepine action at molecular level, *Science* **205**:821–823.

BARKER, S. A., HARRISON, R. E., BROWN, G. B., and CHRISTIAN, S. T., 1979, Gas chromatographic/mass spectrometric evidence for the identification of 1,2,3,4-tetrahydro-β-carboline as an *in vivo* constituent of rat brain, *Biochem. Biophys. Res. Commun.* **87:**146–154.

Barker, J. L., and Mathers, D. A., 1981, GABA analogues activate channels of different duration on cultured mouse spinal neurons, *Science* **213:**358–361.

Battersby, M. K., Richards, J. G., and Moehler, H., 1979, Benzodiazepine receptor: photoaffinity labeling and localization, *Eur. J. Pharmacol.* **57:**277–278.

Beaumont, K., Chilton, W. S., Yamamura, H. I., and Enna, S. J., 1978, Muscimol binding in rat brain: association with synaptic GABA receptors, *Brain Res.* **148:**153–162.

Beer, B., Klepner, C. A., Lippa, A. S., and Squires, R. F., 1978, Enhancement of [^{3}H]-diazepam binding by SQ 65,396: a novel anti-anxiety agent, *Pharmacol. Biochem. Behav.* **9:**849–851.

Bellantuono, C., Reggi, V., Tognoni, G., and Garattini, S., 1980, Benzodiazepines: clinical pharmacology and therapeutic use, *Drugs* **19:**195–219.

Bidder, T. G., Shoemaker, D. W., Boettger, H. G., Evans, M., and Cummins, J. T., 1979, Harman in human platelets, *Life Sci.* **25:**157–164.

Biggio, G., Corda, M. G., Lamberti, C., and Gessa, G. L., 1979, Changes in benzodiazepine receptors following GABAergic denervation of substantia nigra, *Eur. J. Pharmacol.* **58:**215–216.

Biggio, G., Corda, M. G., Montis, G., de, and Gessa, G. L., 1980, Differential effects of kainic acid on benzodiazepine receptors, GABA receptors, and GABA-modulin in the cerebellar cortex, in: *Receptors for Neurotransmitters and Peptide Hormones* (G. Pepeu, M. J. Kuhar, and S. J. Enna, eds.), pp. 265–270, Raven Press, New York.

Billingsley, M. L., and Kubena, R. K., 1978, The effects of naloxone and picrotoxin on the sedative and anticonflict effects of benzodiazepines, *Life Sci.* **22:**897–906.

Bjerrum, O. J., 1977, Immunochemical investigation of membrane proteins: a methodological survey with emphasis placed on immunoprecipitation in gels, *Biochim. Biophys. Acta* **472:**135–195.

Blanchard, J. C., Boireau, A., Garret, C., and Julou, L., 1979, *In vitro* and *in vivo* inhibition by zopiclone of benzodiazepine binding to rodent brain receptors, *Life Sci.* **24:**2417–2420.

Blas, A., de, and Mahler, H. R., 1978, Studies on nicotinic acetylcholine receptors in mammalian brain. Characterization of a microsomal subfraction enriched in receptor function for different neurotransmitters, *J. Neurochem.* **30:**563–577.

Borbe, H. O., Mueller, W. E., and Wollert, U., 1980, The identification of benzodiazepine receptors with brain-like specificity in bovine retina, *Brain Res.* **182:**466–469.

Bosmann, H. B., Case, K. R., and DiStefano, P., 1977, Diazepam receptor characterization: specific binding of a benzodiazepine to macromolecules in various areas of rat brain, *FEBS Lett.* **82:**368–372.

Bosmann, H. B., Penney, D. P., Case, K. R., DiStefano, P., and Averill, K., 1978, Diazepam receptor, specific binding of [^{3}H]diazepam and [^{3}H]flunitrazepam to rat brain subfractions, *FEBS Lett.* **87:**199–202.

Bosmann, H. B., Penney, D. P., Case, K. R., and Averill, K., 1980, Diazepam receptor: specific nuclear binding of [^{3}H]flunitrazepam, *Proc. Natl. Acad. Sci. USA* **77:**1195–1198.

Bowery, N. G., Hill, D. R., and Hudson, A. L., 1980, (−)-Baclofen decreases neurotransmitter release in the mammalian CNS by an action at a novel GABA receptor, *Nature (London)* **283:**92–94.

Braestrup, C., and Nielsen, M., 1978, Ontogenetic development of benzodiazepine receptors in the rat brain, *Brain Res.* **147:**170–173.

Braestrup, C., and Nielsen, M., 1980*a*, Benzodiazepine receptors, *Arzneim.-Forsch.* **30:**852–857.

Braestrup, C., and Nielsen, M., 1980*b*, Strychnine as a potent inhibitor of the brain GABA/benzodiazepine receptor complex, *Brain Res. Bull.* **5:**681–684.

Braestrup, C., and Nielsen, M., 1980*c*, Seaching for endogenous benzodiazepine receptor ligands, *Trends in Pharmacological Sciences*, Nov. pp. 424–427.

Braestrup, C., and Nielsen, M., 1981*a*, Modulation of benzodiazepine receptors, *Adv. Biosciences* **31**:221–227.

Braestrup, C., and Nielsen, M., 1981*b*, [^{3}H]Propyl-β-carboline-3-carboxylate ([^{3}H]-PrCC) as a selective radioligand for the BZ-1 benzodiazepine receptor subclass, *J. Neurochem.* **37**:333–341.

Braestrup, C., and Nielsen, M., 1981*c*, GABA reduces binding of [^{3}H]methyl-β-carboline-3-carboxylate to brain benzodiazepine receptors, *Nature (London)* **294**:472–474.

Braestrup, C., and Squires, R. F., 1977, Specific benzodiazepine receptors in rat brain characterized by high-affinity [^{3}H]diazepam binding, *Proc. Natl. Acad. Sci. USA* **74**:3805–3809.

Braestrup, C., and Squires, R. F., 1978*a*, Brain Specific benzodiazepine receptors, *Br. J. Psychiatry* **133**:249–260.

Braestrup, C., and Squires, R. F., 1978*b*, Pharmacological characterization of benzodiazepine receptors, *Eur. J. Pharmacol.* **48**:263–270.

Braestrup, C., Albrechtsen, R., and Squires, R. F., 1977, High densities of benzodiazepine receptors in human cortical areas, *Nature (London)* **269**:702–704.

Braestrup, C., Nielsen, M., Squires, R. F., and Laurberg, S., 1978*a*, Benzodiazepine receptor in brain, *Acta Psychiatr. Scand. Suppl.* **274**:27–32.

Braestrup, C., Squires, R. F., Bock, E., Pedersen, C. T., and Nielsen, M., 1978*b*, Benzodiazepine receptors: cellular and subcellular localization in brain, in: *Advances in Pharmacology and Therapeutics*, Vol. 7 (J. P. Tillement, ed.), pp. 173–185, Pergamon Press, Oxford.

Braestrup, C.,, Nissen, C., Squires, R. F., and Schousboe, A., 1978*c*, Lack of brain specific benzodiazepine receptors on mouse primary astroglial cultures, *Neurosci. Lett.* **9**:45–49.

Braestrup, C., Nielsen, M., Nielsen, E. B., and Lyon, M., 1979*a*, Benzodiazepine receptors in the brain as affected by different experimental stresses: the changes are small and not undirectional, *Psychopharmacology* **65**:273–277.

Braestrup, C., Nielsen, M., and Squires, R. F., 1979*b*, No changes in rat benzodiazepine receptors after withdrawal from continuous treatment with lorazepam and diazepam, *Life Sci.* **24**:347–350.

Braestrup, C., Nielsen, M., Biggio, G., and Squires, R. F., 1979*c*, Neuronal localization of benzodiazepine receptors in cerebellum, *Neurosci. Lett.* **13**:219–224.

Braestrup, C., Nielsen, M., Krogsgaard-Larsen, P., and Falch, E., 1979*d*, Partial agonists for brain GABA/benzodiazepine receptor complex, *Nature (London)* **280**:331–333.

Braestrup, C., Nielsen, M., and Olsen, C. E., 1980*a*, Urinary and brain β-carboline-3-carboxylates as potent inhibitors of brain benzodiazepine receptors, *Proc. Natl. Acad. Sci. USA* **77**:2288–2292.

Braestrup, C., Nielsen, M., Krogsgaard-Larsen, P., and Falch, E., 1980*b*, Two or more conformations of benzodiazepine receptors depending on GABA receptors and other variables, in: *Receptors for Neurotransmitters and Peptide Hormones* (G. Pepeu, M. J. Kuhar, and S. J. Enna, eds.), pp. 301–311, Raven Press, New York.

Braestrup, C., Nielsen, M., Skovbjerg, H., and Gredal, O., 1981, β-Carboline-3-carboxylates and benzodiazepine receptors, in: *GABA and Benzodiazepine Receptors* (E. Costa, G. DiChiara and G. L. Gessa, eds.), pp. 147–155, Raven Press, New York.

Braestrup, C., Schmiechen, R., Neef, G., Nielsen, M., and Petersen, E. N., 1982*a*, Interaction of convulsive ligands with benzodiazepine receptors, *Science* **216**:1241–43.

Braestrup, C., Schmiechen, R., Nielsen, M., and Petersen, E. N., 1982*b*, Benzodiazepine receptor ligands, receptor occupancy, pharmacological effect, and GABA receptor coupling, in: *Pharmacology of Benzodiazepine* (S. M. Paul, P. Skolnick D., Greenblatt, F. Tallman, and E. Usdin, eds.), Macmillan, New York and London, in press.

Brennan, M. J. W., and Cantrill, R. C., 1979, δ-Aminolaevulinic acid is a potent agonist for GABA autoreceptors, *Nature (London)* **280**:514–515.

BRILEY, M. S., and LANGER, S. Z., 1978, Influence of GABA receptor agonists and antagonists on the binding of [^{3}H]diazepam to the benzodiazepine receptor, *Eur. J. Pharmacol.* **52:**129–132.

BURNSTOCK, G., 1979, Past and current evidence for the purinergic nerve hypothesis, in: *Physiological and Regulatory Functions of Adenosine and Adenine Nucleotides* (H. P. Baer and G. I. Drummond, eds.), pp. 3–32, Raven Press, New York.

BYMASTER, F. P., and WONG, D. T., 1980, Solubilization of γ-aminobutyric acid (GABA) and benzodiazepine (BZ) binding sites by detergents, *Soc. Neurosci. Abstr.* 252.

CACCIA, S., GUISO, G., and GARATTINI, S., 1980, Brain concentrations of clobazam and *N*-desmethylclobazam and antieptazol activity, *J. Pharm. Pharmacol.* **32:**295–296.

CANDY, J. M., and MARTIN, I. L., 1979, The postnatal development of the benzodiazepine receptor in the cerebral cortex and cerebellum of the rat, *J. Neurochem.* **12:**655–658.

CEPEDA, C., TANAKA, T., BESSELIÈVRE, R., POTIER, P., NAQUET, R., and ROSSIER, J., 1981, Proconvulsant effects in baboons of β-carboline, a putative endogenous ligand for benzodiazepine receptors, *Neurosci. Lett.* **24:**53–57.

CHANG, R. S. L., and SNYDER, S. H., 1978, Benzodiazepine receptors: labeling in intact animals with [^{3}H]flunitrazepam, *Eur. J. Pharmacol.* **48:**213–218.

CHANG, R. S. L., and TRAN, V. T., and SNYDER, S. H., 1980, Neurotransmitter receptor localizations: brain lesion induced alterations in benzodiazepine, GABA, β-adrenergic, and histamine H_1-receptor binding, *Brain Res.* **190:**95–110.

CHANG, L-R., BARNARD, E. A., LO, M. M. S., and DOLLY, J. O., 1981, Molecular sizes of benzodiazepine receptors and the interacting GABA receptors in the membrane are identical, *FEBS Lett.* **126:**309–312.

CHIU, T. H., and ROSENBERG, H. C., 1978, Reduced diazepam binding following chronic benzodiazepine treatment, *Life Sci.* **23:**1153–1158.

CHIU, T. H., and ROSENBERG, H. C., 1979*a*, GABA receptor-mediated modulation of [^{3}H]-diazepam binding in rat cortex, *Eur. J. Pharmacol.* **56:**337–345.

CHIU, T. H., and ROSENBERG, H. C., 1979*b*, Differential effects of Triton X-100 on benzodiazepine and GABA binding in a frozen–thawed synaptosomal fraction of rat brain, *Eur. J. Pharmacol.* **58:**335–338.

CHIU, T. H., DRYDEN, D. M., and ROSENBERG, H. C., 1982, Kinetics of [^{3}H]Flunitrazepam binding to membrane-bound benzodiazepine receptors, *Mol. Pharmacol.* **21:**57–65.

COLELLO, G. D., CASE, K., and BOSMANN, H., 1978*a*, Benzodiazepine binding in rat brain: assay parameters, *Pharmacologist* **20:**273.

COLELLO, G. D., HOCKENBERY, D. M., BOSMANN, H. B., FUCHS, S., and FOLKERS, K., 1978*b*, Competitive inhibition of benzodiazepine binding by fractions from porcine brain, *Proc. Natl. Acad. Sci. USA* **75:**6319–6323.

COMAR, D., MAZIERE, M., GODOT, J. M., BERGER, G., SOUSSALINE, F., MENINI, C., ARFEL, G., and NAQUET, R., 1979, Visualisation of [^{11}C]flunitrazepam displacement in the brain of the live baboon, *Nature (London)* **280:**329–331.

COOK, L., and SEPINWALL, J., 1975, Behavioral analysis of the effects and mechanisms of action of benzodiazepines, in: *Mechanism of Action of Benzodiazepines* (E. Costa and P. Greengard, eds.), pp. 1–28, Raven Press, New York.

COOK, L., and SEPINWALL, J., 1980, Relationship of anticonflict activity of benzodiazepines to brain receptor binding, serotonin, and GABA, *Psychopharmacol. Bull.* **16:**30–32.

COSTA, E., 1980, Benzodiazepines and neurotransmitters, *Arzneim.-Forsch.* **30:**858–861.

COSTA, E., and GUIDOTTI, A., 1979, Molecular mechanisms in the receptor action of benzodiazepines, *Annu. Rev. Pharmacol. Toxicol.* **19:**531–545.

COSTA, E., GUIDOTTI, A., MAO, C. C., and SURIA, A., 1975, New concepts on the mechanism of action of benzodiazepines, *Life Sci.* **17:**167–186.

COSTA, E., GUIDOTTI, A., and TOFFANO, G., 1978, Molecular mechanisms mediating the action of diazepam on GABA receptors, *Br. J. Psychiatry* **133:**239–248.

COSTA, T., PERT, A., and PERT, C. B., 1979*a*, Lesions of the substantia nigra reveal benzodiazepine receptors with and without γ-aminobutyric acid receptor linkage, *Soc. Neurosci. Abstr.* 552.

COSTA, T., RODBARD, D., and PERT, C. B., 1979*b*, Interaction between the benzodiazepine receptor, the GABA receptor, and a common anion channel, in: *Physical Chemical Aspects of Cell Surface Events in Cellular Regulation* (C. DeLisi and R. Blumenthal, eds.), pp. 37–52, Elsevier/North-Holland, New York.

COSTA, T., RODBARD, D., and PERT, C. B., 1979*c*, Is the benzodiazepine receptor coupled to a chloride anion channel? *Nature* (*London*) **277:**315–317.

COWEN, P. J., GREEN, A. T., NUTT, D. J., and MARTIN, I. L., 1981, Ethyl-β-carboline carboxylate lowers seizure threshold and antagonizes flurazepam-induced sedation in rats, *Nature* (*London*) **290:**54–55.

CURTIS, D. R., and JOHNSTON, G. A. R., 1974, Convulsant alkaloids, in: *Neuropoisons* (L. L. Simpson, ed.), pp. 207–248, Plenum Press, New York.

CURTIS, D. R., DUGGAN, A. W., FELIX, D., and JOHNSTON, G. A. R., 1971, Bicuculline, an antagonist of GABA and synaptic inhibition of the spinal cord of the cat, *Brain Res.* **32:**69–96.

CZERNIK, A. J., PETRACK, B., KALINSKY, H. J., PSYCHOYOS, S., CASH, W. D., TSAI, C., RINEHART, R. K., GRANAT, F. R., LOVELL, R. A., BRUNDISH, D. E., and WADE, R., 1982, CGS 8216: Receptor binding characteristics of a potent benzodiazepine antagonist, *Life Sci.* **30:**363–372.

DAMM, H. W., MUELLER, W. E., SCHLAEFER, U., and WOLLERT, U., 1978, [^{3}H]flunitrazepam: its advantages as a ligand for the identification of benzodiazepine receptors in rat brain membranes, *Res. Commun. Chem. Pathol. Pharmacol.* **22:**597–600.

DAMM, H. W., MUELLER, W. E., and WOLLERT, U., 1979, Is the benzodiazepine receptor purinergic? *Eur. J. Pharmacol.* **55:**331–333.

DARRAGH, A., LAMBE, R., SCULLY, M., BRICK, I., O'BOYLE, C., and WILSON DOWNIE, W., 1981*a*, Investigation in man of the efficacy of a benzodiazepine antagonist, Ro 15-1788, *Lancet* **ii:**8–10.

DARRAGH, A., LAMBE, R., and BRICK, I. 1981*b*, Reversal of benzodiazepine-induced sedation by intravenous Ro 15-1788, *Lancet* **ii:**1042.

DAVIDOFF, R. A., 1972, Penicillin and presynaptic inhibition in the amphibian spinal cord, *Brain Res.* **36:**218–222.

DAVIES, L. P., COOK, A. F., POONIAN, M., and TAYLOR, K. M., 1980, Displacement of [^{3}H]-diazepam binding in rat brain by dipyridamole and by 1-methylisoguanosine, a marine natural product with muscle relaxant activity, *Life Sci.* **26:**1089–1095.

DAVIS, L. G., and COHEN, R. K., 1980*a*, Identification of an endogenous peptide-ligand for the benzodiazepine receptor, *Biochem. Biophys. Res. Commun.* **92:**141–148.

DAVIS, L. G., and COHEN, K., 1980*b*, Inhibition of [^{3}H]diazepam binding by an endogenous fraction from rat brain synaptosomes, *J. Pharm. Pharmacol.* **32:**218–219.

DILIBERTO, E. J., and AXELROD, J., 1974, Characterization and substrate specificity of a protein carboxymethylase in the pituitary gland, *Proc. Natl. Acad. Sci. USA* **71:**1701–1704.

DILLON, J., SPECTOR, A., and NAKANISHI, K., 1976, Identification of β-carbolines isolated from fluorescent human lens proteins, *Nature* (*London*) **259:**422–423.

DISTEFANO, P., CASE, K. R., COLELLO, G. D., and BOSMANN, H. B., 1979, Increased specific binding of [^{3}H]diazepam in rat brain following chronic diazepam administration, *Cell Biol. Int. Rep.* **3:**163–167.

DOBLE, A., and IVERSEN, L. L., 1982, Molecular size of benzodiazepine receptor in rat brain *in situ:* evidence for a functional dimer? *Nature* (*London*) **295:**522–523.

DREWES, L. R., and GILBOE, D. D., 1973, Glycolysis and the permeation of glucose and lactate in the isolated, perfused dog brain during anoxia and postanoxic recovery, *J. Biol. Chem.* **248:**2489–2496.

DUDAI, Y., 1979, Modulation of benzodiazepine binding sites in calf cortex by an endogenous factor and GABAergic ligands, *Brain Res.* **167:**422–425.

DUDAI, Y., and SHERMAN-GOLD, R., 1980, Studies on the properties of benzodiazepine binding sites from calf cortex, *Prog. Biochem. Pharmacol.* **16**:95–108.

DUDAI, Y., YAVIN, Z., and YAVIN, E., 1979, Binding of [^{3}H]flunitrazepam to differentiating rat cerebral cells in culture, *Brain Res.* **177**:418–422.

DUDEL, J., 1965, Presynaptic and postsynaptic effects of inhibitory drugs on the crayfish neuromuscular junction, *Pfluegers Arch.* **283**:104–118.

DUKA, T., HOELLT, V., and HERZ, A., 1979, *In vivo* receptor occupation by benzodiazepines and correlation with the pharmacological effect, *Brain Res.* **179**:147–156.

EDVINSSON, L., HAAKANSON, R., ROENNBERG, A.-L., and SUNDLER, F., 1973, Tryptophyl-polypeptides in rat brain, *J. Neurochem.* **20**:897–899.

EHLERT, F. J., ROESKE, W. R., and BRAESTRUP, C., YAMAMURA, S. H., and YAMAMURA, H. I., 1981, γ-Aminobutyric acid regulation of the benzodiazepine receptor: biochemical evidence for pharmacologically different effects of benzodiazepines and propyl β-carboline-3-carboxylate, *Eur. J. Pharmacol.* **70**:593–596.

EHLERT, F. J., RAGAN, P., CHEN, A., ROESKE, W. R., and YAMAMURA, H. I., 1982, Modulation of benzodiazepine receptor binding: insight into pharmacological efficacy, *Eur. J. Pharmacol.* **78**:249–253.

ENNA, S. J., and SNYDER, S. H., 1977, Influences of ions, enzymes, and detergents on γ-aminobutyric acid receptor binding in synaptic membranes of rat brain, *Mol. Pharmacol.* **13**:442–453.

FELTZ, A., 1971, Competitive interaction of β-guanidinopropionic acid and γ-aminobutyric acid on the muscle fibre of the crayfish, *J. Physiol.* **216**:391–401.

FERNHOLM, B., NIELSEN, M., and BRAESTRUP, C., 1979, Absence of brain-specific benzodiazepine receptors in cyclostomes and elasmobranchs, *Comp. Biochem. Physiol.* **62C**:209–211.

FILE, S. E., and HYDE, J. R. G., 1977, The effects of *p*-chlorophenyllanine and ethanolamine *O*-sulphate in an animal test of anxiety, *J. Pharm. Pharmacol.* **29**:735–738.

FONG, J., and GOLDSTEIN, M., 1980, Purification of the solubilized benzodiazepine receptor, *Soc. Neurosci. Abstr.* 255

FREUND, G., 1980, Benzodiazepine receptor loss in brains of mice after chronic alcohol consumption, *Life Sci.* **27**:987–992.

FRITZ, L. C., WANG, C. C., and GORIO, A., 1979, Avermectin B_{1a} irreversibly blocks postsynaptic potential at the lobster neuromuscular junction by reducing muscle membrane resistance, *Proc. Natl. Acad. Aci. USA* **76**:2062–2066.

FUHRMAN, F. A., FUHRMAN, G. J., KIM, Y. H., PAVELKA, L. A., and MOSHER, H. S., 1980, Doridosine: a new hypotensive *N*-methylpurine riboside from the nudibranch *Anisodoris nobilis, Science* **207**:193–195.

FUJIMOTO, M., TSUKINOKI, Y., HIROSE, K., KURUMA, K., KONAKA, R., and OKABAYASHI, T., 1980*a,* Detection and determination of pharmacologically active benzodiazepines in rat brain after the administration of 2-*o*-chlorobenzoyl-4-chloro-*N*-methyl-*N'*-glycylglycinanilide, using a combination of high-pressure liquid chromatography and radioreceptor assay, *Chem. Pharm. Bull.* **28**:1378–1386.

FUJIMOTO, M., TSUKINOKI, Y., HIROSE, K., HIRAI, K., and OKABAYASHI, T., 1980*b,* Interaction of peptide-aminobenzophenones with benzodiazepine receptors, *Chem. Pharm. Bull.* **28**:1374–1377.

FUJIMOTO, M., HIRAI, K., and OKABAYASHI, T., 1982, Comparison of the effects of GABA and chloride ion on the affinities of ligands for the benzodiazepine receptor, *Life Sci.* **30**:51–57.

FUXE, K., KOEHLER, C., AGNATI, L. F., ANDERSSON, K., OEGREN, S.-O., ENEROTH, P., PÉREZ DE LA MORA, M., KAROBATH, M., and KROGSGAARD-LARSEN, P., 1981, GABA and benzodiazepine receptors. Studies on their localization in the hippocampus and their interaction with central dopamine neurons in the rat brain, in: *GABA and Benzodiazepine Receptors* (E. Costa, GiDichiara, and G. L. Gessa, eds.), pp. 61–76, Raven Press, New York.

GALLAGER, D. W., and MALLORGA, P., 1980, Diphenylhydantoin: pre- and postnatal administration alters diazepam binding in developing rat cerebral cortex, *Science* **208**:64–66.

GALLAGER, D. W., THOMAS, J. W., and TALLMAN, F., 1978, Effect of GABAergic drugs on benzodiazepine binding site sensitivity in rat cerebral cortex, *Biochem. Pharmacol.* **27**:2745–2749.

GALLAGER, D. W., MALLORGA, P., and TALLMAN, J. F., 1980, Interaction of diphenylhydantoin and benzodiazepines in the CNS, *Brain Res.* **189**:209–220.

GALLAGER, D. W., MALLORGA, P., OERTEL, W., HENNEBERRY, R., and TALLMAN, J., 1981, [^{3}H]-Diazepam binding in mammalian central nervous system: a pharmacological characterization, *J. Neurosci.* **1**:218–225.

GARATTINI, S., MUSSINI, E., MARCUCCI, F., and GUAITANI, A., 1973, Metabolic studies on benzodiazepines in various animal species, in: *The Benzodiazepines* (S. Garattini, E. Mussini, and L. O. Randall, eds.), pp. 75–97, Raven Press, New York.

GAVISH, M., and SNYDER, S. H., 1980*a*, Soluble benzodiazepine receptors: GABAergic regulation, *Life Sci.* **26**:579–582.

GAVISH, M., and SNYDER, S. H., 1980*b*, Benzodiazepine recognition sites on GABA receptors, *Nature (London)* **287**:651–652.

GAVISH, M., and SNYDER, S. H., 1981, γ-Aminobutyric acid and benzodiazepine receptors: copurification and characterization, *Proc. Natl. Acad. Sci. USA* **78**:1939–1942.

GAVISH, M., CHANG, R. S. L., and SNYDER, S. H., 1979, Solubilization of histamine H-1, GABA, and benzodiazepine receptors, *Life Sci.* **25**:783–790.

GEE, K. W., and YAMAMURA, H. I., 1982, Benzodiazepine receptor heterogeneity: a consequence of multiple conformational states of a single receptor or multiple populations of structurally distinct macromolecules? Pharmacological effect and GABA receptor coupling, in: *Pharmacology of Benzodiazepine* (S. M. Paul, P. Skolnick, D. Greenblatt, F. Tallman, and E. Usdin, eds.), Macmillan, New York and London, in press.

GERVASI, G. B., BOSELLI, P., CALIARI, S., and BIANCHETTI A., 1978, Pharmacological activities and high-affinity binding to rat brain synaptosomes of pinazepam and other 4-propargyl-substituted benzodiazepines, Abstract, 2nd World Congress of Biological Psychiatry, Barcelona, p. 148.

GLOWINSKI, J., and IVERSEN, L. L., 1966, Regional studies of catecholamines in the rat brain. I. Disposition of [^{3}H]norepinephrine, [^{3}H]dopamine, and [^{3}H]-dopa in various regions of the brain, *J. Neurochem.* **13**:655–669.

GREENLEE, D. V., and OLSEN, R. W., 1979, Solubilization of gamma-aminobutyric acid receptor protein from mammalian brain, *Biochem. Biophys. Res. Commun.* **88**:380–387.

GUIDOTTI, A., 1980, Pharmacological and biochemical evidence for an interaction between the benzodiazepine and GABA receptors, in: *Receptors for Neurotransmitters and Peptide Hormones* (G. Pepeu, M. J. Kuhar, and S. J. Enna, eds.), pp. 271–275, Raven Press, New York.

GUIDOTTI, A., TOFFANO, G., and COSTA, E., 1978, An endogenous protein modulates the affinity of GABA and benzodiazepine receptors in rat brain, *Nature (London)* **275**:553–555.

HAEFELY, W., KULCSÁR, A., MOEHLER, H., PIERI, L., POLC, P., and SCHAFFNER, R., 1975, Possible involvement of GABA in the central actions of benzodiazepines, in: *Mechanism of Action of Benzodiazepines* (E. Costa and P. Greengard, eds.), pp. 131–151, Raven Press, New York.

HAEFELY, W., POLC, P., SCHAFFNER, R., KELLER, H. H., PIERI, L., and MOEHLER, H., 1979, Facilitation of GABAergic transmission by drugs, in: *GABA-Neurotransmitters* (P. Krogsgaard-Larsen, J. Scheel-Krueger, and H. Kofod, eds.), pp. 357–375, Munksgaard, Copenhagen, and Academic Pres, New York.

HEMMINGSEN, R., BRAESTRUP, C., and NIELSEN, M., and BARRY, D. I., 1982, The GABA/benzodiazepine receptor complex during severe ethanol intoxication and withdrawal in the rat, *Acta Psychiat. Scand.* **65**:120–126.

Henn, F. A., and Henke, D. J., 1978, Cellular localization of [^{3}H]diazepam receptors, *Neuropharmacology* **17**:985–988.
Henn, F. A., Deering, J., and Anderson, D., 1980, Receptor studies on isolated astroglial cell fractions prepared with and without trypsin, *Neurochem. Res.* **5**:459–464.
Hertz, L., and Mukerji, S., 1980, Diazepam receptors on mouse astrocytes in primary cultures: displacement by pharmacologically active concentrations of benzodiazepines or barbiturates, *Can. J. Physiol. Pharmacol.* **58**:217–220.
Hertz, L., Wu, P. H., and Phillis, J. W., 1979, Benzodiazepines and purinergic depression of central neurones, *Soc. Neurosci. Abstr.* **5**:404.
Hill, R. G., Simmons, M. A., and Straughan, D. W., 1972, Convulsive properties of *d*-tubocurarine and cortical inhibition, *Nature* (*London*) **240**:51–52.
Hirsch, J. D., and Lydigsen, J. L., 1981, Binding of β-carboline-3-carboxylic acid ethyl ester to mouse brain benzodiazepine receptors *in vivo, Eur. J. Pharmacol.* **72**:357–360.
Honecker, H., and Rommelspacher, H., 1978, Tetrahydronorharmane (tetrahydro-β-carboline), a physiologically occurring compound of indole metabolism, *Naunyn-Schmiedeberg's Arch. Pharmacol.* **305**:135–141.
Howells, R., and Simon, E. J., 1980, Benzodiazepine binding in chicken retina and its interaction with γ-aminobutyric acid, *Eur. J. Pharmacol.* **67**:133–137.
Howells, R., Hiller, J. M., and Simon, E. J., 1979, Benzodiazepine binding sites are present in retina, *Life Sci.* **25**:2131–2136.
Huang, A., Barker, J. L., Paul, S. M., Moncada, V., and Skolnick, P., 1980, Characterization of benzodiazepine receptors in primary cultures of fetal mouse brain and spinal cord neurons, *Brain Res.* **190**:485–491.
Hunkeler, W., Moehler, H., Pieri, L., Polc, P., Bonetti, E. P., Cumin, R., Schaffner, R., and Haefely, W., 1981, Selective antagonists of benzodiazepines, *Nature* (*London*) **209**:514–516.
Hunt, P., and Clement-Jewery, S., 1981, A steroid derivative, R 5135, antagonizes the GABA/benzodiazepine receptor interaction, *Neuropharmacology* **20**:357–361.
Hunt, P., Husson, J.-M., and Raynaud, J.-P., 1979, A radioreceptor assay for benzodiazepines, *J. Pharm. Pharmacol.* **31**:448–451.
Iversen, L. L., 1978, GABA and benzodiazepine receptors, *Nature* (*London*) **275**:477.
Iversen, L. L., 1980*a*, The present status of benzodiazepines in psychopharmacology, *Arzneim.-Forsch.* **30**:907–910.
Iversen, S. D., 1980*b*, Animal models of anxiety and benzodiazepine actions, *Arzneim.-Forsch.* **30**:862–868.
James, V. A., Krogsgaard-Larsen, P., and Walker, R. J., 1978, The action of conformationally restricted analogues of GABA on *Limulus* and *Helix* central neurones, *Experientia* **34**:1630–1631.
Johnston, G. A. R., and Willow, M., 1981, Barbiturates and GABA receptors, in: *GABA and Benzodiazepine Receptors* (E. Costa, G. DiChiara and G. L. Gessa, eds.), pp. 191–198, Raven Press, New York.
Johnson, R. W., and Yamamura, H. I., 1979, Photoaffinity labeling of the benzodiazepine receptor in bovine cerebral cortex, *Life Sci.* **25**:1613–1620.
Kamioka, T., Takagi, H., Kobayashi, S., and Suzuki, Y., 1972, Pharmacological Studies on 10-chloro-11*b*-(2-chlorophenyl)-2,3,5,6,7,11*b*-hexahydrobenzo-(6,7)-1,4-diazepinol(5,4-*b*)-oxazol-6-one (CS-370), a new psychosedative agent, *Arzneim.-Forsch.* **22**:884–891.
Kari, I., Peura, P., and Airaksinen, M. M., 1979, Mass fragmentographic determination of tetrahydro-β-carboline in human blood platelets and plasma, *Med. Biol.* **57**:412–414.
Karobath, M., and Lippitsch, M., 1979, THIP and isoguvacine are partial agonists of GABA-stimulated benzodiazepine receptor binding, *Eur. J. Pharmacol.* **58**:485–488.
Karobath, M., and Sperk, G., 1979, Stimulation of benzodiazepine receptor binding by γ-aminobutyric acid, *Proc. Natl. Acad. Sci. USA* **76**:1004–1006.
Karobath, M., and Supavilai, P., 1982, Distinction of benzodiazepine agonists from

antagonists by photoaffinity labeling of benzodiazepine receptors *in vitro, Neurosci. Lett.* **31**:65–69.

Karobath, M., Sperk, G., and Schoenbeck, G., 1978, Evidence for an endogenous factor interfering with [^{3}H]diazepam binding to rat brain membranes, *Eur. J. Pharmacol.* **49**:323–326.

Karobath, M., Placheta, P., Lippitsch, M., and Krogsgaard-Larsen, P., 1979, Is stimulation of benzodiazepine receptor binding mediated by a novel GABA receptor? *Nature (London)* **278**:748–749.

Karobath, M., Rogers, J., and Bloom, F. E., 1980, Benzodiazepine receptors remain unchanged after chronic ethanol administration, *Neuropharmacology* **19**:125–128.

Karobath, M., Drexler, G., and Supavilai, P., 1981*a*, Modulation by picrotoxin and IPTBO of [^{3}H]flunitrazepam binding to the GABA/benzodiazepine receptor complex of rat cerebellum, *Life Sci.* **28**:307–313.

Karobath, M., Supavilai, P., Placheta, P., and Sieghart, W., 1981*b*, Interactions of anxiolytic drugs with benzodiazepine receptors, *Adv. Biosciences* **31**:229–238.

Kelly, J. S., and Beart, P. M., 1975, Amino acid receptors in CNS. II. GABA in supraspinal regions, in: *Handbook of Psychopharmacology,* Vol. 4, *Amino Acid Neurotransmitters* (L. L. Iversen, S. D. Iversen, and S. H. Snyder, eds.), pp. 129–209, Plenum Press, New York.

Kim, S., and Li, C. H., 1979, Enzymatic methyl esterification of specific glutamyl residue in corticotropin, *Proc. Natl. Acad. Sci. USA* **76**:4255–4257.

Kleihues, P., Kobayashi, K., and Hossmann, K.- A., 1974, Purine nucleotide metabolism in the cat brain after one hour of complete ischemia, *J. Neurochem.* **23**:417–425.

Klepner, C. A., Lippa, A. S., Benson, D. I., Sano, M. C., and Beer, B., 1979, Resolution of two biochemically and pharmacologically distinct benzodiazepine receptors, *Pharmacol. Biochem. Behav.* **11**:457–462.

König, J. F. R., and Klippel, R. A., 1963, The rat brain, *A Sterotaxic Atlas of the Forebrain and Lower Parts of the Brain Stem,* Williams and Wilkins, Baltimore, Maryland.

Krogsgaard-Larsen, P., and Arnt, J., 1980, Pharmacological studies of interactions between benzodiazepines and GABA receptors. *Brain Res. Bull. Suppl. 5* **2**:867–872.

Krogsgaard-Larsen, P., Hjeds, H., Curtis, D. R., Lodge, D., and Johnston, G. A. R., 1979, Dihydromuscimol, thiomuscimol, and related heterocyclic compounds as GABA analogues, *J. Neurochem.* **32**:1717–1724.

Laduron, P. M., Janssen, P. F. M., and Leysen, J. E., 1978, Spiperone: a ligand of choice for neuroleptic receptors, *Biochem. Pharmacol.* **27**:323–328.

Landis, S., 1973, Ultrastructural changes in the mitochondria of cerebellar Purkinje cells of nervous mutant mice, *J. Cell Biol.* **57**:782–797.

Landis, D. M., and Reese, T. S., 1977, Structure of the Purkinje cell membrane in staggerer and weaver mutant mice, *J. Comp. Neurol.* **171**:247–260.

Lang, B., Barnard, E. A., Chang, L.-R., and Dolly, J. O., 1979, Putative benzodiazepine receptor: a protein solubilized from brain, *FEBS Lett.* **104**:149–153.

Lapin, I. P., 1980, Dissimilar effects of nicotinamide and inosine, putative endogenous ligands of the benzodiazepine receptors, on pentylenetetrazol seizures in four strains of mice, *Pharmacol. Biochem. Behav.* **13**:337–341.

Leeb-Lundberg, F., Snowman, A., and Olsen, R. W., 1980, Coupling of barbiturate–picrotoxinin receptors to benzodiazepine receptors in rat brain, *Soc. Neurosci. Abstr.* p. 236.

Leeb-Lundberg, F., Snowman, A., and Olsen, R. W., 1981, Perturbation of benzodiazepine receptor binding by pyrazolopyridines involves picrotoxinin/barbiturate receptor sites, *J. Neurosci.* **1**:471–477.

Lippa, A. S., Klepner, C. A., Yunger, L., Sano, M. C., Smith, W. V., and Beer, B., 1978*a*, Relationship between benzodiazepine receptors and experimental anxiety in rats, *Pharmacol. Biochem. Behav.* **9**:853–856.

Lippa, A. S., Sano, M. C., Coupet, J., Klepner, C. A., and Beer, B., 1978*b*, Evidence that benzodiazepine receptors reside on cerebellar Purkinje cells: studies with "nervous" mutant mice, *Life Sci.* **23:**2213–2218.

LIPPA, A. S., CRITCHETT, D., SANO, M. C., KLEPNER, C. A., GREENBLATT, E. N., COUPET, J., and BEER, B., 1979*a*, Benzodiazepine receptors: cellular and behavioral characteristics, *Pharmacol. Biochem. Behav.* **10:**831–843.

LIPPA, A. S., COUPET, J., GREENBLATT, E. N., KLEPNER, C. A., and BEER, B., 1979*b*, A synthetic non-benzodiazepine ligand for benzodiazepine receptors: a probe for investigating neuronal substrates of anxiety, *Pharmacol. Biochem. Behav.* **11:**99–106.

LIPPA, A. S., KLEPNER, C. A., BENSON, D. I., CRITCHETT, D. J., SANO, M. C., and BEER, B., 1980, The role of GABA in mediating the anticonvulsant properties of benzodiazepines, *Brain Res. Bull. Suppl. 5* **2:**861–865.

LIPPA, A. S., BEER, B., SANO, M. C., VOGEL, R. A., and MEYERSON, L. R., 1981, Differential ontogeny of type 1 and type 2 benzodiazepine receptors, *Life Sci.* **28:**2343–2347.

LUND, J., 1981, Radioreceptor assay for benzodiazepines in biological fluids using a new dry and stable receptor preparation, *Scand. J. Clin. Lab. Invest.* **41:**275–280.

MACDONALD, J. F., BARKER, J. L., PAUL, S. M., MARANGOS, P. J., and SKOLNICK, P., 1979, Inosine may be an endogenous ligand for benzodiazepine receptors on cultured spinal neurons, *Science* **205:**715–717.

MACKERER, C. R., and KOCHMAN, R. L., 1978, Effects of cations and anions on the binding of [^{3}H]diazepam to rat brain, *Proc. Soc. Exp. Biol. Med.* **158:**393–397.

MACKERER, C. R., KOCHMAN, R. L., BIERSCHENK, B. A., and BREMNER, S. S., 1978, The binding of [^{3}H]diazepam to rat brain homogenates, *J. Pharmacol. Exp. Ther.* **206:**405–413.

MAH, H. D., and DALY, J. W., 1976, Adenosine-dependent formation of cyclic AMP in brain slices, *Pharmacol. Res. Commun.* **8:**65–69.

MALICK, J. B., and ENNA, S. J., 1979, Comparative effects of benzodiazepines and non-benzodiazepine anxiolytics on biochemical and behavioral tests predictive of anxiolytic activity, *Commun. Psychopharmacol.* **3:**245–252.

MALLORGA, P., HAMBURG, M., TALLMAN, J. F., and GALLAGER, D. W., 1980*a*, Ontogenetic changes in GABA modulation of brain benzodiazepine binding, *Neuropharmacology* **19:**405–408.

MALLORGA, P., TALLMAN, J. F., HENNEBERRY, R. C., and GALLAGER, D. W., 1980*b*, [^{3}H]-diazepam binding in dissociated primary cortical culture: a pharmacological characterization, *Soc. Neurosci. Abstr.*, p. 500.

MANN, E., and ENNA, S. J., 1980, Phylogenetic distribution of bicuculline-sensitive γ-aminobutyric acid (GABA) receptor binding, *Brain Res.* **184:**367–373.

MARANGOS, P. J., and PATEL, J., 1981, Properties of [^{3}H]-β-carboline-3-carboxylate ethyl ester binding to the benzodiazepine receptor, *Life Sci.* **29:**1705–1714.

MARANGOS, P. J., PAUL, S. M., GRENNLAW, P., GOODWIN, F. K., and SKOLNICK, P., 1978, Demonstration of an endogenous, competitive inhibitor(s) of [^{3}H]diazepam binding in bovine brain, *Life Sci.* **22:**1893–1900.

MARANGOS, P. J., CLARK, R., MARTINO, A. M., PAUL, S. M., and SKOLNICK, P., 1979*a*, Demonstration of two new endogenous "benzodiazepine-like" compounds from brain, *Psychiatry Res.* **1:**121–130.

MARANGOS, P. J., PAUL, S. M., PARMA, A. M., GOODWIN, F. K., SYAPIN, P., and SKOLNICK, P., 1979*b*, Purinergic inhibition of diazepam binding to rat brain (*in vitro*), *Life Sci.* **24:**851–858.

MARANGOS, P. J., PAUL, S. M., GOODWIN, F. K., and SKOLNICK, P., 1979*c*, Putative endogenous ligands for the benzodiazepine receptor, *Life Sci.* **25:**1093–1102.

MARTIN, I. L., 1980, Endogenous ligands for benzodiazepine receptors, *Trends in Neuroscience*, Dec., pp. 299–301.

MARTIN, I. L., and CANDY, J. M., 1978, Facilitation of benzodiazepine binding by sodium chloride and GABA, *Neuropharmacology* **17:**993–998.

MARTIN, I. L., and CANDY, J. M., 1980, Facilitation of specific benzodiazepine binding in rat brain membrane fragments by a number of anions, *Neuropharmacology* **19:**175–179.

MARTINI, C., LUCACCHINI, A., RONCA, G., HRELIA, S., and ROSSI, C. A., 1982, Isolation of putative benzodiazepine receptors from rat brain membranes by affinity chromatography, *J. Neurochem.* **38:**15–19.

MASSOTTI, M., and GUIDOTTI, A., 1980, Endogenous regulators of benzodiazepine recognition sites, *Life Sci.* **27:**847–854.

MASSOTTI, M., ALLEVA, F. R., BALAZS, T., and GUIDOTTI, A., 1980, GABA and benzodiazepine receptors in the offspring of dams receiving diazepam: ontogenetic studies, *Neuropharmacology* **19:**951–956.

MASSOTTI, M., GUIDOTTI, A., and COSTA, E., 1981, Characterization of benzodiazepine and γ-aminobutyric recognition sites and their endogenous modulators, *J. Neurosci.* **1:**409–418.

MATTHEWS, G., and WICKELGREN, W. O., 1979, Glycine, GABA, and synaptic inhibition of reticulospinal neurones of lamprey, *J. Physiol.* **293:**393–415.

MAURER, R., 1979, the GABA agonist THIP, a muscimol analogue, does not interfere with the benzodiazepine binding site on rat cortical membranes, *Neurosci. Lett.* **12:**65–68.

MAZIÉRE, M., GODOT, J. M., BERGER, G., BARON, J. C., COMAR, D., CEPEDA, C., MENINI, C., and NAQUET, R., 1981, Positron tomography. A new method for *in vivo* brain studies of benzodiazepine in animal and in man, in: *GABA and Benzodiazepine Receptors* (E. Costa, G. DiChiara and G. L. Gessa, eds.), pp. 273–286, Raven Press, New York.

McDONALD, R. L., and BARKER, J. L., 1977, Pentylenetetrazol and penicillin are selective antagonists of GABA-mediated post-synaptic inhibition in cultured mammalian neurones, *Nature (London)* **267:**720–721.

McGEER, E. G., OLNEY, J. W., and McGEER, P. L., eds., 1978, *Kainic Acid as a Tool in Neurobiology,* Raven Press, New York.

McISAAC, W. M., 1961, Formation of 1-methyl-6-methoxy-1,2,3,4-tetrahydro-2-carboline under physiological conditions, *Biochim. Biophys. Acta* **52:**607–609.

McNAMARA, J. O., PEPER, A. M., and PATRONE, V., 1980, Repeated seizures induce long-term increase in hippocampal benzodiazepine receptors, *Proc. Natl. Acad. Sci. USA* **77:**3029–3032.

MEINERS, B. A., and SALAMA, A. J., 1982, Enhancement of benzodiazepine and GABA binding by the novel anxiolytic, tracazolate, *Eur. J. Pharmacol.* **78:**315–322.

MELDRUM, B. S., and HORTON, R. W., 1979, Anticonvulsant activity in photosensitive baboons, *Papio papio,* of two new 1,5-benzodiazepines, *Psychopharmacology* **60:**277–280.

MIMAKI, T., DESMUKH, P. P., and YAMAMURA, H. I., 1980, Decreased benzodiazepine receptor density in rat cerebellum following neurotoxic doses of phenytoin, *J. Neurochem.* **35:**1473–1475.

MITCHELL, R., and MARTIN, I., 1980, Ethyl β-carboline-3-carboxylate antagonizes the effect of diazepam on a functional GABA receptor, *Eur. J. Pharmacol.* **68:**513–514.

MOEHLER, H., and OKADA, T., 1977*a,* Properties of [^{3}H]diazepam binding to benzodiazepine receptors in rat cerebral cortex, *Life Sci.* **20:**2101–2110.

MOEHLER, H., and OKADA, T., 1977*b,* Benzodiazepine receptors: demonstration in the central nervous system, *Science* **198:**849–851.

MOEHLER, H., and OKADA, T., 1978, The benzodiazepine receptor in normal and pathological human brain, *Br. J. Psychiatry* **133:**261–268.

MOEHLER, H., and RICHARDS, J. G., 1981, Agonist and antagonist benzodiazepine receptor interaction *in vitro, Nature (London)* **294:**763–765.

MOEHLER, H., and RIEDERER, P., 1979, Benzodiazepine-Rezeptor-(BR)-binding bei der Parkinsonschen Krankheit, *Wien. Klin. Wochenschr.* **91:**355.

MOEHLER, H., OKADA, T., and BIRD, E., 1978*a,* Huntington's chorea: decrease in benzodiazepine receptor binding, *The Seventh International Pharmaceutical Congress in Paris,* France, p. 819.

MOEHLER, H., OKADA, T., and ENNA, S. J., 1978*b,* Benzodiazepine and neurotransmitter receptor binding in rat brain after chronic administration of diazepam or phenobarbital, *Brain Res.* **156**:391–395.

MOEHLER, H., OKADA, T., HEITZ, P., and ULRICH, J, 1978*c*, Biochemical identification of the site of action of benzodiazepines in human brain by [^{3}H]diazepam binding, *Life Sci.* **22**:985–996.

MOEHLER, H., POLC, P., CUMIN, R., PIERI, L., and KETTLER, R., 1979, Nicotinamide is a brain constituent with benzodiazepine-like actions, *Nature (London)* **278**:563–565.

MOEHLER, H., BATTERSBY, M. K., and RICHARDS, J. G., 1980, Benzodiazepine receptor protein identified and visualized in brain tissue by a photoaffinity label, *Proc. Natl. Acad. Sci. USA* **77**:1666–1670.

MOEHLER, H., RICHARDS, J. G., and WU, J.-Y., 1981, Autoradiographic localization of benzodiazepine receptors in immunocytochemically identified GABAergic synpases, *Proc. Natl. Acad. Sci. USA* **78**:1935–1938.

MUELLER, W., and WOLLERT, U., 1975, High stereospecificity of the benzodiazepine binding site on human serum albumin, *Mol. Pharmacol.* **11**:52–60.

MUELLER, W., SCHLAEFER, U., and WOLLERT, U., 1978, Benzodiazepine receptor binding: the interactions of some nonbenzodiazepine drugs with specific [^{3}H]diazepam binding to rat brain synaptosomal membranes, *Naunyn-Schmiedeberg's Arch. Pharmacol.* **305**:23–26.

MULLEN, R. J., EICHER, E. M., and SIDMAN, R. L., 1976, Purkinje cell degeneration, a new neurological mutation in the mouse, *Proc. Natl. Acad. Sci. USA* **73**:208–212.

NAPIAS, C., BERGMAN, M. O., NESS, P. C., VAN, GREENLEE, D. V., and OLSEN, R. W., 1980, GABA binding in mammalian brain: inhibition by endogenous GABA, *Life Sci.* **27**:1001–1011.

NIELSEN, M., and BRAESTRUP, C., 1980, Ethyl β-carboline-3-carboxylate shows differential benzodiazepine receptor interaction, *Nature (London)* **286**:606–607.

NIELSEN, M., BRAESTRUP, C., and SQUIRES, R. F., 1978, Evidence for a late evolutionary appearance of brain-specific benzodiazepine receptors: an investigation of 18 vertebrate and 5 invertebrate species, *Brain Res.* **141**:342–346.

NIELSEN, M., GREDAL, O., and BRAESTRUP, C., 1979, Some properties of [^{3}H]diazepam displacing activity from human urine, *Life Sci.* **25**:679–686.

NIELSEN, M., SCHOU, H., and BRAESTRUP, C., 1981, [^{3}H]Propyl β-carboline-3-carboxylate binds specifically to brain benzodiazepine receptors, *J. Neurochem.* **36**:276–285.

NISTRI, A., CONSTANTI, A., and KRNJEVIĆ, K., 1980, Electrophysiological studies of the mode of action of GABA on vertebrate central neurons, in: *Receptors for Neurotransmitters and Peptide Hormones* (G. Pepeu, M. J. Kuhar, and S. J., Enna, eds.), pp. 81–90, Raven Press, New York.

NUTT, D. J., COWEN, P. J., and LITTLE, H. J., 1982, Unusual interactions of benzodiazepine receptor antagonists, *Nature (London)* **295**:436–438.

OAKLEY, N. R., and JONES, B. J., 1980, The proconvulsant and diazepam-reversing effects of ethyl β-carboline-3-carboxylate, *Eur. J. Pharmacol.* **68**:381–382.

O'BRIEN, R. A., and SPIRT, N. M., 1980, The inhibition of GABA-stimulated benzodiazepine binding by a convulsant benzodiazepine, *Life Sci.* **26**:1441–1445.

O'BRIEN, R. A., SCHLOSSER, W., SPIRTS, N. M., FRANCO, S., HORST, W. D., POLC, P., and BONETTI, E. P., 1981, Antagonism of benzodiazepine receptors by beta-carbolines, *Life Sci.* **29**:75–82.

OLSEN, R. W., and LEEB-LUNDBERG, F., 1981, Convulsant and anticonvulsant drug binding sites related to GABA-regulated chloride ion channels, in: *GABA and Benzodiazepine Receptors* (E. Costa, G. DiChiara, and G. L. Gessa, eds.), pp. 93–102, Raven Press, New York.

OLSEN, R. W., LEEB-LUNDBERG, F., SNOWMAN, A., and STEPHENSON, F. A., 1982, Barbiturate interactions with the benzodiazepine–GABA receptor complex in mamalian brain, in: *Pharmacology of Benzodiazepines* (S. M. Paul, P. Skolnick, D. Greenblatt, F. Tallman, and E. Usdin, eds.), Macmillan, New York, in press.

OSSOLA, L., DEFEUDIS, F. V., and MANDEL, P., 1980, Lack of Na^+-independent binding of [^{3}H]-GABA or [^{3}H]muscimol to particulate fractions of cultured astroblasts, *J. Neurochem.* **34:**1026–1029.

OSTROVSKAYA, R. U., and VORONINA, T. A., 1977, Antagonistic effects of bicuculline and thiosemicarbazide on diazepam tranquilizing action, *Bull. Exp. Biol. Med.* (*Russ.*) **83:**293–295.

OWEN, F., LOFTHOUSE, R., and BOURNE, R., 1979, A radioreceptor assay for diazepam and its metabolites in serum, *Clin. Chim. Acta* **93:**305–310.

PALACIOS, J. M., NIEHOFF, D. L., and KUHAR, M. J., 1979, Ontogeny of GABA and benzodiazepine receptors: effects of Triton X-100, bromide, and muscimol, *Brain Res.* **179:**390–395.

PALACIOS, J. M., UNNERSTALL, J. R., YOUNG, W. S., III, and KUHAR, M. J., 1981, Radiohistochemical studies of benzodiazepine and GABA receptors and their interactions, in: *GABA and Benzodiazepine Receptors* (E. Costa, G. DiChiara, and G. L. Gessa, eds.), pp. 53–60, Raven Press, New York.

PAPEZ, J. W., 1937, A proposed mechanism of emotion, *Arch. Neurol. Psychiatry* **38:**725–743.

PAUL, S. M., and SKOLNICK, P., 1978, Rapid changes in brain benzodiazepine receptors after experimental seizures, *Science* **202:**892–894.

PAUL, S. M., ZATZ, M., and SKOLNICK, P., 1980, Demonstration of brain-specific benzodiazepine receptors in rat retina, *Brain Res.* **187:**243–246.

PAUL, S. M., MARANGOS, P., BROWNSTEIN, M., and SKOLNICK, P., 1981*a*, Demonstration and characterization of an endogenous inhibitor of GABA-enhanced [^{3}H]diazepam binding from bovine cerebral cortex, in: *GABA and Benzodiazepine Receptors* (E. Costa, G. DiChiara, and G. L. Gessa, eds.), pp. 103–110, Raven Press, New York.

PAUL, S. M., KEMPNER, E. S., and SKOLNICK, P., 1981*b*, *In situ* molecular weight determination of brain and peripheral benzodiazepine binding sites, *Eur. J. Pharmacol.* **76:**465–466.

Peroutka, S. J., MOSKOWITZ, M. A., REINHARD, J. F., Jr., and Snyder, S. H., 1980, Neurotransmitter receptor binding in bovine cerebral microvessels, *Science* **208:**610–612.

PETERSEN, E. N., and BUUS LASSEN, J., 1981, A water lick conflict paradigm using drug-experienced rats, *Psychopharmacology* **75:**236–239.

PHILLIS, J. W., 1979, Diazepam potentiation of purinergic depression of central neurons, *Can. J. Physiol. Pharmacol.* **57:**432–435.

PHILLIS, J. W., and WU, P. H., 1980, Interactions between the benzodiazepines, methylxanthines, and adenosine, *Can. J. Neurol. Sci.* **7:**247–249.

PHILLIS, J. W., EDSTROM, J. P., ELLIS, S. W., and KIRKPATRICK, J. R., 1979, Theophylline antagonizes flurazepam-induced depression of cerebral cortical neurons, *Can. J. Physiol. Pharmacol.* **57:**917–920.

PHILLIS, J. W., BENDER, A. S., and WU, P. H., 1980, Benzodiazepines inhibit adenosine uptake into rat brain synaptosomes, *Brain Res.* **195:**494–498.

PICKLES, H. G., 1979, Presynaptic γ-aminobutyric acid responses in the olfactory cortex, *Br. J. Pharmacol.* **65:**223–228.

PICKLES, H. G., and SIMMONDS, M. A., 1980, Antagonism by penicillin of γ-aminobutyric acid depolarizations at presynaptic sites in rat olfactory cortex and cuneate nucleus *in vitro*, *Neuropharmacology* **19:**35–38.

PLACHETA, P., and KAROBATH, M., 1979, Regional distribution of Na^+-independent GABA and benzodiazepine binding sites in rat CNS, *Brain Res.* **178:**580–583.

PLACHETA, P., and KAROBATH, M., 1980, *In vitro* modulation by SQ 20.009 and SQ 65.396 of GABA receptor binding in rat CNS membranes, *Eur. J. Pharmacol.* **62:**225–228.

POLC, P., 1980, Antagonism by caffeine and theophylline of diazepam effects on the cat spinal cord, *Experientia* **36:**713.

POLC, P., ROPERT, N., and WRIGHT, D. M., 1981, Ethyl-β-carboline-3-carboxylate antagonizes the action of GABA and benzodiazepines in the hippocampus, *Brain Res.* **217:**216–220.

PUESCHMANN, S., VON, and ENGELHORN, R., 1973, Pharmakologische Untersuchungen ueber eine Substanz mit antitussiven und atmungsanregenden Eigenschaften, *Arzneim.-Forsch.* **23:**296–305.

Rakic, P., and Sidman, R. L., 1973, Sequence of developmental abnormalities leading to granule cell deficit in cerebellar cortex of weaver mutant mice, *J. Com. Neurol.* **152:**103–132.

Randall, L. O., Schallek, W., Sternbach, L. H., and Ning, R. Y., 1974, Chemistry and pharmacology of the 1,4-benzodiazepines, in: *Psychopharmacological Agents* (M. Gordon, ed.), pp. 175–281, Academic Press, London.

Regan, J. W., Roeske, W. R., and Yamamura, H. I., 1980*a,* The benzodiazepine receptor: its development and its modulation by γ-aminobutyric acid, *J. Pharmacol. Exp. Ther.* **212:**137–143.

Regan, J. W., Roeske, W. R., and Yamamura, H. I., 1980*b,* [^{3}H]Flunitrazepam binding to bovine retina and the effect of GABA thereon, *Neuropharmacology* **19:**413–414.

Reisine, T. D., Nagy, J. I., Beaumont, K., Fibiger, H. C., and Yamamura, H. I., 1979*a,* The localization of receptor binding sites in the substantia nigra and striatum of the rat, *Brain Res.* **177:**241–252.

Reisine, T. D., Wastek, G. J., Speth, R. C., Bird, E. D., and Yamamura, H. I., 1979*b,* Alterations in the benzodiazepine receptor of Huntington's diseased human brain, *Brain Res.* **165:**183–187.

Rice, K. C., Brossi, A., Tallman, J., Paul, S. M., and Skolnick, P., 1979, Irazepine, a noncompetitive, irreversible inhibitor of [^{3}H]diazepam binding to benzodiazepine receptors, *Nature (London)* **278:**854–855.

Robertson, H. A., 1979, Benzodiazepine receptors in "emotional" and "non-emotional" mice: comparison of four strains, *Eur. J. Pharmacol.* **56:**163–166.

Robertson, H. A., 1980*a,* Audiogenic seizures: increased benzodiazepine receptor binding in a susceptible strain of mice, *Eur. J. Pharmacol.* **66:**249–252.

Robertson, H. A., 1980*b,* Harmaline-induced tremor: the benzodiazepine receptor as a site of action, *Eur. J. Pharmacol.* **67:**129–132.

Robertson, H. A., Martin, I. L., and Candy, J. M., 1978, Differences in benzodiazepine receptor binding in maudsley reactive and maudsley non-reactive rats, *Eur. J. Pharmacol.* **50:**455–457.

Rommelspacher, H., Nanz, C., Borbe, H. O., Fehske, K. J., Mueller, W. E., and Wollert, U., 1980*a,* 1-Methyl-β-carboline (harmane), a potent endogenous inhibitor of benzodiazepine receptor binding, *Naunyn-Schmeideberg's Arch. Pharmacol.* **314:**97–100.

Rommelspacher, H., Strauss, S., and Lindeman, J., 1980*b,* Excretion of tetrahydroharmane and harmane into the urine of man and rat after a load with ethanol, *FEBS Lett.* **109:**209–212.

Rosenberg, H. C., and Chiu, T. H., 1979*a,* Benzodiazepine binding after *in vivo* elevation of GABA, *Neurosci. Lett.* **15:**277–281.

Rosenberg, H. C., and Chiu, T. H., 1979*b,* Decreased [^{3}H]diazepam binding is a specific response to chronic benzodiazepine treatment, *Life Sci.* **24:**803–808.

Rosenberg, H. C., and Chiu, T. H., 1981, Tolerance during chronic benzodiazepine treatment associated with decreased receptor binding, *Eur. J. Pharmacol.* **70:**453–460.

Rosenblatt, J. E., Bridge, T. P., and Wyatt, R. J., 1979, A novel method for measuring benzodiazepines in saliva, *Commun. Psychopharmacol.* **3:**49–53.

Rudzik, A. D., Hester, J. B., Tang, A. H., Straw, R. N., and Friis, W., 1973, Triazolobenzodiazepines, a new class of central nervous system-depressant compounds, in: *The Benzodiazepines* (S. Garattini, E. Musini, and L. O. Randall, eds.), pp. 285–297, Raven Press, New York.

Schlosser, W., and Franco, S., 1979, Reduction of γ-aminobutyric acid (GABA)-mediated transmission by a convulsant benzodiazepine, *J. Pharmacol. Exp. Ther.* **211:**290–295.

Schou, J., 1980, Indberetninger af benzodiazepinbivirkninger, *Ugeskr. Laeg.* **142:**1695–1696.

Sepinwall, J., and Cook, L., 1978, Behavioral pharmacology of antianxiety drugs, in: *Handbook of Psychopharmacology,* Vol. 13 (L. L. Iversen, S. D. Iversen, and S. H. Snyder, eds.), pp. 345–393, Plenum Press, New York.

Sherman-Gold, R., and Dudai, Y., 1980, Solubilization and properties of a benzodiazepine receptor from calf cortex, *Brain Res.* **198:**485–490.

SHIBUYA, H., GALE, K., and PERT, C. B., 1980, Supersensitivity to GABA's effect on benzodiazepine receptors develops after striatonigral lesions, *Eur. J. Pharmacol.* **62:**243–244.

SCHOEMAKER, H., BLISS, M., and YAMAMURA, H. I., 1981, Specific high-affinity saturable binding of [^{3}H]-Ro 5-4864 to benzodiazepine binding sites in the rat cerebral cortex, *Eur. J. Pharmacol.* **71:**173–175.

SHOEMAKER, D. W., CUMMINS, J. T., BIDDER, T. G., BOETTGER, H. G., and EVANS, M., 1980, Identification of harman in the rat arcuate nucleus, *Naunyn-Schmiedeberg's Arch. Pharmacol.* **310:**227–230.

SIEGHART, W., and KAROBATH, M., 1980, Molecular heterogeneity of benzodiazepine receptors, *Nature (London)* **286:**285–287.

SIMONSEN, H., NIELSEN, M., and BRAESTRUP, C., 1982, Peripheral metabolism of β-carboline-carboxylic acid esters, *Acta Pharmacol. Toxicol.* **50:**89–93.

SKOLNICK, P., MARANGOS, P. J., GOODWIN, F. K., EDWARDS, M., and PAUL, S., 1978, Identification of inosine and hypoxanthine as endogenous inhibitors of [^{3}H]diazepam binding in the central nervous system, *Life Sci.* **23:**1473–1480.

SKOLNICK, P., SYAPIN, P. J., PAUGH, B. A., and PAUL, S. M., 1979*a*, Reduction in benzodiazepine receptors associated with Purkinje cell degeneration in "nervous" mutant mice, *Nature (London)* **277:**397–399.

SKOLNICK, P., SYAPIN, P. J., PAUGH, B. A., MONCADA, V., MARANGOS, P. J., and PAUL, S. M., 1979*b*, Inosine, an endogenous ligand of the brain benzodiazepine receptor, antagonizes pentylenetetrazole-evoked seizures, *Proc. Natl. Acad. Sci. USA* **76:**1515–1518.

SKOLNICK, P., GOODWIN, F. K., and PAUL, S. M., 1979*c*, A rapid and sensitive radioreceptor assay for benzodiazepine in plasma, *Arch. Gen. Psychiatry* **36:**78–80.

SKOLNICK, P., LOCK, K.-L., PAUGH, B., MARANGOS, P., WINDSOR, R., and PAUL, S., 1980*a*, Pharmacologic and behavioral effects of EMD 28.422: a novel purine which enhances [^{3}H]diazepam binding to brain benzodiazepine receptors, *Pharmacol. Biochem. Behav.* **12:**685–689.

SKOLNICK, P., PAUL, S. M., and BARKER, J. L., 1980*b*, Pentobarbital potentiates GABA-enhanced [^{3}H]diazepam binding to benzodiazepine receptors, *Eur. J. Pharmacol.* **65:**125–127.

SKOLNICK, P., PAUL, S., ZATZ, M., and ESKAY, R., 1980*c*, "Brain-specific" benzodiazepine receptors are localized in the inner plexiform layer of rat retina, *Eur. J. Pharmacol.* **66:**133–136.

SKOLNICK, P., LOCK, K.-L., PAUL, S. M., MARANGOS, P., JONES, R., and IRMSCHER, K., 1980*d*, Increased benzodiazepine receptor number elicited *in vitro* by a novel purine, EMD 28.422, *Eur. J. Pharmacol.* **67:**179–186.

SKOLNICK, P., MONCADA, V., BARKER, J. L., and PAUL, S. M., 1981 Pentobarbital has dual actions to increase brain benzodiazepine receptor affinity, *Science* **211:**1448–1450.

SKOLNICK, P., SCHWERI, M. M., WILLIAMS, E. F., MONCADA, V. Y., and PAUL, S. M., 1982, An *in vitro* binding assay which differentiates benzodiazepine "agonists" and "antagonists," *Eur. J. Pharmacol.* **78:**133–136.

SLATER, P., and LONGMAN, D. A., 1979, Effects of diazepam and muscimol on GABA-mediated neurotransmission: interactions with inosine and nicotinamide, *Life Sci.* **25:**1963–1967.

SNYDER, S. H., and GOODMAN, R. R., 1980, Multiple neurotransmitter receptors, *J. Neurochem.* **35:**5–15.

SNYDER, S. H., PASTERNAK, G. W., and PERT, C. B., 1975, Opiate receptor mechanisms, in: *Handbook of Psychopharmacology* (L. L. Iversen, S. D. Iversen, and S. H. Snyder, eds.), Vol. 5, pp. 329–360, Plenum Press, New York.

SONAWANE, B. R., YAFFE, S. J., and SHAPIRO, B. H., 1980, Changes in mouse diazepam receptor binding after phenobarbital administration, *Life. Sci.* **27:**1335–1338.

SPEEG, K. V., JR., AVANT, G. R., and SCHENKER, S., 1979, Inhibition of cerebral benzodiazepine binding by cimetidine and other antihistamines, *Clin. Res.* **27:**684A.

SPEEG, K. V., JR., WANG, S., AVANT, G. R., BERMAN, M. L., and SCHENKER, S., 1980,

Antagonism of benzodiazepine binding in brain by antilirium, benzyl alcohol, and physostigmine, *J. Neurochem.* **34:**856–865.

SPERK, G., and SCHLOEGL, E., 1979, Reduction of number of benzodiazepine binding sites in the caudate nucleus of the rat after kainic acid injections, *Brain Res.* **170:**563–567.

SPETH, R. C., and YAMAMURA, H. I., 1979, Benzodiazepine receptors: alterations in mutant mouse cerebellum, *Eur. J. Pharmacol.* **54:**397–399.

SPETH, R. C., WASTEK, G. J., JOHNSON, P. C., and YAMAMURA, H. I., 1978, Benzodiazepine binding in human brain: characterization using [^{3}H]flunitrazepam, *Life Sci.* **22:**859–966.

SPETH, R. C., WASTEK, G. J., and YAMAMURA, H. I., 1979*a*, Benzodiazepine receptors: temperature dependence of [^{3}H]flunitrazepam binding, *Life Sci.* **24:**351–358.

SPETH, R. C., BRESOLIN, N., and YAMAMURA, H. I., 1979*b*, Acute diazepam administration produces rapid increases in brain benzodiazepine receptor density, *Eur. J. Pharmacol.* **59:**159–160.

SPETH, R. C., BRESOLIN, N., MIMAKI, T., DESHMUKH, P. P., and YAMAMURA, H. I., 1981, Neuronal localization of benzodiazepine receptors in the murine cerebellum, in: *GABA and Benzodiazepine Receptors* (E. Costa, G. DiChiara, and G. L. Gessa, eds.), pp. 27–39, Raven Press, New York.

SQUIRES, R. F., 1981, GABA receptors regulate the affinities of anions required for brain specific benzodiazepine binding, in: *GABA and Benzodiazepine Receptors* (E. Costa, G. DiChiara, and G. L. Gessa, eds.), pp. 129–138, Raven Press, New York.

SQUIRES, R. F., and BRAESTRUP, C., 1977, Benzodiazepine receptors in rat brain, *Nature (London)* **266:**732–734.

SQUIRES, R. F., BENSON, D. I., BRAESTRUP, C., COUPET, J., KLEPNER, C. A., MYERS, V., and BEER, B., 1979a, Some properties of brain-specific benzodiazepine receptors: new evidence for multiple receptors, *Pharmacol. Biochem. Behav.* **10:**825–830.

SQUIRES, R., NAQUET, R., RICHE, D., and BRAESTRUP, C., 1979*b*, Increased thermolability of benzodiazepine receptors in cerebral cortex of a baboon with spontaneous seizures: a case report, *Epilepsia* **20:**215–221.

SQUIRES, R. F., KLEPNER, C. A., and BENSON, D. I., 1980, Multiple benzodiazepine receptor complexes: some benzodiazepine recognition sites are coupled to GABA receptors and ionophores, in: *Receptors for Neurotransmitters and Peptide Hormones* (G. Pepeu, M. J., Kuhar, and S. J. Enna, eds.), pp. 285–293, Raven Press, New York.

STEIN, L., 1980, Behavioral neurochemistry of benzodiazepines, *Arzneim.-Forsch.* **30:**868–873.

STEIN, L., WISE, C. D., and BELLUZZI, J. D., 1975, Effects of benzodiazepines on central serotonergic mechanisms, in: *Mechanism of Action of Benzodiazepines* (E. Costa and P. Greengard, eds.), pp. 29–44, Raven Press, New York.

STEPHENSON, R. P., 1956, A modification of receptor theory, *Br. J. Pharmacol.* **11:**379–393.

STERNBACH, L. H., 1979, The benzodiazepine story, *J. Med. Chem.* **22:**1–7.

STRITTMATTER, W. J., HIRATA, F., AXELROD, J., MALLORGA, P., TALLMAN, J. F., and HENNEBERRY, R. C., 1979, Benzodiazepine and β-adrenergic receptor ligands independently stimulate phospholipid methylation, *Nature (London)* **282:**857–859.

STUDY, R. E., and BARKER, J. L., 1981, Diazepam and (−)-pentobarbital: fluctuation analysis reveals different mechanisms for potentiation of γ-aminobutyric acid responses in cultured central neurons, *Proc. Natl. Acad. Sci. USA* **78:**7180–7184.

SULLIVAN, J. W., SEPINWALL, J., and COOK, L., 1978, Anticonflict evaluation of muscimol, A GABA receptor agonist, alone and in combination with diazepam. *Fed. Proc. Fed. Am. Soc. Exp. Biol.* **37:**619.

SUPAVILAI, P., and KAROBATH, M., 1979, Stimulation of benzodiazepine receptor binding by SQ 20.009 is chloride-dependent and picrotoxin-sensitive, *Eur. J. Pharmacol.* **60:**111–113.

SUPAVILAI, P., and KAROBATH, M., 1980*a*, Interaction of SQ 20.009 and GABA-like drugs as modulators of benzodiazepine receptor binding, *Eur. J. Pharmacol.* **62:**229–233.

SUPAVILAI, P., and KAROBATH, M., 1980*b*, Heterogeneity of benzodiazepine receptors in rat cerebellum and hippocampus, *Eur. J. Pharmacol.* **64:**91–93.

SUPAVILAI, P., and KAROBATH, M., 1980c, The effects of temperature and chloride ions on the stimulation of [^{3}H]flunitrazepam binding by THIP and PSA, *Neurosci. Lett.* **19:**337–342.

SYAPIN, P. J., and SKOLNICK, P., 1979, Characterization of benzodiazepine binding sites in cultured cells of neural origin, *J. Neurochem.* **32:**1047–1051.

TAKEUCHI, A., and TAKEUCHI, N., 1975, The structure–activity relationship for GABA and related compounds in the crayfish muscle, *Neuropharmacology* **14:**627–634.

TALLMAN, J. F., THOMAS, J. W., and GALLAGER, D. W., 1978, GABAergic modulation of benzodiazepine binding site sensitivity, *Nature (London)* **274:**383–385.

TALLMAN, J. F., THOMAS, J. W., and GALLAGER, D. W., 1979, Identification of diazepam binding in intact animals, *Life Sci.* **24:**873–880.

TALLMAN, J. F., PAUL, S. M., SKOLNICK, P., and GALLAGER, D. W., 1980, Receptors for the age of anxiety: pharmacology of the benzodiazepines, *Science* **207:**274–281.

TALLMAN, J. F., MALLORGA, P., THOMAS, J. W., and GALLAGER, D. W., 1981, Benzodiazepine binding sites: properties and modulation, in: *GABA and Benzodiazepine Receptors* (E. Costa, G. DiChiara, and G. L. Gessa, eds.) pp. 9–18, Raven Press, New York.

TANIGUCHI, T., WANG, J. K. T., and SPECTOR, S., 1980, Properties of [^{3}H]diazepam binding to rat peritoneal mast cells, *Life Sci.* **27:**171–178.

TENEN, S. S., and HIRSCH, J. D., 1980, β-Carboline-3-carboxylic acid ethyl ester antagonizes diazepam activity, *Nature (London)* **288:**609–610.

THIÉBOT, M. H., JOBERT, A., and SOUBRIÉ, P., 1979, Effects comparés du muscimol et du diazépam sur les inhibitions du comportement induites chez le rat par la nouveauté, la punition et le non-renforcement, *Psychopharmacology* **61:**85–89

TICKU, M. K., 1981, Interaction of depressant, convulsant, and anticonvulsant barbiturates with [^{3}H]diazepam binding site at the benzodiazepine–GABA–receptor–ionophore comples, *Biochem. Pharmacol.* **30:**1573–1579.

TICKU, M. K., and BURCH, T., 1980, Purine inhibition of [^{3}H]-γ-aminobutyric acid receptor binding to rat brain membranes, *Biochem. Pharmacol.* **29:**1217–1220.

TICKU, M. K., and OLSEN, R. W., 1978, Interaction of barbiturates with dihydropicrotoxinin binding sites related to the GABA receptor–ionophore system, *Life Sci.* **22:**1643–1652.

TOFFANO, G., GUIDOTTI, A., and COSTA, E., 1978, Purification of an endogenous protein inhibitor of the high-affinity binding of γ-aminobutyric acid to synaptic membranes of rat brain, *Proc. Natl. Acad. Sci. USA* **75:**4024–4028.

TRAVERSA, U., and NEWMAN, M., 1979, Stereospecific influence of oxazepam hemisuccinate on cyclic AMP accumulation elicited by adenosine in cerebral cortical slices, *Biochem. Pharmacol.* **28:**2363–2365.

TUNNICLIFF, C., SMITH, J. A., and NGO, T. T., 1979, Competition for diazepam receptor binding by diphenylhydantoin and its enhancement by γ-aminobutyric acid, *Biochem. Biophys. Res. Commun.* **91:**1018–1024.

TYE, N. C., EVERITT, B. J., and IVERSEN, S. D., 1977, 5-Hydroxytryptamine and punishment, *Nature (London)* **268:**741–743.

UNNERSTALL, J. R., KUHAR, M. J., NIEHOFF, D. L., and PALACIOS, J. M., 1981, Benzodiazepine receptors are coupled to a subpopulation of γ-aminobutyric acid (GABA) receptors: evidence from a quantitative autoradiographic study, *J. Pharmacol. Exp. Ther.* **218:**797–804.

VOGEL, R. A., FRYE, G. D., WILSON, J. H., KUHN, C. M., MAILMAN, R. B., MUELLER, R. A., and BREESE, G. R., 1980, Attenuation of the effect of punishment by thyrotropin-releasing hormone: comparisons with chlordiazepoxide, *J. Pharmacol. Exp. Ther.* **212:**153–161.

WADDINGTON, J. L., and OWEN, F., 1978, Stereospecific benzodiazepine receptor binding by the enantiomers of oxazepam sodium hemisuccinate, *Neuropharmacology* **17:**215–216.

WASSERMAN, S. K., LEW, B., and PAIK, W. K., 1975, Studies on the natural substrate for protein methylase II in mammalian brain and blood, *J. Neurochem.* **24:**625–629.

WASTEK, G. J., SPETH, R. C., TREISINE, T. D., and YAMAMURA, H. I., 1978, The effect of γ-aminobutyric acid on [^{3}H]flunitrazepam binding in rat brain, *Eur. J. Pharmacol.* **50:**445–447.

WEILAND, G. A., MINNEMAN, K. P., and MOLINOFF, P. B., 1979, Fundamental difference between the molecular interactions of agonists and antagonists with the β-adrenergic receptor, *Nature (London)* **281:**114–117.

WILLIAMS, E. F., RICE, K. C., PAUL, S. M., and SKOLNICK, P., 1980, Heterogeneity of benzodiazepine receptors in the central nervous system demonstrated with kenazepine, an alkylating benzodiazepine, *J. Neurochem.* **35:**591–597.

WILLIAMS, M., and RISLEY, E. A., 1979, Enhancement of the binding of [^{3}H]diazepam to rat brain membranes *in vitro* by SQ 20.009, a novel anxiolytic, γ-aminobutyric acid (GABA), and muscimol, *Life Sci.* **24:**833–842.

WILLIAMS, M., and RISLEY, E. A., 1980, Biochemical characterization of putative central purinergic receptors by using 2-chloro-[^{3}H]adenosine, a stable analog of adenosine, *Proc. Natl. Acad. Sci. USA* **77:**6892–6896.

WILLIAMS, M., and RISLEY, E. A., 1982, Interaction of avermectins with [^{3}H]β-carboline-3-carboxylate ethyl ester and [^{3}H]diazepam binding sites in rat brain cortical membranes, *Eur. J. Pharmacol.* **77:**307–312.

WILLIAMS, M., and YARBOROUGH, G. G., 1979, Enhancement of *in vitro* binding and some of the pharmacological properties of diazepam by a novel anthelmintic agent, avermectin B_{1a}, *Eur. J. Pharmacol.* **56:**273–276.

WILLIAMSON, M. J., PAUL, S. M., and SKOLNICK, P., 1978*a*, Labelling of benzodiazepine receptors *in vivo, Nature (London)* **275:**551–553.

WILLIAMSON, M. J., PAUL, S. M., and SKOLNICK, P., 1978*b*, Demonstration of [^{3}H]diazepam binding to benzodiazepine receptors *in vivo, Life Sci.* **23:**1935–1940.

WONG, G. B. and SELLERS, E. M., 1979, Intravascular factors affecting diazepam binding to human serum albumin, *Biochem. Pharmacol.* **28:**3265–3270.

WOOLF, J. H., and NIXON, J. C., 1981, Endogenous effector of the benzodiazepine binding site: purification and characterization, *Biochem.* **20:**4263–4269.

WU, P. H., PHILLIS, J. W., and BENDER, A. S., 1980, Inhibition of [^{3}H]diazepam binding to rat brain cortical synaptosomal membranes by adenosine uptake blockers, *Eur. J. Pharmacol.* **65:**459–460.

YAMAMURA, H. I., EHLERT, F., MIMAKI, T., ROESKE, W., BRAESTRUP, C., HORST, D., and O'BRIEN, R., 1981, Benzodiazepine receptor binding studies in rat brain using two novel nonbenzodiazepine ligands, Abstract from the XXIXth Colloquium—Protides of the Biological Fluids, Brussels, Belgium, 4th–7th May, 1981.

YAMAMURA, H. I., MIMAKI, T., YAMAMURA, S. H., HORST, W. D., MORELLI, M., BAUTZ, G., and O'BRIEN, R. A., 1982, [^{3}H]-CL 218,872, a novel triazolopyridazine which labels the benzodiazepine receptor in rat brain, *Eur. J. Pharmacol.* **77:**351–354.

YOUNG, W. S., and KUHAR, M. J., 1979, Autoradiographic localization of benzodiazepine receptors in the brains of humans and animals, *Nature (London)* **280:**393–395.

YOUNG, W. S., and KUHAR, M. J., 1980, Radiohistochemical localization of benzodiazepine receptors in rat brain, *J. Pharmacol. Exp. Ther.* **212:**337–346.

YOUNG, W. S., NIEHOFF, D., KUHAR, M. J., BEER, B., and LIPPA, A. S., 1981, Multiple benzodiazepine receptor localization by light-microscopic radiohistochemistry, *J. Pharmacol. Exp. Ther.* **216:**425–430.

YOUSUFI, M. A. K., THOMAS, J. W., and TALLMAN, J. F., 1979, Solubilization of benzodiazepine binding site from rat cortex, *Life Sci.* **25:**463–470.

ZAKUSOV, V. V., OSTROVSKAYA, R. V., KOZHECHKIN, S. N., MARKOVICH, V. V., MOLODAVKIN, G. M., and VORONINA, T. A., 1977, Further evidence for GABAergic mechanisms in the action of benzodiazepines, *Arch. Int. Pharmacodyn. Ther.* **229:**313–326.

8

HISTAMINE RECEPTORS IN BRAIN

Jack Peter Green

1. INTRODUCTION

While mounting evidence is elevating histamine to the category of a putative neurotransmitter, its explicit functions in brain remain as unclear as those of most other putative neurotransmitters. Evidence has been reviewed (Green *et al.*, 1978*a*; Schwartz, 1979, Schwartz *et al.*, 1980*a,b*; Hough and Green, 1980) that it may function in arousal, locomotor activity, regulation of hunger and thirst, emesis, thermoregulation, and the elaboration of hormones, most persuasively, of prolactin and the antidiuretic hormone. There is also evidence, too rich to review here, that histamine may alter the formation, release, or activities of other endogenous substances. As provocative as many of these observations are, they remain to be integrated into coherent hypotheses, as do observations on other biogenic amines. The development of methods to measure histamine metabolism (Hough *et al.*, 1981) offers an opportunity to evaluate further the functions of histamine in brain. Methods to study brain histamine receptors provide additional means of assessing the roles of histamine in brain. A description of these receptors, some of their associated effects, and their response to psychotropic drugs is presented in this review.

Jack Peter Green • Department of Pharmacology, Mount Sinai School of Medicine, The City University of New York, New York, New York 10029.

2. CLASSIFICATION OF HISTAMINE RECEPTORS

Following the rules of "pharmacological taxonomy" (Black, 1976) led directly to the discovery of the histamine H_2-antagonists (Black *et al.*, 1972; Schild, 1981) and of other highly selective drugs, which not only benefit the sick but serve to probe biological processes. Failure to follow the rules has caused confusion in the classification of histamine receptors in brain and in other organs and tissues (Green *et al.*, 1978*b*; Green and Hough, 1980; Hough *et al.*, 1980). One error is to exaggerate the selectivity of agonists that are used in the classification. With the exception of dimaprit (Parsons *et al.*, 1977; Bertaccini *et al.*, 1979) and impromidine (Durant *et al.*, 1978), which are extraordinarily specific for the H_2-receptor, the agonists used to define histamine receptors are full agonists on both histamine H_1- and H_2-receptors (see, for example, Black *et al.*, 1972). Hence, at high enough concentrations, they will stimulate both receptors. Table 1 shows ED_{50} values, necessarily obtained from complete dose–response curves, and the relative activities (derived from the ED_{50} values) of some agonists on the H_1- and H_2-receptors. This hierarchy of agonists, having gradational affinities for each receptor, is very useful in receptor classification as long as a relative selectivity is not transmogrified into an absolute "specificity." The need to consider dose is evident when the percent activation of each receptor by any concentration of agonist is calculated from

$$\text{Effect (\% of maximum)} = \frac{e[\mathrm{D}]}{\mathrm{ED}_{50} + [\mathrm{D}]} \times 100 \qquad (1)$$

where e is the intrinsic activity (here equal to one, as all are full agonists) and [D] is the concentration of agonist. For example, at 10^{-4} M, 2-methylhistamine produces 97% of the maximum response on the H_1-receptor; but it also produces 80% of the maximum response on the H_2-

TABLE 1

ED_{50} Values and Relative Activities of Some Histamine Agonists on Histamine H_1- and H_2 Receptors[a]

	ED_{50} (μM)		Relative activity	
Agonist	H_1	H_2	H_1	H_2
Histamine	0.5	1.1	100.0	100.0
4-Methylhistamine	250.0	2.6	0.2	43.0
2-Methylhistamine	3.0	25.0	16.5	4.4
Thiazolylethylamine	1.9	366.7	26.0	0.3

[a] Black *et al.* (1972); Durant *et al.* (1975).

receptor. A concentration of agonist of 10^{-4} M is often used in experiments to classify histamine receptors. A response to 10^{-4} M of 2-methylhistamine has been attributed to the H_1-receptor. Classification of receptors with agonists requires estimates of ED_{50} values relative to histamine, and these may then be compared with the values on known and defined histamine receptors.

Even when all cautions are observed, agonists may produce ambiguous or even misleading results in classification (see Green and Hough, 1980). The use of competitive antagonists to classify receptors provides less ambiguity if these agents are judiciously used. They too have caused confusion because they too have been used at concentrations high enough to vitiate selectivity. Most H_1-antagonists have K_B values of 10^{-8} to 10^{-10} M; e.g., for pyrilamine (also called mepyramine) and tripellenamine these values are respectively 4×10^{-10} and 3.2×10^{-9} M, yet in some studies aimed at defining the receptor associated with a histamine response, antagonists have been used at concentrations of 10^{-6} and 10^{-4} M, and occasionally at 10^{-2} M. Even at 10^{-6} M, these H_1-antagonists block histamine at the H_2-receptor (Green *et al.*, 1977, 1978*b*; Kanof and Greengard, 1979; Green and Hough, 1980). The H_2-antagonists, notably tiotidine (Yellin *et al.*, 1979), metiamide, and cimetidine (Ganellin, 1978), have high selectivity for the H_2-receptor, but at high concentrations they block other receptors (Brimblecombe *et al.*, 1975; McCulloch *et al.*, 1979). These H_2-antagonists have lower affinities for the H_2-receptor than do the commonly used H_1-antagonists for the H_1-receptor; e.g., for the H_2-receptor in guinea pig atria, the K_B values of cimetidine and tiotidine are, respectively, 7.9×10^{-7} (Gannellin, 1978) and 1.5×10^{-8} M (Yellin *et al.*, 1979). Therefore, higher concentrations of H_2-antagonists are required to block the H_2-receptor than the concentrations of H_1-antagonists needed to block the H_1-receptor. Yet in some work designed to define a receptor, equal concentrations of cimetidine and mepyramine were used, e.g., 10^{-6} M, which is near the K_B of the H_2-antagonist but 1000 times the K_B of the H_1-antagonist. Not surprisingly, at these concentrations the H_2-antagonist had little or no effect while the H_1-antagonist blocked histamine, prompting the (false) inference that the activity is linked to the H_1-receptor. Confusion has also resulted from the use of only one concentration of agonist, e.g., histamine; as the antagonism is competitive and surmountable by increasing doses of agonist, one dose of agonist is unrevealing, for if its concentration is too high, it will surmount the effect of the antagonist. Also confounding are additional effects of an antagonist that can obscure its receptor-blocking effects, as exemplified in Section 5.1. Other complexities that present difficulties in receptor classification have been reviewed (Green *et al.*, 1978*b*; Green and Hough, 1980; Hough *et al.*, 1980). Not all biological systems can readily fulfill the demands of rigorous receptor taxonomy; in systems where less rigor must be accepted, the inferences have to be invested with less confidence.

The latest source of error in classifying histamine receptors has arisen in the use of binding techniques that do not consider well-documented artifacts (Hollenberg and Cuatrecasas, 1975; Bennett, 1978; Burt, 1978; Hollenberg and Cuatrecasas, 1979). Cimetidine binding to sites that had been inferred to be H_2-receptors was shown to be resistant to many H_2-antagonists (Smith *et al.*, 1980; Rising *et al.*, 1980). In one of the studies used to define H_2-binding sites, radioactive cimetidine was used, not at a concentration near its K_B (6.03×10^{-7} M), but at 6.25×10^{-9} M, a concentration that could occupy no more than a fraction of a percent of the H_2-receptors, as calculated from

$$\% \text{ occupancy} = \frac{1}{K_B/[D] + 1} \times 100 \tag{2}$$

Availability of labeled tiotidine, the thiazolyl derivative, which has greater H_2-affinity than cimetidine, presented hope of specifically labeling the receptor; but denatured brain membranes bound tiotidine almost as effectively as the native membranes, showing, along with other experiments, that the tiotidine binding sites were unrelated to the H_2-receptor (Maayani *et al.*, 1981).

At this time, evidence shows only two types of histamine receptors in any organ or tissue.

3. THE HISTAMINE H_1-RECEPTOR

3.1 The H_1-Receptor Characterized by Binding Studies

Of all organs and tissues examined, the density of H_1-receptors revealed by [^{3}H]mepyramine binding appears to be highest in brain in the species studied—guinea pig, rat, and rabbit (Chang *et al.*, 1979*a*). It is interesting that the regional distribution in brain of the receptor differs among species, and surprising that the K_B of mepyramine and some other H_1-antagonists for the receptor in whole brain homogenates differed between guinea pig and other species.

Binding of labeled mepyramine, in the presence of triprolidine, to the H_1-receptor in membranes from whole brain of guinea pig (Chang *et al.*, 1979*b*) has a Hill coefficient not different from 1.0, and Scatchard analysis indicates a homogeneous population of binding sites with K_B = 0.5 nM. The comparable K_B values of rat, rabbit, and mouse brains were between 3 and 4 nM, and that of human frontal cortex, 1.0 nM; these, too, showed saturable and homogeneous binding sites. In guinea pig membranes, the rate of association is faster and the rate of dissociation slower than in rat membranes (Chang *et al.*, 1979*b*). The K_B values of some other H_1-antagonists differ among species (Chang *et al.*, 1979*b*). All membranes studied discriminate between the stereoisomers of chlorpheniramine, but the relative K_B values significantly differ, the guinea pig brain membranes showing higher affinity than either the rat or human preparations for *d*-chlorpheniramine, triprolidine, chlorpromazine, and promazine—for the last two compounds, differences of 10-fold in K_B values. Other organs and tissues differ among species in their affinities for H_1-antagonists, and within the same species, tissues varied in these affinities (Chang *et al.*, 1979*a*). The affinities of H_1-antagonists for the H_1-receptor in rat membranes are better correlated with affinities for the H_1-receptor linked to histamine contraction of the guinea pig ileum than they are for the H_1-receptor in guinea pig brain membranes (cf. Chang *et al.*, 1979*b*).

In whole brain membranes of guinea pig, rat, and mouse and in membranes of human frontal cortex, the B_{max} values were similar, 3–10 pmol/g wet weight, rabbit preparations perhaps showing somewhat less density (Chang *et al.*, 1979*b*). Guinea pigs showed highest density in the cerebellum, and lowest in the corpus striatum (Table 2); in the rat the hypothalamus was highest and the cerebellum lowest (Table 2); in calf the parietal and occipital cortices were richest, and the pons, medulla, and midbrain, lowest; and in man, the frontal parietal, cingulate, and temporal cortices were high, and the pons, medulla, and cerebellum, low. In those species for which data are available, there was no relationship between the regional distribution of receptors and of histamine and histidine decarboxylase activity, an independence common to other systems (Chang *et al.*, 1979*b*). Information is not available to relate species differences in histamine sensitivity to the K_B or B_{max} values; it is known that guinea pig and man are especially sensitive to H_1-agonists (see Naranjo, 1966).

The H_1-receptors are sensitive to proteolytic enzymes and to phospholipases A and C (Chang *et al.*, 1979*b*). Binding of H_1-agonists to the receptor has sensitivity to the medium that is reminiscent of other receptors Chang and Snyder, 1980). Sodium ions, especially, decrease affinity of histamine for the H_1-receptor by 10-fold, as measured by mepyramine binding. GTP, its analog, and GDP decrease affinity. The effects of sodium and GTP are additive. Affinity of H_1-agonists is enhanced by manganese, and to a lesser extent by magnesium, but not by calcium. None of these chemicals influenced the K_B values of antagonists (Chang and Synder, 1980). A soluble digitonin extract of guinea pig brain retained sensitivities to sodium and divalent cations but not to GTP (Toll *et al.*, 1980).

Another group using [^{3}H]mepyramine and promethazine to label the H_1-receptors employed different methods, notably using membranes that

TABLE 2
Regional Distribution of the Histamine H_1-Receptor and the H_2-Linked Adenylate Cyclase Activity in Brain

	H_1-Receptor		H_2-Linked adenylate cyclase activity[c] (guinea pig)
	Rat[a]	Guinea pig[b]	
Telencephalon			
Neocortex	3.0	3.4	117[d]
Corpus striatum	2.2	1.7	28[d]
Hippocampus	2.2	4.5	104[d]
Diencephalon			
Thalamus	2.9	5.7	46[d]
Hypothalamus	4.2	3.8	15[d]
Mesencephalon	3.0	3.8	
Central gray			8[d]
Cerebellum	1.7	9.2	0[e]
Pons and Medulla		2.9	0[e]
Other			
Optic chiasma			0[e]

[a] [^{3}H]Mepyramine binding, pmol/g wet weight (Tan-Tran *et al.*, 1978*a*).
[b] [^{3}H]Mepyramine binding, pmol/g wet weight (Chang *et al.*, 1979*b*).
[c] Not a measure of H_2-receptor density.
[d] Formation of cAMP by membranes. Percent increase over basal activity by 10^{-3} M histamine (Green *et al.*, 1978*a*).
[e] Formation of cAMP by homogenates. Percent increase over basal activity by 10^{-4} M histamine (Hegstrand *et al.*, 1976).

sedimented at 6,000 *g* (Hill *et al.*, 1978) rather than at 50,000 *g* (Chang *et al.*, 1979*b*): the H_1-receptors are predominantly in the synaptosomal membranes (Tan-Tran *et al.*, 1980). Another salient difference is in collecting the membranes after incubation with the ligand by centrifugation rather than by filtering. The proportion of promethazine-insensitive binding sites was higher in this system than was the triprolidine-insensitive binding sites used by the other group (Chang *et al.*, 1979*b*), a difference that is more likely attributable to the membrane fraction than to the masking ligand. On these guinea pig brain membranes, a second H_1-binding site with lower affinity was revealed when the concentration of labeled mepyramine was increased (Hill and Young, 1980).

The K_B of the high-affinity site in brain membranes, 0.83 nM (Hill and Young, 1980), was almost identical to that of guinea pig intestinal membranes and slightly higher than that, 0.50 nM, obtained by others (Chang *et al.*, 1979*b*); the B_{max} (Hill *et al.*, 1978) was considerably higher than that found by Chang *et al.* (1979*b*). Binding of antagonists showed stereoselectivity. Most antagonists showed Hill coefficients differing from unity (Hill *et al.*, 1978), in contrast with the other results (Chang *et al.*, 1979*b*). The affinities (actually, IC_{50} values in absence of simple competition) of these antagonists for the binding sites in guinea pig membranes correlated with their affinities for binding to membranes from guinea pig small intestine and for blocking histamine-induced contractions of the guinea pig ileum (Hill *et al.*, 1978). With these membranes, too, rat brain showed less affinity for H_1-antagonists than did guinea pig brain (Hill and Young, 1980), and the rat and guinea pig brain differed in the regional distribution of H_1-binding sites, the cerebellum showing highest density in the brain of guinea pig and lowest in the brain of rat (Hill *et al.*, 1978; Hill and Young, 1980).

Slices of rat and guinea pig brains show binding parameters, K_B and B_{max} values, similar to those measured on homogenates (Palacios *et al.*, 1979, 1981*a*), a concordance that gives confidence to radioautography studies. In guinea pig brain (Palacios *et al.*, 1979), the cerebellum showed high density of H_1-receptors in the molecular layer; few densities were seen in this layer on the rat cerebellum. In hippocampus of the guinea pig, the dentate gyrus was especially rich in H_1-receptors, and CA4 had greater densities than CA1, CA2, or CA3 (Palacios *et al.*, 1979). High densities in structures of the rat brain that are associated with the auditory system (Palacios *et al.*, 1981*a*) may relate to the impairment of auditory vigilance occurring with H_1-blockers in man (Peck *et al.*, 1975). In rat hippocampus, high densities were seen in CA3 and in the subiculum (Palacios *et al.*, 1981*a*); perhaps pertinent is that promethazine, 1 mM, but not metiamide, 1 mM, blocked the slow depolarization produced by histamine on rat hippocampal CA3 pyramidal cells *in vitro* (Segal, 1981). A tribute to this method of radioautography is that binding of membranes isolated from rat and guinea pig dentate gyrus and hippocampus proper (Chang *et al.*, 1980) could reveal none of these subtleties.

In rat, the hypothalamus showed high density in the supraoptic and suprachiasmatic nuclei, ventromedial nucleus, and the nucleus premammilaris ventralis; some of these may be the sites where intraventricularly injected histamine acts to alter food and water intake, thermal regulation, hormone secretion, and other processes, as summarized by Palacios *et al.* (1981*a*).

Neurons in rat hypothalamus show either increased or decreased firing rate when exposed to histamine (Haas, 1974; Renaud, 1976). Iontophoretic studies precluded unambiguous identification of the presumed histamine receptor associated with the excitation because H_1-antagonists are anesthetic and react with numerous sites in high doses. But observations on perfused rat hypothalamic cultures showed that excitation by histamine is blocked by promethazine, an H_1-antagonist lacking local anesthetic activity in low doses, but not by the H_2-antagonist, metiamide (Geller, 1981). [Analogous results were obtained on neurons of two species of molluscs (Carpenter and Gaubatz, 1975; Gruol and Weinreich, 1979; Gotow *et al.*, 1980).]

Kainate injections into the guinea pig cerebellum produced a decrease in H_1-receptors in the molecular layer, suggesting that the receptors are associated with neurons (Palacios *et al.*, 1981*b*). It is still not certain that all H_1-receptors are neuronal, for the H_1-receptor, of all receptors studied, best sustained lesions in rats (Chang *et al.*, 1980). H_1-binding by striatal homogenates were not altered by injection of 6-hydroxydopamine into the substantia nigra, by hemisection of the brainstem, or ablation of the cerebral cortex. Binding by hippocampal homogenates was unaffected by medial forebrain lesions or transection of the fimbria and fornix. Kainate injection into the hippocampus was followed in four days by a fall in H_1-binding, but at 15 days or longer no difference from untreated rats was seen (Chang *et al.*, 1980). The receptors are not likely to be associated with blood vessels, as the H_1-receptor distribution in brain regions is too discrete; they are not associated with mast cells, which have a distribution (Edvinsson *et al.*, 1976) and development in brain very different from those of H_1-receptors. Some of the H_1-receptors in brain may be associated with glia, as suggested by Chang *et al.* (1980).

The H_1-receptor density in rat brain at birth is about 10–15% of that in the adult, which it attains 15–20 days after birth (Tan-Tran *et al.*, 1980; Subramanian *et al.*, 1981). The K_B remains the same. In all regions studied, the ontogenesis is similar (Tan-Tran *et al.*, 1980). H_1-receptor development occurs earlier than development of histidine decarboxylase which, along with subcellular fractionation studies (Tan-Tran *et al.*, 1980) and parallel measurements of the developmental pattern of brain histamine content (Subramanian *et al.*, 1981), allows these authors to affirm that the receptors are not on mast cells. The development of the H_1-receptor paralleled the development of the capacity to incorporate inorganic phosphate (administered by intracisternal injection) into phospholipids at 5 min after intracisternal injection of histamine (Subramanian *et al.*, 1980), which enhances incorporation by acting on the H_1-receptor (Friedel and Schanberg, 1975; Subramanian *et al.*, 1980).

Prolonged treatment of guinea pigs with mepyramine did not alter the B_{max} (or K_B) of mepyramine binding by membranes from different parts of the brain or from intestinal smooth muscle (Hill *et al.*, 1981). Tolerance to an H_1-antagonist (Heinrich, 1953) may not be related to an effect on the H_1-receptor.

The distribution of labeled mepyramine in brain after intravenous injections in mice was similar to the B_{max} of different brain regions, and the potency (ED_{50}) of H_1-antagonists in competing for labeled mepyramine levels in brain did not differ greatly from their affinities for the H_1-receptor (Quach *et al.*, 1979, 1980), affinities that may relate to the sedative properties of the drugs. It is interesting that atropine and scopolamine were inactive (Quach *et al.*, 1979, 1980). In similar experiments *in vivo*, the H_1-antagonists reduced the amount of labeled mepyramine in the particulate fraction, the potency of some H_1-antagonists in this system again reflecting, though inexactly, their affinities for the H_1-receptor (Diffley *et al.*, 1980).

Early studies (e.g., Heinrich, 1953) were not able to establish a clear relationship between affinities for the H_1-receptor and potencies in sedation. This is not surprising in light of the complexities of sedation; of the affinities of H_1-antagonists for other receptors, including muscarinic sites and sites of uptake for biogenic amines; and of all the interactions a drug undergoes after administration before it reacts with a receptor. Even in an isolated tissue bath, the effect of a drug on a well-defined receptor may be obscured by additional actions (see Section 5). For example, *in vitro* the apparent affinity of metiamide for the H_2-receptor is altered in the presence of a phosphodiesterase inhibitor (Angus and Black, 1980), and adrenergic agonist potencies are changed in the presence of uptake blockers (Kenakin, 1980); other such examples abound, some of which are given in the references accompanying these last two papers. In a whole animal, it would be optimistic to expect a correlation between receptor affinities and pharmacological effect such as sedation, especially with a group of drugs that have affinities for so many sites as the H_1-antagonists have. The correlation between competition for mepyramine binding sites *in vivo* (Quach *et al.*, 1979, 1980) and sedation may, as the authors state, present a useful model to predict sedative properties of these drugs, even if inferences about causal relationships and the mechanism of sedation are limited.

Perhaps because of their affinities for many sites, the H_1-antagonists may sedate as well as excite; some are also anticonvulsant, a property that is independent of sedation (Dashputra *et al.*, 1966). No simple, consistent relationship has been found (in concomitant measurements) between peripheral response to H_1-blockers and central effects including sedation in man (Peck *et al.*, 1975; Carruthers *et al.*, 1978). Potency of drugs in preventing motion sickness in man was not correlated with potency as H_1- (or muscarinic) antagonists (Brand, 1968). The hallucinations observed in toxic reactions to H_1-antagonists may be due to muscarinic blockade (see Nigro, 1968). Potency in antagonizing isolation-induced fighting of mice more nearly paralleled antimuscarinic potency than potency for the H_1-receptor (Barnett *et al.*, 1971). The electrical activity produced in cats by intravenous infusion of racemic chlorpheniramine was similar to that produced by an equal dose of the *l*-isomer, despite differences in their affinities for the H_1-receptor (Faingold and Berry, 1972*a*). It is interesting that a low dose, 50 mg, of diphenhydramine heightened mood while decreasing psychomotor activity; with increasing dose, no effect on mood was seen (Jäätelä *et al.*, 1971). At a dose of 100 mg, diphenhydramine altered time perception in man (subjects thought time passed more slowly), an uncommon effect of a drug and similar to the effect of cannabis; mood and cognitive processes were unchanged (Mohs *et al.*, 1978). This effect of diphenhydramine is not necessarily a characteristic of all H_1-antagonists; they differ, for example, in the electrical activity they produce in cat brain; e.g., low doses of diphenhydramine increase cortical synchrony, unlike tripelennamine (Faingold and Berry, 1972*b*).

These paradoxes, maybe contradictions, that have become clear in studies of H_1-antagonists urge caution in making extrapolations from affinities of drugs for receptors to their effects on any complicated pharmacological or therapeutic event, including, perhaps especially, central events. There is no reason to presume that these cautions do not apply to other drugs. The H_1-receptor and the drugs that block it are distinct only in having been studied for a long time (longer than any other group of drugs and other receptors, excepting anticholinergic agents) in many systems and with many techniques.

3.2. The H_1-Linked Guanylate Cyclase

In a cultured mouse neuroblastoma clone, stimulation of the H_1-receptor enhances cGMP formation (Richelson, 1978; Taylor and Richelson, 1979). The activation, which peaks around 30 sec after exposure to 0.1 mM histamine, increases cGMP formation as much as 50-fold. The ED_{50} of histamine was that for the H_1-receptor in peripheral tissue, and 4-methylhistamine, an H_2-agonist, at a concentration of 1 mM, did not stimulate cGMP formation but the same concentration of 2-methylhistamine did (see Table 1). Other evidence that the response is coupled to the H_1-receptor is that mepyramine, at appropriate concentrations, 3 and 6 nM, shifted the dose–response curve to the right without altering the slope. The affinities of other H_1-antagonists for this receptor were similar to their affinities for the H_1-receptor in guinea pig ileum (Richelson, 1980). Histamine and 2-methylhistamine, but not 4-methylhistamine, each at a concentration of 0.1 mM, desensitized the receptor; pyrilamine, 50 nM, but not metiamide, 1.5×10^{-5} M, prevented sensitization induced by 0.1 mM histamine, (Taylor and Richelson, 1979). Cells desensitized to histamine were still responsive to carbamylcholine; the converse also held. The rate of desensitization decreased with increased histamine concentration. Desensitization was not accompanied by a change in the ED_{50} of histamine but rather in the reduction of the maximal response. Also indicating that the receptor itself is desensitized is that omission of calcium from the preincubation medium, which reduces cGMP formation, did not affect development of desensitization. These observations recall early studies of histamine-stimulated cAMP formation in guinea pig cerebral cortical slices which also showed desensitization to histamine which was partly attributable to enhanced phosphodiesterase activity (Schultz and Daly, 1973).

Histamine increases the cGMP content of cerebral cortical slices of rabbit (Kuo *et al.*, 1972) and of guinea pig (Schwabe *et al.*, 1978) but not of cerebellar slices of rabbit (Kuo *et al.*, 1972), guinea pig (Ohga and Daly, 1977), or mouse (Ferrendelli *et al.*, 1975). Histamine also increases the cGMP content of rat pineal body (O'Dea and Zatz, 1976) and bovine superior ganglion (Study and Greengard, 1978). The histamine receptor mediating these responses remains to be classified. This effect of histamine requires calcium ions (Schwabe *et al.*, 1978). It may be pertinent that suggestive evidence has been presented that histamine may increase calcium influx in membranes of smooth muscle containing the H_1-receptor but not in membranes of smooth muscle lacking this receptor (Uchida, 1980). Calcium ions may then activate guanylate cyclase (see Schwabe *et al.*, 1978) and/or, as suggested by Schwartz *et al.* (1980*a*), glycogenolysis.

3.3. The H_1-Linked Adenylate Cyclase

One of the first systems used to show histamine receptors in brain was the stimulation of adenylate cyclase activity by histamine in brain slices of different species (Kakiuchi and Rall, 1968; Daly, 1977). For reasons described in Section 2, notably that H_1-antagonists, at concentrations of 10^{-6} M and higher, block the H_2-receptor (Green *et al.*, 1977; Kanof and Greengard, 1979), ambiguities resulted. The use of guinea pig brain homogenates, as described in Section 4, unequivocally showed that the H_2-receptor is coupled to adenylate cyclase. Moreover, in the homogenates, only an H_2-receptor linked to adenylate cyclase could be discerned (Hegstrand *et al.*, 1976; Green *et al.*, 1977, 1978*a*) (Section 4), and modeling the system showed that if homogenates contain an H_1-receptor linked to the cyclase, it could contribute no more than 15% to the activation (Hough *et al.*, 1980).

The finding that an H_2-receptor is coupled to adenylate cyclase in brain *homogenates* does not rule out the presence of an H_1-linked adenylate cyclase in brain *slices*. Conditions used to evoke it in homogenates may be inadequate, or homogenization could destroy the coupling. There is in fact, evidence for an H_1-linked adenylate cyclase in slices, as previously summarized (Hough *et al.*, 1980). The histamine stimulation of the cAMP formation in slices is about fivefold (see Daly, 1977), whereas in homogenates the maximal stimulation is about one-fold (e.g., Hegstrand *et al.*, 1976; Green *et al.*, 1977). The activity could be elicited in guinea pig cerebellar slices but not in cerebellar homogenates (Chasin *et al.*, 1974). The relative potencies of a series of drugs in blocking histamine-stimulated adenylate cyclase in rabbit cerebral cortical slices differed from their potencies in blocking histamine-stimulated adenylate cyclase in comparable homogenates (Spiker *et al.*, 1976). In a specially prepared homogenate (Chasin *et al.*, 1974; Psychoyos, 1978), which may more nearly resemble a mince or a slice than a cell-free homogenate, there was evidence that the H_1-receptor may be linked to adenylate cyclase in brain. Evidence that both the H_2- and H_1-receptors are linked to adenylate cyclase in guinea pig brain slices was provided by the use of antagonists in appropriate concentrations and over a range sufficient to do a Schild plot (on the H_1-antagonist) which showed competitive antagonism (Palacios *et al.*, 1978). In these slices, the effects of H_1-antagonists on the dose–response curve to histamine suggests that activation of the H_2-receptor may be necessary before the H_1-receptor can be activated (Palacios *et al.*, 1978). Adenosine, which enhances the effect of histamine (and other amines) on adenylate cyclase activity, appears to affect the H_1-receptor rather than the H_2-receptor in the slices. A study of the dose–response curves to an H_1-agonist and an H_2-agonist in the presence and absence of adenosine showed that only for H_1-stimulation was there a clear decrease in the ED_{50} value of adenosine (25 μM as opposed to 250 μM). The use of H_1- and H_2-antagonists also suggested that the effect of adenosine is mediated mainly by the H_1-receptor (Dismukes *et al.*, 1976*a*).

4. THE HISTAMINE H_2-RECEPTOR

As binding with labeled H_2-antagonists, both cimetidine (Smith *et al.*, 1980; Rising *et al.*, 1980) and tiotidine (Maayani *et al.*, 1981), failed to label selectively the H_2-receptor (Section 2), information about the receptor in brain rests on studies of the H_2-linked adenylate cyclase, which is present in many other tissues (Green and Hough, 1980; Karppanen, 1980). In guinea pig brain homogenates, histamine-stimulated adenylate cyclase activity was convincingly shown to be linked to the H_2-receptor. The

TABLE 3
ED_{50} Values of Histamine Agonists on Broken Cell Preparations of Guinea Pig Hippocampus

Agonist	ED_{50} (μM)	
Histamine	8[a]	14[b]
Dimaprit		6.4[b]
N^{α},N^{α}-Dimethyltryptamine		21[b]
4-Methylhistamine	23[a]	24[b]
2-Methylhistamine	120[a]	120[b]

[a] Stimulation of cAMP formation in homogenates of dorsal hippocampus (Hegstrand *et al.*, 1976).
[b] Stimulation of cAMP formation in membranes of whole hippocampus (Green *et al.*, 1977).

TABLE 4
Apparent Dissociation Constants (K_B) and pA_2 ($-\log K_B$) Values of Histamine H_2-Antagonists on Histamine-Stimulated Adenylate Cyclase in Guinea Pig Hippocampus Homogenates Compared with pA_2 Values on Histamine H_2-Receptors in Peripheral Tissue

Antagonists	Adenylate cyclase activity, guinea pig hippocampus		H_2-receptor in peripheral tissue
	K_B, M	pA_2	(pA_2)
Imidazolylpropylmethylthiourea	4.68×10^{-4}[a]	3.33[a]	3.50[b]
N^{α}-Guanylhistamine	7.94×10^{-5}[a]	4.14[a]	3.90[b]
Imidazolylpropylguanidine	3.16×10^{-6}[a]	5.50[a]	4.65[b]
Thiaburimamide	2.39×10^{-6}[a]	5.62[a]	5.50[b]
Metiamide	8.71×10^{-7}[a,c]	6.06[a]	6.03[b]
Cimetidine	6.03×10^{-7}[a,d]	6.22[a]	6.10[b]
Tiotidine	2.57×10^{-8}[e]	7.59[e]	7.80[f]

[a] Green *et al.* (1977).
[b] Guinea pig atrium and rat uterus. From Ganellin (1978).
[c] Hegstrand *et al.* (1976).
[d] Kanof and Greengard (1979).
[e] Maayani *et al.* (1981).
[f] Yellin *et al.* (1979).

relative activities of agonists on this preparation (Table 3) were similar to activities on peripheral H_2-receptors (Black *et al.*, 1972). Seven compounds with affinities ranging over four orders of magnitude for the H_2-receptor in peripheral tissues had affinities for the histamine-linked cyclase similar to those on the peripheral H_2-receptor (Table 4). Criteria for competitive antagonism were met. All compounds caused a parallel dextral shift in the dose–response curve. Some of the values in Table 4 were obtained from Schild plots (i.e., log (dose ratio-1) vs. log [antagonist]) (Schild, 1947; Arunlakshana and Schild, 1959) in which the slopes did not significantly differ from 1.0 as predicted from simple competitive antagonism. Also as expected of simple competitive antagonism, the same K_B values were obtained with different agonists, including dimaprit (Green *et al.*, 1977) which activates the H_2-receptor with negligible or no activation of the H_1-receptor (Parsons *et al.*, 1977; Bertaccini *et al.*, 1979). The same Schild plot resulted when cimetidine was tested on either hippocampal or cortical membranes (Green *et al.*, 1977), implying that the receptors in the two regions were not different.

For H_2-activation of adenylate cyclase (Kanof *et al.*, 1977), GTP (or an analog) is very nearly obligatory. The guanine nucleotides increase V_{max} for histamine (at 10^{-4} M histamine). The ED_{50} for histamine is decreased by the guanine nucleotides. Implicit in this work (Kanof *et al.*, 1977) is suggestive evidence that just as the guanine nucleotides influence the action of histamine on the H_2-linked cyclase (and the binding of agonists to the receptor), histamine may influence the action of the guanine nucleotides on the cyclase: a Michaelis–Menten plot of their data suggests that the ED_{50} of the GTP analog [viz. p(NH)ppg] is decreased threefold (from 3.5×10^{-6} to 1×10^{-6} M) in the presence of histamine, 10^{-4} M.

Another consideration in the H_2-linked adenylate cyclase is magnesium ion. Mg^{2+} complexes with ATP^{4-} to form magnesium ATP (MgATP), which is recognized as the substrate for the enzyme. Magnesium ions may also stimulate by chelating free ATP, which inhibits the enzyme. Because more Mg^{2+} is required than necessary for stoichiometric formation of MgATP, it was suggested that magnesium ions may act directly on an allosteric site on the enzyme. A plot of the data of Kanof *et al.* (1977) shows that increasing magnesium ions increased the affinity of the cyclase for its substrate, MgATP (the apparent K_m value falls). This plot also suggests that the cyclase is inhibited at high concentrations of MgATP. In brain homogenates, the H_2-linked adenylate cyclase has a sensitivity toward magnesium ions in the presence of histamine (10^{-3} M) that is tenfold greater than in its absence. Analogously, in the presence of histamine, ATP is a more potent inhibitor of the cyclase than in the absence of histamine; cf. K_i = 0.16 mM and 0.33 mM (Kanof *et al.*, 1977).

The H_2-receptor-linked adenylate cyclase is also found in monkey brain homogenates, cortex and hippocampal formation showing greatest activity. Within the hippocampal formation, the subiculum was most responsive, CA1–3 was less active, and the dentate gyrus showed least activity (Newton *et al.*, 1982).

Experiments on chicks also showed adenylate cyclase to be linked to the H_2-receptor (Nahorski *et al.*, 1974, 1977). Subcutaneous injection of histamine (10 mg/kg of body weight) to neonates, which lack a blood–brain barrier to histamine, increased the cAMP content of cerebral hemispheres by three to four times. Mepyramine, in doses as high as 20 mg/kg, did not block this rise, but the H_2-antagonists, burimamide (10 and 20 mg/kg) and metiamide (2 and 5 mg/kg), produced a dose-related reduction in the response. *In vitro* the histamine-stimulated adenylate cyclase of chick slices (more responsive than brain slices of any other species reported) was antagonized by 1 and 5 μM, metiamide, and 10 and 50 μM, burimamide, but not by 50 μM mepyramine. The cerebral hemispheres showed more activity than did the cerebellum, medulla-diencephalon, or optic lobes (Nahorski and Smith, 1976).

Histamine stimulated the adenylate cyclase activity in homogenates of rabbit frontal cortex, anterior limbic cortex, and hypothalamus, in all of which activity diminished with age (Makman *et al.*, 1980), and in rat neocortex (Hegstrand *et al.*, 1976) and hippocampus (Green *et al.*, 1977), but the receptor with which the activity is associated was not defined.

Histamine evoked no stimulation of adenylate cyclase in homogenates of whole brain of gerbil or hamster (Hough and Green, 1981), rat corpus striatum or cerebellum (Hegstrand *et al.*, 1976), and either monkey (Ahn and Makman, 1987) or rat hypothalamus (Ahn and Makman, 1977; Hough and Green, 1981). At least under the assay conditions used, rat brain homogenates are far less responsive in this regard than are guinea pig brain homogenates (Hegstrand *et al.*, 1976; Green *et al.*, 1977); rat is also relatively insensitive to histamine-provoked gastric acid secretion (Ivy and Bachrach, 1966), which is mediated by the H_2-receptor.

Table 2 shows that the highest "densities" of H_2-receptor in guinea pig brain, as measured by histamine-stimulated cAMP formation by brain homogenates, are found in cortex and hippocampus (Hegstrand *et al.*, 1976; Green *et al.*, 1978*b*). The "density" is greater in dorsal than in ventral hippocampus (Hegstrand *et al.*, 1976). No H_2-linked cyclase activity could be elicited from homogenates of guinea pig cerebellum (Hegstrand *et al.*, 1976), which is rich in H_1-receptors (Tan-Tran *et al.*, 1978). Failure to show histamine-stimulated cAMP formation in homogenates does not imply that the H_2-receptor is not present. Abundant and mindful electrophysiological studies of rat brain leave no doubt that the H_2-receptor is there and responding, including in hypothalamus (see below) whose homogenates produce no increase in cAMP when treated with histamine (Ahn and Makman, 1977; Hough and Green, 1981).

Mast cells, which contain the H_2-receptor (Lichtenstein and Gillespie, 1973), cannot account for this receptor in brain homogenates because in all species examined mast cells are scarce or not detectable in cerebral cortex whereas they are especially prominent in the hypothalamus (Edvinsson *et al.*, 1977), a distribution opposite to that of the H_2-linked adenylate cyclase (Table 2). Analogously, although mast cells with H_2-receptors are associated with blood vessels in brain (Edvinsson *et al.*, 1976), mast cells cannot account for the H_2-receptors observed in the capillary-rich fraction from rat (Joó *et al.*, 1975), rabbit (Palmer, 1981), and guinea pig cerebral cortex (Karnushina *et al.*, 1980). Mast cells may contribute to or be responsible for the histamine-stimulated adenylate cyclase observed in choroid plexus (Hammers *et al.*, 1977).

The subcellular fraction of guinea pig cerebral cortex containing synaptic membranes was especially rich in the H_2-receptor (Kanof *et al.*, 1977). From this finding and the nonuniform regional distribution (Table 2), it was suggested that the receptor is associated with neurons rather than glia (Kanof *et al.*, 1977). Nahorski *et al.* (1977), noting that the H_2-receptor is associated with glycogenolysis in chick brain slices (Nahorski *et al.*, 1975), that glycogen is associated with glial cells, and that histamine stimulates cAMP formation in astrocytoma cells (Clark and Perkins, 1971), suggested that the H_2-receptor may be located on the surface of glial cells.

However, kainate lesions in the guinea pig hippocampus almost completely eliminated histamine-stimulated cAMP formation (Garbarg *et al.*, 1978).

Although the histamine-stimulated adenylate cyclase is present in homogenates of rat hippocampus, the maximal cAMP formed is far less than that from comparable preparations from guinea pig (Green *et al.*, 1977). Other kinds of experiments have shown H_2-receptors in rat hippocampus. Injection of dimaprit or histamine directly into rat dorsal hippocampus produced the syndrome of morphine withdrawal, which persisted for hours, as though dimaprit were acting indirectly. The effect was blocked by cimetidine (Glick and Crane, 1978). The rat hippocampus has been a favorite object for electrophysiological studies. Histamine depresses its firing rate (Haas, 1974), as it does that of many neurons in mammalian brain [and molluscan neurons as well (Carpenter and Gaubatz, 1975; Gruol and Weinreich, 1979; Gotow *et al.*, 1980)], an effect that is mediated by H_2-receptors (in contrast with the H_1-mediated increase in firing rate mentioned in Section 3.1). The depression is produced by H_2-agonists and blocked by H_2-antagonists (Haas, 1974; Renaud, 1976). Further, in slices of rat hippocampus, histamine and impromidine, the H_2-agonist, hyperpolarize about half of the CA1-pyramidal cells and dentate granule cells when applied by iontophoresis or pressure, with a concomitant reduction in excitability and spontaneous firing rate (Haas, 1981*a,b*). Some cells are depolarized by histamine and impromidine, the probability being higher on application by microdrops (Haas, 1981*b*). While depolarization is usually accompanied by an increase in spontaneous firing rate, decreases were also observed (Haas, 1981*a,b*). In a calcium-deficient or magnesium-enriched medium, which depresses synaptic transmission, the hyperpolarizations were not blocked, but the depolarizations were. The hyperpolarizations were attributed to a postsynaptic action, maybe to increased K^+-conductance in dendrites. Histamine also reduces homocysteate-evoked firing (Haas, 1981*a*). Compared with the action of homocysteate, the hyperpolarizations produced by histamine (as by 5-hydroxytryptamine) are slow in onset and long in duration, properties that may imply modulation rather than transmission (Haas, 1981*a*). The depolarizations are probably indirect actions of histamine on other cells (Haas, 1981*a,b*; Segal, 1981). The depolarizations coupled with reduced firing rate, Haas (1981*b*) suggested, are consistent with dendritic inhibition and somatic disinhibition. It should be noted that all these electrophysiological effects of histamine were blocked by metiamide (Haas, 1981*a*). In rat hypothalamic neurons in culture the H_2-mediated depression also persists in calcium-free medium, again suggesting that this H_2-effect is postsynaptic (Geller, 1981).

Electrical stimulation of the rat medial forebrain bundle depresses firing of cortical neurons (Sastry and Phillis, 1976), and stimulation of fornix depresses hippocampal CA1 pyramidal cells (Haas and Wolf, 1977);

metiamide reduced these depressions, in accord with histaminergic pathways proposed by Schwartz (1979). It is clear that this depressant effect of histamine on neurons is mediated by the H_2-receptor; however, it is not certain that this H_2-effect is linked to adenylate cyclase. H_2-effects are readily elicited in these electrophysiological experiments on neurons in brain regions where very little or no H_2-linked cyclase can be evoked, e.g., rat hypothalamus and hippocampus. Destruction of the medial forebrain bundle in guinea pigs produced hypersensitivity to iontophoretically applied histamine in both the ipsilateral (but not contralateral) cerebral cortex and hippocampus without altering the ED_{50} or V_{max} of histamine-stimulated adenylate cyclase activity in slices from both these regions or in hippocampal homogenates, results implying that in these neurons, the H_2-cyclase is not related to the electrophysiological response (Dismukes *et al.*, 1976*b*; Haas *et al.*, 1978). But in rat, the same lesion resulted in hypersensitivity of histamine-stimulated adenylate cyclase in cortical and hippocampal slices (Dismukes *et al.*, 1975). In rat hypothalamic cultures, Geller (1979) found that the H_2-linked depressant response was enhanced by phosphodiesterase inhibitors. Kainic acid injections into guinea pig hippocampus abolished histamine-stimulated cAMP synthesis in hippocampal slices (Garbarg *et al.*, 1978), where the stimulation is partly linked to the H_2-receptor (Palacios *et al.*, 1978).

Yet additional and pointed evidence that the H_2-receptor is associated with neural events are the central effects described in adverse reactions to cimetidine. Cimetidine does not enter the brain as readily as do lipid-soluble drugs, but measurable levels were present in all of seven post-mortem brains from patients that had been treated with cimetidine, the brain:serum ratio averaging about 1 (Schentag *et al.*, 1981). In patients with advanced renal or hepatic disease, the blood levels rise, and in patients with diseases of both organs, the half-time of cimetidine is prolonged to 10 times the normal value (Schentag *et al.*, 1981). Adverse reactions to cimetidine are uncommon (even though it is the most widely prescribed of all drugs today), but the most frequently reported adverse reaction to cimetidine are dose-related, reversible mental symptoms (see Schentag *et al.*, 1979, 1981; Kimelblatt *et al.*, 1980) which, in order of appearance, are restlessness, confusion, disorientation, agitation, and visual and auditory hallucinations (Schentag *et al.*, 1979). Mental changes have been reported with serum through concentrations as low as 0.25–0.5 μg/ml (Kimelblatt *et al.*, 1980), probably because of alterations in the blood–brain barrier with hepatic disease. The usual cimetidine cerebrospinal:serum ratio is 0.29 (Schentag *et al.*, 1981), but one agitated, confused, and paranoid patient with hepatic failure but with normal serum levels had a ratio of 0.51 (Kimelblatt *et al.*, 1980). Patients with mental symptoms had high cerebrospinal fluid levels of cimetidine (Table 5). With the (fair) assumption that spinal fluid levels are in equilibrium with brain

TABLE 5
Cimetidine Levels in Cerebrospinal Fluid (CSF) and Estimates of H_2-Receptor Occupation

	Cimetidine CSF levels			Percent Occupation of
Mental status[a]	μg/ml[a]	μM	K_B units[b]	H_2-receptor[c]
Unchanged	0.18	0.7	0.9	48
Unchanged	0.25	1.0	1.3	57
Unchanged	0.25	1.0	1.3	57
280% Change	0.87	3.4	4.5	82
430% Change	1.40	5.5	7.2	88

[a] Schentag *et al.*, (1979).
[b] Calculated from $[D]/K_B$ (Green *et al.*, 1977).
[c] Equation (2), Section 2.

receptors, and that most histamine in brain is held in nerve endings, we calculate that in the patients with changes in mental status, over 80% of the H_2-receptors are occupied (Table 5). These concentrations of cimetidine are too low to affect significantly other known receptors for which it has affinities (Brimblecombe *et al.*, 1975; Green *et al.*, 1977; McCulloch *et al.*, 1979; Barker, 1981).

Another more uncommon adverse effect of cimetidine is hyperthermia (see Hough and Green, 1980, for references). This probably also rests on H_2-receptor blockade, for in rats H_2-agonists produce hypothermia on intraventricular injection, and H_2-antagonists, including cimetidine, block this effect (Green *et al.*, 1975; Pilc and Nowak, 1980). Another effect of cimetidine in man is increased serum prolactin levels (Hough and Green, 1980; Ferrari *et al.*, 1979; Scarpignato *et al.*, 1979; Gonzalez-Villapando *et al.*, 1980), which is also seen in rats (Falaschi *et al.*, 1980). Stimulation of the H_2-receptor by intraventricular administration of dimaprit in rat reduces plasma prolactin levels, dimaprit likely acting at a suprapituitary level, perhaps the hypothalamus (see Falaschi *et al.*, 1980).

5. PSYCHOTROPIC DRUGS AND HISTAMINE RECEPTORS

Histamine in an adventitious way initiated the development of modern psychopharmacology and, concomitantly, modern biological psychiatry when phenothiazines—first designed to treat parasitic infections—were shown to be antihistamines (i.e., H_1-antagonists) and promethazine observed to produce a "euphoric quietude," observations that also stimulated studies on the central effects of H_1-antagonists (Swazey, 1974; Kety, 1978). From the phenothiazines evolved the antidepressant drugs which were also designed as H_1-antagonists (Kuhn, 1970). Consideration of histamine as having a role in psychotropic drug activity ebbed, and not until the 1960s was interest renewed (Green, 1964, 1970; Snyder and Taylor, 1972; Taylor, 1975; Schwartz, 1975). Histamine levels were shown to be reduced by drugs that influence behavior such as reserpine, which lowers histamine levels in brain (Adam and Hye, 1966; Pollard *et al.*, 1973) and in stomach (Kim and Shore, 1963; Isaac *et al.*, 1971), and monoamine oxidase inhibitors which alter histamine metabolism in patients too (Fram and Green, 1968). Histamine metabolism is also altered by ethanol (Subramanian *et al.*, 1978) and barbiturates (Pollard *et al.*, 1974), and morphine-induced changes in locomotor activity inversely correlate with brain histamine levels in the mouse (Lee and Fennessy, 1976). Thus, histamine may have a role in the actions of diverse centrally acting drugs, apart from those drugs that react directly with histamine receptors.

5.1. Antidepressant Drugs

The apparent dissociation constants of antidepressant drugs for the histamine H_1- and H_2-receptors are shown in Table 6. Blockade of the H_1-receptor was demonstrated to be competitive by Schild plots in antagonism to histamine on the H_1-receptor linked to contraction of the guinea pig ileum (Figge *et al.*, 1979) and by parallel shifts in the dose–response curve in the guanylate cyclase system (Richelson, 1980); the K_B values for these preparations were highly correlated (Figge *et al.*, 1979). Several differing K_B values, e.g., in the rat studies, have been shown to be due to different concentrations of tissue used in the binding assay (Taylor and Richelson, 1980), but the relative order of affinities of the drugs for the H_1-receptor is in good agreement. As emphasized by Richelson (1980), doxepin and amitriptyline have greater affinities for the H_1-receptor than have those drugs that are classified as H_1-antagonists. The affinities of some antidepressants, e.g., amitriptyline, mianserin, and doxepin, for the H_1-receptor are high enough for them to be used as labeling ligands in binding studies (Green *et al.*, 1978*c*; Rehavi and Sokolovsky, 1978; Peroutka and Snyder, 1981; Tan-Tran *et al.*, 1981). Treating mice with antidepressants reduced the amount of injected [^{3}H]mepyramine in whole brain (Quach *et al.*, 1979; Schwartz *et al.*, 1980*c*; 1981) and in brain particulate fraction (Diffley *et al.*, 1980). Since antidepressants, like H_1-antagonists, are sedative in untreated laboratory animals and normal man, these measurements may be an additional means of testing drugs for their sedative properties.

Similarities in the pharmacological effects of many H_1-antagonists and antidepressant have been noted, e.g., they antagonize reserpine-like drugs, potentiate amphetamine-like drugs, and block muricide behavior

TABLE 6

Apparent Dissociation Constants, K_B (nM), for Antidepressant Drugs and Histamine Receptors

	H_1-Receptor						H_2-Receptor
			Brain, [^{3}H]mepyramine binding				
Drug	Guinea pig ileum[a]	Mouse neuroblastoma cells[b]	Guinea pig[c]	Rat[d]	Rat[e]	Man[f]	Guinea pig hippocampus[g]
Doxepin	0.06	0.03	0.5	0.7	0.03	0.4	150[h]
Mianserin				3.3			65[i]
Amitriptyline	0.08	0.13	1.6	4.1	0.10	1.1	45[h]
Clomimpramine							200[j]
Nortriptyline	3.2	7	30	46	13	10	760[j]
Imipramine	13	10	16	26	27	11	240[j]
Protriptyline	69	35	80	60	66	31	2000[j]
Desipramine	150	260	250	250	230	200	1100[j]
Dibenzepin							1500[h]
Iprindole			460	100		670	3200[h]

[a] Blockade of histamine-stimulated contractions (Figge *et al.*, 1979).
[b] Blockade of H_1-linked guanylate cyclase activity (Richelson, 1980).
[c] Whole brain homogenates (Chang *et al.*, 1979*b*).
[d] Whole brain homogenates. All values are from Chang *et al.* (1979*b*) except for mianserin (Peroutka and Snyder, 1981).
[e] Homogenates of whole brain except for cerebellum (Taylor and Richelson, 1980).
[f] Frontal cortex homogenates (Chang *et al.*, 1979*b*).
[g] Blockade of H_2-linked adenylate cyclase in homogenates.
[h] Green and Maayani (1977).
[i] Kanof and Greengard (1978).
[j] Maayani *et al.* (1978).

and isolation-induced fighting (Barnett *et al.*, 1969, 1971). These effects of H_1-antagonists were not correlated with H_1-antagonist activity (Barnett *et al.*, 1969, 1971). Treatment of depressed patients with high doses of diphenhydramine yielded ambiguous results (Hankoff *et al.*, 1964). Even if the clinical results were more successful (Barnett *et al.*, 1969) than the authors assessed (Hankoff *et al.*, 1964), they cannot be securely attributed to H_1-antagonism, for 69% of the patients showed side effects, most of which can be attributed to muscarinic blockade. Of interest is that a low dose, 50 mg, of diphenhydramine was reported to heighten mood while decreasing psychomotor activity; with increased dose, no effect on mood was seen (Jäätelä *et al.*, 1971).

The antidepressants competitively antagonized histamine at the H_2-receptor as well, as shown initially for four tricyclic antidepressants with different ring structures (Green and Maayani, 1977) and then for other tertiary and secondary amino compounds (Maayani *et al.*, 1978; Kanof and Greengard, 1978), including the tetracyclic compounds mianserin (Kanof and Greengard, 1978). Hydroxyl derivatives and hydroxychloro derivatives of imipramine, like imipramine itself, inhibited histamine-stimulated adenylate cyclase activity in both slices and homogenates of rabbit cerebral cortex (Palmer *et al.*, 1977). Figure 1 shows dextrad parallel shifts in the dose–response curves to histamine produced by four drugs, each at 10^{-5} M, on the same preparation. Schild plots (Green and Maayani, 1977) and other analyses (Kanof and Greengard, 1978) revealed competitive antagonism.

The H^2-receptor in rabbit atria is blocked by amitriptyline and imipramine (Hughes and Coret, 1974) and in the artery of the rabbit ear by desipramine (McCulloch and Story, 1972), but not competitively. On the H_2-receptor in guinea pig atrium, rat uterus, or mouse stomach, amitriptyline did not antagonize histamine (Angus and Black, 1980). But on the H_2-receptor associated with contraction of the guinea pig papillary muscle, amitriptyline competitively antagonized the effect of histamine (Angus and Black, 1980). The authors suggested that tissue factors explain

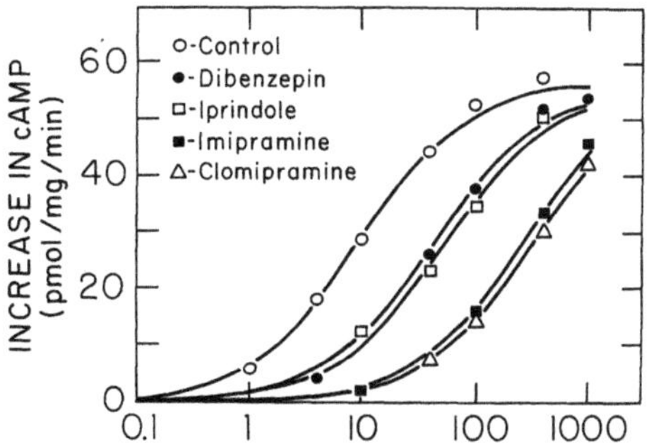

FIG. 1. Inhibition of histamine-stimulated adenylate cyclase activity in the same preparation of homogenates of guinea pig hippocampus by dibenzepin, iprindole, imipramine, and clomimpramine each at 10 μM (unpublished data of Saul Maayani).

the differences between atrium and papillary muscle rather than receptor differences. One factor could be phosphodiesterase. An inhibitor of this enzyme had greater stimulant activity on atrium than on papillary muscle (Angus and Black, 1980). As amitriptyline is a competitive inhibitor of phosphodiesterase (Berndt and Schwabe, 1973; Levin and Weiss, 1976), its block of both the enzyme and the H_2-receptor in the atrium may have "self-cancelled" (Angus and Black, 1980). Consilient with this idea is that a phosphodiesterase inhibitor altered the effect of metiamide on the H_2-receptor in the atrium (Angus and Black, 1980). Similar factors could explain the deviation from competitive blockade by the antidepressants on the H_2-receptor in rabbit atrium (Hughes and Coret, 1974) and artery (McCulloch and Story, 1972) and other H_2-receptor systems (Angus and Black, 1980).

It is interesting that the rank orders of affinities of these drugs for the H_2-receptor and the H_1-receptor are similar as measured on both the guinea pig ileum and brain, with the exception of doxepin (Table 6). Affinities for the H_1-receptor are far higher than for the H_2-receptor (Table 6), and almost all these drugs have higher affinities for the H_1-receptor than for the muscarinic receptor (Richelson, 1980) or the α_1-adrenergic receptor (Snyder, 1980).

Even though the H_2-receptor has less affinity for these drugs, the concentrations of antidepressant drugs in the brain of rats receiving a pharmacological dose, and the concentrations of amitriptyline in the plasma of patients on therapy (Ziegler *et al.*, 1977), are sufficient to occupy the H_2-receptor (see Green and Maayani, 1977). No experiments appear to have been reported yet to show antagonism by antidepressant of a brain response associated with the H_1-receptor, but there is presuasive experimental evidence that at least some antidepressants block effects associated with the H_2-receptor in brain. Amitriptyline reduced the H_2-receptor-linked depression of firing rate of rat cortical neurons (Haas, 1979). This effect, similar to that of metiamide, was seen after direct application of amitriptyline as well as after treating rats with amitriptyline, as little as 3 mg/kg body weight (Haas, 1979). The hypothermia produced in rats by intraventricular injection of dimaprit and other histamine agonists was significantly reduced by pretreatment with imipramine or amitriptyline. The blockade occurred after intraperitoneal injection, 10 and 20 mg/kg body weight, or after intraventricular injection (Nowak *et al.*, 1979).

As antidepressants have greater affinities for the H_1-receptor (Table 6) and the muscarinic (Richelson, 1980) and α_1-adrenergic (Snyder, 1980) receptors than for the H_2-receptor, they too almost certainly must be occupied by antidepressant drugs after treatments that blocked the H_2-receptor. It therefore follows, as emphasized (Green and Maayani, 1977; Green *et al.*, 1980), that the final, gross behavioral effect of these drugs is likely to be the sum of effects of all these blockades in addition to their effects on uptake processes, presynaptic receptors, and, probably, additional sites not yet delineated.

5.2. Neuroleptic Drugs

The phenothiazines originated as H_1-antagonists (Swazey, 1974; Kety, 1978), as noted above, and new neuroleptic agents like butyrophenones and benzamides also block H_1-receptors noncompetitively (Greene, 1972; Van Nueten *et al.*, 1977; Fontaine and Reuse, 1978; Fjalland and Boeck, 1978).

Table 7 shows that neuroleptic drugs block the H_1-receptor in neuroblastoma cells and in brain. Rat brain homogenates had less affinity for the phenothiazines than had comparable homogenates from guinea pig and human brain, but the rank order of potencies in all systems is similar. Many of these drugs competed *in vivo* for [^{3}H]mepyramine binding (Quach *et al.*, 1979).

These drugs block histamine-stimulated adenylate cyclase activity in rabbit cortical slices and homogenates (Spiker *et al.*, 1976; Palmer *et al.*, 1978) and the H_2-linked adenylate cyclase activity in homogenates of guinea pig hippocampus or cortex (Maayani *et al.*, 1978; Kanof and Greengard, 1978). Stereoselectivity was exhibited with both flupenthixol and butaclamol, the clinically active isomer showing much greater potency

TABLE 7

Apparent Dissociation Constants, K_B (nM), for Neuroleptic Drugs and the Histamine H_1-Receptor

Drug	Mouse neuroblastoma cells[a]	Brain, [^{3}H]mepyramine binding		
		Guinea pig[b]	Rat[b]	Man[c]
Clozapine	0.2	4.0	36	1.9
Mesoridazine	1.3			
Promazine	1.4	2.4	22	4.2
Chlorpromazine		4.5	36	9
Pherphenazine	2.4			
Prochlorperazine	5.9			
Fluphenazine	6	19	67	70
Thioridazine	19	34	20	22
Trifluoperazine	580			
Spiroperidol		2700	670	667
Haloperidol	800	800	3300	2500
d-Butaclamol		3100	1300	2100

[a] Blockade of H_1-linked guanylate cyclase activity (Richelson, 1980).
[b] Whole brain homogenates (Chang *et al.*, 1979*b*).
[c] Frontal cortex homogenates (Chang *et al.*, 1979*b*).

TABLE 8
Blockade of the H_2-Receptor by Neuroleptic Drugs[a]

Drug	Potency
Clorprothixene	81
d-Butaclamol	80
cis-Flupenthixol	60
Spiroperidol	56
Trifluperidol	46
Haloperidol	27
Thiethylperazine	20
Clozapine	10
Chlorpromazine	9[b]
Fluphenazine	8
trans-Flupenthixol	7
Sulpiride	3
l-Butaclamol	2

[a] Blockade of H_2-linked adenylate cyclase in homogenates of guinea pig hippocampus (Maayani *et al.*, 1978, and unpublished experiments). Since all compounds were noncompetitive antagonists, inhibitory potency was calculated as inhibition at 10 μM histamine/drug concentration, μM.
[b] This was the only one of the compounds tested that was competitive, K_B = 0.24 μM.

than the other isomer. Of all drugs tested (Table 8) only chlorpromazine showed, like the antidepressants, competitive antagonism. All other neuroleptic agents tested by Maayani *et al.*, (1978) shifted the dose–response curve to histamine in a nonparallel manner, as exemplified in Fig. 2. The rank orders of potencies in two independent studies were in agreement (Maayani *et al.*, 1978; Kanof and Greengard, 1978).

The affinities of these drugs for the H_1- and H_2-receptors were inversely related, i.e., r = -0.95 (cf. Tables 7 and 8), in contrast with the antidepressants (Table 6) in which the rank orders of affinities for the two receptors were similar. This inverse relationship suggests that judicious choices of these neuroleptic drugs in combination with relatively selective histamine agonists can be used to reveal the effects of these drugs that may be attributable to a specific histamine receptor. The inverse relationship may also be exploited to help design selective H_2-antagonists.

5.3. Hallucinogenic Substances

D-LSD is a competitive antagonist of histamine at both the H_1- (Frederickson and Richelson, 1979) and H_2-receptors (Green *et al.*, 1977). The inactive stereoisomer, L-LSD, showed less activity than D-LSD on the H_1-receptor and no measurable activity on the H_2-receptor (Table 9).

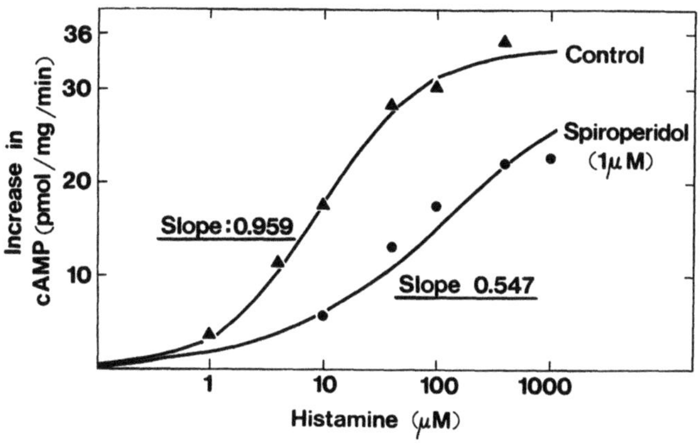

FIG. 2. Inhibition of histamine-stimulated adenylate cyclase activity in homogenates of the guinea pig hippocampus by spiroperidol, 1 μm (unpublished data of Saul Maayani).

Other hallucinogens showed affinities for the H_1-receptor but not for the H_2-receptor. Also inactive on the H_2-receptor were many other types of drugs such as opiates, anticonvulsants, minor tranquilizers, Δ^9-THC, and barbiturates (Green *et al.*, 1978*b*).

For the H_1-receptor, 2-bromo-LSD and D-LSD have about the same affinity, but 2-bromo-LSD has greater affinity than D-LSD for the H_2-receptor (Table 9). Angus and Black (1980) showed that 2-bromo-LSD was a competitive antagonist of histamine on the H_2-receptor in guinea pig papillary muscle.

2-Bromo-LSD is not inactive, as commonly asserted. It has greater

TABLE 9

Apparent Dissociation Constants, K_B (μM), for Hallucinogenic Substances and Histamine Receptors

Drug	H_1[a]	H_2[b]
D-LSD	0.12	1.0
2-Bromo-LSD	0.14	0.07
Psilocybin	1.8	N.A.[c]
Psilocin	1.9	N.A.[c]
L-LSD	25	N.A.[c]
Mescaline	500	N.A.[c]

[a] Blockade of H_1-linked guanylate cyclase activity in mouse neuroblastoma cells (Frederickson and Richelson, 1979).

[b] Blockade of H_2-linked adenylate cyclase activity in homogenates of guinea pig hippocampus (Green *et al.*, 1977).

[c] Not active at 10^{-4} M.

affinity than D-LSD for haloperidol binding sites (Creese *et al.*, 1976), and it is more potent than D-LSD in increasing levels of DOPA (Persson 1977*a*) and DOPAC in rat striatum (Persson, 1977*b*). 2-Bromo-LSD blocks head twitches induced in mice by D-LSD (Corne and Pickering, 1967). 2-Bromo-LSD lacks the agonist activity of D-LSD, but its affinity for postsynaptic sites is similar or greater than that of D-LSD (Green *et al.*, 1977, 1978*b*). Whatever may be the group of receptors that gives rise to the hallucinogenic effect of D-LSD, it is most likely that 2-bromo-LSD shares them, for 2-bromo-LSD blocks the hallucinogenic effect of D-LSD in man (Ginzel and Mayer-Gross 1956; Bertino *et al.*, 1959) at a dosage interval before cross-tolerance is seen. The effect of a dose of D-LSD of 1 μg/kg was blocked by a dose of 2-bromo-LSD of 32 to 654 μg/kg (Bertino *et al.*, 1959).

Of greater interest is that 2-bromo-LSD causes behavioral changes in laboratory animals and in man. Larger doses of BrLSD than LSD are needed to show changes in animal behavior, but they occur (e.g., Uyeno and Mitoma, 1969). In one clinical study in which the dose of 2-bromo-LSD was increased to 32 μg/kg, it produced all the central effects of D-LSD except hallucinations (Schneckloth *et al.*, 1957; Bertino, *et al.*, 1959). What contribution blockade of the H_2-receptor makes to the behavioral effects of D-LSD is not known, but it may be relevant that the adverse reactions to cimetidine in man include hallucinations and hyperpyrexia (Section 4), which D-LSD produces as well.

5.4. Concluding Comment

Each of the two histamine receptors has affinities for psychotropic drugs of three different classes—antidepressant, neuroleptic, and hallucinogenic drugs (notably D-LSD). The histamine receptors are not distinct in this regard, for dopamine receptors have high affinity for both neuroleptic agents and D-LSD (Section 5.3). Since the behavioral effects of these classes of drugs are very different, it follows that these effects cannot be ascribed to interaction with histamine or dopamine receptors alone. Moreover, none of the relatively selective antagonists of any one biogenic amine can produce the behavioral and neural effects of any one of these drugs. Since, at least for antidepressant drugs, usual plasma and brain levels are sufficient to block even that receptor that has least affinity for the drugs, the H_2-receptor, it follows that many receptors are blocked after usual dosing.

The extraordinary effects of these drugs may indeed rest on their affinities for so many sites, the aggregation of interactions producing the behavioral and therapeutic response (Green and Maayani, 1977; Green *et al.*, 1978*b*; Green *et al.*, 1980; Maayani *et al.*, 1981). It is hard to believe that the events by which psychotropic drugs alter human behavior are less complex than how the visual cortex processes a spot of light of known wavelength, intensity, and duration, focused directly on a receptor; this response has been described as "hypercomplex," for the "response properties of most neurons depend on subtle interactions within an intricate system of more and less specific excitatory and inhibitory influences, whose cooperative details determine most aspects of neuronal behavior" (Movshon, 1978). Perhaps specific and tractable questions should be asked about the pharmacology of psychotropic drugs before asking how they alter mood and behavior (for which there are poor laboratory models) or how they cause hallucinations—such seemingly mundane questions as: in altering appetite, on what part or parts of the brain do the antidepressants act, and on what receptor or receptors in these discrete places, and what change in the receptor-coupled responses ensues. A combination of techniques in which the drugs are precisely localized (see Palacios *et al.*, 1981*a*) and in which the events associated with them are precisely defined with a careful choice of agonists and antagonists (see Haas, 1981*a,b*) could, in sum, account for the precise, pharmacological effect. The sum of all the effects could then account for the behavioral effects. The histamine H_1- and H_2-receptors are among the receptors, and the responses that the drugs produce serve to reduce the excitatory and inhibitory effects of histamine. The consequence of these effects are slowly becoming clear, especially for the H_2-receptor, because highly selective agonists and antagonists are available.

Acknowledgments

This work was supported by research grants from the National Institute of Mental Health (MH-31805) and the National Institute on Drug Abuse (DA-01875). Plotting and statistical analyses of results were done on the PROPHET System, a national computer resource sponsored by the National Institutes of Health through the Chemical/Biological Information-Handling Program, Division of Research. I am grateful to Mr. David Allen for patience and sensible and excellent typing, to Dr. Joseph Goldfarb for informative lectures on electrophysiology, and to Dr. Saul Maayani for the original figures and all his other help.

6. REFERENCES

Adam, H. M., and Hye, H. K. A., 1966, Concentration of histamine in different parts of brain and hypophysis of cat and its modification by drugs, *Br. J. Pharmacol.* **28:**137–152.

AHN, H. S., and MAKMAN, M. H., 1977, Neurotransmitter-sensitive adenylate cyclase in the hypothalami of guinea pig, rat, and monkey, *Brain Res.* **138:**125–138.

ANGUS, J. A., and BLACK, J. W., 1980, Pharmacological assay of cardiac H_2-receptor blockade by amitriptyline and lysergic acid diethylamide, *Circ. Res. Suppl. I* **46:**64–69.

ARUNLAKSHANA, O., and SCHILD, H. O., 1959, Some quantitative uses of drug antagonists, *Br. J. Pharmacol.* **14:**48–59.

BARKER, L. A., 1981, Histamine H_1- and muscarinic receptor antagonist activity of cimetidine and tiotidine in the guinea pig isolated ileum, *Agents Actions* **11:**699–705.

BARNETT, A., TABER, R. I., and ROTH, F. E., 1969, Activity of antihistamines in laboratory antidepressant tests, *Int. J. Neuropharmacol.* **8:**73–379.

BARNETT, A., MAKICK, J. B., and TABER, R. I., 1971, Effects of antihistamines on isolation-induced fighting in mice, *Psychopharmacologia* **19:**359–365.

BENNETT, J. P. JR., 1978, Methods in binding studies, in: *Neurotransmitter Receptor Binding* (H. I. Yamamura, S. J. Enna, and M. J. Kuhar, eds.), pp. 57–90, Raven Press, New York.

BERNDT, S., and SCHWABE, U., 1973, Effect of psychotropic drugs on phosphodiesterase and cyclic AMP level in rat brain *in vivo, Brain Res.* **63:**303–312.

BERTACCINI, G., MOLINA, E., ZAPPIA, L., and ZSELI, J., 1979, Histamine receptors in the guniea pig ileum, *Naunyn-Schmiedeberg's Arch. Pharmacol.* **309:**65–68.

BERTINO, J. R., KLEE, G. D., and WEINTRAUB, M. D., 1959, Cholinesterase, *d*-lysergic acid diethylamide, and 2-bromolysergic acid diethylamide, *J. Clin. Exp. Psychopathol.* **20:**218–227.

BLACK, J. W., 1976, Histamine receptors, *Proceedings of the Sixth International Congress of Pharmacology*, Vol. I, *Receptors and Cellular Pharmacology* (E. Klinge, ed.), pp. 3–16, Plenum Press, Oxford.

BLACK, J. W., DUNCAN, W. A. M., DURANT, C. J., GANELLIN, C. R., and PARSONS, E. M., 1972, Definition and antagonism of histamine H_2-receptors, *Nature (London)* **236:**385–390.

BRAND, J., 1968, The pharmacologic basis for the control of motion sickness by drugs, *Pharmacol. Physicians* **2:**1–5.

BRIMBLECOMBE, R. W., DUNCAN, W. A. M., DURANT, G. J., EMMETT, J. C., GANELLIN, C. R., and PARSONS, M. E., 1975, Cimetidine, a non-thiourea H_2-receptor antagonist, *J. Int. Med. Res.* **3:**86–92.

BURT, D. R., 1978, Criteria for receptor identification, in: *Neurotransmitter Receptor Binding* (H. I. Yamamura, S. H. Enna, and M. J. Kuhar, eds.), pp. 41–55, Raven Press, New York.

CARPENTER, D. O., and GAUBATZ, G. L., 1975, H_1 and H_2 histamine receptors on Aplysia Neurones, *Nature (London)* **254:**343–344.

CARRUTHERS, S. G., SHOEMAN, D. W., HIGNITE, C. E., and AZARNOFF, D. L., 1978, Correlation between plasma diphenhydramine level and sedative and antihistamine effects, *Clin. Pharmacol. Ther.* **23:**375–382.

CHANG, R. S. L., and SNYDER, S. H., 1980, Histamine H_1-receptor binding sites in guinea pig brain membranes: regulation of agonist interactions by guanine nucleotides and cations, *J. Neurochem.* **34:**916–922.

CHANG, R. S. L., TRAN, V. T., and SNYDER, S. H., 1979*a*, Characteristics of histamin H_1-receptors in peripheral tissues labeled with [^{3}H]mepyramine, *J. Pharmacol. Exp. Ther.* **209:**437–442.

CHANG, R. S. L., TRAN, V. T., and SNYDER, S. H., 1979*b*, Heterogeneity of histamine H_1-receptors: species variations in [^{3}H]mepyramine binding of brain membranes, *J. Neurochem.* **32:**1653–1663.

CHANG, R. S. L., TRAN, V. T., and SNYDER, S. H., 1980, Neurotransmitter receptor localizations: brain lesion induced alterations in benzodiazepine, GABA, β-adrenergic, and histamine H_1-receptor binding, *Brain Res.* **190:**95–110.

CHASIN, M., MAMRAK, F., and SAMANEIGO, S. G., 1974, Preparation and properties of a cell-free, hormonally responsive adenylate cyclase from guinea pig brain, *J. Neurochem.* **22:**1031–1038.

Clark, R. B., and Perkins, J. P., 1971, Regulation of adenosine 3′,5′-cyclic monophosphate concentration in cultured human astrocytoma cells by catecholamines and histamine, *Proc. Natl. Acad. Sci. USA* **68:**2757–2760.

Corne, S. J., and Pickering, R. W., 1967, A possible correlation between drug-induced hallucinations in man and a behavioural response in mice, *Psychopharmacologia* **11:**65–78.

Creese, I., Burt, D. R., and Snyder, S. H., 1976, The dopamine receptor: differential binding of *d*-LSD and related agents to agonist and antagonist states, *Life Sci.* **17:**1715–1720.

Daly, J., 1977, *Cyclic Nucleotides in the Nervous System*, pp. 97–179, Plenum Press, New York.

Dashputra, P., Sharma, M., Jagtap, M., Khapre, M., and Rajapurkar, M., 1966, Modification of metrazol-induced convulsions in rats by antihistamines, *Arch. Int. Pharmacodyn. Ther.* **160:**106–112.

Diffley, D., Tran, V. T., and Snyder, S. H., 1980, Histamine H_1-receptors labeled *in vivo*: antidepressant and antihistamine interactions, *Eur. J. Pharmacol.* **64:**177–181.

Dismukes, R. K., Ghosh, P., Creveling, C. R., and Daly, J. W., 1975, Altered responsiveness of adenosine 3′,5′-monophosphate-generating systems in rat cortical slices after lesions of the medial forebrain bundle, *Exp. Neurol.* **49:**725–735.

Dismukes, R. K., Rogers, M., and Daly, J. W., 1976*a*, Cyclic adenosine 3′,5′-monophosphate formation in guinea pig brain slices: effect of H_1- and H_2-histaminergic agonists, *J. Neurochem.* **26:**785–790.

Dismukes, R. K., Ghosh, P., Creveling, C. R., and Daly, J. W., 1976*b*, Norepinephrine depletion and responsiveness of norepinephrine-sensitive cyclic AMP generating systems in guinea pig brain, *Exp. Neurol.* **52:**206–215.

Durant, G. J., Ganellin, C. R., and Parsons, M. E., 1975, Chemical differentiation of histamine H_1- and H_2-receptor agonists, *J. Med. Chem.* **18:**905–909.

Durant, G. J., Duncan, W. A. M., Ganellin, G. R., Parsons, M. E., Blakemore, R. C., and Rasmussen, A. C., 1978, Impromidine (SKF 91676) is a very potent and specific agonist for histamine H_2-receptors, *Nature (London)* **276:**403–405.

Edvinsson, L., Owman, C., and Sjoberg, N.-O., 1976, Autonomic nerves, mast cells, and amine receptors in human brain vessels. A histochemical and pharmacological study, *Brain Res.* **115:**377–393.

Edvinsson, L., Cervos-Navarro, J., Larsson, L.-I., Owman, C., and Ronnberg, A.-L., 1977, Regional distribution of mast cells containing histamine, dopamine, or 5-hydroxytryptamine in the mammalian brain, *Neurology* **27:**878–883.

Faingold, C. L., and Berry, C. A., 1972*a*, A comparison of the EEG effects of the potent antihistaminic (DL-chlorpheniramine) with a less potent isomer (L-chlorpheniramine), *Arch. Int. Pharmacodyn.* **199:**213–218.

Faingold, C. L., and Berry, C. A., 1972*b*, Effects of antihistaminic agents upon the electrographic activity of the cat brain: a power spectral density study, *Neuropharmacology* **11:**491–498.

Falaschi, P., Ruggieri, S., D'Urso, R., Rocco, A., Scarnati, E., Forchetti, C., Boghen, M., Frajese, G., and Agnoli, A., 1980, Further evidence for a role of histamine-H_2 receptor in controlling prolactin secretion, in: *Central and Peripheral Regulation of Prolactin Function* (R. M. MacLeod and U. Scapagnini, eds.), pp. 361–364, Raven Press, New York.

Ferrari, C., Caldara, R., Barbieri, C., Cambielli, M., Bierti, L, and Romussi, M., 1979, Prolactin release by intravenous cimetidine in man: evidence for a suprapituitary locus of action, *Clin. Endocrinol.* **11:**619–623.

Ferrendelli, J. A., Kinscherf, D. A., and Chang, M.-M., 1975, Comparison of the effects of biogenic amines on cyclic GMP and cyclic AMP levels in mouse cerebellum *in vitro*, *Brain Res.* **84:**63–73.

Figge, J., Leonard, P., and Richelson, E., 1979, Tricyclic antidepressants: potent blockade of histamine H_1 receptors of guinea pig ileum, *Eur. J. Pharmacol.* **58:**479–483.

Fjalland, B., and Boeck, V., 1978, Neuroleptic blockade of the effect of various neurotransmitter substances, *Acta Pharmacol. Toxicol.* **42:**206–211.

FONTAINE, J., and REUSE, J., 1978, Pharmacological analysis of the effects of substituted benzamides on the isolated guinea pig ileum: study of metoclopramide, sulpiride, bromopride, tiapride, and sultopride, *Arch. Int. Pharmacodyn.* **235**:51–61.

FRAM, D. H., and GREEN, J. P., 1968, Methylhistamine excretion during treatment with monoamine oxidase inhibitor, *Clin. Pharmacol. Ther.* **9**:355–357.

FREDERICKSON, P. A., and RICHELSON, E., 1979, Hallucinogens antagonize histamine H_1 receptors of cultured mouse neuroblastoma cells, *Eur. J. Pharmacol.* **56**:261–264.

FRIEDEL, R. O., and SCHANBERG, S. M., 1975, Effects of histamine on phospholipid metabolism of rat brain *in vivo*, *J. Neurochem.* **24**:819–820.

GANELLIN, C. R., 1978, Chemistry and structure–activity relationships of H_2-receptor antagonists, in: *Handbook of Experimental Pharmacology*, Vol. 18, Part 2, *Histamine II and Anti-Histaminics* (M. Rocha e Silva, ed.), pp. 251–294, Springer-Verlag, Berlin.

GARBARG, M., BARBIN, G., PALACIOS, J. M., and SCHWARTZ, J. C., 1978, Effects of kainic acid on histaminergic systems in guinea pig hippocampus, *Brain Res.* **150**:638–641.

GELLER, H. M., 1979, Are histamine H_2-receptor depressions of neuronal activity in tissue cultures from rat hypothalamus mediated through cyclic adenosine monophosphate?, *Neurosci. Lett.* **14**:49–53.

GELLER, H. M., 1981, Histamine actions on activity of cultured hypothalamic neurons: evidence for mediation by H_1- and H_2-histamine receptors, *Dev. Brain Res.* **1**:89–101.

GINZEL, K. H., and MAYER-GROSS, W., 1956, Prevention of psychological effects of *d*-lysergic acid diethylamide (LSD 25) by its 2-bromo derivative (BOL 148), *Nature (London)* **178**:210.

GLICK, S. D., and CRANE, L. A., 1978, Opiate-like and abstinenece-like effects of intracerebral histamine administration in rats, *Nature (London)* **273**:547–549.

GONZALEZ-VILLAPANDO, C., SZABO, M., and FROHMAN, L. A., 1980, Central nervous system mediated stimulation of prolactin secretion by cimetidine, a histamine H_2-receptor antagonist: impaired responsiveness in patients with prolactin-secreting tumors and idiopathic hyperprolactinemia, *J. Clin. Endocrinol. Metab.* **51**:1417–1424.

GOTOW, T., KIRKPATRICK, C. T., and TOMITA, T., 1980, Excitatory and inhibitory effects of histamine on molluscan neurons, *Brain Res.* **196**:151–167.

GREEN, J. P., 1964, Histamine and the nervous system, *Fed. Proc., Fed. Am. Soc. Exp. Biol.* **23**:1095–1102.

GREEN, J. P., 1970, Histamine, in: *Handbook of Neurochemistry* (A. Lajtha, ed.), Vol. 4, pp. 221–249, Plenum Press, New York.

GREEN, J. P., and HOUGH, L. B., 1980, Histamine receptors, in: *Cellular Receptors for Hormones and Neurotransmitters* (D. Schulster and A. Levitzki, eds.), pp. 287–305, Wiley, New York.

GREEN, J. P., and MAAYANI, S., 1977, Tricyclic antidepressant drugs block histamine H_2 receptor in brain, *Nature (London)* **269**:163–165.

GREEN, M. D., COX, B., and LOMAX, P., 1975, Histamine H_1- and H_2-receptors in the central thermoregulatory pathways of the rat, *J. Neurosci. Res.* **1**:353–359.

GREEN, J. P., JOHNSON, C. L., WEINSTEIN, H., and MAAYANI, S., 1977, Antagonism of histamine-activated adenylate cyclase in brain by *d*-lysergic acid diethylamide, *Proc. Natl. Acad. Sci. USA* **74**:5697–5701.

GREEN, J. P., JOHNSON, C. L., and WEINSTEIN, H., 1978*a*, Histamine as a neurotransmitter, in: *Psychopharmacology: A Generation of Progress* (M. A. Lipton, A. DiMascio, and K. F. Killam, eds.), pp. 319–332, Raven Press, New York.

GREEN, J. P., WEINSTEIN, H., and MAAYANI, S., 1978*b*, Defining the histamine H_2-receptor in brain: the interaction with LSD in: *Quantitative Structure–Activity Relationships of Analgesics, Narcotic Antagonists, and Hallucinogens* (G. Barnett, M. Trsic, and R. Willette, eds.), "QuaSAR" Research Monograph 22, pp. 38–58, National Institute on Drug Abuse, U.S. Government Printing Office, Washington, D.C.

GREEN, J. P., MAAYANI, S., and WEINSTEIN, H., 1978*c*, Histamine receptors in brain: high-affinity binding and the histamine-sensitive adenylate cyclase, 7th International Congress of Pharmacology, July, p. 339.

GREEN, J. P., MAAYANI, S., WEINSTEIN, H., and HOUGH, L. B., 1980, Histamine and psychotropic drugs, *Psychopharmacol. Bull.* **16:**36–38.

GREENE, M. J., 1972, Some aspects of the pharmacology of droperidol, *Br. J. Anaesth.* **44:**1272–1276.

GRUOL, D. L., and WEINREICH, D., 1979, Cooperative interactions of histamine and competitive antagonism by cimetidine at neuronal histamine receptors in the marine mollusc, *Aplysia californica, Neuropharmacology* **18:**415–421.

HAAS, H. L., 1974, Histamine: action on single hypothalamic neurones, *Brain Res.* **76:**363–366.

HAAS, H. L., 1979, Histamine and noradrenaline are blocked by amitriptyline on cortical neurones, *Agents Actions* **9:**83–84.

HAAS, H. L., 1981*a*, Histamine and the central nervous system: analysis of histamine actions by intra- and extracellular recording in hippocampal slices of the rat, *Agents Actions* **11:**125–128.

HAAS, H. L., 1981*b*, Histamine hyperpolarizes hippocampal neurones *in vitro, Neurosci. Lett.* **22:**75–78.

HAAS, H. L., and WOLF, P., 1977, Central actions of histamine: microelectrophoretic studies, *Brain Res.* **122:**269–279.

HAAS, H. L., WOLF, P., PALACIOS, J. M., GARBARG, M., BARBIN, G., and SCHWARTZ, J. C., 1978, Hypersensitivity to histamine in the guinea pig brain: microiontophoretic and biochemical studies, *Brain Res.* **156:**175–291.

HAMMERS, R., CLARENBACH, P., LINDL, T., and CRAMER, H., 1977, Uptake and metabolism of cyclic AMP in rabbit choroid plexus *in vitro, Neuropharmacology* **16:**135–141.

HANKOFF, L. D., GUNDLACH, R. H., PALEY, H. M., and RUDORFER, L., 1964, Diphenhydramine as an antidepressant, *Dis. Nerv. Syst.* **25:**547–553.

HEGSTRAND, L. R., KANOF, P. D., and GREENGARD, P., 1976, Histamine-sensitive adenylate cyclase in mammalian brain, *Nature* (*London*) **260:**163–165.

HEINRICH, M. A., 1953, The effect of the antihistaminic drugs on the central nervous system in rats and mice. *Arch. Int. Pharmacodyn.* **92:**444–463.

HILL, S. J., and YOUNG, J. M., 1980, Histamine H_1-receptors in the brain of the guinea pig and the rat: differences in ligand binding properties and regional distribution, *Br. J. Pharmacol.* **68:**687–696.

HILL, S. J., EMSON, P. C., and YOUNG, J. M., 1978, The binding of [^{3}H]mepyramine to histamine H_1 receptors in guinea pig brain, *J. Neurochem.* **31:**997–1004.

HILL, S. J., HILEY, C. R., and YOUNG, J. M., 1981, Extended mepyramine treatment and histamine H_1-receptors in guinea pig brain, *Eur. J. Pharmacol.* **71:**421–428.

HOLLENBERG, M. D., and CUATRECASAS, P., 1975, Biochemical identification of membrane receptors: principles and techniques, in: *Handbook of Psychopharmacology*, Section I: *Basic Neuropharmacology*, Vol. 2, *Principles of Receptor Research* (L. L. Iversen, S. D. Iversen, and S. H. Snyder, eds.), pp. 129–177, Plenum Press, New York.

HOLLENBERG, M. D., and CUATRECASAS, P., 1979, Distinction of receptor from nonreceptor interactions in binding studies, in: *The Receptors. A Comprehensive Treatise*, Vol. I, *General Principles and Procedures* (R. D. O'Brien, ed.), pp. 193–214, Plenum Press, New York.

HOUGH, L. B., and GREEN, J. P., 1980, Possible functions of brain histamine, *Psychopharmacol. Bull.* **16:**42–44.

HOUGH, L. B., and GREEN, J. P., 1981, Histamine-activated adenylate cyclase in brain homogenates of several species, *Brain Res.* **219:**363–370.

HOUGH, L. B., WEINSTEIN, H., and GREEN, J. P., 1980, One agonist and two receptors mediating the same effect: histamine receptors linked to adenylate cyclase in the brain, *Adv. Biochem. Psychopharmacol.* **21:**183–192.

HOUGH, L. B., KHANDELWAL, J. K., MORRISHOW, A. M., and GREEN, J. P., 1981, An improved GCMS method to measure tele-methylhistamine, *J. Pharmacol. Methods* **5:**143–148.

HUGHES, M. J., and CORET, I. A., 1974, Effects of tricyclic compounds on the histamine response of isolated atria, *J. Pharmacol. Exp. Ther.* **191:**252–261.

ISAAC, L., CHO, A. K., and BEAVEN, M. A., 1971, Decline of histidine decarboxylase activity and histamine levele in rat stomach after reserpine, *Biochem. Pharmacol.* **20**:1453–1461.
IVY, A. C., and BACHRACH, W. H., 1966, Effect of histamine on gastric secretion, in: *Handbook of Experimental Pharmacology*, Vol. 18, Part 1, *Histamine and Antihistamines* (M. Rocha e Silva, ed.), pp. 302–313, Springer-Verlag, New York.
JÄÄTELÄ, A., MANNISTO, P., PAATERO, H., and TUOMISTO, J., 1971, The effects of diazepam or diphenhydramine on healthy human subjects, *Psychopharmacologia* **21**:202–211.
JOÓ, F., RAKONCZAY, Z., and WOLLEMAN, M., 1975, cAMP-mediated regulation of the permeability in the brain capillaries, *Experientia* **31**:582–583.
KAKIUCHI, S., and RALL, T. W., 1968, The influence of chemical agents on the accumulation of adenosine 3',5'-phosphate in slices of rat cerebellum, *Mol. Pharmacol.* **4**:367–378.
KANOF, P. D., and GREENGARD, P., 1978, Brain histamine receptors as targets for antidepressant drugs, *Nature (London)* **272**:329–333.
KANOF, P. D., and GREENGARD, P., 1979, Pharmacological properties of histamine-sensitive adenylate cyclase from mammalian brain, *J. Pharmacol. Exp. Ther.* **209**:87–96.
KANOF, P. D., HEGSTRAND, L. R., and GREENGARD, P., 1977, Biochemical characterization of histamine-sensitive adenylate cyclase in mammalian brain, *Arch. Biochem. Biophys.* **182**:321–334.
KARNUSHINA, I. L., PALACIOS, J. M., BARBIN, G., DUX, E., JOÓ, F., and SCHWARTZ, J. C., 1980, Studies on a capillary-rich fraction isolated from brain: histaminic components and characterization of the histamine receptors linked to adenylate cyclase, *J. Neurochem.* **34**:1201–1208.
KARPPANEN, H., 1980, Histamine receptors and cyclic nucleotides, *Prog. Pharmacol.* **4**:167–184.
KENAKIN, T. P., 1980, Errors in the measurement of agonist potency ratios produced by uptake processes: a general model applied to β-adrenoceptor agonists, *Br. J. Pharmacol.* **71**:407–417.
KETY, S. S., 1978, Strategies of basic research, in: *Psychopharmacology: A Generation of Progress* (M. A. Lipton, A. DiMascio, and K. F. Killam, eds.), pp. 7–11, Raven Press, New York.
KIM, K. S., and SHORE, P. A., 1963, Mechanism of action of reserpine and insulin on gastric amines and gastric acid secretion, and the effect of monoamine oxidase inhibition, *J. Pharmacol. Exp. Ther.* **141**:321–325.
KIMELBLATT, B. J., CERRA, F. B., CALLER, G., BERG, M. J., MCMILLEN, M. A., and SCHENTAG, J. J., 1980, Dose and serum concentration relationships in cimetidine-associated mental confusion, *Gastroenterology* **78**:791–795.
KUHN, R., 1970, The imipramine story, in: *Discoveries in Biological Psychiatry* (F. J. Ayd, Jr., and B. Blackwell, eds.), pp. 205–217, Lippincott, Philadelphia.
KUO, J.-F., LEE, T.-P., REYES, P. L., WALTON, K. G., DONNELLY, T. E., JR., and GREENGARD, P., 1972, Cyclic nucleotide-dependent protein kinases. X. An assay method for the measurement of guanosine 3',5'-monophosphate in various biological materials and a study of agents regulating its levels in heart and brain, *J. Biol. Chem.* **247**:16–22.
LEE, J. R., and FENNESSY, M. R., 1976, Effects of morphine on brain histamine, antinociception, and activity in mice, *Clin. Exp. Pharmacol. Physiol.* **3**:179–189.
LEVIN, R. M., and WEISS, B., 1976, Mechanism by which psychotropic drugs inhibit adenosine cyclic-3'5'-monophosphate phosphodiesterase of brain, *Mol. Pharmacol.* **12**:581–589.
LICHTENSTEIN, L. M., and GILLESPIE, E., 1973, Inhibition of histamine release by histamine controlled by H_2 receptor, *Nature (London)* **244**:287–288.
MAAYANI, S., GREEN, J. P., and WEINSTEIN, H., 1978, LSD, tricyclic antidepressants, and neuroleptics inhibit histamine-stimulated adenylate cyclase in brain, *Fed. Proc.* **37**:612.
MAAYANI, S., HOUGH, L. B., WEINSTEIN, H., and GREEN, J. P., 1981, Response of the histamine H_2-receptor in brain to antidepressant drugs, in: *Typical and Atypical Antidepressants*, vol. 31, *Advances in Biochemical Psychopharmacology*, (G. Racagni, and E. Costa, eds.), pp. 133–147, Raven Press, New York.

Makman, M. H., Ahn, H. S., Thal, L. J., Sharpless, N. S., Dvorkin, B., Horowitz, S. G., and Rosenfeld, M., 1980, Evidence for selective loss of brain dopamine and histamine-stimulated adenylate cyclase activities in rabbits with aging, *Brain Res.* **192:**177–183.

McCulloch, M. W., and Story, D. F., 1972, Antagonism of noradrenaline and histamine by desipramine in the isolated artery of the rabbit ear, *Br. J. Pharmacol.* **46:**140–150.

McCulloch, M. W., Medgett, I. C., and Rand, M. J., 1979, Effects of the histamine H_2-receptor blocking drugs burimamide and cimetidine on noradrenergic transmission in the isolated aorta of the rabbit and atria of the guinea pig, *Br. J. Pharmacol.* **67:**535–543.

Mohs, R. C., Tinklenberg, J. R., Roth, W. T., and Kopell, B. S., 1978, Methamphetamine and diphenhydramine effects on the rate of cognitive processing, *Psychopharmacology* **59:**13–19.

Movshon, J. A., 1978, Hypercomplexities in the visual cortex, *Nature (London)* **272:**305–306.

Nahorski, S. R., and Smith, B. M., 1976, Stimulated formation of cyclic AMP in different areas of chick brain, *Eur. J. Pharmacol.* **40:**273–278.

Nahorski, S. R., Rogers, K. J., and Smith, B. M., 1974, Histamine H_2 receptors and cyclic AMP in brain, *Life. Sci.* **15:**1887–1894.

Nahorski, S. R., Rogers, K. J., and Edwards, C., 1975, Cerebral glycogenolysis and stimulation of B-adrenoceptors and histamine H_2 receptors, *Brain Res.* **92:**529–533.

Nahorski, S. R., Rogers, K. J., and Smith, B. M., 1977, Stimulation of cyclic adenosine 3′,5′-monophosphate in chick cerebral hemisphere slices: effects of H_1 and H_2 histaminergic agonists and antagonists, *Brain Res.* **126:**387–390.

Naranjo, P., 1966, Toxicity of histamine: lethal doses, in: *Handbook of Experimental Pharmacology*, Vol. 18, Part 1, *Histamine and Antihistamines* (M. Rocha e Silva, ed.), pp. 179–201, Springer-Verlag, New York.

Newton, M. V., Hough, L. B., and Azimitia, E. C., 1982, Histamine sensitive adenylate cyclase in monkey brain, *Brain Res.* **239:**639–643.

Nigro, S. A., 1968, Toxic psychosis due to diphenhydramine hydrochloride, *J. Am. Med. Assoc.* **203:**301–302.

Nowak, J. Z., Bielkiewicz, B., and Lebrecht, U., 1979, Dimaprit-induced hypothermia in normal rats: its attenuation by cimetidine and by tricyclic antidepressant drugs, *Neuropharmacology* **18:**783–789.

O'Dea, R. F., and Zatz, M., 1976, Catecholamine-stimulated cyclic GMP accumulation in the rat pineal: apparent presynaptic site of action, *Proc. Natl. Acad. Sci. USA* **73:**3398–3402.

Ohga, Y., and Daly, J. W., 1977, The accumulation of cyclic AMP and cyclic GMP in guinea pig brain slices. Effect of calcium ions, norepinephrine, and adenosine, *Biochem. Biophys. Acta* **498:**46–60.

Palacios, J. M., Garbarg, M., Barbin, G., and Schwartz, J. C., 1978, Pharmacological characterization of histamine receptors mediating the stimulation of cyclic AMP accumulation in slices from guinea pig hippocampus, *Mol. Pharmacol.* **14:**971–982.

Palacios, J. M., Young, W. S., III, and Kuhar, M. J., 1979, Autoradiographic localization of H_1-histamine receptors in brain using [^{3}H]mepyramine: preliminary studies, *Eur. J. Pharmacol.* **58:**295–304.

Palacios, J. M., Wamsley, J. K., and Kuhar, M. J., 1981*a*, The distribution of histamine H_1-receptors in the rat brain: an autoradiographic study, *Neuroscience* **6:**15–37.

Palacios, J. M., Wamsley, J. K., and Kuhar, M. J., 1981*b*, GABA, benzodiazepine, and histamine-H_1 receptors in the guniea pig cerebellum: effects of kainic acid injections studied by autoradiographic methods, *Brain Res.* **214:**155–162.

Palmer, G. C., 1981, Regional distribution of neurohumoral responsive adenylate cyclase in rabbit brain microvessels, *Neurosci. Lett.* **21:**207–210.

Palmer, G. C., Wagner, H. R., Palmer, S. J., and Manian, A. A., 1977, Histamine-stimulated adenylate cyclase: blockade by imipramine and its analogues, *Commun. Pharmacol.* **1:**61–69.

PALMER, G. C., WAGNER, H. R., PALMER, S. J., and MANIAN, A. A., 1978, Histamine-, norepinephrine-, and dopamine-sensitive central adenylate cyclases: effects of chlorpromazine derivatives and butaclamol, *Arch. Int. Pharmacodyn.* **233**:314–325.

PARSONS, M. E., OWEN, D. A. A., GANELLIN, C. R., and DURANT, C. J., 1977, Dimaprit [*S*-[3-(*N,N*-dimethylamino)propyl]isothiourea]—a highly specific histamine H_2-receptor agonist, Part 1, Pharmacology, *Agents Actions* **7**:31–38.

PECK, A. W., FOWLE, A. S. E., and BYE, C., 1975, A comparison of triprolidine and clemastine on histamine antagonism and performance tests in man: implications for the mechanism of drug-induced drowsiness, *Eur. J. Clin. Pharmacol.* **8**:455–463.

PEROUTKA, S. J., and SNYDER, S. H., 1981, [^{3}H]Mianserin: differential labeling of serotonin and histamine receptors in rat brain, *J. Pharmacol. Exp. Ther.* **216**:142–148.

PERSSON, S.-A., 1977*a*, The effect of LSD and 2-bromo-LSD on the striatal dopa accumulation after decarboxylase inhibition in rats, *Eur. J. Pharmacol.* **43**:73–83.

PERSSON, S.-A., 1977*b*, Effects of LSD and 2-bromo-LSD on striatal dopac levels, *Life Sci.* **20**:1199–1206.

PILC, A., and NOWAK, J. Z., 1980, The influence of 4-methylhistamine, an agonist of histamine H_2 receptors, on body temperature in rats, *Neuropharmacology* **19**:773–775.

POLLARD, H., BISCHOFF, S., and SCHWARTZ, J.-C., 1973, Increased synthesis and release of ^{3}H-histamine in rat brain by reserpine, *Eur. J. Pharmacol.* **24**:399–401.

POLLARD, H., BISCHOFF, S., and SCHWARTZ, J.-C., 1974, Turnover of histamine in rat brain and its decrease under barbiturate anesthesia, *J. Pharmacol. Exp. Ther.* **190**:88–99.

PSYCHOYOS, S., 1978, H_1- and H_2-histamine receptors linked to adenylate cyclase in cell-free preparations of guinea pig cerebral cortex, *Life Sci.* **23**:2155–2162.

QUACH, T. T., DUCHEMIN, A. M., ROSE, C., and SCHWARTZ, J. C., 1979, *In vivo* occupation of cerebral histamine H_1-receptors evaluated with [^{3}H]mepyramine may predict sedative properties of psychotropic drugs, *Eur. J. Pharmacol.* **60**:391–392.

QUACH, T. T., DUCHEMIN, A. M., ROSE, C., and SCHWARTZ, J. C., 1980, Labeling of histamine H_1-receptors in the brain of the living mouse, *Neurosci. Lett.* **17**:49–54.

REHAVI, M., and SOKOLOVSKY, M., 1978, Multiple binding sites of tricyclic antidepressant drugs to mammalian brain receptors, *Brain Res.* **149**:525–529.

RENAUD, L. P., 1976, Histamine microiontophoresis on identified hypothalamic neurons: three patterns of response in the ventromedial nucleus of the rat, *Brain Res.* **115**:339–344.

RICHELSON, E., 1978, Histamine H_1 receptor-mediated guanosine 3′,5′-monophosphate formation by cultured mouse neuroblastoma cells, *Science* **201**:69–71.

RICHELSON, E., 1980, Psychotherapeutic drugs, histamine H_1 and muscarinic acetylcholine receptors, in: *Psychopharmacology and Biochemistry of Neurotransmitter Receptors* (H. I. Yamamura, R. W. Olsen, and E. Usdin, eds.), pp. 263–277, Elsevier/North-Holland, New York.

RISING, T. J., NORRIS, D. B., WARRANDER, S. E., and WOOD, T. P., 1980, High-affinity [^{3}H]-cimetidine binding in guinea pig tissues, *Life Sci.* **27**:199–206.

SASTRY, B. S. R., and PHILLIS, J. W., 1976, Evidence for an ascending inhibitory histaminergic pathway to the cerebral cortex, *Can. J. Physiol. Pharmacol.* **54**:782–786.

SCARPIGNATO, C., VALENTI, G., CEDA, G. P., and BERTACCINI, 1979, Effects of cimetidine on the secretion of some pituitary hormones, *Pharmacology* **19**:111–115.

SCHENTAG, J. J., CERRA, F. B., CALLERI, G., DEGLOPPER, E., ROSE, J. G., and BERNHARD, H., 1979, Pharmacokinetic and clinical studies in patients with cimetidine-associated mental confusion, *Lancet* **1**:177–181.

SCHENTAG, J. J., CERRA, F. B., CALLERI, G. M., LEISING, M. E., FRENCH, M. A., and BERNHARD, H., 1981, Age, disease, and cimetidine disposition in healthy subjects and chronically ill patients, *Clin. Pharmacol. Ther.* **29**:737–743.

SCHILD, H. O., 1947, p*A*, a new scale for the measurement of drug antagonism, *Br. J. Pharmacol.* **2**:189–206.

SCHILD, H. O., 1981, The multiple facets of histamine research, *Agents Actions* **11**:12–19.

SCHNECKLOTH, R., PAGE, I. H., DEL GRECO, F., and CORCORAN, A. C., 1957, Effects of serotonin antagonists in normal subjects and patients with carcinoid tumors, *Circulation* **16:**523–532.

SCHULTZ, J., and DALY, J. W., 1973, Cyclic adenosine 3′-5′-monophosphate in guinea pig cerebral cortical slices, III: Formation, degradation, and reformation of cyclic adenosine 3′-5′-monophosphate during sequential stimulations by biogenic amines and adenosine, *J. Biol. Chem.* **248:**860–866.

SCHWABE, U., OHGA, Y., and DALY, J. W., 1978, The role of calcium in the regulation of cyclic nucleotide levels in brain slices of rat and guinea pig, *Naunyn-Schmiedeberg's Arch. Pharmacol.* **302:**141–151.

SCHWARTZ, J.-C., 1975, Histamine as a transmitter in brain, *Life Sci.* **17:**503–518.

SCHWARTZ, J.-C., 1979, Minireview: histamine receptors in brain, *Life Sci.* **25:**895–912.

SCHWARTZ, J.-C., BARBIN, G., TUONG, M. D. T., DUCHEMIN, A.-M., GARBARG, M., QUACH, T. T., RODERGAS, E., and ROSE, C., 1980*a*, Pharmacology and biochemistry of histamine receptors in brain, in: *Psychopharmacology and Biochemistry of Neurotransmitter Receptors* (H. I. Yamamura, R. W. Olsen, and E. Usdin, eds.), pp. 279–300, Elsevier/North-Holland, New York.

SCHWARTZ, J.-C., POLLARD, H., and QUACH, T. T., 1980*b*, Histamine as a neurotransmitter in mammalian brain: neurochemical evidence, *J. Neurochem.* **35:**26–33.

SCHWARTZ, J.-C., BARBIN, G., DUCHEMIN, A. M., GARBARG, M., PALACIOS, J. M., QUACH, T. T., and ROSE, C., 1980*c*, Histamine receptors in the brain: characterization by binding studies and biochemical effects, *Adv. Biochem. Psychopharmacol.* **21:**169–182.

SCHWARTZ, J.-C., GARBARG, M., and QUACH, T. T., 1981, Histamine receptors in brain as targets for tricyclic antidepressants, Trends in Pharmacological Sciences, **3:**122–125.

SEGAL, M., 1981, Histamine modulates reactivity of hippocampal CA3 neurons to afferent stimulation *in vitro, Brain Res.* **213:**443–448.

SMITH, I. R., CLEVERLEY, M. T., GANELLIN, C. R., and METTERS, K. M., 1980, Binding of [^{3}H]-cimetidine to rat brain tissue, *Agents Actions* **10:**422–426.

SNYDER, S. H., 1980, Tricyclic antidepressant drug interactions with histamine and adrenergic receptors, *Pharmakopsychiatr./Neuro-Psychopharmakol.* **13:**62–67.

SNYDER, S. H., and TAYLOR, K. M., 1972, Histamine in the brain: a neurotransmitter? in: *Perspectives in Pharmacology. A Tribute to Julius Axelrod* (S. H. Snyder, ed.), pp. 43–73, Oxford University Press, New York.

SPIKER, M. D., PALMER, G. C., and MANIAN, A. A., 1976, Action of neuroleptic agents on histamine-sensitive adenylate cyclase in rabbit cerebral cortex, *Brain Res.* **104:**401–406.

STUDY, R. E., and GREENGARD, P., 1978, Regulation by histamine of cyclic nucleotide levels in sympathetic ganglia, *J. Pharmacol. Exp. Ther.* **207:**767–778.

SUBRAMANIAN, N., MITZNEGG, and ESTLER, C.-J., 1978, Ethanol-induced alterations in histamine content and release in the rat hypothalamus, *Naunyn-Schmiedeberg's Arch. Pharmacol.* **302:**119–121.

SUBRAMANIAN, N., SEIDLER, F. J., WHITMORE, W. L., and SLOTKIN, T. A., 1980, Histamine stimulates brain phospholipid turnover through a direct, H-1 receptor mediated mechanism, *Life Sci.* **27:**1315–1319

SUBRAMANIAN, N., WHITMORE, W. L., SEIDLER, F. J., and SLOTKIN, T. A., 1981, Ontogeny of histaminergic neurotransmission in the rat brain: concomitant development of neuronal histamine, H-1 receptors, and H-1 receptor-mediated stimulation of phospholipid turnover, *J. Neurochem.* **36:**1137–1141.

SWAZEY, J. P., 1974, *Chlorpromazine in Psychiatry: A Study of Therapeutic Innovation*, MIT Press, Cambridge, Massachusetts.

TAN-TRAN, V., CHANG, R. S. L., and SNYDER, S. H., 1978, Histamine H_1 receptors identified in mammalian brain membranes with [^{3}H]mepyramine, *Proc. Natl. Acad. Sci. USA* **75:**6290–6294.

TAN-TRAN, V., FREEMAN, A. D., CHANG, R. S. L., and SNYDER, S. H., 1980, Ontogenetic development of histamine H_1-receptor binding in rat brain, *J. Neurochem.* **34:**1609–1613.

TAN-TRAN, V., LEBOVITZ, R., TOLL, L., and SNYDER, S. H., 1981, [^{3}H]Doxepin interactions with histamine H_1-receptors and other sites in guinea pig and rat brain homogenates, *Eur. J. Pharmacol.* **70:**501–509.

TAYLOR, K. M., 1975, Brain histamine, in: *Handbook of Psychopharmacology* (L. L. Iversen, S. D. Iversen, and S. H. Snyder, eds.), Vol. 3, pp. 327–379, Plenum Press, New York.

TAYLOR, J. E., and RICHELSON, E., 1979, Desensitization of histamine H_1 receptor-mediated cyclic GMP formation in mouse neuroblastoma cells, *Mol. Pharmacol.* **15:**462–471.

TAYLOR, J. E., and RICHELSON, E., 1980, High-affinity binding of tricyclic antidepressants to histamine H_1-receptors: fact and artifact, *Eur. J. Pharmacol.* **67:**41–46.

TOLL, L., TRAN, V. T., GAVISH, M., and SNYDER, S. H., 1980, Properties of soluble histamine H_1-receptors in the brain, in: *Psychopharmacology and Biochemistry of Neurotransmitter Receptors* (H. I. Yamamura, R. W. Olsen, and E. Usdin, eds.), pp. 301–311, Elsevier/North-Holland, New York.

UCHIDA, M., 1980, Histamine-induced decrease of membrane-bound calcium ions in the membrane fraction of rabbit taenia coli, *Eur. J. Pharmacol.* **64:**357–630.

UYENO, E. T., and MITOMA, C., 1969, The relative effectiveness of several hallucinogens in disrupting maze performance by rats, *Psychopharmacologia* **16:**73–80.

VAN NUETEN, J. M., RENEMAN, R. S., and JANSSEN, P. A. J., 1977, Specific α-adrenoceptor blocking effect of droperidol on isolated smooth muscles, *Eur. J. Pharmacol.* **44:**1–8.

YELLIN, T. O., BUCK, S. H., GILMAN, D., JONES, D. F., and WARDLEWORTH, J. M., 1979, ICI 125,211: a new gastric antisecretory agent acting on histamine H_2-receptors, *Life Sci.* **25:**2001–2009.

ZIEGLER, V. E., CLAYTON, P. J., and BIGGS, J. T., 1977, A comparison study of amitriptyline and nortriptyline with plasma cells, *Arch. Gen. Psychiatry* **34:**607–612.

INDEX

MIX
Papier aus verantwortungsvollen Quellen
Paper from responsible sources
FSC® C105338

If you have any concerns about our products,
you can contact us on
ProductSafety@springernature.com

In case Publisher is established outside the EU,
the EU authorized representative is:
Springer Nature Customer Service Center GmbH
Europaplatz 3, 69115 Heidelberg, Germany

Printed by Libri Plureos GmbH
in Hamburg, Germany